FUZZY ENGINEERING

FUZZY ENGINEERING

Bart Kosko
University of Southern California

PHI Learning Private Limited
Delhi-110092
2015

₹ 550.00

FUZZY ENGINEERING (with CD-ROM)
by Bart Kosko

ISBN-978-81-203-5053-3

Published by Asoke K. Ghosh, PHI Learning Private Limited, Rimjhim House, 111, Patparganj Industrial Estate, Delhi-110092 and Printed by Rajkamal Electric Press, Plot No. 2, Phase IV, HSIDC, Kundli-131028, Sonepat, Haryana.

For Lotfi Zadeh

CONTENTS

PART VII: COMPUTING IN FUZZY CUBES 401

PART VIII: FEEDBACK IN FUZZY CUBES 465

PREFACE

FUZZY ENGINEERING: FUZZY LOGIC AS FUNCTION APPROXIMATION

Fuzzy systems approximate functions. They are universal approximators if they use enough fuzzy rules. In this sense fuzzy systems can model any continuous function or system. Those systems can just as well come from physics or sociology as from control theory or signal processing.

The quality of the fuzzy approximation depends on the quality of the rules. In practice experts guess at the fuzzy rules. Or neural schemes learn the rules from data and tune the rules with new data. The result always approximates some unknown nonlinear function that can change in time. Better brains and better neural networks give better function approximations. This is not the standard view of fuzzy systems but it is the view we take in this text and call *fuzzy engineering*: function approximation with fuzzy systems.

The standard view is that fuzzy systems theory or "fuzzy logic" is a linguistic theory that models how we reason with vague rules of thumb and common sense. Fuzzy sets and systems serve as means to this linguistic end. This no doubt holds in many cases. It tends to hold in practice when the number of inputs and outputs in a problem is small enough and when the time scale is slow enough for a human to find some solution paths as when we focus a camera lens or back up a car or grill a steak. And it reflects the kinds of issues the first fuzzy engineers addressed and shows the kinds of tools they often used in their work and the language they used to defend it.

But both the name *fuzzy* and the linguistic view behind it have often sent the wrong message to outsiders and to laymen and to the field's critics. The view suggested to some that fuzzy logic was an obscure branch of linguistics or mathematical logic or artificial intelligence or that it was not a branch of science or math at all. In the worst case the linguistic view suggested that fuzzy logic was no more than a clever mix of software programming and wordplay. The nonsystems branches of fuzzy theory compounded this perception with what sometimes looked like fuzzification for fuzzification's sake: Just add a continuous 0-to-1 index to a model and call it fuzzy. Before the 1980s many fuzzy journal articles were of this form and a few still are. The new names *fuzzy function approximation* and *fuzzy engineering* should help keep critics from painting fuzzy systems with that brush.

Engineers seemed to have had a special distrust for the linguistic view of fuzzy systems. The irony is that the linguistic view grew out of the new surge in systems engineering in the 1960s that began with the Kalman filter. Some control engineers were openly hostile to the field and its linguistic cloak. Fuzzy engineers dealt with many of the same control problems that they did but the fuzzy engineers spoke not just in a new language but spoke *in* language and not in math. Behind the fuzzy language lay the new algebra of fuzzy sets cast in the abstract math of multivalued set theory and logic. Engineers found much of the field twice removed from their formal training in closed-form math models. The fuzzy systems looked more like verbal guidelines than exact algorithms and few engineers had taken courses in topology or the more obscure branches of symbolic logic.

Critics had a further reason to distrust fuzzy logic: It stood for a type of mathematical ruthlessness. When fuzzy engineers needed to break a "law" of binary logic to model a concept like cool air or tall men or small change they just went ahead and broke it. They worried about the breaking later if they worried about it at all. Physicists bend math this way as a matter of course and few seem to mind. Fuzzy engineers used words to bend math and that was harder to accept. You can fit a delta pulse into your math scheme if you work at it. But to give up Aristotle's either–or law of excluded middle means to give up the whole binary math scheme except as a limiting case. The reasons and results had better be good to pay so high a cost. Few thought the payoff was worth it.

THE RISE OF FUZZY ENGINEERING

The tide turned when the new "smart" gadgets and fuzzy control systems appeared in mass in Japan and Korea in the 1990s. Even then the words often got in the way of the success of these new fuzzy systems embedded in consumer electronics. Critics found more smoke than fire when they read how fuzzy rules and fuzzy chips raised the machine IQ of washing machines and camcorders and transmissions. Few mathematicians could see how a multivalued set theory itself led to smarter machines. They saw some new set math and saw rules cast as sentences. They did not see how it led to a system in the math terms of an input–output map or function. And often the fuzzy engineers did not see this either.

As a result many who read about fuzzy logic in an airline magazine or heard it invoked in a talk at a conference ignored the field as oversold engineering or dismissed it as hype. A fuzzy washing machine might save energy and wear and tear on the clothes. And the Sendai subway system might run smoother and more cheaply with a fuzzy controller than with a human or automatic controller. That was not enough to give up Aristotle.

The engineering view of fuzzy logic as function approximation does not depend on words. It does not rest on the latest cognitive theory or linguistic paradigm and has no ties to modal theories of possibility and necessity. It rests on the math of function approximation and statistical learning theory. Most of this math is well known. To some the math may come as a letdown. It shows there is no magic in fuzzy systems despite the colorful language of some of their designers and critics. Fuzzy systems are just a natural way to turn speech and measured action into functions that approximate hard tasks.

Words still map to math in fuzzy engineering. But often there are no such words to be had. Words can have great value in practice when we first need to initialize the function approximation or later when we need to tune it.. Brains and the words they emit are too rich a resource to ignore. Neural nets bypass words and focus on behavior. They get us closer to brains. Words are just a tool or ladder we climb down on to get to the task of function approximation.

On this view fuzzy language is a means to the end of computing and not the other way around. To compute is to turn inputs into outputs or to turn causes into effects or to turn questions into answers. Systems do that and we model them as functions. Systems compute and so do fuzzy systems. What matters is how well they do it and at what cost. Any match-up with natural language is a bonus and not a goal. The goal is to model systems so that we can control them or shape them or at least predict them. In this sense fuzzy engineering is a branch of function approximation.

FUZZY EPISTEMOLOGY AND THE CONCEPTUAL ANARCHY

The basic unit of fuzzy function approximation is the if-then rule: "If the wash water is dirty then add more detergent" or "If the price is high then the demand is low." A fuzzy system is a set of if-then rules that maps inputs to outputs. Each rule maps some of the inputs to some of the outputs. It maps input fuzzy sets like *dirty washwater* to output fuzzy sets like *more detergent*. Dirty washwater is fuzzy because all washwater is dirty and not dirty to some degree. A lone rule like "Dirty washwater implies more detergent" can act as a minimal fuzzy system. By itself it defines a constant function or line segment. Overlapping rules define polynomials or still richer functions.

Rules also show how natural language can lead to a fuzzy system and thus how words can map to math. The fuzzy system grows from words to sentences to paragraphs. Words stand for sets. The noun *washwater* stands for a set or subset of water. The *washwater* set may have fuzzy or gray boundaries but it need not. The adjectives *dirty* and *very dirty* stand for fuzzy subsets of the washwater set. So the noun phrase *dirty washwater* stands for a fuzzy set and so does *more detergent*. We can model these fuzzy sets with curves other than rectangles like triangles or trapezoids or bell curves. A full sentence like "Dirty washwater implies more detergent" stands for a fuzzy rule or fuzzy association between the if-part fuzzy set *dirty washwater* and the then-part fuzzy set *more detergent*. Below we note that such a rule defines a fuzzy patch or subset of the input–output state space and acts as the minimal unit of function approximation. Paragraphs or lists of such sentences define a fuzzy system or set of fuzzy rule patches that cover some function or some family of functions. We can add or delete rules or sentences at will in a type of macrotuning. Neural systems tend to do this only at the start of the training cycle. They microtune from there as they slowly change the shape of fuzzy sets like *dirty washwater* or *more detergent* and thus change their "meaning."

The meaning of a concept is just the fuzzy set it stands for. All objects belong to the fuzzy set and its complement to some degree. So meaning is fuzzy. Change the fuzzy set even a little and you change the meaning of the term. One person might draw *cool air* as a triangle centered at 70° Fahrenheit. He might find 70° air 100% cool and find 72° air or 68° air only 80% cool. But meaning is also relative and brains and neural nets capture it by moving and tuning the fuzzy sets. Someone else might draw or mean *cool air* as a thinner or fatter triangle or as one centered at 65° or at 60°. What is 100% cool for one person may be 80% cool for the next. The fuzzy rule "If the air is cool then set the motor speed to slow" has the same word form for all meanings of its terms. The fuzzy sets differ and so do the patch structure of the rule and the function such rules define.

The epistemology of fuzzy systems is that of a conceptual anarchy. We all use the same terms but do not mean the same things by them. The set of circles or other math objects may not be fuzzy or relative in the language of math but they may be both in a brain-based mind or neural network. Outside of math all patterns are fuzzy and relative to at least some degree. The humble sets of *dirty washwater* and *cool air* give a first glimpse of our conceptual anarchy. The fuzz and relativity grows as we move from sensory terms like *blue* or *large* or *cool* to the more abstract terms of head and heart like *similar* or *smart* or *fear* and on up to the broad terms like *season* or *progress* or *war* that describe complex physical or social processes. The same newspaper prints all the fuzzy terms in the same clear type as it states the "facts" of the world. It is a surprise that we can communicate at all since no two persons mean the same thing with what they think or say. Fuzzy systems maintain a common rule structure while each user tunes the sets and finds his niche in the conceptual anarchy. This turns the anarchy into a search space for user friendliness. Each person digs his own niche in the fuzzy state space.

ADDITIVE FUZZY SYSTEMS AND THE STANDARD ADDITIVE MODEL

The fuzzy system $F: R^n \rightarrow R^p$ itself is a shallow but wide rule tree. It is a feedforward approximator. There are m rules of the form "If X is fuzzy set A then Y is fuzzy set B." At this level the system depends less and less on words. Each input x partially fires all rules in parallel. So the system acts as an associative processor as it computes the output $F(x)$. Rule firing is a type of pattern matching of the input x with the if-part fuzzy sets of the m rules. The system then combines the partially fired then-part fuzzy sets in a sum and converts this sum to a scalar or vector output. We can also view a match-and-sum fuzzy approximator as a generalized artificial intelligence (AI) expert system or as a neural-like fuzzy associative memory.

Fuzzy systems can extend their rule tree to a rule digraph by letting the rules feed back to one another and to themselves. Then the feedback fuzzy system models a dynamical system of the form $\frac{dx}{dt} = F(x)$ or of some more complex form. Inputs swirl into equilibrium attractors that range from fixed points and limit cycles to chaos. The rare case of stable systems may admit a simple Lyapunov function to assure that all nearby inputs converge to a fixed point. Other feedback fuzzy systems define fuzzy cognitive maps and act as adaptive knowledge networks. Feedback fuzzy systems model functions or systems that change in time. They may use far fewer rules than feedforward systems would use for such tasks. But their feedback and nonlinearity make them hard to analyze or control.

This text looks at how *additive* fuzzy systems approximate continuous functions. These systems *add* the then-parts of the if-then rules when the rules fire. The systems can be feedforward or feedback approximators. We apply additive fuzzy systems to a wide range of problems in engineering and machine intelligence: curve fitting, chaos modeling, control, signal processing, communications, hardware design, pattern recognition, knowledge engineering, and virtual reality. There are fuzzy systems other than additive ones. Chapter 2 on additive systems gives some of the technical reasons why we ignore them. So far only additive systems are proven universal approximators for rules that use fuzzy sets of any shape. The computational simplicity of additive systems is a further reason why this text focuses on them.

The broader reason is that this text seeks to push additive fuzzy systems as far as we can now push them in modern engineering. That task rightly falls on the author of the new architecture. He can criticize other fuzzy systems to distinguish or explain his own system but in the end he bears the burden of showing how it works. Universal approximation theorems are hollow if you do not show the old things and new things the approximators can do and how well they can do them. That task filled this volume and could easily have filled a second.

Additive fuzzy systems have the geometry of eggs on a string. This geometry runs throughout the text. Each rule defines a fuzzy patch in the input–output state space. In many cases the patch may have the form of an egg or ellipsoid but it need not. The patches geometrize the words or knowledge. Precise rules define small patches. Vague or noisy rules define large patches. A rule relates fuzzy patterns: "If input X is fuzzy pattern A then output Y is fuzzy pattern B". The Cartesian product $A \times B$ defines the fuzzy rule patch or fuzzy subset of the input-output state space $X \times Y$. An additive fuzzy system covers the graph of the function with these rule patches and adds or averages rule patches that overlap:

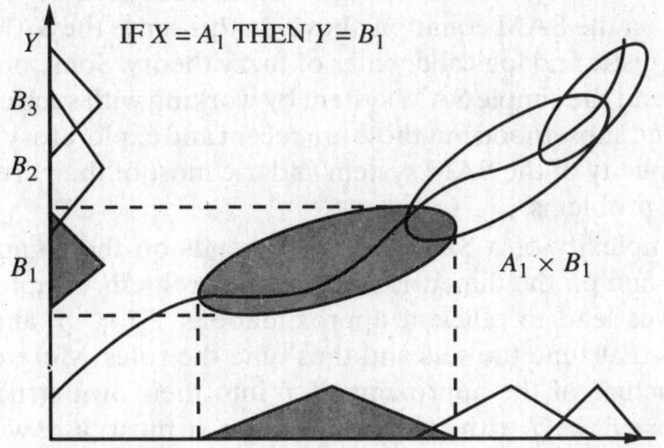

The function approximation tends to get finer and finer as the patches grow in number and shrink in size. The patch covering in the figure is the emblem of this text. Each chapter deals with it in some way.

The class of additive systems is vast. In practice most fuzzy engineers use some form of what we call the *standard additive model* or SAM. A SAM defines a function $F: R^n \rightarrow R^p$. Its additive structure leads by theorem to a form simple enough to state in a preface:

$$F(x) = \frac{\sum_{j=1}^{m} w_j \, a_j(x) V_j c_j}{\sum_{j=1}^{m} w_j \, a_j(x) V_j}.$$

The SAM stores m rules of the patch form $A_j \times B_j$ or the word form "If input variable X equals fuzzy set A_j then output variable Y equals fuzzy set B_j" each with scalar rule weight $w_j > 0$. The vector input x belongs to the if-part fuzzy set $A_j \subset R^n$ to degree or percentage $a_j(x)$ in [0,1]. The SAM need not use the fuzz or multivaluedness of then-part fuzzy set $B_j \subset R^p$. It just uses the volume or area V_j of set B_j and its centroid or center of mass c_j. In practice we replace B_j with a nonfuzzy rectangle or box with the same volume and centroid. We can even dispense with the volumes if the then-part sets B_j all have the same volume or area. Then the SAM reduces to the COG or center-of-gravity fuzzy systems found in most Japanese consumer products. In turn Gaussian COGs coincide with radial basis function neural networks.

The SAM equation departs from the linguistic context of earlier fuzzy models and lies at the heart of fuzzy engineering. It converts an input x to an output $F(x)$ and need not use fuzzy concepts or break the either–or laws of binary logic. For to compute $F(x)$ it takes just a few additions and multiplications and just one division. The if-part fuzzy set A_j does break the either-or laws if we view it along side its set complement A_j^c. For then $A_j \cap A_j^c \neq \emptyset$ and $A_j \cup A_j^c \neq X$ hold and they contradict the binary assumptions of ancient and modern logic and set theory. But most fuzzy systems work with just A_j and not its complement. The engineer works with *dirty washwater* and not explicitly with *not-dirty washwater*. What counts is the accuracy of the patch cover and that depends on the SAM equation above. In this sense the SAM equation sidesteps the older linguistic and logical debates of fuzzy theory. Some of the later chapters in this book extend the simple SAM system by working with such concepts as measures of fuzziness and subsethood but these are recent and exploratory paths. Most chapters keep the simplicity of the SAM system and use most of their creative effort to apply it to complex problems.

The complexity of a SAM system depends on the complexity of the if-part fuzzy sets A_j and on the dimensionality of the problem. Simple sets like trapezoids and bell curves lead to efficient approximations $F \approx f$ and simple supervised learning laws that tune the sets and thus tune the rules. More complex sets A_j can code the structure of the approximand f into their own structure and give exact representations $F = f$ with as few as two rules. In most cases we have no knowledge of the approximand f. We face blind approximation and have only expert guesses at the rules or the raw and noisy input–output data that come from f. In practice we assume simple shapes for the fuzzy sets and do our best to tune them and relate them in rules. This limits the set complexity at the price of approximation accuracy. It does not affect the more severe complexity problem of dimension that all fuzzy systems face in blind approximation: rule explosion.

THE CURSE OF DIMENSIONALITY AND LEARNING: PATCH THE BUMPS

Fuzzy systems face exponential rule growth in high dimensions. This "curse of dimensionality" stems from the patch cover of the fuzzy function approximation. The above figure shows that a scalar fuzzy system $F\colon R \rightarrow R$ needs on the order of k rule patches to cover some finite section of the input–output plane. One large rule patch would cover the graph of the approximand f and would give F as a flat-line constant system. The fuzzy system $F\colon R^n \rightarrow R$ needs on the order of k^n rule patches. The full vector system $F\colon R^n \rightarrow R^p$ needs on the order of k^{n+p-1} rules. The exponential rule explosion stems far more from the inputs than outputs since $n > p$ holds in practice.

Most fuzzy applications have had small dimension. They have used just two or three inputs and one output and a k equal to 5 or 7. So they have used less than 150 rules. Simple fuzzy systems may work poorly if at all with the more complex systems of finance or medicine or manufacturing. Such fuzzy experts need not only a lot of rules but a lot of good ones.

Learning can help deal with a tight rule budget. Learning moves the rule patches to tune the fuzzy system. Learning can come from a math algorithm or a software search or an expert's guesses or trials and errors. It is a means to the end of new and better rules. Learning changes the shape or place of the if-part sets or changes that of the then-part sets or changes them both. New sets give new rules. New rules give a new system and thus a new function approximation. The fuzzy system F moves as a point in the space of continuous functions and tries to move to the point f. The goal of learning is optimality or exact representation $F = f$.

Learning moves the rule patches as F moves toward f. The number of rules m will in general be much smaller than the exponential number of rules k^{n+p-1} the system needs to cover the graph of f. The best place to put the rules depends on the extrema or bumps of the error function. The number of error bumps may well exceed the number m of rules. Each time one adds a rule it should lie at the largest bump of the error curve. The next rule should lie at the largest bump of the residual error curve and so on. This does not remove the curse of dimensionality but it does the best job in the face of it. The error bumps often depend on the bumps in f. This holds for lone rules in local regions of the input–output state space. Large lone rules are the rule and not the exception in problems of high dimension. Then the best rule patches lie at the bumps or extrema of the function. They *patch the bumps*.

In practice we do not know the shape of the approximand f or the error surface between the fuzzy system F and f. So we ask an expert or use neural or other learning schemes to find and move the fuzzy rule patches as close as they can to the unknown bumps. Expertise often shows itself at the turning points. Extra rules can fill in between the bumps to improve the interpolation or can fall outside the far bumps to improve the extrapolation. The best learning schemes converge quickly to the optimal rule patches. To date almost all adaptive fuzzy systems have learned blindly without the goal of moving fixed or changing rule patches to their optimal positions. This often adds the exponential complexity of many learning schemes to the exponential rule complexity with little payoff in approximation accuracy. Patching the bumps is the best way to spend a rule budget. It also reduces much of fuzzy function approximation to the old task of finding the zeroes of a derivative map.

The learning scheme itself need not depend on a neural network. The learning scheme can map sample data to new values for the fuzzy parameters. These schemes tend to be either unsupervised clustering of data or supervised gradient descent. Unsupervised learning is faster but less accurate. Supervised learning requires more knowledge of the approximand f to compute an error signal and often requires orders of magnitudes more training cycles. Both schemes can tune the values in a fuzzy system or in a neural network. Indeed the line between neural and fuzzy systems is itself fuzzy.

Fuzzy systems and neural networks are both black-box approximators and suffer all the good and ill that come with that model-free status. They apply to any input–output process but may not admit model-based proofs of stability or robustness. Feedforward versions offer more control and less power than feedback versions. And we cannot be sure of what functions or patterns they have learned without checking a large number of input–out test cases. The fuzzy black box does let the user open it to see the fuzzy rules and tune them or remove them as independent modules. Neural networks offer only "connectionist glop" on their insides. Neural nets also suffer from the distributed structure of their web of synapses. Most nets forget part of what they have learned each time they learning something new. New patterns crowd out and distort some of the old patterns. Neural nets model brains too well in this sense. They cannot recall what they have forgotten and so we never know what they have learned or unlearned. The last word on model-free systems will always come from computer simulation.

The fuzzy systems in this text learn with the same learning laws that a neural network uses. These are adaptive fuzzy systems but we often call them "neural fuzzy" or "fuzzy neural" systems. Here "neural" just means "adaptive" or "learning." There are a few systems that use fuzzy math to model a neuron or that combine neural and fuzzy systems in master–slave or slave–master hybrids. These systems lie outside the scope of this book and of the mainstream research on black-box approximators. This book applies the SAM system and its variants to a wide range of engineering problems and tunes those systems with unsupervised or supervised learning laws. Neural nets might behave the same in many cases if tuned with the same learning laws and the same training data. The same result may hold for other classes of black-box approximators such as polynomials or blind nonlinear regressors. The purpose of this book is to explore the power and limits of that new class of function approximators known as additive fuzzy systems.

ABOUT THIS TEXT

This book is both a unified textbook and to some degree an edited volume. It aims at the college audience in engineering. Each chapter contains detailed homework problems though some of the later research chapters contain fewer problems than do earlier chapters. And each chapter fits in the larger topics scheme that runs from theory to applications and the topological scheme that runs from feedforward systems to feedback systems. I worked on some chapters with my graduate students and asked colleagues to prepare a third of the chapters and wrote the rest myself. We worked on the book for over two years and both students and colleagues had to endure my editorial control. In that time I added and deleted many chapters to best fit and extend the shifting frontiers of fuzzy engineering. The result has the collaborative strength of an edited volume and the structure and tone that only a sole author can give.

The book starts with a review and moves from feedforward fuzzy systems to feedback fuzzy systems. The first two chapters reflect this global structure and the reader should master them and their many homework problems before moving on to the application chapters. The first chapter extends the article "Fuzzy Logic" that ran in the July 1993 issue of *Scientific American*. Its final section lists the core mathematics of fuzzy sets and systems. Its homework problems cover older topics in multivalued logic and fuzzy sets that help sharpen the students skill when working with the more advanced fuzzy systems.

Chapter 2 is the main chapter in the text and by far the longest. It presents the theory of fuzzy engineering in a sequence of theorems and lesser results. Chapters 3–15 apply and extend the theory in Chapter 2. The homework problems consist largely of more theorems and extensions. The serious student should work these problems to fully grasp the concepts in the chapter. Chapter 3 also focuses on fuzzy function approximation but in the special case of ellipsoidal rules that factor onto coordinate axes.

Chapters 4 and 5 apply fuzzy systems to control and chaos modeling. Chapter 4 describes a large research effort to use fuzzy systems to control platoons of smart cars both in simulations and in road tests. Chapter 5 shows how fuzzy rule patches can recursively partition a state space to model chaotic attractors. This chapter shows how exponential rule explosion can overwhelm a feedforward fuzzy system and points to the need for feedback fuzzy systems that might model the same attractors with a fixed set of rules.

Chapters 6 and 7 apply fuzzy systems to signal and image processing. Chapter 6 introduces the important new concept of alpha-stability to model impulsive noise. Alpha-stable statistics have bell curves with thicker tails than the Gaussian bell curve so common in modern science and engineering. The thicker tails often give a better model of noise but do not admit much analysis since all variances and higher moments are infinite. This chapter shows how a fuzzy system can predict time-series data in the presence of massive impulsive noise and can filter such noise from signals. The system uses ellipsoidal rules but does not factor them. It instead uses the joint set function based on the metrical structure of the rules. Chapter 7 uses a fuzzy system to find a high-energy subtree of a tiling tree in image subband coding. Fuzzy rules split frequency bands to act as a bank of wavelet filters or wavelet packets.

Chapters 8 and 9 apply fuzzy systems to modern communications. Chapter 8 recasts the standard additive fuzzy model to work with modular arithmetic. The fuzzy system learns "random rules" by watching the output of a random number generator. The rules are part of a wireless system that spreads and despreads a signal with frequency hopping spread spectrum. Chapter 9 shows how a weighted sum of fuzzy systems can detect signals in the presence of both Gaussian and impulsive alpha-stable noise.

Chapters 10 and 11 show how to capture additive fuzzy systems in hardware. The designer of the first digital fuzzy processor shows in Chapter 10 how to design SAM chips and adaptive SAM chips that use the supervised learning laws of Chapter 2 to tune the fuzzy system. Chapter 11 shows how to design optical systems that capture the essential structure of additive fuzzy systems.

The last four chapters deal with fuzzy cubes or unit hypercubes of high dimension that house all fuzzy subsets of some discrete space of objects. Chapter 12 presents the theory of fuzzy cubes and mappings between cubes and real spaces. The chapter embeds Shannon information theory in a more general fuzzy information theory and presents wavelike equations of how fuzzy information fluid may unfold in time. Chapter 13 applies the key fuzzy idea of subsethood or partial set containment in fuzzy cubes to learning and function approximation. It uses a Gaussian additive fuzzy system that has the same structure as the popular radial basis function networks of neural theory.

The last two chapters deal with feedback in fuzzy cubes. The founders of adaptive resonance theory present a fuzzy adaptive resonance theory in Chapter 14 based on the fuzzy measure of subsethood. They apply the model to pattern recognition and function approximation. Chapter 15 presents the fuzzy cognitive map of Chapter 2 as a framework for virtual reality systems and as a tool to help animate dynamical systems. The size of the fuzzy sets or concepts defines the conceptual granularity of the feedback virtual world. The chapter applies the theory to the swirling undersea world of dolphins and sharks and shows how to use unsupervised learning to encode virtual-world dynamics and to help map them to a cartoon-like animation. The Appendix shows how to use the software for fuzzy function approximation that comes with the textbook.

Bart Kosko
Director
Signal and Image Processing Institute
Department of Electrical Engineering
University of Southern California

PART I

INTRODUCTION

Vague or fuzzy logic has a long history in mathematics and philosophy. It begins with the insight that not all statements are true or false to the same degree. Some claims are more true than others and so truth is a matter of degree. This means that the old laws or axioms of either-or logic do not apply except in the limiting case.

Fuzzy logic extends these vague or continuous logics to reasoning with vague concepts or sets. This requires a new set algebra for the vague concepts. The fuzzy set algebra allows words to map to fuzzy sets and allows sentences to map to fuzzy rules or associations among the fuzzy sets. The rules combine to form systems or maps from an input domain to an output range. Most fuzzy applications depend on a fuzzy system and not a mere fuzzy or gray-scale index of some term.

Fuzzy engineering applies fuzzy systems to the broad task of function approximation. It shifts the focus more toward the mathematical structure of fuzzy concepts and away from their ties to language. Fuzzy engineering seeks the best way to model complex nonlinear systems with the math tools of fuzzy systems and their many extensions.

Chapter 1 reviews fuzzy logic and fuzzy engineering and some of the applications that have helped lead from the former to the latter. It also reviews supervised and unsupervised learning in neural networks as a way to learn and shape fuzzy rules. The last section presents the basic mathematics of both fuzzy logic and fuzzy engineering. It presents the tools one needs to work the detailed homework problems that follow.

1

FUZZY LOGIC AND ENGINEERING

Logicians have too much neglected the study of vagueness, not suspecting the important part it plays in mathematical thought.

Charles Sanders Peirce
Collected Works

Everything is vague to a degree you do not realize till you have tried to make it precise.

Bertrand Russell
The Philosophy of Logical Atomism

The vagueness of the word chair is typical of all terms whose application involves the use of the senses. In all such cases "borderline cases" or "doubtful objects" are easily found to which we are unable to say either that the class name does or does not apply.

Max Black
"Vagueness: An Exercise in Logical Analysis"
Philosophy of Science, volume 4, 1937

PREVIEW

This chapter reviews the basic concepts of fuzzy sets and systems. Statements are true or false to only some degree in a continuous or vague logic. Objects belong to fuzzy sets to only some degree. Fuzzy rules associate if-part fuzzy sets with then-part fuzzy sets. Fuzzy systems are sets of fuzzy rules that map inputs to outputs. Most applied fuzzy systems map sensor measurements to control actions. Neural systems can help find and tune the rules in fuzzy systems. But all fuzzy systems face exponential rule explosion in high dimensions. The last section of this chapter reviews the basic concepts of multivalued logic and set theory. The homework problems extend these concepts.

1.1 FUZZINESS AS VAGUENESS: EVERYTHING IS A MATTER OF DEGREE

Brains do not reason as computers do. Computers reason in clear steps with statements that are black or white. They reason with strings of 0s and 1s. We reason with the vague terms of common sense as in "The air is cool" or "The speed is fast" or

"He is young." These fuzzy or gray facts are true only to some degree between 0 and 1 and they are false to some degree. Brains work with these fuzzy patterns with ease and computers may not work with them at all. Fuzzy logic and engineering try to change that.

The key idea of fuzziness comes from the multivalued logic of the 1920s: *Everything is a matter of degree.* A statement of fact like "The sky is blue" or "The angle is small" or "$e = mc^2$" does not have a binary truth value. It has a *vague* or "fuzzy" truth value between 0 and 1. And so does its negation "The sky is not blue." So the sky is both blue and not blue to some degree. This simple point of fact violates the either-or laws of logic that extend from the first formal logic of ancient Greece to the foundations of modern math and science.

No one has put forth a binary statement fact and no one is likely to. Yet modern science describes its subjects with binary statements and equations. The best we can do is to get the science right to a few decimal places. We cannot get it right to the infinite decimal places that binary truth requires. Factual truth is just accuracy of description and accuracy is a matter of degree [24]. Some logicians and philosophers still deny that and insist on bivalence no matter what the cost [25]:

> If the term "table" is to be reconciled with bivalence, we must posit an exact demarcation, exact to the last molecule, even though we cannot specify it. We must hold that there are physical objects, coincident except for one molecule, such that one is a table and the other is not.

No one has yet put forth the molecular definition of a table or chair or mountain or any other "thing."

Fuzzy logic builds gray truth into complex schemes of formal reasoning. It is a new branch of machine intelligence that tries to make computers reason with our gray common sense. The earlier uses of the term *fuzzy logic* were the same as *continuous truth* or *vagueness*. It meant matters of degree and gray borders and thus breaking the either-or law of binary logic. Today fuzzy logic refers to a fuzzy system or mapping from input to output that depends on fuzzy rules. The rules in turn depend on fuzzy sets or vague concepts like *cool air* or *blue sky* or *small angle* and these terms depend on fuzzy degrees of truth or set membership. Fuzzy logic means reasoning with vague concepts. In practice it can mean computing with words.

The fuzzy system itself is a function or mapping. It is a set of fuzzy if-then rules that maps inputs to outputs. It converts stimuli to responses or sensor measurements to control actions. The rules might have the verbal form "If the wash water is very dirty then add much more detergent" or "If the error is small and positive then turn the wheel a little to the left" or "If the air is cool then set the motor speed to slow." The inputs and outputs can be numbers or vectors of numbers. These rule-based systems can in theory model any system. Fuzzy approximation theorems [16] ensure that and we review them in the next chapter. But these theorems do not ensure that we can always find the rules with a given pool of experts or with a fixed set of data.

Fuzzy engineering is fuzzy function approximation. It often uses fuzzy systems to approximate or control or predict systems in problems of control or signal processing or communication or optimization. It searches for fuzzy systems that come as close as possible to a known or unknown system or function.

A fuzzy system can help model or control a system when we do not have a math model of how the system's output depends on its input. The fuzzy system uses commonsense rules in place of the math model or the so-called plant model. This builds a bridge from the input space to the output space. Until recently [30] there was no known math model to back up a truck-and-trailer in a parking lot to a loading dock if the truck-and-trailer starts from any position. Both humans and fuzzy systems can perform this nonlinear control even for a truck with five trailers [29]. An adaptive fuzzy system can learn its rules from the human as the human gives examples of backing up the truck-and-trailer [13]. Most experts can state fuzzy rules more easily and more accurately than they can state a math model.

A new wave of commercial fuzzy products has fueled interest in fuzzy logic even though the field has grown in some form for most of this century [15, 18]. In 1980 the contracting firm of F. L. Smidth & Co. in Copenhagen first used a fuzzy system to control a cement kiln [8]. The fuzzy system used rules to map the temperature and oxygen content of exhaust gas and the kiln drive torque and clinker temperature to the rate of air flow and coal feed in the kiln. In 1988 Hitachi put a subway under fuzzy control in the Japanese city of Sendai [34]. Since then Japan has produced hundreds of "smart" products that use fuzzy logic. They range from camcorders and washing machines and microwave ovens and TVs to car transmissions and computer disc drives and robot graspers.

Computer chips stud these devices. Toasters and washing machines may have just one or two chips in them while a new car may have 50 to 100 chips in it. Engineers need only reprogram these chips to encode a fuzzy system that smoothly matches control actions to changing sensor measurements. Most control problems do not require new hardware. Communication problems may require special hardware because they work on much faster time scales and may involve more system variables.

Japan's Ministry of International Trade and Industry (MITI) estimates that in 1992 Japan produced about $2 billion worth of fuzzy products. MITI also helped sponsor the fuzzy industrial consortium LIFE or Laboratory for International Fuzzy Research in Yokohama. The $70 million LIFE project ran from March 1989 to March 1995. The Japanese island of Kyushu still sponsors the Fuzzy Logic Systems Institute at the Kyushu Institute of Technology. Now many firms in the East and West sell products that house fuzzy systems in their microprocessors. Many Americans drive GM Saturns that use fuzzy rules to smoothly downshift their automatic transmissions. The large European firms Thomson and Siemens sponsor research teams in fuzzy systems. The U.S. firms Motorola and National Semiconductor build fuzzy chips and support fuzzy research groups. The Institute for Electrical and Electronics Engineers (IEEE) began its annual international fuzzy conference series in 1992 and began the journal *IEEE Transactions on Fuzzy Systems* in 1993.

1.2 THE HISTORY OF VAGUE LOGIC

Fuzziness began as vagueness in the late nineteenth century. Pragmatist philosopher Charles Sanders Peirce seems the first logician to have dealt with vagueness [23]: "Vagueness is no more to be done away with in the world of logic than friction in mechanics." A concept is vague just in case it has blurred boundaries. The concept *mountain* is vague because we do not know where a mountain ends and a hill begins.

Logician Bertrand Russell first identified vagueness at the level of symbolic logic [27]. Concept *A* is vague if and only if it breaks Aristotle's "law" of excluded middle—if and only if *A or not-A* fails to hold. This law fails to hold just to the extent that the "contradiction" *A and not-A* tends to hold as homework problem 1.3 shows. Statements of logic or math obey Aristotle's laws: "$1 + 1 = 2$" is 100% true and 0% false. "$1 + 1 = 3$" is 0% true and 100% false. But statements of fact are vague and have truth values between these binary extremes: "Grass is green" may be true only 80% and so "Grass is not green" is true 20%. Russell first saw this mismatch between gray fact and binary math and then looked for it in math itself.

Russell revised the *sorites* paradox of the ancient Greek philosopher Zeno to show that induction was not binary: Am I bald? No. Pluck out a hair. Am I bald now? No. You keep plucking and asking but do not find that one hair (or molecule) that takes you from not-bald to bald. Yet you are bald if you pluck out all or most of your 100,000 or so head hairs. Each plucked hair slightly increases your degree of baldness and decreases the strength of the inductive chain [5]. The final vague inference of not-baldness has the near-zero truth value that comes from multiplying a large number of values between 0 and 1. This would not affect a pure binary inference because it would multiply a string of 1s. It would imply that you are still 100% not-bald even after you have plucked out all your hair.

Russell found a deeper paradox in math itself as he worked with Alfred North Whitehead on the pioneering volumes of *Principia Mathematica* [31]. He found the ancient paradox of the liar from Crete. The Cretan says that all Cretans lie. Does he lie or tell the truth? If he lies then he tells the truth and so does not lie. If he does not lie then he tells the truth and so he lies. Both cases lead to the contradiction *A and not-A*.

Russell found the same paradox in set theory. The set of all sets is a set. So it is a member of itself. But many sets are not members of themselves. The set of apples is not a member of itself since its members are apples and not sets. But what about the set of all sets that are not members of themselves? Is it a member of itself? If it is then it is not. And if it is not then it is. Here *A and not-A* holds not in the gray world of things but in the formal system of binary mathematics.

Russell at first put forth his "theory of types" to ban such paradoxes but the paradoxes still emerged in other forms despite the ban. There is something deeply counterintuitive about denying that the set of all sets is not itself a set as many have done by calling it a "class." Russell saw by the time of his 1923 article "Vagueness" that we might have to relax if not reject Aristotle's law of excluded middle both to deal with paradoxes and to account for the vagueness of factual statements. Formal fuzzy logic begins with this 1923 article.

The math section 1.6 shows that fuzziness resolves the liar paradox with the midpoint truth value of 1/2 or 50%. The paradoxes are half truths. The Cretan lies 50% and does not lie 50%. Bivalent logic gets stuck because it cannot round off 1/2 to 0 or to 1. They are both equally close corners or vertices in the unit interval [0,1] of truth values. More complex paradoxes pick out the midpoints of fuzzy "cubes" of higher dimension [13]. Figure 1.1 shows the fuzzy 1-cube and the fuzzy 2-cube or square and the paradox that can lie at their midpoints. Grim [7] has shown that we can also view the liar paradox as a dynamical system of truth values that flip back and forth from 0 to 1 or that converge to the lone fixed-point attractor of the midpoint 1/2. Multidimensional paradoxes take place in multidimensional fuzzy cubes. A fuzzy 2-cube houses the medieval dualist paradox:

Socrates: "What Plato is about to say is true."
Plato: "Socrates lies."

The math review section 1.6 shows one of many ways to convert the dualist paradox into a dynamical system in a fuzzy 2-cube.

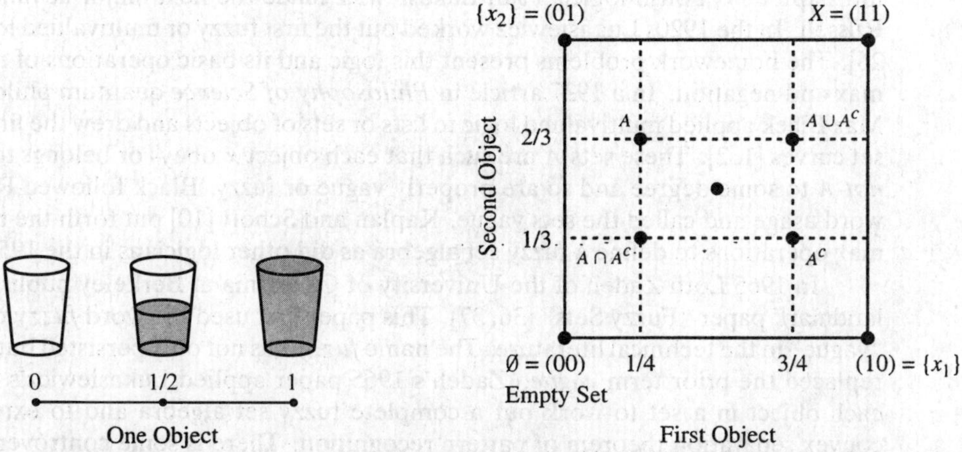

Figure 1.1 Discrete fuzzy sets and paradoxes. One object can vary from 100% present to 0% present (or 100% absent) and so defines the unit interval or 1-D fuzzy cube [0,1]. Two objects define a 2-D fuzzy cube in the same way. Each cube contains a unique midpoint that is equidistant to the binary corners. The point $A = (1/4,\ 2/3)$ defines a fuzzy subset of the two objects. It has the fuzzy set complement $A^c = (3/4,\ 1/3)$. These two points lie at the corners of an interior square with the intersection corner $A \cap A^c = (1/4,\ 1/3)$ and the union corner $A \cup A^c = (3/4,\ 2/3)$. Long diagonals connect both nonfuzzy and fuzzy set complements. The four fuzzy points contract to the midpoint as A gets fuzzier and expand out to the binary corners as A gets less fuzzy.

A set X of n objects $x_1, \ldots, x_n$ gives rise to the fuzzy n-cube or unit hypercube $[0, 1]^n$ and the view of *sets as points* [13]. The fuzzy n-cube has 2^n corners or binary subsets and just one midpoint. One corner of a 6-cube is the bit list (1 0 0 1 1 0) where a 1 means an object is present and a 0 means it is absent. The bit list (1 0 0 1 1 0) means that the corner set contains the first, fourth, and fifth objects and so defines the set $\{x_1,\ x_4,\ x_5\}$. The points inside the cube define fuzzy subsets. The *fit* (fuzzy unit) list (1, 3/4, 2/3, 1/3, 1/4, 0) defines a fuzzy set that contains only five objects and contains four of them to only some degree.

The deeper inside the cube a fuzzy set A lies the closer it is to the midpoint $(1/2, \ldots, 1/2)$ and the more A resembles *not*-A and the more it breaks the law of excluded middle. The midpoint is the fuzziest set of all since it and it alone obeys the "paradoxical" relation $A =$ *not*-A. There is no binary way to round off a cube midpoint to one of the 2^n corners since it is just that set equidistant to all 2^n binary corners. *A or not-A* holds 100% at the cube corners or for binary sets. *A and not-A* holds 100% at the cube midpoint. Both relations hold only partially between these two extremes. Homework problem 3 shows that a theorem of multivalued logic constrains this relation between excluded middle and contradiction. Probability logics [5, 26] always obey both the law of excluded middle and the law of noncontradiction [$p(A \text{ or not-}A) = 1$ and $p(A \text{ and not-}A) = 0$] as discussed in the next chapter. This theorem also shows that some degree of paradox in logic is the rule and not the exception.

The paradoxes motivated much of the early work in vague or fuzzy logic as did the uncertainty principle of quantum mechanics (and of signal processing as discussed in Chapter 7). Polish logician Jan Lukasiewicz made the next major advance after Russell. In the 1920s Lukasiewicz worked out the first fuzzy or multivalued logic [17, 26]. The homework problems present this logic and its basic operations of min and max and negation. In a 1937 article in *Philosophy of Science* quantum philosopher Max Black applied multivalued logic to lists or sets of objects and drew the first fuzzy set curves [1, 2]. These sets A are such that each object x obeys or belongs to *A and not-A* to some degree and so are properly vague or fuzzy. Black followed Russell's word usage and called the sets vague. Kaplan and Schott [10] put forth the min and max operations to define a fuzzy set algebra as did other logicians in the 1950s [4].

In 1965 Lotfi Zadeh of the University of California at Berkeley published the landmark paper "Fuzzy Sets" [36, 37]. This paper first used the word *fuzzy* to mean "vague" in the technical literature. The name *fuzzy* has not only persisted but largely replaced the prior term *vague*. Zadeh's 1965 paper applied Lukasiewicz's logic to each object in a set to work out a complete fuzzy set algebra and to extend the convex separation theorem of pattern recognition. There is some controversy here [15, 18] because Zadeh did not refer to the works of Lukasiewicz or any of the other multivalued logicians who had long since defined the vague concepts and pointwise operators (*min*, *max*, and $1 - x$) at the heart of fuzzy set theory. This has often led to the perception that multivalued logic begins with fuzzy sets despite over a half century of prior work [4]. Still Zadeh brought about the second wave of multivalued research under the banner and language of fuzzy logic and did so almost singlehandedly. The IEEE awarded him its medal of honor in 1995 for his work in fuzzy sets.

In the mid–1970s Ebrahim H. Mamdani of Queen Mary College in London first applied fuzzy sets to systems in the form of a rule-based fuzzy system [19]. Mamdani designed a fuzzy system to control a steam engine and used a fuzzy system that differed only in detail from the fuzzy systems used today. Mamdani's work marks the start of fuzzy engineering.

1.3 FUZZY SETS AND SYSTEMS

A set contains objects. A set is the theoretical primitive of mathematics just as the symbol is the theoretical primitive of logic [31]. Set A contains an object x to some degree. Bivalent set A contains x all or none. A fuzzy set A contains x to some degree in [0,1]. The discrete set (1 0 0 1 1 0) contains just three of six possible objects. The discrete fuzzy set (1, 3/4, 2/3, 1/3, 1/4, 0) contains four elements only partially. We can extend these set distinctions to the continuous case where the sets are subsets of the real line.

Consider a simple air conditioner. The air temperature defines a fuzzy variable or "linguistic" variable [37]. The air can take on the fuzzy-set values cold, cool, just right, warm, or hot. We can define fuzzy-set curves for these sets as in Figures 1.2 and 1.3. Figure 1.2 compares bivalent and multivalent sets of cool-air degrees. The fuzzy set lists partial contradictions that sum to unity for each object x. The air is both cool and not-cool to some degree. If the air is 60% cool then it must be 40% not cool. The cube midpoint now becomes the midpoint line at 1/2. It gives the only fuzzy set such that $A = not\text{-}A$. A fuzzy set A always intersects its set complement *not-A* at the midpoint line. Bivalent sets jump over this midpoint line as they switch abruptly from A to *not-A*.

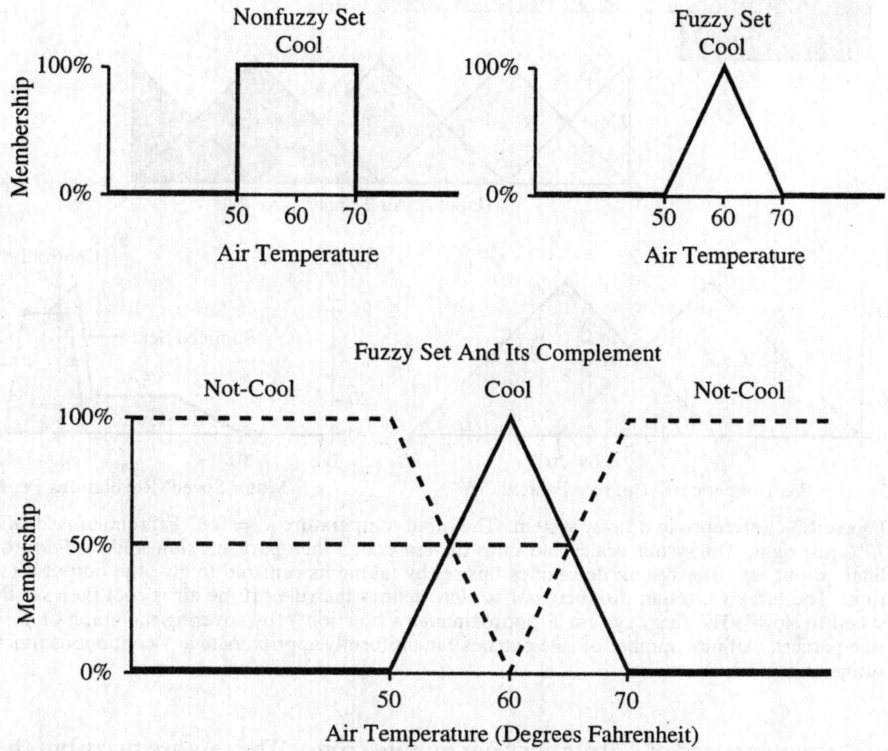

Figure 1.2 Bivalent versus fuzzy continuous subsets of the real line. Each curve defines a set function.

Fuzzy sets form the building blocks for fuzzy if-then rules such as "If the air is cool then set the motor speed to slow" or "If the air is hot then set the motor speed to blast." The rules have the form "If X is A then Y is B" where A and B are fuzzy sets. A fuzzy system is a set of fuzzy rules that converts inputs to outputs. So a fuzzy system is a mapping or function from an input space of alternatives to an output space. The fuzzy air conditioner might map the input temperature 68° Fahrenheit to

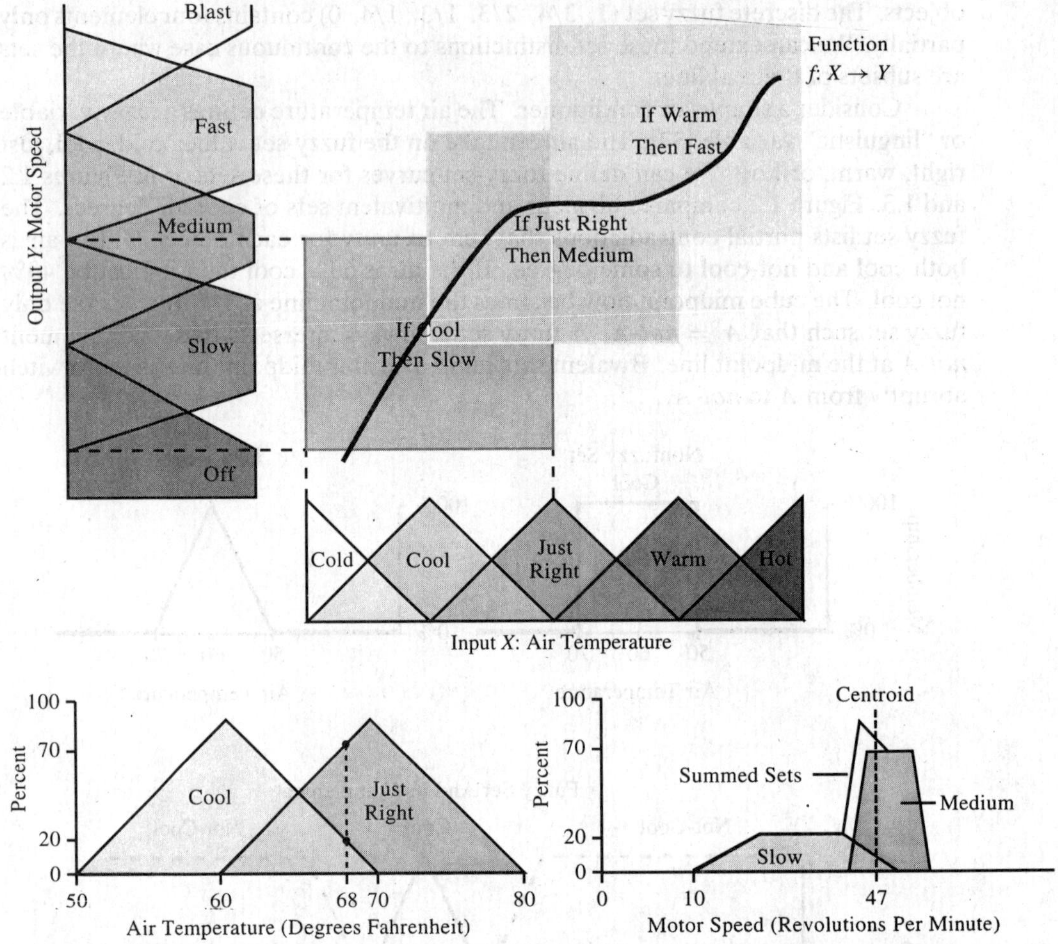

Figure 1.3 Inference in a fuzzy system. The input temperature $x = 68°$ Fahrenheit is 20% cool and 70% just right. The system scales and sums the associated then-part sets *slow* and *medium* to give the final output set. The system defuzzifies this set by taking its centroid to give the output $F(x) = 47$ rpm. The fuzzy Cartesian product $cool \times slow$ defines the rule "If the air is cool then set the motor speed to slow." The fuzzy system F approximates a function f by covering the graph of f with such rule patches. A finite number of rule patches can uniformly approximate any continuous function on a compact domain.

a motor speed of 47 rotations per minute (rpm). The sets are fuzzy but the inputs can be exact. The inputs can also be fuzzy sets themselves in a *set* system as discussed in Chapter 2 (Section 2.7).

Fuzzy engineering begins with three steps. The first step picks the input and output variables X and Y. Here the input variable X is air temperature. We might add to it the variables humidity or light intensity. The output variable Y is a set of numerical motor speeds. The second step picks fuzzy subsets of these variables. Figure 1.3 shows the if-part triangular sets for cold, cool, just right, warm, and hot air and the then-part trapezoidal sets for stop, slow, medium, fast, and blast motor speeds. The third step relates the output sets to the input sets in fuzzy rules:

Rule 1: If the air is cold then set the motor speed to stop.

Rule 2: If the air is cool then set the motor speed to slow.

Rule 3: If the air is just right then set the motor speed to medium.

Rule 4: If the air is warm then set the motor speed to fast.

Rule 5: If the air is hot then set the motor speed to blast.

This gives the first-cut fuzzy system. An expert might give the rules or we might guess at them ourselves or use an adaptive algorithm to grow them from training data. Much of fuzzy engineering deals with tuning these rules and perhaps adding new rules and deleting old ones.

The fuzzy system F maps an input x to an output $F(x)$ in three steps. The first step matches the input x to all the if-part fuzzy sets in parallel. This step "fires" or "activates" the rules by how much the input x belongs to each if-part set A. Each input x fires at most two rules for the if-part sets in Figure 1.3. Most applications use such overlapping sets along each input axis. So an input n-vector x fires 2^n rules. This reflects the rule explosion that all fuzzy systems suffer in higher dimensions. Then each fired if-part set A scales its then-part set B and B shrinks down to this height. The second step adds all the scaled or shrunken then-part sets into a final output set. The third step is "defuzzification." The system computes the output $F(x)$ as the centroid or center of gravity of this final output set. Sometimes the system picks the mode or maximum value of this output set as $F(x)$. Section 1.6 states the SAM (standard additive model) equation that computes the output $F(x)$ given the input x. Figure 1.3 shows the fuzzy inference process for the input temperature 68° Fahrenheit.

A fuzzy chip repeats this parallel inference process thousands or millions of times each second. We measure a chip's inference rate in FLIPS or fuzzy logical inferences per second. Masaki Togai and Hiroyuki Watanabe built the first digital fuzzy chip at Bell Laboratories in 1985. It processed 16 simple rules in 12.5 milliseconds or at a rate of 1.28 kilo-FLIPs. Takeshi Yamakawa of Kyushu Institute of Technology built the first analog fuzzy chip in 1987. It ran at 0.2 mega-FLIPs. Analog chips [20] tend to run faster than digital chips run but users can more easily reprogram a digital chip to change the shape of its fuzzy sets and rule patches. Most modern chips are digital and process many rules at the mega-FLIPs rate. Motorola and National Semiconductor make fuzzy chips in the United States. Chapter 10 reviews the design of fuzzy chips and shows how to design a learning or adaptive fuzzy chip. Few small-scale problems require fuzzy hardware. Instead software runs some form of the SAM equation (1.18) in Section 1.6.

Figure 1.3 also shows how fuzzy rule patches geometrize knowledge. Each rule like "If the air is cool then set the motor speed to slow" defines a fuzzy rule patch or Cartesian product *cool* × *slow* [13, 16, 37]. Patch size can measure the vagueness or uncertainty in the rule. Smaller patches tend to reflect more precise knowledge or less noisy data. A patch shrinks to a point in the precise limit when both *cool* and *slow* define spike sets. Larger patches are less certain and often stem from noisy data. Chapter 3 works with egg-shaped or ellipsoidal rule patches and shows how noisy data can both grow them and adjust their size.

Rule patches lead to the fundamental fuzzy approximation theorem [14–16] of fuzzy engineering. A fuzzy system F approximates a function f by covering the graph of f with rule patches and averaging patches that overlap. The next chapter explores this result in detail and many other chapters apply it to real problems. The approximation is uniform and that allows the user in theory to pick the approximation error level in advance. The search for such rules may not be so easy in practice and makes up one of the main research areas of fuzzy engineering. The approximation theorem has a constructive proof that suggests that data clusters can define rule patches if there are enough data and if the data reflect the unknown system or process f.

The patch covering of fuzzy systems also leads to their greatest weakness: exponential rule explosion [14]. The number of rules a fuzzy system F needs to cover the graph of a function f grows exponentially with the number of input and output dimensions n and p. More input variables can give a better causal model of a process but at a higher cost in knowledge acquisition and computation. Suppose n input variables each have m fuzzy subsets on their axes. Then it takes m^n rules just to cover the input space. Optimal rules can make the best of a fixed rule budget but may be hard to find. Lone optimal rule patches cover the turning points or extrema of f but that knowledge may be of no help if the user does not know at least the rough shape of the approximand f. The goal of fuzzy learning theory is to shape and move the rule patches to optimal locations. But the neural learning schemes discussed in Section 1.5 often involve their own exponential computational complexity and require a stream of accurate data that the user may not have.

1.4 FUZZY SYSTEMS IN COMMERCIAL PRODUCTS

Fuzzy products use fuzzy systems in their microprocessors. Some use a fuzzy chip but most just reprogram their existing chip. The fuzzy inference scheme discussed previously and in Section 1.6 takes up only a few lines of software code. Japan leads the world in fuzzy products and holds over a thousand patents in Japan on fuzzy designs. Most applications have had few inputs and outputs and this has helped keep the rule explosion manageable. Fuzzy systems are also highly nonlinear and model-free. So it often takes extensive computer simulations or real tests to study their sensitivity and stability in control applications.

The most famous fuzzy application is the fuzzy control of the subway system in the Japanese city of Sendai. The subway runs on a 13.6-kilometer route with 16 stations. Hitachi [35] programmed rules in a fuzzy system to brake the subway and a fuzzy system to speed and slow it: If the train speed exceeds the limit speed then slow the speed. If the train is in the allowed zone then brake slightly (rather than accelerate). The fuzzy system gives a smoother ride than did human control and it outperforms standard PID (proportional-integrative-differential) controllers in smoothness of braking and acceleration and in electric power consumption. It also stops the subway with greater accuracy.

The Japanese government required extensive testing of the fuzzy system. In the 1980s Hitachi Corporation ran over 300,000 simulations of the fuzzy system and ran over 3,000 runs on real but empty subways before the government let the fuzzy system replace human controllers in 1987. Today the fuzzy system runs the subway during peak hours. Humans still control the subway in nonpeak hours to keep up their operating skills.

Several firms in Japan and Korea manufacture fuzzy washing machines. A fuzzy washing machine gives a finer wash than does a "dumb" washing machine with fixed commands. It tailors the wash to each set of clothes and changes the wash strategy as the clothes clean. That helps prevent cloth damage and under- or overwashing. A fuzzy washing machine turns sensor data into wash times and wash cycles. A pulsing optical sensor measures the murk or clarity of the wash water and measures how long it takes a stain to dissolve or saturate in the wash water. Mud and dirt break down quickly. Oil stains break down more slowly. Some machines use a load sensor to change the agitation rate or water temperature. Some shoot bubbles into the wash to help dissolve dirt and detergent.

The simplest systems map water clarity and saturation time to wash time. They may use as few as 10 fuzzy rules: If the water clarity is low and the saturation time is short then wash long. If the water clarity is low and the saturation time is long then wash very long. If the water clarity is high and the saturation time is short then wash short. Most wash water is clear to some degree and saturates quickly or slowly to some degree. So each second the machine's microprocessor fires most of the rules to some degree. Engineers run thousands of experiments in advance to tune the fuzzy sets *short*, *medium*, *high*, *long*, and others to new shapes. Some systems use neural or statistical schemes to tune the fuzzy-set curves.

Fuzzy cameras and camcorders map image data to lens settings. The first fuzzy camera was the Canon handheld H800 model in 1990. It tuned the autofocus with 13 fuzzy rules. Sensors measure the image clarity and the change in the image clarity in six parts of the image. The 13 rules take up only 1.1 kilobytes of memory and convert the 12 types of sensor data to new lens settings. The first fuzzy camcorder was the Sanyo-Fisher 8mm FVC–880 model. It tuned the autofocus with only 9 fuzzy rules. The rules tune the lens setting with the relative contrast between six regions of the image. The center region counts most and the border regions count least. Nonfuzzy camcorders just weigh and add the image intensity in the six regions.

The fuzzy rules adjust the camcorder lens with the relative contrast between the six regions. Some regions are much brighter or a little less brighter than others or are almost equal in brightness. Matsushita (Panasonic) adds more rules to cancel the image jitter that a shaking hand causes in small camcorders. The fuzzy rules tell movement within the frame from movement of the whole frame: If all the image points move in the same direction then the hand shakes. The system compares the current frame with past frames to look for global movement. Math models of image jitter tend to model and cancel only a few types of jitter.

A few fuzzy rules map sensor data to control settings in many household products and car systems. A fuzzy dryer converts load size and fabric type and the flow of hot air to drying times and drying strategies. A fuzzy vacuum sweeper uses changes in dust flow to judge if the floor is bare or carpeted. A pulsing infrared light emitting diode measures the dust flow. A simple 4-bit microprocessor converts the dust flow to suction power and beater-bar settings. Mitsubishi and Korea's Samsung report energy savings of 40% to 100% with their fuzzy vacuum sweepers over their nonfuzzy designs.

Hitachi, Sanyo, Sharp, and Toshiba have designed fuzzy microwave ovens. A fuzzy microwave oven measures infrared light patterns, temperature, humidity, and change in food shape. Fuzzy rules associate these conditions with whether the food is frozen or thawed and with how well cooked it is. This in turn maps to power and cooking times and control actions like hot-air blowing or roasting.

Nissan has patented a fuzzy antiskid braking system, fuzzy transmission system, and fuzzy fuel injector. The antiskid system tries to brake right up to the point of locking when the car slows down too fast. The fuel injector adjusts the fuel flow with fuzzy rules in an on-board microprocessor. Sensors measure the throttle setting, manifold pressure, water temperature, and car rpm. A second set of fuzzy rules times the ignition with sensor data on the car rpm, water temperature, and oxygen concentration. Mitsubishi has developed an omnibus fuzzy system that controls a car's suspension, transmission, steering, traction, and air conditioner.

Michio Sugeno of the Tokyo Institute of Technology has designed one of the most complex fuzzy systems. It uses fuzzy voice commands like "up" and "land" and "hover" to control an unmanned helicopter [28]. Pilots must train for weeks to hover in a helicopter. The first version was a scaled helicopter of length 3.58 meters and that weighed 20 kilograms and had a 1-meter rotor. The new version has a 3-meter rotor. The system takes 13 inputs and gives four control outputs in terms of the elevator, aileron, throttle, and rudder. One set of fuzzy rules controls the higher-level task of navigation: If the flight mode is hover and if the flight state is forward then the standard trim is such and such mix of pitch angle, roll angle, elevator offset, and aileron offset. A second set of rules controls the lower-level task of stabilization: If the body pitches then control the elevator in reverse. An Omron 16-bit microprocessor houses the entire fuzzy system. Cavalcante [3] has designed a like fuzzy system to control a simulated helicopter.

Fuzzy systems can also manage information systems. Omron uses a fuzzy system to manage five medical databases in a health management system for large firms. The fuzzy system uses 500 rules to give each of up to 10,000 patients a health diagnosis and a personal health plan to help prevent disease and to help each patient stay fit and reduce stress. Hitachi uses 150 or so fuzzy rules to trade in Japanese bonds and bond futures. Yamaichi Securities uses hundreds of rules to manage a stock fund. The rules model private firms and the Japanese economy and large industries such as textiles and automobiles and consumer electronics.

In most fuzzy products and systems an expert has given rules to the engineer. A few fuzzy rules give a quick approximation of the control system. Then the hard task is to tune the rules and fuzzy sets. That can take days or weeks of software trial and error. There are now many software packages that help users tune fuzzy systems and port them to microprocessors. Often there are no experts to ask for rules. Then the user must turn to statistical or neural schemes to learn the rules from training data.

1.5 ADAPTIVE FUZZY SYSTEMS AND THE SEARCH FOR BETTER RULES

The Achilles' heel of a fuzzy system is its rules. Smart rules give smart systems and other rules give less smart or even dumb systems. Some rules can conflict with other rules or contradict them outright. Experts give and help tune most of the rules in commercial fuzzy systems. Adaptive fuzzy systems can help automate this process by using neural networks or neural-like learning algorithms to tune rules and thus to move the fuzzy rule patches in the state space.

Neural networks are sets of neurons and synapses that map inputs to outputs. A node or neuron adds up signals from other nodes and then emits its own signal. These signals are just numbers. Simple neurons emit an on (1) or off (0) signal and act as switches. Smooth neurons emit a signal that rises smoothly from off (0) to on (1). The signals travel over the edges or synapses that connect the nodes. Each edge has a numerical synaptic value. A neural net with n neurons needs n^2 synapses to interconnect it fully.

The neural nodes and edges change with time and so define a dynamical system. At any time the state of the network is either a transient state or an equilibrium state. Once trained a neural net quickly "settles down" or converges to an equilibrium state when fresh input data perturbs it. The input data act as a question or stimulus. The equilibrium state acts as an answer or system response. More complex neural systems consist of networks of networks [22] as in the mammalian brain [6]. The goal of neural engineering is to tie the equilibrium states to answers to hard questions or solutions to hard tasks like pattern recognition or signal filtering or cost minimization. An adaptive fuzzy system ties the equilibrium states to the choosing or tuning of fuzzy rules.

Most neural nets run in software. Then the neurons and synapses reduce to equations that update their state. Digital neural chips run the neural equations at high speed and can update the synapses millions of times per second. In analog neural chips resistors act as synapses and amplifiers act as neurons [20]. In optical neural systems light beams act as synapses and special mirrors or other electro-optical devices act as neurons.

A neural net learns by changing its synapses. The numerical synaptic values change when input data make the neurons fire. The net can learn to recognize patterns like a face in an image or a scrawled zip code on an envelope or a high-risk loan applicant or a bomb in an X-ray scan of a suitcase. In general a neural net maps inputs to outputs. So like a fuzzy system it learns or approximates a function or relation of cause and effect and can act as a universal approximator.

Neural systems resemble fuzzy systems in other ways. Radial basis function networks have the same form as Gaussian fuzzy systems and so suffer from the same rule explosion in high dimension. Multilayer feedforward "perceptron" networks [13] do not suffer from rule explosion since they can approximate functions with a fixed number of "hidden" neurons in their interior layers or fields of neurons. But they do suffer from the "connectionist glop" of massive webs of synapses. These distributed webs have no direct meaning and give no audit trail of how the neural net infers an output. Fuzzy rules in contrast are modular and the user can put them in the system or take them out at will. Distributed synaptic webs have the further problem of forgetting old things when they learn new things. A user has no way to know what a neural net has learned or forgotten except to test the net with input-output pairs.

Neural learning is supervised or unsupervised depending on what data the neural net uses to learn. A supervised net learns by trial and error. It needs a teacher to tell it when it has made an error. The error is the difference between the desired output and the net's output. The user runs the training samples through the net over and over until the net gives the right response for each input. This statistical method can take thousands or hundreds of thousands of training cycles.

Supervised nets tune the rules of a fuzzy system as if they were synapses in the net. They slightly vary the size or shape of one fuzzy set and see how well the system performs with that change. The net tends to keep those changes that improve how the system performs and tends to ignore the rest. The user must give the fuzzy system its first set of rules. Then the neural net tests hundreds of thousands of cases as it tries to fine tune those first fuzzy rules. The final tuned fuzzy system still depends on how well humans or some math algorithm guesses at the first set of rules.

In Japan Sanyo uses supervised neural learning to tune the fuzzy rules in its microwave oven. Several firms use supervised learning to tune the rules in their washing machines. Sharp uses it to tune the rules in its refrigerator and to learn how often the user opens the door. The neural net learns off-line before the customer buys the product. Once sold the tuned rules are fixed. Future research will tune the rules on-line as the user uses the product.

The goal is to shape the fuzzy sets to the user in real time: Data in and rules out. Consider again the fuzzy air conditioner. Each person means something similar but different when they say the air temperature is "cool" or "warm" or "just right." The same person may mean different things by these words at different times of the day or in different seasons. The words stand for fuzzy-set curves. Each person means a slightly different curve for each word. The overall shape may look like a triangle or trapezoid or bell curve or any other figure. The different shades of meaning in *warm* mean that up close the curves differ in length or width or position or slope or wiggle. A user could train a supervised net by telling it how an air temperature feels or by ranking different air temperatures. Then each user would have to help retune the system just as now each driver has to readjust the rear-view mirror.

Supervised learning schemes suffer from suboptimality as well as intense computational complexity. They tend to get stuck in local minima of the "error surface" [22]. Each set of network synaptic values defines one error value. All possible such sets define a bumpy error surface. The net starts at one point on the error surface and tries to move down the surface to the point of least error as if the net were a marble that rolled down a large bumpy sheet of paper. The neural net gets stuck in the nearest local error minimum just as the marble gets stuck in the bottom of the nearest pit in the paper.

Neural researchers have proposed hundreds of schemes to lessen the problem of local error minima but none has removed it [9]. In practice that means it can take hours or days of computer time to tune a fuzzy system with a supervised neural net.

And the final result may not be much better than the first set of rules. That leads to trial and error with different choices of fuzzy rules and different nets. At present there is no closed-form scheme to tune a fuzzy system optimally with supervised learning. Local optimality may be the best neural-fuzzy systems can ever achieve in general.

Unsupervised learning clusters data. The clusters then approximate the fuzzy rule patches. No teacher tells the neural net to which class a piece of data belongs or whether the net gives a good or bad response to the input. There is no error since there is no desired response. The net must blindly cluster the data into groups whose members resemble one another. Unsupervised learning is much faster than supervised learning. The math algorithms are simpler and in theory use just one pass of the data. In some cases when data are sparse it must recycle through the system more than once.

Unsupervised learning can find the first set of rules for a fuzzy system. All it needs is numerical input-output data from an expert or from a physical process or from a math algorithm. Unsupervised learning can convert most math algorithms into a set of fuzzy rules that behaves much as the math algorithm behaves. Supervised learning can tune the rules once unsupervised learning has found the first set of rules. Then unsupervised learning acts as a human expert who converts his experience into a rough set of rules. Chapter 3 uses this hybrid scheme to tune ellipsoidal fuzzy rules. Unsupervised competitive learning finds the first rule patches to initialize the gradient descent of the supervised learning. Competitive learning drives quantization vectors that move in the state space to best approximate the pattern of sample data. A covariance matrix measures the error ball or ellipsoidal about each quantization vector. These ellipsoids define the fuzzy rules. Chapter 6 extends this neural technique to a like measure of covariation for statistics that have infinite variance and suffer from highly impulsive noise.

Future neural-fuzzy research must deal with large-scale nonlinear systems with many variables. These systems can arise in plant control or airline scheduling or economic modeling. Users might not have experts or common sense to give the rules. Neural nets may have no data or only sparse data to use to learn rules.

More variables also increase the complexity of fuzzy systems. Again the number of fuzzy rules grows exponentially with the number of system variables. This rule explosion or "curse of dimensionality" faces all math models and computer schemes. Large rule patches reduce the rule count but give less precise control or approximation. So fuzzy systems must trade some accuracy for ease of computation. So far most fuzzy applications have had few variables and have been in control. This is as much due to social chance as to the ease of designing small-scale fuzzy control systems. Many of the first fuzzy engineers were control theorists and a control loop lies inside most products in consumer electronics.

Future fuzzy systems may extend well beyond problems in control and machine intelligence. Already they have spread to new designs in signal processing and communications and multimedia. The fuzzy approximation theorem applies to any model of cause and effect in any branch of science or engineering.

1.6 OVERVIEW OF FUZZY LOGIC, SET THEORY, AND SYSTEMS

A *statement* S is a claim that has a truth value $t(S)$. So truth defines a mapping from the set of statements to the set of truth values: t: {*Statements*} $\rightarrow$ {*Truth Values*}. Classical or Aristotelian truth allows only the two truth values *true* and *false* or 1 and 0: t: {*Statements*} $\rightarrow$ {0, 1}. An n-valued logic maps to n truth values. A common 3-valued logic maps to truth and falsehood and midpoint: t: {*Statements*} $\rightarrow$ {0, 1/2, 1}. A continuous-valued logic maps to the unit interval: t: {*Statements*} $\rightarrow$ [0, 1]. All such logics are fuzzy logics and include the classical logic as a special case.

Truth functions give the truth of compound statements in terms of the truth of their component statements [5, 26]:

$$\begin{aligned} t(A \text{ AND } B) &= \min(t(A),\ t(B)) \\ t(A \text{ OR } B) &= \max(t(A),\ t(B)) \\ t(\textit{not-}A) &= 1 - t(A). \end{aligned} \tag{1.1}$$

Suppose t(*Grass is green*) = .8 and t(*Snow is white*) = .9. Then t(*Grass is green AND Snow is white*) = .8, t(*Grass is green OR Snow is white*) = .9, t(*Grass is NOT green*) = .2, t(*Snow is both white and NOT white*) = .1, and t(*Snow is either white or NOT white*) = .9. The homework problems show that functions [21] other than min and max can define *and* [11, 33] and *or* [34] and even *not*. Other problems use fuzzy knowledge combiners [12, 32] to fill in the operators that range between *and* and *or* and thus between *min* and *max*. The homework problems review these operators.

The truth function for implication can have many forms. The Lukasiewicz form comes from the truth gap $t(A) - t(B)$ and is most common [5]:

$$t_L(A \rightarrow B) = \min(1,\ 1 - t(A) + t(B)). \tag{1.2}$$

Statements A and B are logically equivalent or $A = B$ if and only if A implies B and B implies A. Then homework problem 6 shows that Lukasiewicz equivalence has the form

$$t_L(A = B) = 1 - |t(A) - t(B)|. \tag{1.3}$$

So two fuzzy statements A and B are (100%) equivalent if and only if they have the same truth values: $t(A) = t(B)$.

Now consider Russell's paradox. It ends up with two implications of the form $A \rightarrow \textit{not-}A$ and $\textit{not-}A \rightarrow A$. So A and *not-*A are logically equivalent: $A = \textit{not-}A$. So they have the same truth values [5]:

$$\begin{aligned} t(A) &= t(\textit{not-}A) \\ &= 1 - t(A). \end{aligned} \tag{1.4}$$

Then bivalent logic leads to the contradictions $1 = 0$ or $0 = 1$. Note that this argument assumes no form for implication or equivalence. It uses only the order reversal of negation. There is no logical reason to insist that either $t(A) = 0$ or $t(A) = 1$ must hold for all statements A. Instead we can let the structure of the paradox tell its own story and solve the equation $t(A) = 1 - t(A)$ for the "paradoxical" truth value $t(A)$. The result is the midpoint of the 1-dimensional fuzzy hypercube [0,1]: $t(A) = 1/2$.

Grim [7] has shown that we can view Russell's paradox and other "liar" paradoxes as a discrete dynamical system of truth values that changes with n:

$$t(S_{n+1}) = 1 - t(S_n). \tag{1.5}$$

The binary case gives the oscillating sequence T, F, T, F, T, F, . . . or 1, 0, 1, 0, 1, 0, . . . where the truth values flip back and forth as a type of dialogue with a self-referential liar. The midpoint truth value 1/2 gives the only fixed-point attractor of the dynamical system. The Lukasiewicz equality operator also leads to the liar dynamical system

$$t(S_{n+1}) = 1 - |(1 - t(S_n)) - t(S_n)| \tag{1.6}$$

to model such self-referential statements as "This sentence is true if and only if it is false" or "This sentence is as true as it is false." Here the state space is the fuzzy 2-cube or $[0, 1]^2$. Graphs of the 2-cube trajectories often show the liar falling into a chaotic attractor. The squaring function can model the adjective *very* in fuzzy logic and leads to the dynamical system

$$t(S_{n+1}) = (1 - |(1 - t(S_n)) - t(S_n)|)^2 \tag{1.7}$$

to model such statements as "This sentence is very true if and only if it is false." The dualist paradox

Socrates: "What Plato is about to say is true."
Plato: "Socrates lies."

might give rise to a nonlinear 2-D discrete dynamical system with state equations

$$\begin{aligned} t(S_{n+1}) &= 1 - |t(S_n) - t(U_n)| \\ t(U_{n+1}) &= 1 - |t(U_n) - (1 - t(S_n))^2|. \end{aligned} \tag{1.8}$$

Here the midpoint of the 2-cube does not act as a fixed-point attractor as in the 1-D case.

A set A contains an object x to degree $a(x)$:

$$a(x) = Degree(x \in A). \tag{1.9}$$

The map $a: X \rightarrow$ {*Membership Degrees*} is a set function or "membership" function. A classical or bivalent set A has a binary set function $a: X \rightarrow \{0, 1\}$ that maps all objects x to 1 or 0 or to *in* or *out*:

$$a(x) = \begin{cases} 1 & \text{if } x \in A \\ 0 & \text{if } x \notin A. \end{cases} \tag{1.10}$$

A multivalued or vague or "fuzzy" set A has a set function that maps to 3 or more *fit* or fuzzy unit values. Here we take the range of fit values as the continuum of the 1-D unit hypercube: $a: X \rightarrow [0, 1]$.

The fit status of each object x in a fuzzy set A acts as a statement of multivalued truth: $a(x) = t(x \in A)$. We can also view it as the value of a discrete conditional probability: $a(x) = \text{prob}(x \in A | X = x)$. Suppose $A \subset R$ is the set of *cool* air temperatures. The fuzzy view casts A as a locus of fit values. The term $X = A$ means "The temperature is cool" or the fuzzy variable X takes on the fuzzy-set value A. The fit value $a(x)$ might mean "The temperature value of 68° Fahrenheit is cool to degree $a(x)$." On the probability view $a(x)$ means "The probability that the air is cool is $a(x)$ given that the temperature is x degrees Fahrenheit or given that $X = x$." So the probability view casts A not as a set but as a locus of two-point conditional densities. We work only with the set view [36] in this text.

Fuzzy truth functions define fuzzy-set operations pointwise through set functions. Suppose X is the ground space that contains the fuzzy subsets A and B. Then

$$a^c(x) = 1 - a(x)$$

$$a \cap b(x) = \min(a(x), b(x)) \tag{1.11}$$

$$a \cup b(x) = \max(a(x), b(x)).$$

Again other functions can define these operations and the homework problems explore some of them. At the systems level we often use product instead of min to define pairwise intersection: $a \cap b(x) = a(x)b(x)$. This holds in the SAM or standard additive fuzzy model below and that we use throughout this text. A set $A \subset X$ is fuzzy if and only if it breaks the "law" of noncontradiction $A \cap A^c \neq \emptyset$ and the "law" of excluded middle $A \cup A^c \neq X$. A set is bivalent if equality holds in both set relations.

Suppose the space X contains n objects x_i: $X = \{x_1, \ldots, x_n\}$. Then the power set 2^X contains 2^n bivalent sets. The power set 2^X contains all subsets of X and is isomorphic to the Boolean n-cube: $2^X = \{0, 1\}^n$. Each nonfuzzy set defines a bit vector of 1s and 0s of length n. The two extreme bit vectors are X itself and the empty set: $X = (1, \ldots, 1)$ and $\emptyset = (0, \ldots, 0)$. These 2^n bivalent sets lie at the corners or vertices of the fuzzy n-cube or unit hypercube $I^n = [0, 1]^n$.

Fuzzy sets $A \subset X$ fill in the unit hypercube. The fuzzy power set $F(2^X)$ is the nonfuzzy set of all fuzzy subsets $A \subset X$. It is isomorphic to the unit hypercube: $F(2^X) = I^n$. On this geometric view [13] sets are points in a fuzzy cube and define fit vectors: $A = (a(x_1), \ldots, a(x_n)) = (a_1, \ldots, a_n)$. The cube midpoint $M = (1/2, \ldots, 1/2)$ is the only set equidistant to all 2^n vertices or bit vectors. The cube midpoint is also unique in that it alone maximally violates the "laws" of noncontradiction and excluded middle: $A = A \cap A^c = A \cup A^c = A^c$ when min and max define intersection and union pointwise.

Suppose $A = (1/3, 1/2, 1)$ is a point in the fuzzy 3-cube. Then the previous pointwise operations give the fit vectors

$$\begin{aligned} A &= \left(\frac{1}{3}, \frac{1}{2}, 1\right) \\ A^c &= \left(\frac{2}{3}, \frac{1}{2}, 0\right) \\ A \cap A^c &= \left(\frac{1}{3}, \frac{1}{2}, 0\right) \\ A \cup A^c &= \left(\frac{2}{3}, \frac{1}{2}, 1\right) \end{aligned} \tag{1.12}$$

These 4 points lie at 4 of the 8 vertices of an interior subcube centered at the cube midpoint. The other 4 vertex points have fit values made up of those of A and A^c. Each fuzzy set A in a fuzzy cube I^n defines such an interior subcube that is unique up to the 2^n-fold symmetry of the fuzzy cube. The subcube shrinks toward the midpoint M as A gets fuzzier and its fit values move closer to 1/2. The subcube expands toward the entire fuzzy cube I^n as A gets less fuzzy and its fit values move closer to 0 or 1.

The subset relation $A \subset B$ at the set level extends the logical implication $A \rightarrow B$ at the statement level. Binary containment $A \subset B$ holds for binary sets A and B if every object x in A belongs to B iff A belongs to B's power set 2^B iff $a(x) \leq b(x)$ holds for all $x \in X$. The set-function relation also defines when containment holds 100% between fuzzy sets A and B. In general both A and B belong only partially to each other. A *subsethood* operator measures this partial containment or inclusion: $S(A, B) = Degree(A \subset B)$. Again suppose the space of objects X is discrete: $X = \{x_1, \ldots, x_n\}$. Then Chapter 12 derives the ratio subsethood operator [13]

$$S(A, B) = \frac{c(A \cap B)}{c(A)} \tag{1.13}$$

for the counting measure

$$c(A) = \sum_{i=1}^{n} a_i$$

and for A non-null. The continuous case of $A \subset R^n$ gives the like ratio

$$S(A, B) = \frac{\int_{R^n} \min(a(x), b(x))dx}{\int_{R^n} a(x)dx} \tag{1.14}$$

provided the integrals exist (which they do when we take X as a compact or closed and bounded subset of R^n). Chapter 13 uses the subsethood measure to aid in learning with a Gaussian SAM system or radial-basis-function neural network. Chapter 14 uses the subsethood measure as a matching criterion for the fuzzy ART (adaptive resonance theory) model.

The subsethood measure gives back many of the set operations of fuzzy theory. The whole-in-the-part measure $S(X, A) = Degree(X \subset A)$ reduces even the counting operation to subsethood:

$$c(A) = nS(X, A) \tag{1.15}$$

with a like result in the continuous case. This in turn gives back the defining measure of "relative frequency" or the success ratio in discrete probability theory:

$$\begin{aligned} S(X, A) &= \frac{c(A)}{n} \\ &= \frac{n_A}{n} \end{aligned} \tag{1.16}$$

for n_A binary wins out of n binary trials. The whole-in-the-part containment term $S(A \cup A^c, A \cap A^c)$ gives the measure of fuzziness or entropy $E(A)$ of Chapter 12:

$$\begin{aligned} S(A \cup A^c, A \cap A^c) &= \frac{c(A \cap A^c)}{c(A \cup A^c)} \\ &= E(A) \end{aligned} \tag{1.17}$$

with again a like result in the continuous case. Chapter 9 uses the *clarity* measure $1 - E(A)$ to weight the rule firings of combined fuzzy systems. Note that the fuzziness of A depends on just its counted violations of the laws of noncontradiction and excluded middle. Only the min and max operators achieve this equality.

Fuzzy systems $F: R^n \rightarrow R^p$ use m rules to map vector inputs x to vector or scalar outputs $F(x)$. The jth rule has the form "If $X = A_j$ then $Y = B_j$" as in "If the air temperature is *cool* then set the motor speed to *slow*." The rule defines a patch or fuzzy Cartesian product $A_j \times B_j \subset R^n \times R^p$. If-part fuzzy set $A_j \subset R^n$ has set function $a_j: R^n \rightarrow [0, 1]$. Many systems factor the if-part set function a_j into n scalar set functions $a_j^1, \ldots, a_j^n$ but at the expense of ignoring the correlations among the input components. Then-part fuzzy set $B_j \subset R^p$ has set function $b_j: R^p \rightarrow [0, 1]$. Some systems factor these sets as well. In general most

n-D or p-D sets do not factor uniquely into n or p components. Hyper-rectangles factor but ellipsoids do not. Product is the simplest and most powerful way to factor a joint if-part set function:

$$a_j(x) = \prod_{i=1}^{n} a_j^i(x_i).$$

It does not ignore $n - 1$ of the fit values for each x as does the older *min* combiner. The product of fit values gives a small number but only the relative values matter in the SAM ratio below in (1.18).

The rule patch $A_j \times B_j$ can use any pointwise *AND* operation such as min or product or the t-norms discussed in the homework problems. In practice and in this text we use product to define the rule-patch fuzzy set: $a_j \times b_j(x, y) = a_j(x)b_j(y)$. This is the "standard" part of a SAM or standard additive model. The "additive" part means that input x "fires" or activates the jth rule to degree $a_j(x)$ and the system *sums* [13] the "fired" or scaled then-part sets $a_j(x)B_j$:

$$B(x) = \sum_{j=1}^{m} w_j a_j(x) B_j \subset R^p$$

for scalar rule weights $w_j > 0$. Earlier fuzzy systems used max instead of sum to combine the fired rules. Modern fuzzy systems dispense with min and max. They consist of only multiplies, adds, and divisions as in the SAM Theorem discussed next.

Then the output $F(x)$ is the defuzzified result of the generalized fuzzy set $B(x)$: $F(x) = \text{Defuzzied}(B(x))$. The centroid and mode are the most common defuzzifiers. Chapter 9 uses the mode to defuzzify a fused or combined additive system for signal detection. The centroid uses all and only the information in the output set $B(x)$ in a Bayesian sense. SAMs defuzzify with the centroid: $F(x) = \text{Centroid}(B(x))$. Chapter 2 shows that this leads to the SAM Theorem [13] that is the central result of this text:

$$F(x) = \frac{\sum_{j=1}^{m} w_j a_j(x) V_j c_j}{\sum_{j=1}^{m} w_j a_j(x) V_j} \tag{1.18}$$

with m bounded then-part volumes or areas

$$V_j = \int_{R^p} b_j(y_1, \ldots, y_p) dy_1 \ldots dy_p > 0 \tag{1.19}$$

and m then-part set centroids

$$c_j = \frac{\int_{R^p} y b_j(y_1, \ldots, y_p) dy_1 \ldots dy_p}{\int_{R^p} b_j(y_1, \ldots, y_p) dy_1 \ldots dy_p}. \tag{1.20}$$

In practice the then-part sets B_j are often scalar fuzzy sets. Then the volumes reduce to areas and the then-part centroids involve only integrals over the real line or over some compact subset of it. The user can pick these areas and centroids in advance to simplify the computation.

SAM systems account for almost all fuzzy systems in practice. The special case of independent scalar Gaussian if-part set functions $a_j^1, \ldots, a_j^n$ also reduces to the popular radial basis function model of neural networks for equal rule weights and equal then-part set volumes. Chapter 2 extends the SAM model in many ways to include these cases and others and to study the stability of feedback SAMs. The reader should commit the ratio form of the SAM Theorem to memory. For now we observe that the SAM output $F(x)$ equals a convex sum of the then-part set centroids c_j:

$$F(x) = \sum_{j=1}^{m} p_j(x) c_j \tag{1.21}$$

for the m convex coefficients $p_j(x) \geq 0$ of the form

$$p_j(x) = \frac{w_j a_j(x) V_j}{\sum_{k=1}^{m} w_k a_k(x) V_k}. \tag{1.22}$$

So

$$\sum_{j=1}^{m} p_j(x) = 1$$

for each input x. The SAM output $F(x)$ equals a convex sum or discrete expectation of the m then-part set centroids c_j.

The SAM model reduces to the COG or center of gravity model used in many fuzzy applications [4] if we ignore the relative rule weights and the relative values or volumes of the then-part sets and assume $w_1 = \cdots = w_m > 0$ and $V_1 = \cdots = V_m > 0$. Then the previous ratio of the SAM Theorem reduces to the simpler COG form

$$F(x) = \frac{\sum_{j=1}^{m} a_j(x) c_j}{\sum_{j=1}^{m} a_j(x)} \tag{1.23}$$

which also assumes that at least one rule fires for each x.

The COG model ignores both the fuzzy structure of the then-part sets and how those sets affect the uncertainty or variance of the fuzzy system. The next chapter deals with these issues in detail. Homework problem 1.42 shows how to design a simple COG system. In practice some rules have more weight or value than other rules have and likewise some then-part sets B_j are wider and less certain than are other sets. The COG model can give a good first approximation of a fuzzy system but does not have the flexibility of the more general SAM system. Adaptive fuzzy systems can also tune the weight and volume parameters.

REFERENCES

[1] Black, M., "Vagueness: An Exercise in Logical Analysis," *Philosophy of Science*, vol. 4, 427–455, 1937.

[2] Black, M., "Reasoning with Loose Concepts," *Dialogue*, vol. 2, 1–12, 1963.

[3] Cavalcante, C., Cardosa, J., Ramos, J. G., and Neves, O. R., "Design and Tuning of a Helicopter Fuzzy Controller," *Proceedings of the IEEE International Conference on Fuzzy Systems (FUZZ–95)*, 1549–1554, March 1995.

[4] Dubois, D., Prade, H., and Yager, R. R., editors, *Readings in Fuzzy Sets for Intelligent Systems*, Morgan Kaufmann, 1993.

[5] Gaines, B. R., "Foundations of Fuzzy Reasoning," *International Journal of Man-Machine Studies*, vol. 8, 623–688, 1976.

[6] Gazzaniga, M, S., editor, *The Cognitive Neurosciences*, MIT Press, 1995.

[7] Grim, P., "Self-reference and Chaos in Fuzzy Logic," *IEEE Transactions on Fuzzy Systems*, vol. 1, no. 4, 237–253, November 1993.

[8] Holmblad, L. P., and Ostergaard, J. J., "Control of a Cement Kiln by Fuzzy Logic," *Fuzzy Information and Decision Processes*, M. M. Gupta and E. Sanchez, editors, 389–399, North Holland, 1982.

[9] Hush, D.R., and Horne, B. G., "Progress in Supervised Neural Networks," *IEEE Signal Processing Magazine*, January 1993.

[10] Kaplan, A., and Schott, H. A., "A Calculus for Empirical Classes," *Methodos* III, vol. 11, 165–188, 1951.

[11] Klir, G. J., and Folger, T. A., *Fuzzy Sets, Uncertainty, and Information*, Prentice Hall, 1988.

[12] Kosko, B., "Fuzzy Knowledge Combination," *International Journal of Intelligent Systems*, vol. 1, no, 4, 293–320, Winter 1986.

[13] Kosko, B., *Neural Networks and Fuzzy Systems*, Prentice Hall, 1991.

[14] Kosko, B., and Isaka, S., "Fuzzy Logic," *Scientific American*, vol. 269, no. 1, 76–81, July 1993.

[15] Kosko, B., *Fuzzy Thinking*, Hyperion, New York, 1993.

[16] Kosko, B., "Fuzzy Systems as Universal Approximators," *IEEE Transactions on Computers*, vol. 43, no. 11, November 1994; an earlier version appears in the *Proceedings of the First IEEE International Conference on Fuzzy Systems (IEEE FUZZ–92)*, 1153–1162, March 1992.

[17] Lukasiewicz, J., "Philosophical Remarks on Many-Valued Systems of Propositional Logic," *Selected Works*, Borkowski, editor, Studies in Logic and the Foundations of Mathematics, 153–179, North Holland, 1970.

[18] McNeill, D., and Freiberger, P., *Fuzzy Logic*, Simon & Schuster, 1993.

[19] Mamdani, E. H., "Application of Fuzzy Logic to Approximate Reasoning Using Linguistic Synthesis," *IEEE Transactions on Computers*, vol. C–26, no. 12, 1182–1191, December 1977.

[20] Mead, C., *Analog VLSI and Neural Systems*, Addison-Wesley, 1989.

[21] Menger, K., "Statistical Metrics," *Proceedings of the National Academy of Sciences, USA*, vol. 28, 535–537, 1942.

[22] Minsky, M., *The Society of Mind*, Simon & Schuster, 1985.

[23] Peirce, C. S., *Collected Papers of Charles Sanders Peirce*, C. Hartshorne and P. Weiss, editors, Harvard University Press, 1931.

[24] Putnam, H., *Mind, Language, and Reality*, vol. 2, Cambridge University Press, 1975.

[25] Quine, W. V. O., "What Price Bivalence?" *Journal of Philosophy*, vol. 78, no. 2, 90–95, February 1981.

[26] Resher, N., *Many-Valued Logic*, McGraw-Hill, 1969.

[27] Russell, B., "Vagueness," *Australian Journal of Philosophy*, vol. 1, 1923.

[28] Sugeno, M., Griffin, M. F., and Bastian, A., "Fuzzy Hierarchical Control of an Unmanned Helicopter," *Proceedings of the 17th IFSA World Congress*, 179–182, 1993.

[29] Tanaka, K., and Yoshioka, K., "Design of Fuzzy Controller for Backer-Upper of a Five-Trailers and Truck," *Proceedings of the 1995 IEEE International Conference on Fuzzy Systems (FUZZ–95)*, 1543–1548, March 1995.

[30] Tilbury, D., Murray, M. M., and Sastry, S. S., "Trajectory Generation for the *N*-Trailer Problem Using Goursat Normal Form," *IEEE Transactions on Automatic Control*, vol. 40, no. 5, 802–819, May 1995.

[31] Whitehead, A. N., and Russell, B., *Principia Mathematica*, vol. I, University of Cambridge Press, 1910.

[32] Yager, R. R., "Multiple Objective Decision Making Using Fuzzy Sets," *International Journal of Man-Machine Studies*, vol. 9, 375–382, 1977.

[33] Yager, R. R., "On a General Class of Fuzzy Connectives," *Fuzzy Sets and Systems*, vol. 4, 235–242, 1980.

[34] Yager, R. R., "On Ordered Weighted Averaging Aggregation Operators in Multi-criteria Decision Making," *IEEE Transactions on Systems, Man, and Cybernetics*, vol. 18, 183–190, 1988.

[35] Yasunobu, S., Miyamoto, S., and Ihara, H., "Fuzzy Control for Automatic Train Operation System," *Proceedings of the 4th IFAC/IFIP/IFORS International Conference on Control in Transportation Systems*, Baden-Baden, Germany, 33–39, 1983.

[36] Zadeh, L. A., "Fuzzy Sets," *Information and Control*, vol. 8, 338–353, 1965.

[37] Zadeh, L. A., *Fuzzy Sets and Applications: Selected Papers*, R. R. Yager, S. Ovchinnikov, R. M. Tong, and H. T. Nguyen, editors, Wiley-Interscience, 1987.

HOMEWORK PROBLEMS

1.1. Truth maps statements into truth values. A multivalued or "fuzzy" truth function t: $\mathbf{S} \rightarrow [0, 1]$ maps the set of statements $\mathbf{S}$ into the unit interval [0,1] or other totally ordered set of truth values. Then the Lukasiewicz definitions for conjunction, disjunction, and negation have the form

$$t(A \text{ AND } B) = \min(t(A),\ t(B))$$

$$t(A \text{ OR } B) = \max(t(A),\ t(B))$$

$$t(\textit{not-}A) = 1 - t(A).$$

Show that min and max obey DeMorgan's laws:

$$t(A \text{ AND } B) = t(\textit{not-}(\textit{not-}A \text{ OR } \textit{not-}B))$$

$$t(A \text{ OR } B) = t(\textit{not-}(\textit{not-}A \text{ AND } \textit{not-}B))$$

or

$$\min(x,\ y) = 1 - \max(1 - x,\ 1 - y)$$

$$\max(x,\ y) = 1 - \min(1 - x,\ 1 - y).$$

1.2. Prove the modular equality for Lukasiewicz fuzzy logic:

$$t(A) + t(B) = t(A \text{ AND } B) + t(A \text{ OR } B).$$

1.3. Use the modular equality of problem 1.2 to prove the general multivalued relation between the "laws" of noncontradiction and excluded middle:

$$t(A \text{ AND } \textit{not-}A) + t(A \text{ OR } \textit{not-}A) = 1.$$

Hence the bivalent law of noncontradiction $t(A \text{ AND } \textit{not-}A) = 0$ holds if and only if the bivalent law of excluded middle $t(A \text{ OR } \textit{not-}A) = 1$ holds. What happens in the "paradox" case when $t(A) = t(\textit{not-}A)$?

1.4. The Lukasiewicz implication operator has the form

$$t_L(A \rightarrow B) = \min(1,\ 1 - t(A) + t(B)).$$

Jan Lukasiewicz used this if-then operator to derive the max form for the OR operator. Prove this multivalued theorem:

$$t(A \text{ OR } B) = t_L((A \rightarrow B) \rightarrow B).$$

Then the order-reversing negation operator $t(\textit{not-}A) = 1 - t(A)$ in DeMorgan's law gives back min as the AND operator. So negation and implication generate Lukasiewicz fuzzy logic.

1.5. Material implication holds in binary logic: $A \text{ OR } B = \textit{not-}A \rightarrow B$. Verify this with a binary truth table. Material implication need not hold in a multivalued logic. Suppose it holds for Lukasiewicz implication. Then use DeMorgan's law to show that this would give the fuzzy AND operator as the so-called continuous triangular norm T_c:

$$t(A \text{ AND } B) = T_c(t(A),\ t(B)) = \max(0,\ t(A) + t(B) - 1).$$

Then use the generalized DeMorgan's law for triangular norms

$$t(A \text{ OR } B) = 1 - T_c(1 - t(A), 1 - t(B))$$

to derive the new OR operator

$$t(A \text{ OR } B) = \min(1, t(A) + t(B)).$$

Then show whether these new AND and OR operators obey the modular equality

$$t(A) + t(B) = t(A \text{ AND } B) + t(A \text{ OR } B).$$

1.6. The biconditional relation $A \leftrightarrow B = (A \rightarrow B)$ AND $(B \rightarrow A)$ defines the equality relation $A = B$. So statement A is logically equivalent to statement B if and only if the two statements imply each other. Use the Lukasiewicz implication operator to derive the multivalued equality operator

$$t_L(A = B) = 1 - |t(A) - t(B)|.$$

The *exclusive*-OR operator is negated equality and so $t(A \text{ XOR } B) = 1 - t(A = B)$. Lukasiewicz exclusive-OR is just l^1 or fuzzy Hamming distance:

$$t_L(A \text{ XOR } B) = |t(A) - t(B)|.$$

So in the binary case an exclusive-or statement is true if and only if the two binary truth values differ.

1.7. The binary implication $A \rightarrow B$ is true if the antecedent A is false. Then the truth of the consequent B does not matter. So the statement "If snow is blue then the moon is made of cheese" is true since snow is not blue. The statement "If I am dead then I am alive" is true as long as you are not dead. These statements are vacuously true in the logical sense of binary truth tables but they seem to defy the *causal* sense of implication. Show that this anticausal effect still holds in the multivalued case for the Lukasiewicz implication operator by first showing that $t_L(A \rightarrow B) = 1$ if and only if $t(B) \geq t(A)$. This reflects the medieval scholastic doctrine that "truth cannot imply falsehood." Indeed the threshold implication operator

$$t_T(A \rightarrow B) = \begin{cases} 0 & \text{if } t(B) < t(A) \\ 1 & \text{if } t(B) \geq t(A) \end{cases}$$

defines the minimal implication operator for multivalued truth that obeys the scholastic doctrine. Show that the threshold implication operator has the same truth table as the binary implication $A \rightarrow B$ has when $t(A)$ and $t(B)$ take values only in $\{0, 1\}$.

1.8. Kurt Gödel put forth the conjunctive min operator as an implication operator:

$$t_G(A \rightarrow B) = \min(t(A), t(B)).$$

Explain why this operator avoids the anticausal property of the implication operators in Problem 1.7. The Gödel implication operator and the related product implication operator $t(A \rightarrow B) = t(A)t(B)$ turn out to be the implication operators that carry over pointwise to the case of fuzzy systems where A and B are fuzzy subsets of real vector spaces. The product operator holds for set-level implications or rules in the case of SAMs or standard additive models. (This holds because min or product defines the pointwise value of a fuzzy rule's Cartesian product or patch structure.)

The Gödel implication operator obeys a multivalued form of the *modus ponens* rule of inference but fails to give back the *modus tollens* rule of inference in the binary case. To see this prove the following *modus ponens* inference schema and check that it reduces to the binary *modus ponens* rule of inference when $t_G(A \rightarrow B) = t(A) = 1$:

$$\begin{array}{l} t_G(A \rightarrow B) = c \\ \\ t(A) \geq a \\ \hline \therefore t(B) \geq a. \end{array}$$

So $t(B) \geq a$ if $t(A) \leq t(B)$. Do the same for the *modus ponens* schema when the Lukasiewicz implication replaces the Gödel implication operator:

$$\begin{array}{l} t_L(A \rightarrow B) = c \\ \\ t(A) \geq a \\ \hline \therefore t(B) \geq \max(0,\ a + c - 1). \end{array}$$

Now derive the *modus tollens* inference schema for the Lukasiewicz implication operator:

$$\begin{array}{l} t_L(A \rightarrow B) = c \\ \\ t(A) \leq b \\ \hline \therefore t(A) \leq \min(1,\ 1 - c + b). \end{array}$$

Verify that the binary *modus tollens* schema holds when $t_L(A \rightarrow B) = 1$ and $t(B) = 0$. Explain why no such *modus tollens* schema holds in the binary case if the Gödel implication operator replaces the Lukasiewicz implication operator.

1.9. Repeat the derivations in problem 1.8 of the two inference schemas for the Gaines implication operator:

$$t_{\text{Gaines}}(A \rightarrow B) = \begin{cases} 1 & \text{if } t(A) = 0 \\ \min\left(1, \frac{t(B)}{t(A)}\right) & \text{if } t(A) > 0. \end{cases}$$

The conclusions of the Gaines inference schemas will differ from the conclusions in the Lukasiewicz case. Do the Gaines schemas reduce to the binary *modus ponens* and *modus tollens* schemas for binary truth values? Does the Gaines implication operator avoid the anticausal effect?

1.10. A negation operator N: $[0, 1] \rightarrow [0, 1]$ should obey double negation: *not-*(*not-A*) = A or $N(N(x)) = x$. Show that the Yager negation operator

$$N_p(x) = [1 - x^p]^{\frac{1}{p}}$$

obeys double negation for parameter $p > 0$. Show that the Sugeno negation operator

$$N_p(x) = \frac{1 - x}{1 + px}$$

obeys double negation if $p > -1$.

1.11. Negation operators N act as discrete dynamical systems if they take as input their own output: $x_{n+1} = N(x_n)$. The fit sequence $x_0, x_1, x_2, \ldots$ converges to a negation *fixed-point* equilibrium if there is some fit value x_e such that $x_e = N(x_e)$. Show that the Yager negation operator has the fixed-point fit value

$$x_e = \frac{1}{2^{\frac{1}{p}}}.$$

What is the equilibrium fit value of the Sugeno negation operator in Problem 1.10?

1.12. Triangular norms or t-norms T generalize the AND operator in multivalued logic. Triangular conorms or t-conorms S generalize the OR operator. A t-norm is a between-cube mapping T: $[0, 1] \times [0, 1] \rightarrow [0, 1]$ that satisfies five axioms:

1. $T(0, 0) = 0$
2. $T(x, 1) = T(1, x) = x$
3. $T(x, y) = T(y, x)$
4. $T(u, v) \leq T(x, y)$ if $u \leq x$ and $v \leq y$
5. $T(T(x, y), z) = T(x, T(y, z))$.

A t-conorm S: $[0, 1] \times [0, 1] \rightarrow [0, 1]$ satisfies five like axioms:

1. $S(1, 1) = 1$
2. $S(x, 0) = S(0, x) = x$
3. $S(x, y) = S(y, x)$
4. $S(u, v) \leq S(x, y)$ if $u \leq x$ and $v \leq y$
5. $S(S(x, y), z) = S(x, S(y, z))$.

Verify that min is a t-norm and max is a t-conorm. Use the t-norm and t-conorm axioms to prove the bound

$$T(x, y) \leq \min(x, y) \leq \max(x, y) \leq S(x, y).$$

1.13. Verify that the product operator $T(x, y) = xy$ is a t-norm. Find its DeMorgan dual t-conorm $S(x, y) = 1 - T(1 - x, 1 - y)$. An *Archimedean* t-norm obeys $T(x, x) < x$ for all x in [0,1]. So it is not idempotent. (At the set level this means $A \cap A \subset A$ but $A \not\subset A \cap A$.) Are min and the product t-norm Archimedean?

1.14. Show that the t-norm

$$T_{\text{lower}}(x, y) = \begin{cases} x & \text{if } y = 1 \\ y & \text{if } x = 1 \\ 0 & \text{if else} \end{cases}$$

is the "smallest" t-norm in the sense that $T_{\text{lower}}(x, y) \le T(x, y)$ for all t-norms T. Find the "largest" t-conorm S_{upper} and show that it obeys $S(x, y) \le S_{\text{upper}}(x, y)$ for all t-conorms S.

1.15. The Dubois-Prade family of t-norms

$$D_p(x, y) = \frac{xy}{\max(p, x, y)}$$

depends on a parameter p that ranges from 0 to 1. The Dubois-Prade t-norm generalizes the product t-norm. Show that the DeMorgan dual t-conorm of the Dubois-Prade t-norm is

$$P_p(x, y) = \frac{x + y - xy - \min(1 - p, x, y)}{\max(p, 1 - x, 1 - y)}.$$

Show that the Dubois-Prade norms give back the product t-norm and t-conorm when $p = 1$. What t-norm and t-conorm do the Dubois-Prade norms give as p approaches 0?

1.16. The Yager t-norm has the form

$$Y_p(x, y) = 1 - \min\left(1, [(1 - x)^p + (1 - y)^p]^{\frac{1}{p}}\right)$$

for $p > 0$. Find its DeMorgan dual t-conorm. What familiar t-norm does the Yager t-norm equal when $p = 1$? Show that the Yager t-norm has DeMorgan dual t-conorm

$$Z_p(x, y) = \min(1, [x^p + y^p]^{\frac{1}{p}}).$$

Prove that the Yager t-norm converges to the min operator in the infinite limit:

$$\lim_{p \to \infty} Y_p(x, y) = \min(x, y).$$

To prove this it helps to use L'Hôpital's rule of differential calculus and to show that the Yager t-conorm converges to the max operator.

1.17. Aggregation operators $A: [0, 1]^n \to [0, 1]$ fuzzify the AND-OR region between min and max that t-norms and t-conorms ignore. An aggregation operator must commute among all its n input arguments and must satisfy the min-max bound

$$\min(x_1, \ldots, x_n) \le A(x_1, \ldots, x_n) \le \max(x_1, \ldots, x_n).$$

The paragon aggregation operator is the arithmetic mean:

$$A(x_1, \ldots, x_n) = \frac{1}{n} \sum_{i=1}^{n} x_i.$$

Show that the arithmetic mean satisfies the min-max bound.

1.18. Credibility weights $w_1, \ldots, w_n$ can change the structure of aggregation operators. Suppose the weights are normalized to $0 \leq w_i \leq 1$ and that at least one weight is strictly positive: $w_j > 0$. Then a *simple* weighted mean has the unnormalized form

$$A_w(x_1, \ldots, x_n) = \frac{1}{n} \sum_{i=1}^{n} w_i x_i.$$

Show that a simple weighted mean need not satisfy the min-max bound. A *standard* weighted mean has the normalized form

$$A_w(x_1, \ldots, x_n) = \frac{\sum_{i=1}^{n} w_i x_i}{\sum_{i=1}^{n} w_i}.$$

Show that a standard weighted mean does satisfy the min-max bound.

1.19. Yager OWA (ordered weighted averaging) operators are weighted aggregation operators that are linear and that generalize the arithmetic mean. Their definition depends on the ordering $x_1 \leq x_2 \leq \ldots \leq x_n$ of the components of the *fit* (fuzzy unit) vector $x = (x_1, \ldots, x_n) \in [0, 1]^n$:

$$O_w(x_1, \ldots, x_n) = \sum_{i=1}^{n} w_i x_i.$$

The OWA weight vector w also lies in the fuzzy cube: $w = (w_1, \ldots, w_n) \in [0, 1]^n$. It obeys the further normalization or probability constraint

$$\sum_{i=1}^{n} w_i = 1.$$

This constrains OWA weight vectors to the simplex in the fuzzy cube. Define the n cube vertices or unit bit vectors as $u^1 = (1, 0, \ldots, 0)$, $u^2 = (0, 1, 0, \ldots, 0)$, $u^n = (0, \ldots, 0, 1)$. Are standard weighted means OWA operators? Show that $O_{u^1} = \min(x_1, \ldots, x_n)$ and $O_{u^n} = \max(x_1, \ldots, x_n)$. Show that OWA operators satisfy the min-max bound and hence are proper aggregation operators. What aggregation operator does an OWA give for the uniform weight fit vector $w = (\frac{1}{n}, \ldots, \frac{1}{n})$? Define an OWA weight vector that picks the median value of the inputs.

1.20. OWA operators vary from AND to OR. Define the OR-ness of an OWA vector w as

$$OR(w) = \frac{1}{n-1} \sum_{i=1}^{n} w_i (i-1).$$

Verify that $OR(w)$ lies in the unit interval [0,1]. Define the AND-ness of an OWA vector w as the negation of the OR-ness:

$$\text{AND}(w) = 1 - \text{OR}(w).$$

An OWA vector w is more OR-like the larger its degree of OR-ness. Show that $\text{OR}(u^n) = 1$ and $\text{OR}(u^1) = 0$ and hence that $\text{AND}(u^n) = 0$ and $\text{AND}(u^1) = 1$.

1.21. Knowledge combination operators $K: [0, 1]^n \rightarrow [0, 1]$ are symmetric in their arguments ($K(x) = K(x_\sigma)$ for all $n!$ permutations σ of $(x_1, \ldots, x_n)$) and obey the min-max bound of all aggregation operators: $l \leq K(x) \leq m$ for the *least* or smallest order statistic $l = \min(x_1, \ldots, x_n)$ and largest order statistic $m = \max(x_1, \ldots, x_n)$. The difference $m - l$ defines the *knowledge gap* among the n knowledge sources or experts who give the knowledge response $x = (x_1, \ldots, x_n)$. Knowledge combination operators further depend only on the order statistics l and m: $K(x) = K(l, m)$. Confirm that the knowledge-gap operator

$$K(l, m) = \min(m, 1 - m + l)$$

is a knowledge combination operator. Show that

$$K(l, m) = \begin{cases} 1 - m + l & \text{if } m \geq \dfrac{1 + l}{2} \\ m & \text{if } m \leq \dfrac{1 + l}{2}. \end{cases}$$

Find the value of the combined knowledge for knowledge responses $x = (.2, .6, .7, .6, .8)$, $x = (.8, .5, 0, 1, .3)$, and $x = (.4, .2, .3, .5, .2)$.

1.22. A knowledge combination operator is *conservative* if it tends toward the lower bound l as the knowledge gap $m - l$ grows to unity: $K(l, m) \downarrow l$ as $m - l \uparrow 1$. Show that the knowledge-gap operator is conservative. Is the mean

$$K(l, m) = \frac{l + m}{2}$$

a knowledge combination operator? Is it conservative? A knowledge combination operator is *lenient* if it tends toward the upper bound m as the knowledge gap falls to zero: $K(l, m) \uparrow m$ as $m - l \downarrow 0$. Are the above two operators lenient?

1.23. A weight fit vector $w = (w_1, \ldots, w_n) \in [0, 1]^n$ can weight the n knowledge sources and thus weight an operator K that combines their knowledge. The weight vector w must weight both the least fit value l and the most fit value m:

$$K^w = \min(m^w, 1 - m^w + l^w)$$

for weight functions

$$l^w = \min_i f_i(w_i, x_i)$$

$$m^w = \max_i g_i(w_i, x_i).$$

The weight functions should obey three axioms:

$$\text{(W1)} \quad f_i(w_i, x_i) \to x_i \quad \text{and} \quad g_i(w_i, x_i) \to x_i \quad \text{as} \quad w_i \uparrow 1$$

$$\text{(W2)} \quad f_i(w_i, x_i) \uparrow 1 \quad \text{as} \quad w_i \downarrow 0$$

$$\text{(W3)} \quad g_i(w_i, x_i) \downarrow 0 \quad \text{as} \quad w_i \downarrow 0.$$

Axiom $W1$ tends to take experts at their word as they get more credible. Axiom $W2$ tends to ignore an expert's response from the *min*-based least operator l^w as the expert gets less credible. Axiom $W3$ tends to ignore an expert's response from the *max*-based most operator m^w as the expert gets less credible. Show that the choices

$$f_i(w_i, x_i) = \max(1 - w_i, x_i)$$

$$g_i(w_i, x_i) = \min(w_i, x_i)$$

obey the three weight axioms. Find another pair of t-norms and t-conorms that also obey the three weight axioms. Use the weight fit vector $w = (.3, .8, 1, .7, .9)$ to find the weighted combined knowledge $K^w(l^w, m^w)$ for knowledge responses $x = (.2, .6, .7, .6, .8)$, $x = (.8, .5, 0, 1, .3)$, and $x = (.4, .2, .3, .5, .2)$. What goes wrong as the weight vector w approaches the null vector $(0, \ldots, 0)$? How does this problem increase or decrease with the sample size n?

1.24. This problem and the next few use the fit vectors $F \in [0, 1]^4$:

	x_1	x_2	x_3	x_4
$X =$	(1	1	1	1)
$\emptyset =$	(0	0	0	0)
$A =$	(.3	.6	.4	.2)
$B =$	(.4	.5	.4	.7)
$C =$	(.6	.7	.7	1)
$D =$	(1	1	0	0).
$E =$	(1	0	1	0)
$M =$	(.5	.5	.5	.5).

Compute the overlap set $F \cap F^c$ and the underlap set $F \cup F^c$ for each of the eight sets. Use min and max for intersection and union.

1.25. Repeat problem 1.24 but use the product t-norm and its DeMorgan dual for intersection and union.

1.26. The l^p distance between fuzzy sets A and B has the form

$$l^p(A, B) = \sqrt[p]{\sum_{i=1}^{n} |a_i - b_i|^p}$$

for $p > 0$. An integral (when it exists) replaces the sum for continuous set functions a and b. l^1 is just the sum of absolute fit differences or the fuzzy Hamming distance:

$$l^1(A, B) = \sum_{i=1}^{n} |a_i - b_i|.$$

So fuzzy Hamming distance reduces to fuzzy exclusive *OR* when $n = 1$ and thus when the fuzzy cube of sets reduces to the unit interval of multivalued truth values. Compute the 64 pairwise fuzzy Hamming distances between each of the above 8 fit and bit vectors. It may help to list the values in an 8-by-8 matrix.

1.27. The fit vector $A = (a_1, \ldots, a_n)$ gives rise to 2^n fit vectors of the form $A^* = (a_1 \text{ or } 1 - a_1, \ldots, a_n \text{ or } 1 - a_n)$. Prove that these 2^n fit vectors lie at the corners of an interior fuzzy cube that has the same midpoint as the entire fuzzy cube. Then prove that each of the corners A^* lies the same distance from a nearest corner A^*_{near} and a farthest corner A^*_{far} for *all* l^p metrics. The 2^n fit vectors A^* move in toward the cube midpoint as A moves toward its complement A^c. They move out toward the 2^n binary corners as A moves away from A^c. Plot the 4 sets in the 2-cube for a fuzzy set inside the unit square.

1.28. Confirm that the midpoint $M = (1/2, \ldots, 1/2)$ is equidistant to the 2^n binary vertices V in a fuzzy n-cube for all l^p metrics. What is this distance? Prove that M is unique in that it alone has this equidistant property. The cube midpoint is a metrical invariant.

1.29. The count $c(A)$ of fit vector A is its distance from the origin or empty set $\emptyset$ in the fuzzy cube. Show that for fuzzy Hamming distance the count of A is just the sum of its fit values:

$$c(A) = l^1(A, \emptyset) = \sum_{i=1}^{n} a_i.$$

Compute the counts of the above 8 fit and bit vectors. What does the *product* of the fit values measure in a fuzzy cube?

1.30. The ratio entropy or fuzziness $E(A)$ of fuzzy set A is the ratio of the distance from A to the nearest vertex A_{near} over the distance from A to the farthest vertex A_{far}:

$$E(A) = \frac{l^1(A, A_{\text{near}})}{l^1(A, A_{\text{far}})}.$$

Compute A_{near} and A_{far} for the above 8 sets. Then compute their fuzzy entropies.

1.31. Prove the fuzzy entropy theorem:

$$E(A) = \frac{c(A \cap A^c)}{c(A \cup A^c)}.$$

Use the theorem to compute the fuzzy entropies of the above 8 sets.

1.32. The fuzzy entropy theorem uses min and max for intersection and union. Suppose we use any other t-norm or t-conorm for these operations to give a ratio of counts $F(A)$. Prove that $F(A) \leq E(A)$.

1.33. A probability distribution P on n variables is a fit vector with unit count $c(P) = 1$. Show that $E(P) = \frac{1}{n-1}$ if each fit value obeys $p_i \leq \frac{1}{2}$. Show that

$$E(P) = \frac{2(1 - p_j)}{n - 2(1 - p_j)}$$

if some $p_j > \frac{1}{2}$. So the fuzzy entropy of a probability fit vector P decreases with cube dimension since $E(P) \leq \frac{1}{n-1}$.

1.34. The subsethood theorem gives the degree $S(A, B) = \text{Degree}(A \subset B)$ to which A is a subset of B as a ratio of set counts:

$$S(A, B) = \frac{c(A \cap B)}{c(A)}.$$

Compute the degrees of subsethood between each pair of the above 8 sets. It may help to list the values in an 8-by-8 matrix.

1.35. Prove that fuzzy entropy reduces to subsethood:

$$E(A) = S(A \cup A^c, A \cap A^c).$$

So fuzzy entropy or fuzziness is the degree to which the subset $A \cap A^c$ contains its own superset $A \cup A^c$. Does this equality hold if other t-norms and t-conorms define intersection and union in the fuzzy entropy ratio?

1.36. Prove the inequality

$$S(A, B) + S(A, B^c) \geq 1.$$

This result holds for arbitrary fuzzy sets and limits the types of Bayes theorems fuzzy sets can achieve.

1.37. Prove the odds-form Bayes theorem

$$\frac{S(A \cap H, B)}{S(A \cap H, B^c)} = \frac{S(B \cap H, A)}{S(B^c \cap H, A)} \frac{S(H, B)}{S(H, B^c)}$$

for arbitrary fuzzy sets A, B, and H.

1.38. Prove this subsethood-based fuzzy Bayes theorem:

$$S(E, H_i) = \frac{S(H_i, E)c(H_i)}{\frac{1}{m}\sum_{j=1}^{m} \left[c(E \cup H_j) + c(E \cap H_j) - c(H_j)\right]}$$

for fuzzy "evidence" set E and arbitrary fuzzy "hypothesis" sets $H_1, \ldots, H_m$. The theorem gives the degree to which each new piece of evidence supports each hypothesis. The hypotheses can both overlap and be fuzzy. They need not give a disjoint partition of the space X. Compute the three subsethood degrees for the fit vectors

$$\begin{aligned} E &= (.9 \quad .8 \quad .6 \quad .4 \quad .2 \quad 0) \\ H_1 &= (1 \quad .8 \quad 0 \quad .1 \quad .1 \quad 0) \\ H_2 &= (0 \quad 0 \quad .7 \quad 1 \quad 0 \quad .3) \\ H_3 &= (.1 \quad .4 \quad 1 \quad 1 \quad .4 \quad .1). \end{aligned}$$

1.39. Let the continuous fuzzy subsets A_1 and A_2 of $[0, 1]$ have the set functions

$$a_1(x) = e^{-x} \qquad a_2(x) = \frac{1}{2}$$

on the unit-interval domain $[0, 1]$. Compute the subsethood values $S(A_1, A_2)$ and $S(A_2, A_1)$.

1.40. The exponential set A has the set function

$$a(x) = e^{-x}$$

on the unit-interval domain $[0, 1]$. Compute the entropy or fuzziness value $E(A)$.

1.41. Adding the fit values of $A = (a_1, \ldots, a_n)$ gives the count $c(A)$. *Multiplying* the fit values gives the volume $v(A) = \prod_{i=1}^{n} a_i$. The volume $v(A)$ measures the volume of the hyper-rectangle or fuzzy power set $F(2^A)$ in the fuzzy n-cube. Suppose $v(A) > 0$. This measure suggests the volume subsethood measure $S_v(A, B) = \frac{v(A \cap B)}{v(A)}$. Prove that $S_v(A, B) \leq S(A, B)$ where $S(A, B)$ is the usual or sum-based subsethood measure.

1.42. Suppose a 3-rule SAM $F: [-2, 2] \to [-1, 1]$ has 3 rectangular if-part sets A_1, A_2, A_3 with set functions

$$a_1(x) = \begin{cases} 0 & \text{if } x \notin [-2, 0] \\ 1 & \text{if } x \in [-2, 0] \end{cases}$$

$$a_2(x) = \begin{cases} 0 & \text{if } x \notin [-1, 1] \\ 1 & \text{if } x \in [-1, 1] \end{cases}$$

$$a_3(x) = \begin{cases} 0 & \text{if } x \notin [0, 2] \\ 1 & \text{if } x \in [0, 2]. \end{cases}$$

The 3 then-part sets B_1, B_2, B_3 are triangles of unit area with centers at -1, 0, and 1. State the exact form of the SAM map $F: [-2, 2] \to [-1, 1]$ and plot its graph.

PART II

FUZZY FUNCTION APPROXIMATION

Fuzzy systems can take many forms. Minimal systems act as lone rules or associations. They map fuzzy subsets or concepts of one space to fuzzy subsets of a second space as when degrees of high pressures map to degrees of high temperatures. Most fuzzy systems are more complex. They store a bank of fuzzy rules and use them to map vectors to vectors or to map fuzzy sets to fuzzy sets. The rules define a wide and shallow rule tree from inputs to outputs. The rule tree suffers from exponential rule explosion as the dimensions of the input and output spaces grow.

Fuzzy systems differ in how they combine fired rules. A vector or set input fires the if-part of each rule to some degree. Then these partial and parallel firings scale the then-part fuzzy sets of the rules. Older fuzzy systems are not additive. They form the union of the fired then-part sets to combine them. This takes the pairwise maximum of the set values and so ignores most of the set values. Additive fuzzy systems simply add the then-part set values. The sum values remain sensitive to all changes in input and lead to a simple ratio form for the output. Most fuzzy applications use some form of an additive system.

Chapter 2 presents the theory of additive fuzzy systems and serves as a mathematical prelude for the remaining chapters in the text. The chapter presents feedforward additive systems as universal function approximators and explores their statistical and learning properties. Exponential rule explosion emerges as a structural problem that all fuzzy systems face in high dimensions. Optimal rules help allocate a fixed rule budget and act as the goal or end points of rule learning.

Feedback fuzzy systems take their output as their own input. They can model nonlinear dynamical systems but may not be stable or may lead to chaos. Feedback additive fuzzy systems may be stable if each rule defines a stable linear operator. More complex feedback additive systems can model richer classes of dynamical systems but they may not yield easily to formal analysis.

2

ADDITIVE FUZZY SYSTEMS

PREVIEW

This chapter presents the theory of additive fuzzy systems. Additive fuzzy systems sum the then-part fuzzy sets that inputs fire or activate. They can uniformly approximate any real continuous (or bounded measurable) function on a compact domain. The additive approximator covers the graph of the approximand with rule patches and adds or averages rule patches that overlap. All fuzzy systems face exponential rule explosion in high dimensions.

The standard additive model or SAM includes most fuzzy systems found in practice and in this text. The SAM has the simple form of a convex sum for both vector and set inputs and extends to more complex models that replace the then-part fuzzy sets with nonlinear operators. SAMs can combine any number of weighted fuzzy systems into a common fuzzy system. Additive systems act as conditional expectations with conditional variances that describe their output uncertainty.

Neural or statistical clustering can learn the rule patches from training data. Then gradient descent or supervised learning can tune the rule patches. Lone optimal rule patches cover the extrema or turning points of the approximand. They "patch the bumps" and can help deal with the rule explosion in high dimensions. Better learning schemes move lone rule patches to extrema and then move extra rule patches between the extrema as the rule budget allows.

A Lyapunov analysis shows that feedback SAMs are stable in some cases of practical interest. SAMs can also extend causal knowledge networks or feedback fuzzy cognitive maps that have complex nonlinear dynamics.

2.1 FUZZY FUNCTION APPROXIMATION AS A FUZZY COVER

A fuzzy system F is a set of if-then rules that maps inputs to outputs. The rules define fuzzy patches in the input-output state space $X \times Y$. The fuzzy system $F: X \rightarrow Y$ approximates a function $f: X \rightarrow Y$ by covering its graph with rule patches and averaging patches that overlap. The approximation tends to improve as the fuzzy rule patches grow in number and shrink in size as in Figure 2.1. The rules grow exponentially in number as the dimensions of X and Y grow. The best lone rule patches cover extrema or bumps in the graph of f.

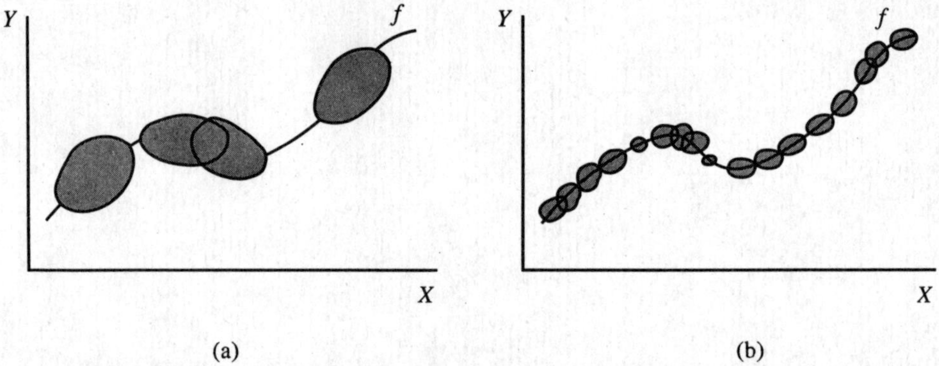

Figure 2.1 In (a) four large fuzzy rule patches cover part of the graph of the unknown function or approximand $f: X \rightarrow Y$. In (b) more smaller patches better cover f but at a greater computational cost. Each rule patch defines a fuzzy subset of the product space $X \times Y$. A large but finite number of fuzzy or precise rules can cover the graph and give a fuzzy system F that approximates f with arbitrary accuracy.

This chapter presents the algebraic details of the patch-covering geometry. All fuzzy systems give some type of patch covering. *Additive* fuzzy systems [18, 21] average the patches that overlap by adding them. Then a centroid or other operation converts the patch cover to the function $F: R^n \rightarrow R^p$. We show that all centroidal fuzzy systems F compute a conditional mean: $F(x) = E[Y|X = x]$. An additive fuzzy system splits this global conditional mean into a convex sum of local (rule) conditional means.

The fuzzy system F does not use a math model of the system or function f that it tries to approximate with the conditional mean $E[Y|X]$. So F is a *model-free* statistical estimator. This model freedom accounts for most of a fuzzy system's modeling power both in practice and in theory. The fuzziness or multivaluedness of the sets can help tie words to math if an expert states the rules. We show below that to first order we can dispense with the then-part fuzzy sets and just use rectangles or nonfuzzy sets.

Each fuzzy rule defines a fuzzy patch or subset of the input-output state space $X \times Y$. Figure 2.2 shows the fuzzy rule "If X is Negative Small then Y is Positive Small" as the Cartesian product $NS \times PS$ of the "fuzzy" [58] or multivalued sets NS and PS. A 3-D plot would show the fuzzy patch $NS \times PS$ as a barn-like structure that rises up from its rectangular base in the plane. Each pair (x, y) is a point in

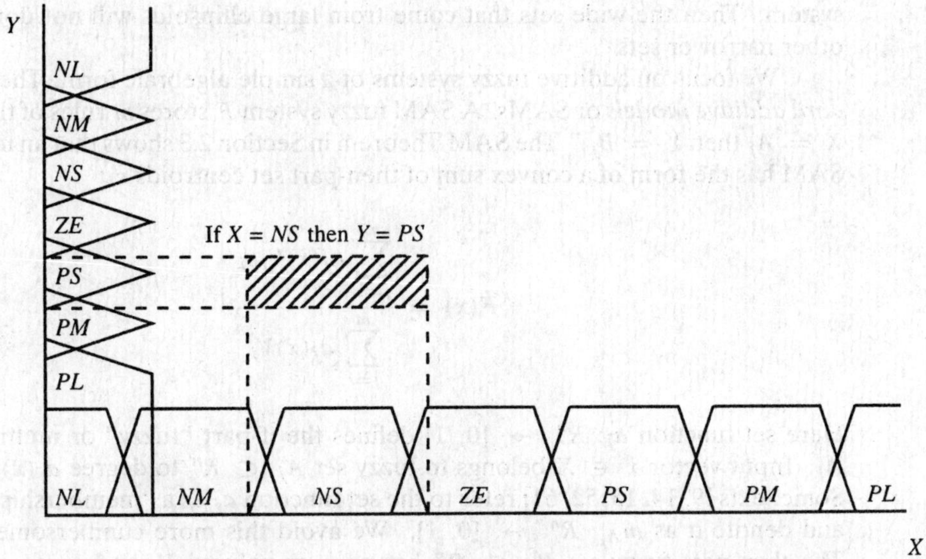

Figure 2.2 Fuzzy rule as a state-space patch or Cartesian product of fuzzy sets. The product patch $NS \times PS$ stands for the fuzzy rule "If X is Negative Small then Y is Positive Small." Here trapezoids and triangles define the fuzzy or multivalued sets.

the plane and belongs to the rule patch to some degree between 0 and 1 just as each input x belongs to each trapezoidal fuzzy set to some degree. Zadeh [59] first observed that planar rule patches can cover both functions and relations. He did not observe the resulting exponential rule explosion or the optimality of rule patches that cover extrema or show how to convert a set of abstract rules or rule patches into a well-defined function $F: R^n \rightarrow R^p$ with output $F(x)$ for each input x.

The patch structure of a rule suggests that a neural or statistical clustering algorithm can learn rules from data. Data clusters can define or center the first set of rules. The simplest scheme centers a rule patch at a sample vector value and in practice centers fuzzy sets at the vector components. Then new data can tune the rules and thus move them in the product space or change their shape. Sparse or noisy data tend to give large patches and thus less certain rules. Dense or filtered data tend to give small patches and thus more certain rules. The rule structure can have many forms. Each form constrains the learning algorithms used to tune the rules.

In the next chapter we use egg-shaped or ellipsoidal rules. These rules trade the generality of the patch structure for rules with simple math. Ellipsoidal rules arise from the covariance matrices of clustering algorithms. They have a simple math structure that leads to simple learning algorithms. The ellipsoids project onto the axes to give back the fuzzy sets as triangles or sets with other shapes. Such projections are not unique because ellipsoids and most solids do not factor into independent components. Other techniques can tune the fuzzy sets themselves and so tune and move the rule patches. These techniques give other fuzzy sets and rules and so give other function approximations. We can also normalize all then-part sets to have the same area or volume. This does not affect the first-order structure of the additive system. Then the wide sets that come from large ellipsoids will not dominate the other narrower sets.

We focus on additive fuzzy systems of a simple algebraic form. They are *standard additive models* or SAMs. A SAM fuzzy system F stores m rules of the form "If $X = A_j$ then $Y = B_j$." The SAM Theorem in Section 2.3 shows that an unweighted SAM has the form of a convex sum of then-part set centroids c_j:

$$F(x) = \frac{\sum_{j=1}^{m} a_j(x) V_j c_j}{\sum_{j=1}^{m} a_j(x) V_j}. \tag{2.1}$$

Here set function a_j: $R^n \rightarrow [0, 1]$ defines the if-part "fuzzy" or multivalued set A_j. Input vector $x \in X$ belongs to fuzzy set $A_j \subset R^n$ to degree $a_j(x) \in [0, 1]$. Some texts [9, 14, 15, 52, 61] refer to the set function a_j as a "membership" function and denote it as m_{A_j}: $R^n \rightarrow [0, 1]$. We avoid this more cumbersome notation. The then-part fuzzy set $B_j \subset R^p$ has area or volume V_j and centroid c_j. The then-part set is often a scalar set in practice. The SAM Theorem lets us compute these volumes and local conditional means or centroids in advance. We explore (2.1) in great detail in later sections and the reader should commit its structure to memory.

SAMs and all additive fuzzy systems are function approximators. A finer and finer rule-patch cover is the thrust of the fuzzy approximation theorem [22] below. We can always find a finite number of rule patches to cover the graph of f to keep the distance $|f(x) - F(x)|$ as small as we please for all x. In practice we must guess at the rules or use a neural or statistical scheme to learn them from data. We must also balance the fineness of the patch approximation with how much it costs to process and tune the rules. In some cases a small number of rules exactly represent [55, 56] a function f in the sense that $F(x) = f(x)$ for all x. But this requires total knowledge of the approximand f and in that case we can just use f itself. We work in general with blind approximation when we do not know f.

A uniform approximation lets us pick the error level ε in advance. Then for all $\varepsilon > 0$ we have that $|f(x) - F(x)| < \varepsilon$ for *all* x in X. The domain X is a compact (closed and bounded) subset of R^n. X may be an interval $[u, v]$ where $u < v$ or a product of intervals $[u_1, v_1] \times \cdots \times [u_n, v_n]$. The uniform approximation of continuous functions allows us in theory to replace each continuous fuzzy set with a finite discretization or a point in a unit hypercube $[0, 1]^n$ or fuzzy space of high dimension. Then we can work with balls around the point sets and use the known structure of fuzzy cubes. The last three chapters apply these cubes to function approximation and pattern recognition and other problems.

The fuzzy approximation would not be of much use if each input x had its own error level $\varepsilon(x)$. Thus we insist on a uniform approximation. Then once we pick the error level $\varepsilon > 0$ we can find a finite set of fuzzy rules that gives a fuzzy system F ε-close to f. These ε-rules "exist" just as in a chess game an optimal set of moves exists from any stage in the game even though no one has found these moves.

In practice we may never find these rules or we may find only some of them in some regions of the product state space $X \times Y$. We can ask experts for the rules or watch experts or real processes to learn them or we can just guess at them. These first rules initialize the learning process. Then we can tune them with gradient descent or random hill climbing ("genetic algorithms") or other search techniques. But the rules learned from a finite stream of data may not give an F ε-close to f. Most supervised learning schemes need thousands or tens of thousands of iterations. Exponential rule complexity only compounds this learning complexity.

2.2 THE CURSE OF DIMENSIONALITY: RULE EXPLOSION AND OPTIMAL RULES

All fuzzy systems suffer from the *curse of dimensionality*: rule explosion. They need too many rules to approximate most functions [23, 25]. The number of rules grows exponentially with the number of input and output variables. Fuzzy systems do not scale up.

The patch geometry in Figure 2.1 shows how rules grow in a patch cover. The approximand is the scalar map $f\colon R \to R$. It takes k rule patches to cover the graph of f on some rectangle in the plane. For $f\colon R^2 \to R$ it takes on the order of k^2 rule patches to cover the surface in some 3-D box. For $f\colon R^n \to R^p$ it takes on the order of k^{n+p-1} rules to cover the graph of f in some hyperbox. Large rule patches can slow the rule explosion because they use a small value k. They cannot lessen its exponential form.

More variables add realism to a math model. They ignore less structure and relate more causes and effects. But they tax our skill at guessing at the rules (or equations) that relate them and they come at the supreme cost of rule explosion. All algorithms face some form of this curse of dimensionality. In the fuzzy case it can in the end defeat any expert who guesses at the rules or any neural system that tries to learn the rules from data. Rules are a scarce good in fuzzy approximation.

Optimal rules may be the best defense. Lone optimal rules cover the extrema or turning points of the function f [25]. They *patch the bumps*. The best place to put one lone rule is at the highest peak or the lowest trough. More rules can patch more bumps and can extend the base of the input fuzzy sets to give a full cover. Still more rules can fill in patches between the bump patches. The best place to put these extra patches is at the extrema of the error curve $(f - F)^2$ or the *residual* error curve after patches have covered the first set of extrema.

The extremal points $\hat{x}$ of f are those points where the derivative map f' is zero: $f'(\hat{x}) = 0$. This reduces much of adaptive fuzzy function approximation to estimating the zeroes of unknown functions f'. Neural or direct methods can estimate f' from the difference of noisy samples $(x_i, f(x_i))$. Newton's method [27] or other iterative or contraction maps can find some or all of the zeroes of f'. Then we can center the input fuzzy sets at these roots and perhaps add fuzzy sets centered between the roots. Or clustering algorithms can estimate the bumps directly from the data as in *product space clustering* [12]. Supervised or unsupervised learning can further tune the rules. The next chapter shows how such learning drives ellipsoidal rule patches to quickly cover the extrema of a bumpy polynomial.

Section 2.13 shows why lone rules should cover extrema. The idea is that the fuzzy system F is constant in the region of the lone patch: $F(x) = c$. On the real line this region is some interval $[u, v]$. Each choice of interval gives a new fuzzy system F over the domain. We must search through a family of fuzzy systems to find the optimal one-rule fuzzy system $\hat{F}$. This task involves some form of the calculus of variations. The case of overlapping input fuzzy sets leads to more complex optimality conditions. We postpone that derivation for this overview and instead present a local argument to see why lone optimal rules involve $f'(\hat{x}) = 0$. This amounts to a search along an error curve rather than a search through a family of error curves. The homework problems extend these results.

Suppose f is a smooth scalar map $f\colon [u, v] \to (0, \infty)$ with isolated critical points. Suppose the fuzzy system F has just one rule and the base of the input fuzzy set is the interval $[u, v]$. Then F is constant and defines the flat line $F(x) = c > 0$ for all x on the interval. The next section and (2.1) show that an additive fuzzy system with one rule "If X is A then Y is B" gives a scaled output fuzzy set aB for some $a > 0$. The scaled set aB has the same centroid and mode as the unscaled set B. So it always "defuzzifies" or maps to the same value c. That constant is the centroid or "center of mass" of B: $F(x) = c$. We here assume these properties of lone rules but prove them in later sections.

Now define the *error function* $e(x)$ as the square of the desired value $f(x)$ minus the actual value $F(x)$:

$$e(x) = (f(x) - F(x))^2 = (f(x) - c)^2. \tag{2.2}$$

The simplest case occurs when f lies above the line segment of the fuzzy system: $f(x) > c$ for all x. For now it may help to think of the line as just the interval $[u, v]$ itself and thus $c = 0$. Each value $e(x)$ measures the distance from $f(x)$ to c. Then the minima of $e(x)$ are the same as the minima of $f(x)$. Calculus confirms this geometric insight:

$$\frac{de(\hat{x})}{dx} = 0 = 2(f - F)(f' - F') \tag{2.3}$$

$$= 2(f - c)f'. \tag{2.4}$$

Then $f'(\hat{x}) = 0$ since $f > c$ and thus $(f - c) > 0$. The extremum $\hat{x}$ is here a local minimum of e if and only if it is a local minimum of f because

$$\frac{d^2e(\hat{x})}{dx^2} = 2(f'(\hat{x}))^2 + 2(f - c)f''(\hat{x}) \tag{2.5}$$

$$= 2(f - c)f''(\hat{x}) \tag{2.6}$$

for $f'(\hat{x}) = 0$. So e'' and f'' have the same sign at $\hat{x}$ since $2(f(\hat{x}) - c) > 0$.

The previous local argument holds for a given c on the interval $[u, v]$. In the global case we must vary the interval to find the best one-rule fuzzy system $\hat{F}$. The solution will contain the extrema of the approximand f and perhaps its endpoint extrema. For a fixed interval we can ask which constant $\hat{c}$ gives the best choice of F and hence the best choice of the then-part set B. The optimal $\hat{c}$ should equal the centroid of B and this in turn should equal the centroid of f on $[u, v]$. To show this we differentiate the integral of the error function $e(x)$ in (2.2) with respect to c and then solve:

$$0 = \frac{d}{dc} \int_u^v e(x)dx \tag{2.7}$$

$$= -\int_u^v 2(f(x) - \hat{c})dx \tag{2.8}$$

$$= -2 \int_u^v f(x)dx + 2\hat{c} \int_u^v dx \tag{2.9}$$

$$= -2 \int_u^v f(x)dx + 2\hat{c}(v - u). \tag{2.10}$$

So the best constant fuzzy system $\hat{F}$ equals the centroid or local mean of the function f:

$$\hat{c} = \frac{1}{v - u} \int_u^v f(x)dx. \tag{2.11}$$

The value $\hat{c}$ minimizes the mean-squared error since from (2.10) the second derivative in (2.7) with respect to c is $2(v - u) > 0$. The homework problems show that the same local result holds for the l^p error function $e(x) = \frac{1}{p}(f - c)^p$.

Lone optimal rules cover extrema and involve local centroids of f for all fuzzy systems F. This holds in general in complex systems since locally just one rule patch may cover any region of the graph of f. The first goal in building or learning any fuzzy system is to find the extremal points $f'(\hat{x}) = 0$ but this may require more knowledge of f than we have. Guessing at the surface bumps remains the best first step when guessing at the surface.

The next section formally defines additive fuzzy systems and states and proves the SAM Theorem (2.1). We return to optimal fuzzy systems in Section 2.13 after we have explored the structure of SAM systems in a wide range of contexts and system extensions.

2.3 ADDITIVE FUZZY SYSTEMS: THE STANDARD ADDITIVE MODEL

An additive fuzzy F system stores m fuzzy rules of the form "If $X = A_j$ then $Y = B_j$" and computes the output $F(x)$ as the centroid (or mode) of the *summed* and partially fired then-part fuzzy sets B'_j. Each input x fires all the rules to some degree and in parallel. So the additive fuzzy system $F: R^n \rightarrow R^p$ acts as an associative processor or global *fuzzy associative memory* (FAM) [21]. The homework problems show that each rule acts as a local FAM. Figure 2.3 shows the FAM signal-flow structure of a feedforward additive system.

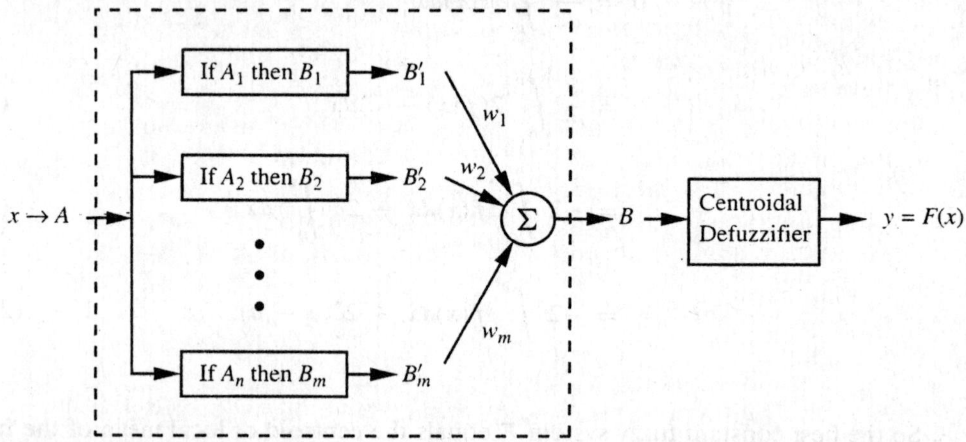

Figure 2.3 Additive fuzzy system architecture. Real input x fires the if-part of all m rules in parallel and to some degree (most to zero degree). Then the system scales or weights the then-parts to give the new fuzzy sets B'_j and then sums these to form the output set B. The system "defuzzifies" B or maps it to a scalar to give the output $y = F(x)$ by taking the centroid of B or by picking its mode or by some other means. In the discrete case x is a unit bit vector or column of the n-by-n identity matrix and each rule is an n-by-p matrix or fuzzy associative memory (FAM). In the continuous case x_i is a delta pulse $\delta(x - x_i)$. Neural or genetic or other adaptive systems can change the rule weights w_j or change the fuzzy sets A_j or B_j with sample data. Both the global fuzzy system F and its local rules act as types of FAMs.

The feedforward *standard additive model* (SAM) is the most important special case of an additive fuzzy system and we use it throughout this chapter and book. In the SAM case the fired then-part set B'_j is just the fit product $a_j(x)B_j$. The additive structure comes from the sum B of fired then-part sets:

$$B = \sum_{j=1}^{m} w_j B'_j \tag{2.12}$$

$$= \sum_{j=1}^{m} w_j a_j(x) B_j. \tag{2.13}$$

The sum is over the m rules that map fuzzy subsets $A_j \subset R^n$ to fuzzy subsets $B_j \subset R^p$. The *fit* (fuzzy unit) value $a_j(x)$ states the degree to which the input x belongs to the if-part fuzzy set A_j. So a_j: $R^n \rightarrow [0, 1]$ and b_j: $R^p \rightarrow [0, 1]$ define *set functions*. We later extend the then-part functions to maps from fuzzy subsets $A \subset R^n$ to vector points in R^p to give a *set* SAM (see (2.97) below). The rule weights w_j can be any real non-negative numbers.

The sum in (2.12) leads to a general additive system [21]. The sum in (2.13) leads to a SAM system. Rule weights w_j can scale each term in the sum to reflect rule credibility or frequency or "usuality" [59] and thus to give an extra term for a learning system to tune. They enter the SAM structure as multiples of the input set values $a_j(x)$. In practice we often ignore the rule weights and take them as all equal or all unity: $w_1 = \ldots = w_m = 1$.

The sum in (2.13) gives a SAM if we further map the output set B to a number $F(x)$ by computing its centroid for each x. We could also map B to $F(x)$ by choosing the mode of B (as we do in Chapter 9) or by choosing any other number that relates to B. The mode gives $F(x)$ as a MAP or maximum *a posteriori* estimate. Section 2.9 shows that the centroid gives a fuzzy system F the structure of a conditional expectation. So it acts as an optimal nonlinear approximator in the mean-squared sense. The centroid choice $F(x) = \text{Centroid}(B)$ uses all the information in the output set B and only the information in B. It is optimal in this Bayesian sense.

The summed output set B in (2.12) may have elements that exceed unity: $b(x, y) > 1$. Earlier fuzzy systems combined with pairwise maximum [30, 31, 45, 52] to avoid this outcome as we discuss in Section 2.6. The centroid uses only the relative information in B. So we require just that the set function $b(x, .)$: $R^p \rightarrow R^+$ be non-negative and integrable with respect to the output variable y for each input x. This gives a generalized set function.

We can now state and prove the simplest form of the *SAM Theorem*. It shows that a fuzzy system maps input x to $F(x)$ as a convex sum of the m then-part set centroids c_j. These centroids can change with each x as in a generalized SAM or can change with time as in an adaptive SAM. The SAM Theorem decomposes the global centroid of B to this convex sum of local centroids. The m coefficients $p_1(x), \ldots, p_m(x)$ are convex in that each term is non-negative $p_j(x) \geq 0$ and they sum to unity:

$$\sum_{j=1}^{m} p_j(x) = 1.$$

So the weights define a discrete probability density function for each x. Much of this chapter explores the structure of these convex coefficients and how they shape centroidal fuzzy systems. Theorem 2.2 proves a more general case and does so in the statistical framework of conditional means and variances.

Theorem 2.1. (*SAM Theorem*) Suppose the fuzzy system $F: R^n \rightarrow R^p$ is a *standard additive model*: $F(x) = \text{Centroid}(B) = \text{Centroid}\left(\sum_{j=1}^{m} w_j\, a_j(x)\, B_j\right)$.

Then $F(x)$ is a convex sum of the m then-part set centroids:

$$F(x) = \frac{\sum_{j=1}^{m} w_j\, a_j(x) V_j\, c_j}{\sum_{j=1}^{m} w_j\, a_j(x)\, V_j} \tag{2.14}$$

$$= \sum_{j=1}^{m} p_j(x)\, c_j. \tag{2.15}$$

The convex coefficients or discrete probability weights $p_1(x), \ldots, p_m(x)$ depend on the input x through the ratios

$$p_j(x) = \frac{w_j\, a_j(x)\, V_j}{\sum_{k=1}^{m} w_k\, a_k(x)\, V_k}. \tag{2.16}$$

V_j is the finite positive volume (or area if $p = 1$ in the range space R^p) and c_j is the centroid of then-part set B_j:

$$V_j = \int_{R^p} b_j(y_1, \ldots, y_p) dy_1 \ldots dy_p > 0 \tag{2.17}$$

$$c_j = \frac{\int_{R^p} y b_j(y_1, \ldots, y_p) dy_1 \ldots dy_p}{\int_{R^p} b_j(y_1, \ldots, y_p) dy_1 \ldots dy_p}. \tag{2.18}$$

The popular scalar case of $p = 1$ reduces (2.17) and (2.18) to

$$V_j = \int_{-\infty}^{\infty} b_j(y) dy \tag{2.19}$$

$$c_j = \frac{\int_{-\infty}^{\infty} y b_j(y) dy}{\int_{-\infty}^{\infty} b_j(y) dy} \tag{2.20}$$

Proof. The theorem follows by expanding the centroid of $B(x)$ and invoking the SAM assumption (2.13) to rearrange terms:

$$F(x) = \text{Centroid}(B) = \frac{\int_{R^p} yb(y)dy}{\int_{R^p} b(y)dy} \tag{2.21}$$

$$= \frac{\int_{R^p} y \sum_{j=1}^{m} w_j b'_j(y)dy}{\int_{R^p} \sum_{j=1}^{m} w_j b'_j(y)dy} \tag{2.22}$$

$$= \frac{\int_{R^p} y \sum_{j=1}^{m} w_j a_j(x) b_j(y)dy}{\int_{R^p} \sum_{j=1}^{m} w_j a_j(x) b_j(y)dy} \tag{2.23}$$

$$= \frac{\sum_{j=1}^{m} w_j a_j(x) \int_{R^p} yb_j(y)dy}{\sum_{j=1}^{m} w_j a_j(x) \int_{R^p} b_j(y)dy} \tag{2.24}$$

$$= \frac{\sum_{j=1}^{m} w_j a_j(x) V_j \frac{\int_{R^p} yb_j(y)dy}{V_j}}{\sum_{j=1}^{m} w_j a_j(x) V_j} \tag{2.25}$$

$$= \frac{\sum_{j=1}^{m} w_j a_j(x) V_j c_j}{\sum_{j=1}^{m} w_j a_j(x) V_j}. \tag{2.26}$$

■

The standard additive model in (2.14) involves light computation. We compute the volumes V_j and centroids c_j in advance. They change only when the system learns or tunes its rules or when a user varies them to hand tune the system. For each input x we need compute just the m fit values $a_j(x)$ and then update the ratio in

(2.14). Simple then-part sets B_j can take the form of triangles or trapezoids or bell curves and so have simple areas and centroids. We now discuss some of the basic properties of SAM systems.

The SAM structure (2.14) lets us replace all then-part fuzzy sets B_j with rectangles or *non*fuzzy sets R_j that have the same volume V_j and centroid c_j. This does not change the output value $F(x)$ but it can change the variance or *uncertainty* of the output as we discuss in Section 2.10. A symmetric set B_j has the same centroid no matter how wide or thin the set. But a thicker set codes less certain knowledge and thus so does a thicker rule patch. Most applications ignore this second level of output uncertainty.

Consider the hypercube R_j centered at c_j that has volume V_j and unit height. Then R_j has a binary set function $r_j\colon R^p \to \{0, 1\}$ and so defines a standard or nonfuzzy set. In the scalar case $F\colon R^n \to R$ the jth rectangle has the 1-D cube base

$$\left[c_j - \frac{V_j}{2},\ c_j + \frac{V_j}{2}\right].$$

In the vector case $F\colon R^n \to R^p$ the jth rectangle has a p-D cube base with p sides of the form

$$\left[c_j^k - \frac{\sqrt[p]{V_j}}{2},\ c_j^k + \frac{\sqrt[p]{V_j}}{2}\right].$$

So we can dispense with half the fuzziness of a fuzzy system F. An exercise shows that we can always find rectangles centered at the then-part centroids c_j that partition the p coordinates of the range space. These rectangles need not have unit height.

Sometimes we normalize the volumes to unity so the terms V_j do not appear in (2.14). Otherwise the volumes act as rule weights. The larger V_j the more it forces the global output $F(x)$ to act like the local rule output c_j:

$$\lim_{V_j\to\infty} F(x) = \lim_{V_j\to\infty} \frac{\sum_{k=1}^{m} a_k V_k c_k}{\sum_{k=1}^{m} a_k V_k} \tag{2.27}$$

$$= \lim_{V_j\to\infty} \frac{a_j V_j c_j + c}{a_j V_j + d} \tag{2.28}$$

$$= c_j. \tag{2.29}$$

Some learning schemes may weight the jth rule with the inverse volume $1/V_j$ to give less weight to large or uncertain rules. The next chapter shows how this can arise when we grow ellipsoidal rules from error covariance matrices. The variance weight $w_j = 1/\sigma_j^2$ has the same effect for Gaussian SAMs.

Learning schemes may also require that F be differentiable. This holds just in case all the if-part sets $a_j(x)$ are differentiable. So the fuzzy system is only *piecewise* differentiable if the if-part sets are triangles, trapezoids, rectangles, or other curves with "corners."

The convex weights $p_1(x), \ldots, p_m(x)$ define a discrete probability density function $\mathbf{p}(x)$ for each input x since the m terms are non-negative and sum to one. So for each input x the fuzzy system F defines the expected value of the m output centroids c_j with respect to $\mathbf{p}(x)$:

$$F(x) = \sum_{j=1}^{m} p_j(x)c_j = E_{\mathbf{p}(x)}[C]. \tag{2.30}$$

Again we will see that each such value $F(x)$ is just one realization of a random conditional mean vector $E[Y|X = x]$. So we can view a SAM fuzzy system as a way to pick a good set of centroid weights for each input x. And again each output value has a distinct uncertainty value. The uncertainty grows as the output interpolates between more centroids c_j. The output has least uncertainty when one rule fires "dead on" and some convex coefficient obeys $p_j(x) = 1$ while others obey $p_k(x) = 0$.

Convexity also implies the key geometric constraint that $F(x)$ lies at or between the least and greatest centroid value for each of the p centroid components in the range space R^p. We can state this for the scalar case $F\colon R^n \to R$ as follows:

$$c_1 \leq F(x) \leq c_m \tag{2.31}$$

This holds for all densities $p(x)$ and for all centroidal additive systems. The proof in Section 2.10 of the fuzzy approximation theorem depends on this fact.

The vector case $F\colon R^n \to R^p$ shows that $F(x)$ lies in a p-dimensional hyperrectangle or centroid box of the form $[c^1_{left}, c^1_{right}] \times \cdots \times [c^p_{left}, c^p_{right}] \subset R^p$. The convex weight $p_j(x)$ scales the jth vector centroid $c_j = (c^1_j, \ldots, c^p_j)$ by scaling its p components: $p_j(x)c_j = (p_j(x)c^1_j, \ldots, p_j(x)c^p_j)$. Then

$$F(x) = (F_1(x), \ldots, F_p(x)) = \sum_{j=1}^{m} p_j(x)c_j \tag{2.32}$$

$$= \left(\sum_{j=1}^{m} p_j(x)c^1_j, \ldots, \sum_{j=1}^{m} p_j(x)c^p_j \right). \tag{2.33}$$

So each fuzzy output component $F_k(x)$ is a convex combination of the m scalars $c^k_1, \ldots, c^k_m$. So $c^k_{left} \leq F_k(x) \leq c^k_{right}$ holds for $c^k_{left} = \min(c^k_1, \ldots, c^k_m)$ and $c^k_{right} = \max(c^k_1, \ldots, c^k_m)$. So $F(x) \in [c^1_{left}, c^1_{right}] \times \cdots \times [c^p_{left}, c^p_{right}] \subset R^p$ holds as claimed. To match the centroid cube to the range of f is a key geometric step in fuzzy function approximation.

So far we have focused on the then-part structure of SAMs. The next section looks at the if-part structure of the set functions $a_j\colon R^n \to [0, 1]$ that shape the convex coefficients $p_j\colon R^n \to [0, 1]$.

2.4 SET FUNCTIONS IN SAMS

The SAM Theorem allows us to pick arbitrary joint if-part set functions $a_j: R^n \to [0, 1]$. Chapter 6 works with metrical joint set functions. In practice we often work with the n scalar set-function factors $a_j^1, a_j^2, \ldots, a_j^n$ even though they ignore the correlations among components in the joint set functions. These factors $a_j^i: R \to [0, 1]$ are often the familiar triangles, trapezoids, and bell curves of fuzzy engineering. Symmetric triangles give a piecewise linear SAM. The most common way to combine set values is with pairwise minimum:

$$a_j(x) = a_j^1(x_1) \wedge a_j^2(x_2) \wedge \ldots \wedge a_j^n(x_n) \tag{2.34}$$

for input vector $x = (x_1, x_2, \ldots, x_n)$. Suppose the input temperature value is 80% warm and the input humidity is 60% high in a simple air conditioner. Then the factor 60% = min(80%,60%) scales the then-part set B_j to give $.6B_j$.

The min combiner is a conjunctive or AND combiner. Both factors must hold to a high degree to "fire" the rule to a high degree. We assume that the if-part factors combine a_j^i with AND though the SAM model allows any logic scheme to arrive at $a_j: R^n \to [0, 1]$. So we could replace the min in (2.34) with max to model OR if-parts. This seldom occurs in practice. In this text we sometimes use (2.34) as the default combiner.

Engineers also often use product in place of min to factor joint if-part set functions:

$$a_j(x) = \prod_{i=1}^{n} a_j^i(x_i). \tag{2.35}$$

The product if-part combiner may arise in digital hardware devices [32] that compute products more easily than they compute minima. Optical devices compute products and sums with ease but have more trouble with mins, maxes, divisions, and differences. Chapter 11 looks at optical SAM systems. The end of this section compares (2.34) and (2.35) in a numerical example of a 3-rule SAM system.

There is one key reason to use the product if-part combiner (2.35) over the min combiner (2.34). The product combiner does not ignore information as does the min combiner. The min gives $a_j(x) = .6$ in the above example but the product combiner gives $a_j(x) = .48$. But suppose we add more dimensions to the input space. Then min(.8,.6,.6,.6) or even min(.8,.6,.9,.9) still gives $a_j(x) = .6$. But the product combiner gives $a_j(x) = .1728$ or $a_j(x) = .3888$ to reflect the diminished or increased *joint* strength of the if-part rule firing. The product tends to get small for large n but that does not affect the SAM output. The convex coefficients in (2.15) normalize the rule firings.

The product if-part combiner (2.35) can also arise when working with radial basis [12, 44] or Gaussian [54] if-part set functions:

$$a_j^i(x_i) = s_i^j \exp\left[-\frac{1}{2}\left(\frac{x_i - \bar{x}_i^j}{\sigma_i^j}\right)^2\right] \tag{2.36}$$

for scaling constant $0 < s_i^j \leq 1$. Then the product in (2.35) gives the exponential of a sum. These radial basis nets center a Gaussian ball at the mean vector $\bar{x}_j = (\bar{x}_1^j, \ldots, \bar{x}_n^j)$. Each scalar mean value $\bar{x}_i^j$ centers a bell curve with variance $(\sigma_i^j)^2$. These n variances make up the diagonal of some covariance matrix K with zero entries off the diagonal. So (2.35) assumes that the n factors in (2.36) are independent Gaussian random variables. The choice of min versus product often makes little difference in practice for systems with only two or three inputs. We can use either one but in theory the product combiner (2.35) has more discriminatory power. It also gives a SAM map that uses only the standard operations of multiplication, addition, and division.

In theory we can also use any conjunctive combiner T to compute $a_j(x)$ from its n factors $a_j^1, a_j^2, \ldots, a_j^n$:

$$a_j(x) = T\left(a_j^1(x_1), \ldots, a_j^n(x_n)\right). \tag{2.37}$$

Here T can be a so-called t-norm or triangular norm [15, 21] with dual disjunctive or OR t-conorm S. T-norms and t-conorms obey $T(x, y) \leq \min(x, y) \leq \max(x, y) \leq S(x, y)$ for x and y in [0,1] as discussed in the homework problems of Chapter 1. The product t-norm $T(x, y) = xy$ has De Morgan dual t-conorm $S(x, y) = 1 - T(1 - x, 1 - y) = x + y - xy$. We can also use a t-norm to form the fired or inferred set B_j' in (2.12): $B_j' = T(a_j(x), B_j)$. Few engineers use t-norms in practice other than min or product. We can also use any other combination scheme to compute the joint if-part set function $a_j(x)$ from its n factors. One such scheme is the arithmetic mean

$$a_j(x) = \frac{1}{n}\sum_{i=1}^{n} a_j^i(x_i).$$

The homework problems in Chapter 1 discuss other combination schemes that fall between min and max and hence between t-norms and t-conorms.

The min clip $B_j' = a_j(x) \wedge B_j$ or *correlation-minimum inference* [21] still occurs in practice but much less so than it once did. It both ignores all information in B_j' above the fit-value threshold $a_j(x)$ and does not lead to as simple a transfer function as the SAM equation (2.14). The same holds for other t-norm schemes of the form $B_j' = T(a_j(x), B_j)$. None work as well as the SAM scheme $B_j' = a_j(x)B_j$ of *correlation-product inference*. T-norms are far more popular among theorists than engineers. They do capture the De Morgan duality of AND and OR among if-part factors. But their popularity in the older fuzzy literature seems more the result of the analytical platform they provide than of their proven value in real systems. We mention them for completeness but discuss them in detail only in the homework problems of Chapter 1.

One question remains of how a SAM model computes output $F(x)$ from input x: How does a numeric input x_0 "pick off" the fit value $a_j(x_0)$ from the if-part set A_j? In practice we just plug the input x_0 into the function a_j and compute the value $a_j(x_0)$. This contrasts with a random view of a_j as a continuous probability density function. Then each value is zero and we would have to integrate to get a nonzero value. One way to deal with this is to view A_j as a *random set* [34] or locus of discrete two-point conditional densities: $a_j(x) = prob\{x \in A_j | X = x\}$ and $a_j^c(x) = 1 - a_j(x) = prob\{x \notin A_j | X = x\}$.

Another approach is to consider the general case of passing an entire fuzzy set A through the fuzzy system F. The input set A matches the if-part A_j of each rule as a fuzzy associative memory and so similar inputs map to similar outputs: $A \approx A_j \to B'_j \approx B_j$. We have discussed this in full detail for the discrete case of fit vectors and FAM rule matrices [21]. The same view holds in the continuous case that we emphasize in this text. Equations (2.92)–(2.99) deal with the general case of an arbitrary fuzzy set A as input to the SAM system. We can view the input x_0 as a binary singleton set $\{x_0\}$. Then the fuzzy set A collapses to a spike or Dirac delta pulse $\delta(x - x_0)$ centered at x_0. (In the discrete case $\{x_0\}$ defines a unit bit vector with a 1 in the present slot and 0s in all other slots.) Then the "fired" fit value $a_j(x_0)$ follows from the sifting or combing property of convolving the set function with the delta pulse:

$$a_j(x_0) = \int_{R^n} \delta(x - x_0) a_j(x) dx \tag{2.38}$$

The pulse view of inputs also shows how an input x_0 "fires" a rule patch $A_j \times B_j$ in a SAM. The patch is a Cartesian product of sets. Each pair (x, y) belongs to the patch to some degree in [0,1]. Some conjunctive operator defines the patch set function in terms of the marginal set functions a_j and b_j. The SAM combines the two set functions with product:

$$R_{A_j \to B_j}(x, y) = a_j(x) b_j(y). \tag{2.39}$$

In the discrete case (2.39) defines a Hebbian or correlation FAM matrix as the outer-product matrix $A_j^T B_j$ of the row-fit vectors $A_j = (a_1, \ldots, a_n)$ and $B_j = (b_1, \ldots, b_p)$. Passing input x_0 through the rule patch gives the multiplicative form $B'_j = a_j(x_0) B_j$ in the additive sum (2.13):

$$b'_j(y) = \int_{R^n} \delta(x - x_0) R_{A_j \to B_j}(x, y) dx \tag{2.40}$$

$$= \int_{R^n} \delta(x - x_0) a_j(x) b_j(y) dx \tag{2.41}$$

$$= b_j(y) \int_{R^n} \delta(x - x_0) a_j(x) dx \tag{2.42}$$

$$= a_j(x_0) b_j(y) \quad \text{for all } y \in R^p. \tag{2.43}$$

The delta-pulse method extends to *any* rule-patch set function $R_{A_j \to B_j}: R^n \times R^p \to [0, 1]$ and thus to any t-norm coding of the rule patch such as pairwise minimum. Put $g_j(x) = R_{A_j \to B_j}(x, y)$ in the integrand of (2.40) to remind us that here the rule patch depends only on x. Then $b'_j(y) = R_{A_j \to B_j}(x_0, y)$. T-norm coding with $R_{A_j \to B_j}(x, y) = T(a_j(x), b_j(y))$ gives $b'_j(y) = T(a_j(x_0), b_j(y))$. The special case of the min t-norm or correlation-minimum coding [21] gives the "min clip" $b'_j(y) = a_j(x_0) \wedge b_j(y)$ or $B'_j = a_j(x_0) \wedge B_j$.

We conclude this section with an example of a simple 3-rule SAM system. The next sections explore some special cases of the SAM system and how it extends to more complex fuzzy mappings.

Consider the numerical example of a SAM map $F: R^2 \to R$ with just 3 rules or rule patches $A_1 \times B_1$, $A_2 \times B_2$, $A_3 \times B_3$. Suppose the 3 then-part sets B_j have the following centroids and volumes (areas):

$$c_1 = -5 \qquad V_1 = 1$$

$$c_2 = 0 \qquad V_2 = 2$$

$$c_3 = 5 \qquad V_3 = 1.$$

Suppose the 3 rules have the same or unity rule weights: $w_1 = w_2 = w_3 = 1$. Then we can ignore the weights.

We will compute the SAM output $F(x)$ for just one vector input $x = (x_1, x_2)$ for both min if-part combination (2.34) and product if-part combination (2.35). Suppose x fires the 3 rules or belongs to the 3 product sets $A_1 = A_1^1 \times A_1^2$, $A_2 = A_2^1 \times A_2^2$, $A_3 = A_3^1 \times A_3^2$ and gives the 3 pairs of fit values:

$$a_1^1(x_1) = .6 \qquad a_1^2(x_2) = .1$$

$$a_2^1(x_1) = .3 \qquad a_2^2(x_2) = .9$$

$$a_3^1(x_1) = .1 \qquad a_3^2(x_2) = .2$$

A glance at the fit values shows that we should expect the SAM output $F(x)$ to favor the middle centroid value $c_2 = 0$. The double-volume weight $V_2 = 2$ further favors this outcome. The first rule fires slightly more than the third rule and so we can expect the output centroid $F(x) = Centroid(B)$ to move slightly in the negative direction. This happens with the product if-part combiner but not with the min combiner.

The min if-part combiner gives

$$a_1(x) = \min\left(a_1^1(x_1),\ a_1^2(x_2)\right) = \min(.6,\ .1) = .1$$

$$a_2(x) = \min\left(a_2^1(x_1),\ a_2^2(x_2)\right) = \min(.3,\ .9) = .3$$

$$a_3(x) = \min\left(a_3^1(x_1),\ a_3^2(x_2)\right) = \min(.1,\ .2) = .1.$$

So the min gives the same weight to the first and third rules. The SAM output $F(x)$ equals

$$F(x) = \frac{\sum_{j=1}^{m} a_j(x) V_j c_j}{\sum_{j=1}^{m} a_j(x) V_j}$$

$$= \frac{\sum_{j=1}^{3} \min\left(a_j^1(x_1),\ a_j^2(x_2)\right)\ V_j c_j}{\sum_{j=1}^{3} \min\left(a_j^1(x_1),\ a_j^2(x_2)\right)\ V_j}$$

$$= \frac{(.1 \times 1 \times (-5)) + (.3 \times 2 \times 0) + (.1 \times 1 \times 5)}{(.1 \times 1) + (.3 \times 2) + (.1 \times 1)}$$

$$= \frac{-.5 + .5}{.62}$$

$$= 0.$$

The product if-part combiner gives a different set of fit values:

$$a_1(x) = a_1^1(x_1) \times a_1^2(x_2) = .6 \times .1 = .06$$

$$a_2(x) = a_2^1(x_1) \times a_2^2(x_2) = .3 \times .9 = .27$$

$$a_3(x) = a_3^1(x_1) \times a_3^2(x_2) = .1 \times .2 = .02.$$

Then the SAM has the output

$$F(x) = \frac{\sum_{j=1}^{m} a_j(x) V_j c_j}{\sum_{j=1}^{m} a_j(x) V_j}$$

$$= \frac{\sum_{j=1}^{3} \left[a_j^1(x_1) a_j^2(x_2)\right] V_j c_j}{\sum_{j=1}^{3} \left[a_j^1(x_1) a_j^2(x_2)\right] V_j}$$

$$= \frac{(.06 \times 1 \times (-5)) + (.27 \times 2 \times 0) + (.02 \times 1 \times 5)}{(.06 \times 1) + (.27 \times 2) + (.02 \times 1)}$$

$$= \frac{-(.06 \times 5) + (.02 \times 5)}{.62}$$

$$= -\frac{10}{31}$$

or $F(x) \approx -0.32$.

2.5 THE "CENTER OF GRAVITY" METHOD AS A CONSTANT-VOLUME SAM

The SAM fuzzy system

$$F(x) = \frac{\sum_{j=1}^{m} w_j a_j(x) V_j c_j}{\sum_{j=1}^{m} w_j a_j(x) V_j} \tag{2.44}$$

reduces to the *center of gravity* or COG fuzzy model

$$F(x) = \frac{\sum_{j=1}^{m} a_j(x) P_j}{\sum_{j=1}^{m} a_j(x)} \tag{2.45}$$

if the modes or "peaks" P_j of the then-part sets $B_j \subset R^p$ equal the then-part set centroids c_j and if the then-part sets B_j all have the same areas or volumes V_j and the same rule weights w_j: $P_j = c_j$ and $V_1 = \ldots = V_m > 0$ and $w_1 = \ldots = w_m > 0$. Sugeno [45] and other Japanese fuzzy engineers [49, 52] have popularized the COG model as an *ad hoc* scheme to "defuzzify" a fuzzy system's set output B. Most digital and analog fuzzy chips use the COG model. In this way many fuzzy engineers have used the SAM model without knowing it. Below we show that the SAM/COG models include the popular radial basis function networks [12, 33, 44] of modern neural network theory.

Some engineers dispense with the fuzzy structure of B_j and replace it with a spike centered at c_j on the real line R or in theory at some point in R^p. Then the set function $b_j: R^p \to [0, 1]$ of then-part set B_j is just a delta pulse: $b_j(y) = \delta(y - c_j)$. Some use the unit-pulse convention that $b_j(c_j) = 1$ but this is not needed. Indeed formally a Dirac delta function obeys $\delta(y - c_j) = \infty$ if $y = c_j$ and $\delta(y - c_j) = 0$ if $y \neq c_j$. In any case the pulse set B_j has unit volume and has c_j as its centroid:

$$V_j = \int_{R^p} \delta(y - c_j)dy = 1 \tag{2.46}$$

$$Centroid(B_j) = \frac{\int_{R^p} y\delta(y - c_j)dy}{\int_{R^p} \delta(y - c_j)dy} = \int_{R^p} y\delta(y - c_j)dy = c_j. \tag{2.47}$$

The proof of the SAM Theorem still holds even if the scalar w_j weights the jth then-part spike or gives the value of its "height":

$$F(x) = Centroid(B(x)) = Centroid\left(\sum_{j=1}^{m} w_j a_j(x)\delta(y - c_j)\right) \tag{2.48}$$

$$= \frac{\int_{R^p} y \sum_{j=1}^{m} w_j a_j(x)\delta(y - c_j)dy}{\int_{R^p} \sum_{j=1}^{m} w_j a_j(x)\delta(y - c_j)dy} \tag{2.49}$$

$$= \frac{\sum_{j=1}^{m} w_j a_j(x) \int_{R^p} y\delta(y - c_j)dy}{\sum_{j=1}^{m} w_j a_j(x) \int_{R^p} \delta(y - c_j)dy} \tag{2.50}$$

$$= \frac{\sum_{j=1}^{m} w_j a_j(x) c_j}{\sum_{j=1}^{m} w_j a_j(x)}. \tag{2.51}$$

So again the SAM system (2.44) or (2.51) reduces to the unweighted COG model (2.45) with $P_j = c_j$ and $V_1 = \ldots = V_m > 0$ and $w_1 = \ldots = w_m > 0$. Note that the spike weights w_j do not change with each input x though they can if we want them to and write them as $w_j(x)$. The weights can also take on negative values to give an "upside down" spike. Most real systems do not use them at all.

Often engineers misuse the SAM/COG model because they vary the structure of the then-part sets B_j while they still assume that the set structure does not vary. They hand tune the sets B_j or let a learning scheme move and reshape them. Then the peaks P_j may differ from the set centroids c_j . More often the then-part set widths or volumes differ. This seems to stem from a desire to keep each then-part set triangle, trapezoid, or bell curve of unit height and yet adjust its width or base to reflect the rule's importance to the system output $F(x)$. We tend to place narrow sets near equilibrium regions for more precise control and place wider sets farther away for rougher control and to quickly bring the system closer to equilibrium. But then the COG model gives more weight to the less important rules since their then-parts B_j have more area or volume V_j. This can lead to needless bouts of tuning and to poor function approximation. The simple solution is either to use the full SAM system (2.44) with variable set volumes V_j or to not insist that the then-part "sets" B_j have unit height and to not normalize them and so to use a proper SAM system. More complex solutions [7] can adjust the rule weights w_j as a function of the volumes V_j.

The misuse of COG models runs so deep in fuzzy engineering because of the model's *ad hoc* nature. Engineers use the COG model (2.45) to compute a quick $F(x)$ and then justify the model by pointing to its practical effects. No doubt these effects would improve somewhat or perhaps a great deal if a SAM processed the same data with the same rules. So why not use a SAM to begin with?

The literature shows that here the new fuzzy school of function approximation meets the older school of linguistic analysis. Papers use the COG model after they first cite a nonadditive scheme for combining fired then-part sets B'_j. This runs from the first work on fuzzy systems of Mamdani in the 1970s [30, 31] to the recent conference proceedings of the IEEE-FUZZ conferences. Engineers cite the "extension principle" combination rule of pairwise maxima of fired then-part sets

$$B = \bigcup_{j=1}^{m} B'_j = \bigcup_{j=1}^{m} a_j(x) B_j \tag{2.52}$$

as a type of homage to the "extension principle" [9, 14, 58, 59]. This principle converts or "extends" a function f that maps points in X to points in Y to a function that maps fuzzy subsets $A \subset X$ to fuzzy subsets $B \subset Y$—from $y = f(x)$ to $B =$

$f(A)$. The extension principle has the form of a supremum or maximum of pairwise minima (or products). The max combiner in (2.52) uses $b(y) = \max\{a_j(x)b_j(y): j = 1, \ldots, m\}$ for product scaling or $b(y) = \max\{\min(a_j(x), b_j(y)): j = 1, \ldots, m\}$ for min clipping. The result B is the *envelope* or silhouette of the "fired" then-part sets B'_j. There is no easy way to compute the centroid of B in the formula $F(x) = \text{Centroid}(B)$ in this scheme that Terano and Sugeno [52] call "inference method 1" (the next section discusses their other inference methods). Hence an engineer must appeal to *ad hoc* but efficient techniques such as (2.45) to arrive at $F(x)$ from the rule set and the input data x. This move in effect assumes the competing additive combination scheme

$$B = \sum_{j=1}^{m} B'_j(x) = \sum_{j=1}^{m} a_j(x)B_j \tag{2.53}$$

and thus rejects the max-min edifice it tries to support. Of course the SAM models (2.44) or (2.45) do not logically imply the sum combiner (2.53). But the written appeal to some sort of then-part combination scheme strongly suggests it. Language further confuses the issue. Fuzzy engineers often call systems that invoke the max combiner (2.52) a "Mamdani system" even if the systems are additive systems like SAMs or COGs. The point again is that few real fuzzy systems actually use the max-combined B in (2.52) to defuzzify with a centroid even though many cite the max scheme in their written system description.

There are two main problems with the extension principle or max combiner (2.52): It is hard to compute and the envelope B tends toward the same rectangular pulse as more rules fire. The max or supremum nature of (2.52) does not lead to a simple closed-form solution if the then-part sets B'_j overlap. The extreme case of disjoint then-part sets B'_j reduces to (2.53) since the real-number equality $x + y = \min(x, y) + \max(x, y)$ reduces to $x + y = \max(x, y)$ iff the pairwise intersection is null: $\min(x, y) = 0$ or $B'_j \cap B'_{j+1} = \emptyset$. We showed previously that the SAM model remains unchanged if we replace all then-part sets B_j with overlapping rectangles R_j (along each axis) of varying height and thus varying area or volume. Then the max combiner (2.52) acts the same as the sum combiner (2.53) but only because we have started with (2.53) to compute B in the output centroid $F(x)$. The max combiner does not let us replace then-part sets B_j with rectangles R_j in the general case when then-part sets overlap.

The max combiner (2.52) ignores the consensus in the "fired" overlap $B'_j \cap B'_{j+1}$. Consider the extreme case when two rules fire and have the same fired then-parts: $B'_j = B'_{j+1}$. The max combiner gives $B'_j \cup B'_{j+1} = B'_j$ while the sum combiner (2.53) gives $B'_j + B'_{j+1} = 2B'_j$. The problem becomes more pronounced as more fired then-part sets have more overlap. This happens both in lone fuzzy systems and in combined fuzzy systems where many experts may tend to agree in their rule structure. It tends to happen more as the input dimension n grows since exponentially more of the if-part sets tend to fire for each input x. Each input fires 2^n rules in the common case of rules with overlapping or contiguous if-part sets along each of the n axes. Hence the

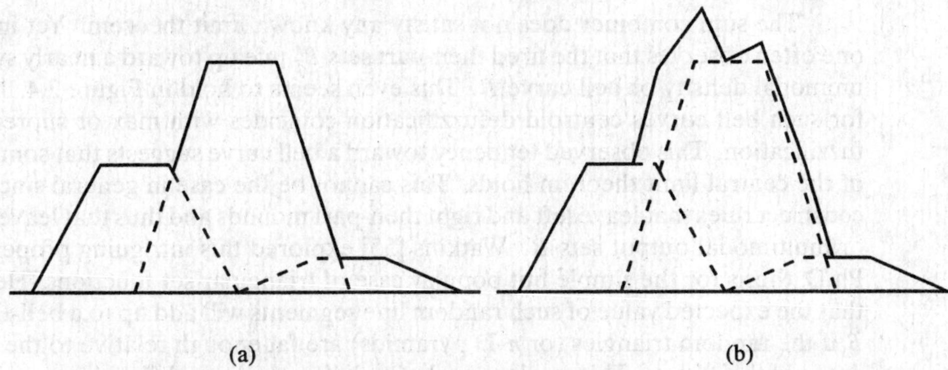

Figure 2.4 Max versus sum combination of fired then-part sets. The max envelope (a) tends toward a rectangle as the number of trapezoids grows. The sum envelope (b) may tend toward a bell shape or symmetric unimodal curve or some more complex curve. It does not tend toward a rectangle. The centroid of the sum envelope in (b) must fall at or between the centroids of the first and third then-part trapezoids: $c_1 \leq F(x) \leq c_3$. The proof of the fuzzy approximation theorem depends on this fact. Centroids of max envelopes can fall outside these bounds.

more rules that fire then the less the centroidal output $F(x) = \text{Centroid}(B)$ tends to change. The system grows less sensitive as its knowledge base grows. Figure 2.4 compares the max combiner with the sum combiner.

A limit theorem [17] lies behind the max combiner's monotonic march to a rectangular pulse or the uniform probability density. The idea is that the max or supremum of a list of random numbers in an interval $[a, b]$ can only grow toward the upper bound b. This still holds if one forms a third sequence from two random sequences by picking their pairwise minima and then taking the global supremum. The same holds with probability one for the supremum of any finite number of pairwise independent sequences:

$$\limsup_{i\to\infty} x_i^1 \wedge x_i^2 \wedge \ldots \wedge x_i^n = b \quad \text{with probability one} \tag{2.54}$$

for n pairwise independent sequences of independent identically distributed random variables $\{x_i^j\}$ that take values in $[a, b]$ and that have sequences of probability density functions $\{p_i^j\}$ that are nondegenerate in the sense that their Lebesgue integral [40] obeys

$$\int_I p_i^j(x)dx > 0$$

for each interval I of the form $[k, 1]$ or $[0, 1-k]$ for all k in $(0,1)$. This theorem follows from the Borel-Cantelli lemma of probability theory and shows that the extension principle does not "extend" a fuzzy set in general. It tends to map it to a binary or near-binary set. In many cases one can just as well pick the value 1 for the induced set value if it is nonzero. An averaging operation might better extend the fuzzy structure from the domain of a function to its range. In any case the max combiner (2.52) seems as troubled in theory as it is ignored in practice.

The sum combiner does not satisfy any known limit theorem. Yet in practice one often observes that the fired then-part sets B'_j pile up toward a nearly symmetric unimodal density or bell curve B. This even seems to hold in Figure 2.4. Note that for such bell curves centroid defuzzification coincides with max or supremum defuzzification. This observed tendency toward a bell curve suggests that some version of the central limit theorem holds. This cannot be the case in general since we can construct rules that leave left and right then-part mounds and thus that leave bimodal or multimodal output sets B. Watkins [55] explored this intriguing property in his Ph.D. thesis for the simple but popular case of triangular set functions. He showed that the expected value of such random line segments will add up to a bell-like curve B if the random triangles (or n-D pyramids) are fat enough relative to the effective range of the B'_j sets. This random analysis of a fuzzy system differs from earlier models of fuzzy random variables [37] and fuzzy laws of large numbers [28] that fuzzify random techniques. It simply applies random techniques to the SAM model F in (2.44). Still we lack both practical and general theoretical guidelines for when or to what degree the combined set B will approach a bell curve. This remains one of many open research problems.

We last show that the *radial basis function networks* or RBFs of neural network theory [12] are a special case of the SAM. A simple Gaussian SAM gives both a COG model and the popular radial-basis-function (RBF) model of neural networks of Moody [33] and Specht [44]. Wang and Mendel [54] have recently restated this RBF model in fuzzy notation as a simple scalar Gaussian SAM F: $R^n \to R$:

$$F(x) = \frac{\sum_{j=1}^{m} \bar{z}^j \left(\prod_{i=1}^{n} \mu_{A_i^j}(x_i) \right)}{\sum_{j=1}^{m} \left(\prod_{i=1}^{n} \mu_{A_i^j}(x_i) \right)}. \tag{2.55}$$

The SAM (2.44) reduces to (2.55) if for Gaussian sets with product combination of if-part set values we have

$$y = z \tag{2.56}$$

$$a_j(x) = \prod_{i=1}^{n} a_i^j(x_i) \tag{2.57}$$

$$= \prod_{i=1}^{n} \mu_{A_i^j}(x_i) \tag{2.58}$$

$$V_j = 1 \tag{2.59}$$

$$c_j = \bar{z}^j. \tag{2.60}$$

The unity volume follows in (2.59) since Wang and Mendel integrate their m then-part Gaussian sets over all of R and thus use the scaling constant in (2.36) in their if-part Gaussian sets to account for the input truncation to a compact set. All that matters is that the then-part sets have the same area. Other *ad hoc* schemes can change the then-part areas. Or the rule weights w_j can reflect the variance or then-part bell-curve uncertainty as when $w_j = 1/(\sigma_j^i)^2$. Then the RBF in (2.55) is still a SAM but no longer a COG. Equation (2.60) follows because the mode of a Gaussian set equals its centroid and Wang and Mendel use the mode definition "$\bar{z}^j$ is the point in R at which $\mu_{B^j}(z)$ achieves its maximum value." They further call the SAM convex coefficients $p_j(x)$ "fuzzy basis functions" in this Gaussian case even though they are not orthogonal. These new names add no new content to the RBF SAM in (2.55).

Moody [33] arrived at (2.55) in his search for a neural network built from m input-ouput pairs $(x_j,\ y_j)$ with light computation. Specht [44] independently arrived at (2.55) from the theory of Parzen density estimators and the use of conditional expectations as mean-squared optimal estimators. They all center a vector Gaussian set or ball at each input vector x_j and center a Gaussian bell curve at each output value y_j. Chapter 13 applies the theory of the fuzzy unit hypercube to the RBF SAM in (2.55).

2.6 GENERALIZED SAMS AND THE TSK MODEL: CONVEX SUMS OF NONLINEAR SYSTEMS

Additive systems include a wide range of fuzzy systems. SAM systems remain the most important class of these systems and they extend in many directions and include many models as special cases. All these systems defuzzify (usually with a centroid) some output set B that changes with each input vector x and that arises from the sum of fired then-part sets:

$$B(x) = \sum_{j=1}^{m} B'_j(x) \tag{2.61}$$

These are general additive systems because the fired then-parts $B'_j(x)$ depend explicitly on x. They are simple *standard* additive models if the input fit value $a_j(x)$ scales a *fixed* then-part set: $B'_j(x) = a_j(x)B_j$.

The fuzzy systems are *generalized* standard additive models if the then-part set maps each x to some nonconstant output function or vector $B_j(x)$: $B'_j(x) = a_j(x)B_j(x)$. Here the then-part set B_j defines a map or operator from the input vector space to an output vector space or output function space. This restricts the sum combiner (2.53) to the still-wide class of fuzzy systems that defuzzify the $B(x)$ in

$$B(x) = \sum_{j=1}^{m} B'_j(x) = \sum_{j=1}^{m} a_j(x)B_j(x) \tag{2.62}$$

to compute $F(x)$. Note again that the structure of such systems does not depend on how or whether we factor the if-part set functions a_j: $R^n \rightarrow [0, 1]$. These set functions also need not map to just the unit interval. They can map to a subset of the real line or to the entire real line.

This section shows that generalized SAMs still obey some form of the SAM Theorem depending on how the map or operator B_j behaves. This includes as a special linear case the so-called TSK or just TS fuzzy models [46, 49–52] of Takagi and Sugeno. All the systems compute the centroidal output $F(x)$ as a convex sum of linear or nonlinear systems.

Consider first the case when B_j maps each input x to a new fuzzy subset $B_j(x) \subset R^p$. Then we can view the then-part set function b_j as a map from a product space to real numbers or b_j: $R^n \times R^p \rightarrow [0, 1]$. Each input x picks out a new then-part fuzzy set $B_j(x)$ through the restricted set function $b_j(x, .)$: $R^p \rightarrow [0, 1]$. Then both the set volumes and centroid depend on x and the proof of the SAM Theorem in (2.21)–(2.26) goes through otherwise with no change:

$$F(x) = \frac{\sum_{j=1}^{m} a_j(x) V_j(x) c_j(x)}{\sum_{j=1}^{m} a_j(x) V_j(x)} = \sum_{j=1}^{m} p_j(x) c_j(x). \tag{2.63}$$

This convex sum forces the user to compute each term in the SAM ratio for each input.

Now suppose B_j is an arbitrary map from the input vector space R^n to the output vector space R^p. The map B_j: $R^n \rightarrow R^p$ is just an n-by-p matrix B_j in the linear case. In general B_j is a nonlinear operator. It maps each x to a new output vector $y_j = B_j(x)$. The output vector y_j depends on x but we omit this subscript for simplicity. This nonlinear operator case is a special case of the above point-to-set case when the then-part fuzzy set is just the singleton set $\{y_j\}$. This gives a unit pulse or binary set function:

$$b_j(x, .) = \begin{cases} 1 & \text{if } B_j(x) = y_j \\ 0 & \text{if } B_j(x) \neq y_j. \end{cases} \tag{2.64}$$

Then the discrete case gives a SAM Theorem as shown in the homework problems. The continuous case requires us to identify the singleton set $\{y_j\}$ with a delta pulse or generalized set function:

$$b_j(x, .) = \delta(y - y_j) = \delta(y - y(x)). \tag{2.65}$$

Then each x gives a generalized "set" $B_j(x)$ with unit volume and with a new spike or range point y_j for its centroid $c_j(x)$:

$$V_j(x) = \int_{R^p} b_j(x, y)dy = \int_{R^p} \delta(y - y_j)dy = 1 \tag{2.66}$$

$$c_j(x) = \frac{\int_{R^p} y b_j(x, y) dy}{\int_{R^p} b_j(x, y) dy} = \int_{R^p} y\delta(y - y_j) dy = y_j(x). \tag{2.67}$$

Then the additive combiner in (2.62) further reduces to

$$B(x) = \sum_{j=1}^{m} B_j'(x) = \sum_{j=1}^{m} a_j(x) B_j(x) = \sum_{j=1}^{m} a_j(x)\delta(y - y_j) \tag{2.68}$$

and leads to a generalized SAM Theorem:

$$F(x) = Centroid(B(x)) \tag{2.69}$$

$$= \frac{\int_{R^p} y b(x, y) dy}{\int_{R^p} b(x, y) dy} \tag{2.70}$$

$$= \frac{\int_{R^p} y \sum_{j=1}^{m} a_j(x)\delta(y - y_j) dy}{\int_{R^p} \sum_{j=1}^{m} a_j(x)\delta(y - y_j) dy} \tag{2.71}$$

$$= \frac{\sum_{j=1}^{m} a_j(x) \int_{R^p} y\delta(y - y_j) dy}{\sum_{j=1}^{m} a_j(x) \int_{R^p} \delta(y - y_j) dy} \tag{2.72}$$

$$= \frac{\sum_{j=1}^{m} a_j(x) y_j}{\sum_{j=1}^{m} a_j(x)} \tag{2.73}$$

$$= \sum_{j=1}^{m} p_j(x) y_j \tag{2.74}$$

$$= \sum_{j=1}^{m} p_j(x) B_j(x). \tag{2.75}$$

Note that this convex sum defines a nonlinear map $F: R^n \rightarrow R^p$ as long as $m > 1$ or the system has 2 or more rules even if each then-part set B_j is a linear map or matrix.

Tanaka [50, 51] has explored the special case of (2.75) when the operator B_j is not only a matrix but an n-by-n square matrix. Then the SAM system can define a discrete autonomous dynamical or feedback system or nonlinear difference equation:

$$x(k+1) = F(x(k)) \tag{2.76}$$

$$= \sum_{j=1}^{m} p_j(x(k)) B_j(x(k)). \tag{2.77}$$

The square matrix B_j can house the coefficients of a piecewise polynomial or more complex set function b_j. Triangles and trapezoids are examples of such continuous piecewise-polynomial set functions. The square matrix B_j can also house a separate control or forcing function $u_j(k)$.

The last section of this chapter discusses this discrete feedback SAM model in detail. The main result is that the SAM map F is stable or has fixed-point equilibria if (but not only if) there exists a positive definite matrix P so that *each* of the m matrices $B_j^T P B_j - P$ is negative definite or so that the quadratic-form inequality

$$x^T B_j^T P B_j x - x^T P x < 0 \tag{2.78}$$

holds for all j and for all non-null n-vectors x. This is the standard result for the linear case of $m = 1$ extended to the convex sum (2.77). There is no general way to find the common positive definite matrix P. The odds of finding such a stabilizer fall as the number of rules grows. The convexity in (2.77) also implies that the SAM system need not be stable even if each linear subsystem or rule is stable. Of course the user may not want fixed-point stability when modeling complex nonlinear processes.

A still less general SAM case is the TSK or TS case. Sugeno [45] and Terano [52] call this the "third inference method." It just replaces B_j with a piecewise linear map f_j or appropriate n-by-p matrix operator B_j in the SAM equation:

$$F(x) = \sum_{j=1}^{m} p_j(x) B_j(x) \tag{2.79}$$

$$= \frac{\sum_{j=1}^{m} a_j(x) f_j(x_1, \ldots, x_n)}{\sum_{j=1}^{m} a_j(x)} \tag{2.80}$$

$$= \frac{\sum_{j=1}^{m} a_j(x)\left[b_0^j + b_1^j x_1 + b_2^j x_2 + \ldots + b_n^j x_n\right]}{\sum_{j=1}^{m} a_j(x)} \tag{2.81}$$

in Sugeno's notation. Then the jth rule has the form

$$\text{IF } X = A_j \text{ THEN } Y = b_0^j + b_1^j x_1 + b_2^j x_2 + \ldots + b_n^j x_n \tag{2.82}$$

where the then-part term describes a piecewise linear set function such as triangle or trapezoid. Strictly speaking the SAM model (2.63) should apply here since the piecewise-linear sets have different areas or volumes and centroids and these volumes and centroids change with each input x. The varying volumes affect both the linguistic "meaning" of the then-part sets and how they weight the SAM output $F(x)$.

Sugeno sees the convex sum of linear forms (2.81) as carving the input space into m subspaces [45]: "A linear relation is built in each fuzzy subspace of the input space . . . Those parameters [coefficients of linear equations] are identified so that the output error is minimized." So least squares [27] can find the coefficients. Takagi and Sugeno [49] published a paper in the same year (1985) that carried out this program of linear parameter estimation. In both çases the if-part set functions are linear forms as are the then-part sets.

Takagi and Sugeno solved the least-squares or mean-squared optimal "normal equations"

$$B = (V^T V)^{-1} V^T Y \tag{2.83}$$

to find the best linear coefficients given l input-output pairs or training samples $(x(1),\ y(1)), \ldots, (x(l),\ y(l))$ for invertible matrix $V^T V$. They applied this scheme to the convex SAM sum in the form

$$F(x) = \sum_{j=1}^{m} p_j(x)\left[b_0^j + b_1^j x_1 + \ldots + b_n^j x_n\right] \tag{2.84}$$

$$= \sum_{j=1}^{m}\left[p_j(x) b_0^j + p_j(x) b_1^j x_1 + \ldots + p_j(x) b_n^j x_n\right] \tag{2.85}$$

for a block $m(n+1)$-vector of parameters

$$B = \left[b_0^1, \ldots, b_0^m | b_1^1, \ldots, b_1^m | \cdots | b_n^1, \ldots, b_n^m\right]^T. \tag{2.86}$$

The l-by-$m(n+1)$ observation matrix V has l rows v_k of the form

$$V = \begin{pmatrix} v_1 \\ \vdots \\ v_l \end{pmatrix} \tag{2.87}$$

$$= \begin{pmatrix} p_1(x(1)), \ldots, p_m(x(1)) | p_1(x(1))x_1(1), \ldots, \\ p_m(x(1))x_1(1)| \cdots |p_1(x(1))x_n(1), \ldots, p_m(x(1))x_n(1) \\ \vdots \\ p_1(x(l)), \ldots, p_m(x(l)) | p_1(x(l))x_1(l), \ldots, \\ p_m(x(l))x_1(l)| \cdots |p_1(x(l))x_n(l), \ldots, p_m(x(l))x_n(l) \end{pmatrix} \tag{2.88}$$

Y is the l-by–1 column vector of observed scalar outputs

$$Y = [y_1, \ldots, y_l]^T. \tag{2.89}$$

Then Takagi and Sugeno used recursive least squares or a simple stable-state form of the Kalman filter [35] to estimate the optimal B values in (2.83):

$$B_{k+1} = B_k + C_{k+1} v_{k+1} [y_{k+1} - v_{k+1} B_k] \tag{2.90}$$

$$C_{k+1} = C_k - \frac{C_k \, v_{k+1}^T \, v_{k+1} \, C_k}{1 + v_{k+1} \, C_k \, v_{k+1}^T} \tag{2.91}$$

for filter index $k = 0, 1, \ldots, l - 1$ and $B_l = B$. The update equation (2.90) has the form of a full predictor-corrector with an adaptive learning or corrector weight. C is an $m(n + 1)$-by-$m(n + 1)$ matrix and part of the Kalman "gain" or learning weight. The recursion starts with $B_0 = \emptyset$ and $C_0 = cI$ where $c > 0$ is some large number and I is the identity matrix. One can also estimate B in the stochastic case with the Widrow-Hoff LMS algorithm [21] or with some other form of stochastic gradient descent.

The key point is that the TSK *model* in (2.80)–(2.82) is a generalized SAM with piecewise linear then-part set functions. The Kalman learning scheme is not part of the TSK model. It is one of the infinitely many ways to use training data to tune some of the parameters of a SAM map F. The next section further extends the above generalized SAMs to the case of fuzzy-set inputs or SAMs F that map fuzzy sets A to point or set outputs $F(A)$.

2.7 GENERALIZED SAMS AS SET MAPPINGS AND CORRELATORS: THE SET SAM

The next level of SAM generalization shows how to map fuzzy sets A in the input space R^n to points y in the output vector space R^p. So the *set SAM F* has as its domain the fuzzy power set $F(2^{R^n})$ or the set of all fuzzy subsets $A \subset R_n$ with arbitrary set functions $a\colon R^n \rightarrow [0, 1]$ or even with arbitrary non-negative set functions $a\colon R^n \rightarrow [0, \infty)$. The SAM map $F\colon F(2^{R^n}) \rightarrow R^p$ gives the vector point $y = F(A)$ for each fuzzy-set input A. Some fuzzy engineers call such a system a "nonsingleton fuzzifier" even though very few engineers use such a system or such a name scheme.

More future systems may pass an entire fuzzy set A through the SAM's bank of m fuzzy rules. An estimation system might define what a user means by *cool* air with a set function and feed the entire waveform to the fuzzy air conditioner. Or a Kalman filter or other system might feed a fuzzy system a probability density $p(x|z)$ of state x conditioned on the observed data z. Here the system might use noisy radar data z to track an airplane x that flies behind a cloud. The idea is to match the input set A with the m stored if-part sets A_j. We now develop a correlation scheme that does this in a simple way and still preserves the SAM structure.

The above SAM systems that map points to points define "singleton fuzzifiers" and suggest how to extend the domain of SAM maps. The lone or "singleton" point input $x_0 \in R^n$ gives rise to a unit bit vector in the discrete case (as in the homework problems) or a generalized set function or delta pulse $\delta(x - x_0)$ in the continuous case. So we can replace the delta pulse with the arbitrary set function a in (2.38):

$$a_j(A) = \int_{R^n} a(x)a_j(x)dx \tag{2.92}$$

if the integral exists. This gives the "fired" then-part set $B'_j(A)$ as

$$b'_j(y) = \int_{R^n} a(x)R_{A_j \to B_j}(x, y)dx \tag{2.93}$$

$$= \int_{R^n} a(x)a_j(x)b_j(y)dx \tag{2.94}$$

$$= b_j(y)\int_{R^n} a(x)a_j(x)dx \tag{2.95}$$

$$= a_j(A)b_j(y). \tag{2.96}$$

This extends the generalized SAM combiner in (2.62) to a set-SAM combiner

$$B(A) = \sum_{j=1}^{m} B'_j(A) = \sum_{j=1}^{m} a_j(A)B_j(A) \tag{2.97}$$

and so extends the generalized SAM theorem in (2.63) to

$$F(A) = \frac{\sum_{j=1}^{m} a_j(A)V_j(A)c_j(A)}{\sum_{j=1}^{m} a_j(A)V_j(A)} = \sum_{j=1}^{m} p_j(A)c_j(A). \tag{2.98}$$

Only the if-part set functions a_j depend on the input A in most real systems. This greatly simplifies (2.97) and (2.98):

$$F(A) = \frac{\sum_{j=1}^{m} a_j(A)V_j c_j}{\sum_{j=1}^{m} a_j(A)V_j} = \sum_{j=1}^{m} p_j(A)c_j. \tag{2.99}$$

The convex sum (2.99) gives the most weight to those then-part centroids c_j whose if-part sets A_j best match the input set A.

We can also define the rule-patch set function $R_{A_j \to B_j}: R^n \times R^p \to [0, 1]$ with min or any other conjunctive or t-norm operation. This gives (2.93) as

$$b'_j(y) = \int_{R^n} a(x) \min(a_j(x), b_j(y))dx \tag{2.100}$$

$$= \int_{R^n} \min(a(x)a_j(x), a(x)b_j(y))dx. \tag{2.101}$$

So min coding no longer factors out the then-part fit value $b_j(y)$ from the integral as it does in (2.40)–(2.43) for the singleton input $a(x) = \delta(x - x_0)$. Again we can replace the min rule-patch set function in (2.93) with any conjunctive or t-norm operator $T(a_j(x), b_j(y))$. Only the multiplicative or t-norm in (2.94) gives the factored SAM form (2.96).

The generalized set function a_j: $F(2^{R^n}) \to R^+$ defines a mathematical *correlation* and measures the area or volume of the intersection $A \cap A_j$. So the SAM (2.98) acts as a type of correlation detector [21] or nearest-neighbor set matcher. To see this first define the L^2 or Euclidean norm [40] of an integrable fuzzy set A as the integral of its square or its self-correlation:

$$\|A\|^2 = \int_{R^n} a^2(x)dx = a(A). \tag{2.102}$$

Now suppose all m if-part sets have nearly the same size or obey the *equi-norm* constraint:

$$\|A_1\|^2 = \cdots = \|A_m\|^2 < \infty. \tag{2.103}$$

The if-part set functions all lie on the surface of the same abstract "sphere." Suppose that the kth if-part set A_k best matches the input fuzzy set A:

$$a_k(A) = \max_j a_j(A) = -\min_j - a_j(A). \tag{2.104}$$

The equi-norm constraint (2.103) lets us add normed if-part sets to both sides of (2.104) to "complete the square" in (2.106) below. The equality (2.104) holds if and only if

$$||A_k||^2 - 2\int_{R^n} a(x)a_k(x)dx = \min_j \left(||A_j||^2 - 2\int_{R^n} a(x)a_j(x)dx\right) \tag{2.105}$$

if and only if

$$\int_{R^n} \left(a^2(x) + a_k^2(x) - 2a(x)a_k(x)\right) dx = \min_j \int_{R^n} \left(a^2(x) + a_j(x) - 2a(x)a_j(x)\right) dx \tag{2.106}$$

if and only if

$$\int_{R^n} (a(x) - a_k(x))^2 dx = \min_j \int_{R^n} (a(x) - a_j(x))^2 \, dx \tag{2.107}$$

if and only if the "stored" if-part set A_k is the nearest neighbor of the input fuzzy set A:

$$||A - A_k||^2 = \min_{1\le j\le m} ||A - A_j||^2. \tag{2.108}$$

Optical processors can compute the correlation set matches by passing light beams of data through systems of lenses, mirrors, beamsplitters, and spatial light modulators. The integrals in (2.104) may favor optics and other analog tools [32] over digital VLSI chip designs or discrete sampling for realtime computation. Chapter 11 shows how to implement simple optical SAM systems.

Older fuzzy systems would use some form of the "extension principle" [9, 59] to map the input fuzzy set A to the output fuzzy set B or to map it to the output vector y. This amounts to a sup combiner of pairwise minima in place of the integrals (or sums) of pairwise products in (2.94) or (2.100):

$$b'_j(y) = \sup a(x) \wedge R_{A_j \to B_j}(x, y) \tag{2.109}$$

$$= \sup a(x) \wedge a_j(x)b_j(y) \tag{2.110}$$

where the supremum ranges over all x in the input space R^n. Or correlation-min can define the rule-patch set function $R_{A_j \to B_j}$: $R^n \times R^p \to [0, 1]$ to give the "fired" then-part set $B'_j(A)$ as

$$b'_j(y) = \sup a(x) \wedge R_{A_j \to B_j}(x, y) \tag{2.111}$$

$$= \sup a(x) \wedge a_j(x) \wedge b_j(y). \tag{2.112}$$

Or again any t-norm can define the rule-patch set function.

The motivation for such supremum schemes seems to be a mix of min-max tradition and a desire to restrict the fit values of the then-part sets $B'_j(A)$ to the unit interval. The cost is high both in practice and in theory. The sup schemes (2.109)–(2.112) force one to solve a nonlinear optimization problem for each of the continuum many fit values $b'_j(y)$ of each then-part set $B'_j(A)$. This poses an uncomputable problem in most cases. And the search complexity grows as the nonlinearity of the input set A grows. This may prevent realtime performance or even exact computation off-line.

The sup schemes also throw away most of the information in the input set A and the stored if-part set A_j. Perhaps the best way to see this is to check the case where the sup schemes work best—the singleton case of input $\{x_0\}$ or its delta-pulse set function $\delta(x - x_0)$. The sup does not apply to the delta pulse $\delta(x - x_0)$ and so we must replace it with the unit pulse δ_{x_0}:

$$a(x) = \delta_{x_0} = \begin{cases} 0 & \textit{if } x \neq x_0 \\ 1 & \textit{if } x = x_0. \end{cases} \tag{2.113}$$

Then (2.110) becomes

$$b'_j(y) = \sup \delta_{x_0}(x) \wedge a_j(x) b_j(y) \tag{2.114}$$

$$= a_j(x_0) b_j(y) \tag{2.115}$$

or $B'_j(x_0) = a_j(x_0) B_j$ and thus defines a proper SAM. The unit pulse also combs the correlation-minimum case (2.112) to give

$$b'_j(y) = \sup \delta_{x_0}(x) \wedge a_j(x) \wedge b_j(y) \tag{2.116}$$

$$= a_j(x_0) \wedge b_j(y) \tag{2.117}$$

or $B'_j(x_0) = a_j(x_0) \wedge B_j$ and thus defines a nonstandard or min-clip additive model. The sup schemes discard structure for all other nonsingleton sets A. These schemes search only for the largest fit value in the intersection $A \cap A_j$ and ignore the rest of the intersection. A low-volume set function A with a large spike or bump in the right place can match a stored if-part set A_j just as well as A_j matches itself. The extension-principle schemes work best in the singleton case where we do not need them and do not work well in the set-valued case where we do need them.

The standard correlation matcher (2.93)–(2.96) computes the input match $a_j(A)$ as a linear operator. Some software packages and hardware devices can compute these terms in realtime. The generalized SAM in (2.98) requires only that the correlations are finite or that the m integrals in (2.73) exist. Then the convex coefficients $p_1(A), \ldots, p_m(A)$ act as then-part centroid normalizers. The correlations $a_j(A)$ measure the entire volume of the intersection $A \cap A_j$. Different t-norm definitions of the intersection can change the value of this volume but not whether it depends on the entire intersection $A \cap A_j$. This too arises from the simple sum-and-product structure of SAMs and gives further reason for using them.

2.8 COMBINING FUZZY SYSTEMS: CONVEX SUMS OF EXPERT RESPONSES

Additive systems offer a natural way to combine any number of fuzzy systems. The systems might come from a pool of human experts or a pool of competing neural systems that try to predict the same input-output relation. The fuzzy systems need not be additive or SAM systems. The combined system is a SAM and so acts as a convex sum of the combined fuzzy systems. The advance lies in the structure of the convex weights.

Suppose we wish to combine q experts or fuzzy systems F_k: $R^n \rightarrow R^p$. The systems may be scalar valued in practice. We can view the q systems as q experts or knowledge sources [17, 57]. Then the combined output $F(x)$ answers a question: What do we conclude from the answers of q experts when we ask them question x? This is an old question that ranges from epistemology in philosophy to knowledge acquisition in expert systems theory to issues of jury and citizen voting in modern legal and political theory. You have a bump on your neck and you get a biopsy and then see two physicians. One says the tumor may turn malignant and the other doubts it. Who do you believe? How do you combine their knowledge responses to a final belief state?

A weighted average or mean is a standard way to combine the fuzzy systems. It combines their *outputs*

$$M(x) = \frac{1}{q} \sum_{k=1}^{q} w_k F_k(x) \tag{2.118}$$

for system or expert credibility weights $w_k \in [0, 1]$ such that at least one expert has nonzero credibility or some $w_i > 0$. A *fully weighted* average normalizes the sum with the sum of credibility weights and has the form of a minimum-variance [36] average:

$$W(x) = \frac{\sum_{k=1}^{q} w_k F_k(x)}{\sum_{k=1}^{q} w_k}. \tag{2.119}$$

The fully weighted mean W is a convex sum of the fuzzy systems' outputs. The inequality

$$q \geq \sum_{k=1}^{q} w_k$$

implies that $W > M$ if some $w_i < 1$ or at least one expert is not maximally credible. So $W = M$ holds if and only if $w_k = 1$ for all k or all the weights are maximally credible.

We can compare W and M by seeing how they behave as the fit vector $w = (w_1, \ldots, w_q)$ moves through the fuzzy cube $I^q = [0, 1]^q$. Chapter 12 explores the structure of fuzzy cubes. Each fuzzy system F_k is an "expert" to some degree and $w = (w_1, \ldots, w_q)$ states this list of q degrees. We can always normalize finite non-negative weight vectors to map them to points in a fuzzy cube.

Suppose first that the q experts have the same weight w but $0 < w < 1$. Then the weights w cancel in (2.119) just as they do in any SAM structure when all the weights or then-part set volumes are the same. This reduces W to the unweighted sample average

$$W(x) = \frac{1}{q} \sum_{k=1}^{q} F_k(x). \tag{2.120}$$

The unweighted average reflects that there is no reason to prefer one expert's response over another's. But w scales the weighted average M in (2.118):

$$M(x) = \frac{w}{q} \sum_{k=1}^{q} F_k(x). \tag{2.121}$$

This shows that the weighted mean M is not invariant on the locus of constant credibility weights in the fuzzy cube $I^q = [0, 1]^q$. The inequality $W > M$ reflects this scaling effect that stems from the constant normalizer q.

Next suppose that the weight vector $w = (w_1, \ldots, w_q)$ is one of the $2^q - 1$ binary vectors or weights in the fuzzy weight cube. We exclude the origin or null vertex $w = (0, \ldots, 0)$ since we assume at least one weight is nonzero. The fuzzy cube I^q has 2^q binary or bit-vector vertices. Suppose that r weights are 1s and the other $q - r$ weights are 0: $w_1 = \ldots = w_r = 1$ and $w_{r+1} = \ldots = w_q = 0$. Then the fully weighted mean W just averages the r nonzero expert outputs:

$$W(x) = \frac{1}{r} \sum_{k=1}^{r} F_k(x). \tag{2.122}$$

But the simple weighted mean M scales down as an artifact of the normalizer q:

$$M(x) = \frac{1}{q} \sum_{k=1}^{r} F_k(x). \tag{2.123}$$

Now look at the q cases where just one weight $w_k = 1$ and all other weights are 0. Then the fully weighted mean W becomes $W(x) = F_k(x)$. We take what the one maximally credible expert says as the answer. But the simple weighted mean M becomes $M(x) = F_k(x)/q$. The size q of the expert pool scales down the fuzzy system output. These cases reflect $W > M$ and show that q does not behave as well as the weight-dependent normalizer in (2.119). SAM combination gives just this desired form.

The idea of an additive combiner is to additively combine *throughputs* rather than *outputs* [26]. This means the system combines before it defuzzifies and not after. Averaging combines what the experts say and thus limits us to combining defuzzified outputs. We would like to combine what the experts *know*. In the extreme case we would like to combine their brains into one common brain. For fuzzy systems the most we can do is combine their rules. This comes to the same thing as combining their rule firings. For additive fuzzy systems this comes to adding the weighted inferred sets $w_k B^k$ as in (2.124) below.

Each expert or fuzzy system F_k: $R^n \rightarrow R^p$ combines the fired then-part sets of his or her rules to give the output set $B^k(x)$ for input vector x. These systems need not use the additive combiner in (2.12) or (2.61) to form the combined set $B^k(x)$. Any combiner will work as long as the "set" stays bounded. Each expert or system computes its output as the centroid of its combined set: $F_k(x) = Centroid(B^k(x))$. Then the additive combiner weights each output set $B^k(x)$ and sums them at a higher level:

$$B(x) = \sum_{k=1}^{q} w_k B^k(x). \tag{2.124}$$

The sum expands if each fuzzy system F_k: $R^n \rightarrow R^p$ is also a SAM. Then (2.124) becomes

$$B(x) = \sum_{k=1}^{q} w_k \sum_{j=1}^{m_k} a_j^k(x) B_j^k = \sum_{k=1}^{q} \sum_{j=1}^{m_k} w_k a_j^k(x) B_j^k. \tag{2.125}$$

The SAM Theorem and (2.124) give the SAM combiner for all centroidal fuzzy systems F_k: $R^n \rightarrow R^p$:

$$F(x) = Centroid(B(x)) \tag{2.126}$$

$$= \frac{\sum_{k=1}^{q} w_k V^k(x) c^k(x)}{\sum_{k=1}^{q} w_k V^k(x)} \tag{2.127}$$

$$= \sum_{k=1}^{q} p_k(x) F_k(x) \tag{2.128}$$

with convex coefficients

$$p_i(x) = \frac{w_i V^i(x)}{\sum_{k=1}^{q} w_k V^k(x)} \tag{2.129}$$

and with $c^k(x) = Centroid(B^k(x)) = F_k(x)$ and $V^k(x) = Volume(B^k(x))$. The credibility weight w_k stays fixed while the volume and centroid terms change with each x. So we cannot fold the weights into the volumes as we can with a simple SAM system.

The SAM Theorem gives a more detailed SAM structure if (2.125) holds and thus if each system is itself a SAM:

$$F(x) = \frac{\sum_{k=1}^{q} \sum_{j=1}^{m_k} w_k \, a_j^k(x) V_j^k c_j^k}{\sum_{k=1}^{q} \sum_{j=1}^{m_k} w_k \, a_j^k(x) V_j^k} \tag{2.130}$$

$$= \sum_{k=1}^{q} \sum_{j=1}^{m_k} p_j^k(x) c_j^k \tag{2.131}$$

with $m_1 + \ldots + m_q$ convex coefficients

$$p_l^i(x) = \frac{w_i a_l^i(x) V_l^i}{\sum_{k=1}^{q} \sum_{j=1}^{m_k} w_k \, a_j^k(x) V_j^k}. \tag{2.132}$$

Here the kth expert or SAM fuzzy system F_k: $R^n \rightarrow R^p$ stores m_k rules or rule patches $A_1^k \times B_1^k, \ldots, A_{m_k}^k \times B_{m_k}^k$ with $c_j^k = Centroid(B_j^k)$ and $V_j^k = Volume(B_j^k)$.

The SAM combiner (2.127) reduces to the fully weighted mean W in (2.119) in the rare equi-volume case $V^1 = \cdots = V^q > 0$. This means in effect that the q experts all answer the "question" x with the same confidence. The volume $V^k(x)$ gives a weight that depends on each input x unlike the constant volumes in the simple or lone SAM in (2.14). A two-expert example shows how these volumes can (and should) affect the combined outcome.

Suppose a cancer patient asks two physicians for a diagnosis. Suppose both doctors have the same credibility: $w_1 = w_2 = 1$. The value of the credibility w does not matter as long as $w > 0$ but the default extremal case $w_1 = w_2 = 1$ simplifies the analysis since then $W = M$. Both experts see the same data x. The first doctor says the patient has only a 30% chance to live: $F_1(x) = .3$. The second doctor gives him a 70% chance: $F_2(x) = .7$. Then the simple weighted means M and W in (2.118)–(2.119) give him a 50% chance to live.

Throughput combination tends to give a different answer. The mean approach assumes that the doctors have output sets B^1 and B^2 centered at .3 and .7 and that the sets have the same volumes V^1 and V^2. But suppose $V^1 < V^2$. Both sets may look like bumpy triangles but B^1 may have less height and width than B^2. So $B^1 + B^2$ will look more like B^2 than like B^1 and will have a centroid $F(x)$ in (2.127) closer to .7 than to .3. So the result might be a predicted 60% chance to live.

The doctors may have the same credibility but they need not respond the same way to all parts x of the input space. After all they are distinct experts. So in general the volume of their combined rule firings will differ for each inference: $V^1 < V^2$. This reveals a second credibility measure that averaging the outputs ignores.

Chapter 9 applies this combination theory to the problem of detecting a signal in noise. Two additive fuzzy systems F_1 and F_2 map noisy signals x to the combined sets B^1 and B^2. The first system F_1 maps the noisy signal x_k at time k to the combined set $B^1(x_k)$. The second system F_2 maps the noisy context signal vectors x_{k-1} and x_{k+1} to the combined set $B^2(x_{k-1}, x_{k+1})$. Then the system weights and sums these sets to give the global combined set $B(x_k) = c(x_k)B(x_k) + c(x_{k-1}, x_{k+1})B(x_{k-1}, x_{k+1})$. The clarity weight c gives less weight to the more fuzzy signal vectors x based on where they fall in a fuzzy hypercube that stores messages at some of its vertices. The system combines before it defuzzifies but does not defuzzify $B(x_k)$ by taking its centroid. Instead the output $F(x_k) = y_{\max}$ picks the mode of $B(x_k)$ over a finite alphabet or library of messages Y. A rounded-off centroid can give the same answer as the mode gives but often does not in detection problems. So the combined fuzzy system in this example cannot use the simpler SAM combiner (2.127) though it remains an additive combiner.

We can write the SAM combiner (2.127) in recursive form for $q + 1$ experts or fuzzy systems as

$$F(x) = \frac{\sum_{k=1}^{q+1} w_k V^k(x) F_k(x)}{\sum_{k=1}^{q+1} w_k V^k(x)} \tag{2.133}$$

$$= F_{q+1}(x) + \frac{1}{\sum_{k=1}^{q+1} w_k V^k(x)} \sum_{k=1}^{q} w_k V^k(x)[F_k(x) - F_{q+1}(x)] \tag{2.134}$$

$$= F_{q+1}(x) + \sum_{k=1}^{q} \left(\frac{w_k V^k(x)}{\sum_{j=1}^{q+1} w_j V^j(x)} \right) [F_k(x) - F_{q+1}(x)] \tag{2.135}$$

$$= F_{q+1}(x) + \sum_{k=1}^{q} p_k(x)[F_k(x) - F_{q+1}(x)]. \tag{2.136}$$

The proof is homework problem 2.21.

The update equation (2.136) states the additive combiner in a weak type of "predictor-corrector" form. Other types let us write the "corrector" term in simpler or more complex forms. This lets us add new experts to the expert pool without a batch computation. It also has the approximate form of a static Kalman filter [21]. The next section explores this connection with statistics and optimal estimation and shows that indeed SAM systems act as model-free conditional expectations or mean-squared optimal estimators:

$$F(x) = E[Y|X = x] = \sum_{j=1}^{m} p_j(x)E[B_j|X = x].$$

This holds in all cases. It does not depend on assuming jointly Gaussian or other jointly distributed random vectors. It also gives a way to compute a conditional variance or dispersion measure of confidence for each fuzzy system output $F(x)$ and shows that all outputs do not have the same uncertainty.

2.9 FUZZY SYSTEMS AS CONDITIONAL MEANS: THE PROBABILITY CONNECTION

The debate over whether fuzziness differs from randomness occurs at both the set level and the system level. The debate at the set level is a lively but largely philosophical debate. The February 1994 issue of the *IEEE Transactions on Fuzzy Systems* covers this debate [24, 29] from many viewpoints. We sketch some of the issues here and explore others in the fuzzy cube framework of Chapter 12.

The debate on fuzziness versus probability often stems from news of fuzzy-system applications. The press has been quick to print and reprint the claims of a few extreme probabilists who claim that fuzziness is randomness in disguise. Many journalists believe this gives "balance" to an article on fuzzy washing machines or subways or computer chips. In most cases the critics confuse notions of shades of gray or multivaluedness with rule-based systems engineering. They put forth philosophical claims about sets to answer questions about systems. But we could replace the if-part triangles and trapezoids in many applications with binary rectangles and with work most systems would still perform well. Then the system would have lost its fuzziness but the philosophical claims would still apply. This holds because the power of fuzzy systems lies far more in the *model-free* nature of their black boxes than in their use of multivalued sets. Indeed the critics have missed the main result at the systems level: Fuzzy systems are probabilistic systems—and it is all the better for fuzzy systems. We now examine these issues in turn.

The set-level debate turns on how we use binary or multivalued sets to model events. Then fuzziness deals with the degree to which an event occurs [21]: "The rain today is light." All rain patterns are both light L and not-light L^c to some degree. In general this breaks the bivalent "laws" of noncontradiction and excluded middle: $L \cap L^c \neq \emptyset$ and $L \cup L^c \neq X$. Randomness deals with whether an event occurs: "There is a 20% chance that it will rain today." Here whether it rains is binary or either-or. It will rain R or it will not rain R^c and not both or neither: $\text{prob}\{R \cap R^c \neq \emptyset\} = 0$ and $\text{prob}\{R \cup R^c = X\} = 1$. Standard probability measures [40] map only binary sets to real numbers. So noncontradiction and excluded middle always hold: $R \cap R^c = \emptyset$ and $R \cup R^c = X$. Fuzzy sets are just those sets that break these binary laws.

These two modes of set or event uncertainty can apply to each other at a higher level if one mode describes the other. Fuzzy probability deals with the vagueness in a random description [59]: "The odds are slight that it will rain today." Here the odds of the event are vague but the raining event itself is binary. The probability of fuzzy events deals with whether vague events occur: "There is a 20% chance of light rain today." Here the odds are exact but the light-rain event is not binary or either-or.

We can view a fuzzy set A as a random set [34] or locus of two-point conditional probability densities. Then the set degree $a(x) = \text{Degree}\{x \in A\}$ becomes the local conditional probability $\text{prob}\{X = A | X = x\}$. The complement fit value $1 - a(x) = \text{Degree}\{x \notin A\}$ becomes the dual probability $\text{prob}\{X \neq A | X = x\}$. Suppose A is the subset of cool air temperatures. On the global set view $a(x)$ is the degree to which air temperature value x is a cool value. On the local random-set view $a(x)$ is the probability that the temperature is cool *given* that the temperature value is x. So we can equally view A as a locus of multivalued set values or as a locus of two-point conditional probabilities. Both views can appeal to linguistic convention but the fuzzy-set view works with a simpler conceptual chunk. Vastly more people speak of air as cool or warm or hot rather than as loci of two-point conditional densities.

We can also view how much one set contains another as a measure of subsethood [21] or conditional probability and from this derive many of the key concepts of both fuzzy sets and probability theory. These views range from the mathematical to the aesthetic. They do not directly affect the nature of fuzzy systems even though they often arose when fuzzy engineers first put forth their systems and the press first tried to describe them. Chapter 12 reviews the theory of subsethood and shows how to apply some of its constructs to information theory.

The debate at the system level $F: R^n \rightarrow R^p$ is not a debate at all. A simple argument shows that *all* centroidal fuzzy systems are probabilistic systems. Indeed the result is stronger: A centroidal fuzzy system F computes a conditional expectation $E[Y|X]$ and thus computes a mean-squared optimal nonlinear estimator [36]. The power of the fuzzy system lies both in this optimality result and in its *model-free* structure. Most popular conditional-mean systems use a math model of the plant. The linear Kalman filter uses a linear Gauss-Markov state model and assumes that all variables are jointly Gaussian. A fuzzy system does its mean-squared best to model a system or approximate a function with its rules or paired sets or densities. The sets A_j can be rectangles and thus nonfuzzy and still the rules may give a good graph cover and thus a good approximation to some sampled or learned approximand f. SAM maps $F: R^n \rightarrow R^p$ do not in practice use the fuzzy overlap terms $A_j \cap A_j^c$ as if-part sets though they could. An input x air temperature that is both cold and cool to nonzero degree ($A_{j-1} \cap A_j \neq \emptyset$) is not the same as one that is both cool and not cool to nonzero degree ($A_j \cap A_j^c \neq \emptyset$). Overlapping rectangles can give the former but only fuzzy sets can give the latter.

The proof that centroidal fuzzy systems are conditional means follows from the ratio structure of the centroid and the boundedness and non-negativity of the set values $b(x, y) \geq 0$ of the combined set B of "fired" then-part sets B_j in (2.52) or (2.53) or in any other combination scheme. Each input x gives its own $B(x)$ and thus its own output $F(x)$:

$$F(x) = Centroid(B(x)) \tag{2.137}$$

$$= \frac{\int_{R^p} yb(x, y)dy}{\int_{R^p} b(x, y)dy} \tag{2.138}$$

$$= \int_{R^p} yp(y|x)dy \tag{2.139}$$

$$= E[Y|X = x] \tag{2.140}$$

for each $x \in R^n$. This holds because the ratio in (2.138) of the joint distribution to the marginal defines a proper conditional probability density

$$p(y|x) = \frac{b(x, y)}{\int_{R^p} b(x, y)dy} \tag{2.141}$$

even though $b(x, y) > 1$ may hold.

The fuzzy system $F\colon R^n \to R^p$ in (2.137) need not be additive. The SAM Theorem shows what happens if F is additive and SAM in structure. Then $F(x)$ is a convex sum of centroids or local conditional (then-part set) means. We now show that this convexity property holds for all additive systems and that the same convex weights decompose the conditional variance of the system [26]. The conditional variance gives a confidence measure for each output $F(x)$ and shows that the structure of the then-part sets B_j matters after all.

In practice we can compute the local conditional means or centroids in advance but not so with the local conditional variances. The result holds for general additive maps $F\colon R^n \to R^p$ that combine fired then-part sets B_j' with the sum combiners (2.12) or (2.61). We prove this for the scalar case $F\colon R^n \to R$ for simplicity and to avoid the matrix notation for the covariance. The result holds for standard additive models when $B_j'(x) = a_j(x)B_j$ or in the generalized case when $B_j'(x) = a_j(x)B_j(x)$.

Theorem 2.2 (*Additive Statistics*) Suppose $F: R^n \rightarrow R$ is an additive fuzzy system such that $F(x) = \text{Centroid}(B)$ and $B(x) = \sum_{j=1}^{m} w_j B'_j(x)$. Then

$$F(x) = E[Y|X = x] = \sum_{j=1}^{m} p_j(x) E_{B'_j}[Y|X = x] \tag{2.142}$$

$$V[Y|X = x] = \sum_{j=1}^{m} p_j(x) V[Y|X = x, B'_j]. \tag{2.143}$$

The convex coefficients $p_j(x)$ are weighted volume ratios of the "fired" sets B'_j:

$$p_j(x) = \frac{w_j V'_j(x)}{\sum_{k=1}^{m} w_k V'_k(x)} \tag{2.144}$$

$$V'_j(x) = \int b'_j(x, y) dy. \tag{2.145}$$

Proof. We first repeat the chain of equalities to show that the additive fuzzy system F computes a realization of the conditional expectation for each input x. Then the same chain of equalities shows that $F(x)$ is a convex sum of local conditional mean realizations or centroids:

$$F(x) = \text{Centroid}(B(x)) \tag{2.146}$$

$$= \frac{\int_{-\infty}^{\infty} y b(x, y) dy}{\int_{-\infty}^{\infty} b(x, y) dy} \tag{2.147}$$

$$= \int_{-\infty}^{\infty} y p_B(y|x) dy \tag{2.148}$$

$$= E[Y|X = x] \tag{2.149}$$

$$= \frac{\sum_{j=1}^{m} w_j \int_{-\infty}^{\infty} y b'_j(x, y) dy}{\sum_{j=1}^{m} w_j \int_{-\infty}^{\infty} b'_j(x, y) dy} \tag{2.150}$$

$$= \frac{\sum_{j=1}^{m} w_j \int_{-\infty}^{\infty} b'_j(x, y) dy \dfrac{\int_{-\infty}^{\infty} y b'_j(x, y) dy}{\int_{-\infty}^{\infty} b'_j(x, y) dy}}{\sum_{j=1}^{m} w_j \int_{-\infty}^{\infty} b'_j(x, y) dy} \tag{2.151}$$

$$= \frac{\sum_{j=1}^{m} w_j V'_j \int_{-\infty}^{\infty} y b'_j(x|y) dy}{\sum_{j=1}^{m} w_j V'_j} \tag{2.152}$$

$$= \sum_{j=1}^{m} p_j(x) E_{B'_j}[Y|X = x] \tag{2.153}$$

$$= \sum_{j=1}^{m} p_j(x) c'_j(x) \tag{2.154}$$

which proves (2.142). The variance (covariance) result (2.143) follows in the same way:

$$V[Y|X = x] = \frac{\int_{-\infty}^{\infty} (y - E[Y|X = x])^2 b(x, y) dy}{\int_{-\infty}^{\infty} b(x, y) dy} \tag{2.155}$$

$$= \sum_{j=1}^{m} p_j(x) \int_{-\infty}^{\infty} (y - E[Y|X = x])^2 p_{B'_j}(y|x) dy \tag{2.156}$$

$$= \sum_{j=1}^{m} p_j(x) V[Y|X = x, B'_j]. \tag{2.157}$$

■

The unconditional output variance $V_Y(Y)$ follows from (2.142) and (2.143) and from the standard variance decomposition

$$V_Y(Y) = E_X(V[Y|X]) + V_X(E[Y|X]) \tag{2.158}$$

in the scalar case $F: R \to R$. A like decomposition holds in the vector case. The unconditional mean E_X and variance V_X depend on an arbitrary probability density function p_X defined on the input space R^1 for the random variable (measurable function) $g(X)$:

$$E_X(g) = \int_{-\infty}^{\infty} g(x)p_X(x)dx \tag{2.159}$$

$$V_X(g) = \int_{-\infty}^{\infty} (x - g(x))^2 p_X(x)dx. \tag{2.160}$$

We do not know the density p_X in most cases. The special and trivial case of the delta-pulse "density" $p(x) = \delta(x - x_0)$ acts as the limiting case of a Gaussian or other symmetric unimodal density and leads us to expect to observe $F(x_0)$ as the system output. There is no simple way to compute the induced probability density of the output $F(x)$ given the input density $p_X(x)$. We can view the set SAM $F(A)$ in (2.98) as one way to map the information in the fuzzy set or unnormalized density A to the output space. The variance term (2.158) offers little information in practice. We must instead compute the conditional variance terms in the theorem and these come at a high computational price.

The variance term $V[Y|X = x, B'_j]$ does not equal the local jth conditional variance $V_{B'_j}[Y|X = x]$ in general since it measures the dispersion about the global centroid $F(x)$ and not about the local centroid $c'_j(x)$ of the jth fired then-part set $B'_j(x)$:

$$V_{B'_j}[Y|X = x] = \int_{-\infty}^{\infty} (y - c'_j)^2 p_{B'_j}(y|x)dy. \tag{2.161}$$

The rare case of $F(x) = c'_j$ holds if $p_j(x) = 1$ and may arise if input x belongs more to if-part fuzzy set A_j than to the near neighbors A_{j-1} and A_{j+1} (or other close sets in n dimensions): $a_j(x) > \max(a_{j-1}(x), a_{j+1}(x))$. The extreme case of $a_j(x) = 1$ and $a_k(x) = 0$ for $k \neq j$ and (2.157) lead to $V[Y|X = x] = V_{B'_j}[Y|X = x]$ as well as to $F(x) = c'_j$. The constant-centroid result $c'_j = c_j$ holds in the standard additive model. Then we can compute the local then-part set centroids c_j in advance. Result (2.143) confirms that we must compute the local variance-like terms for each input x even in the SAM case. The variance term $V[Y|X = x, B'_j]$ is large when $F(x)$ lies far from c'_j. But then a_j tends to be small and so too will be the variance weight V'_j.

So Theorem 2.2 shows that not all additive outputs $F(x)$ are the same. Some outputs are more certain than others. The most certain outputs $F(x) = c'_j$ occur when $p_j(x) = 1$ and thus when the system does not interpolate between then-part centroids. The spread or dispersion about these centroids leads to system uncertainty. The fuzzy system pays for its interpolation in increased uncertainty.

The SAM structure $B'_j(x) = a_j(x)B_j$ simplifies the conditional variance $V_{B'_j}[Y|X = x]$ because the conditional probability density $p_{B'_j}(y|x)$ does not depend on the input x for $a_j(x) > 0$:

$$p_{B'_j}(y|x) = \frac{a_j(x)b_j(y)}{a_j(x)\displaystyle\int_{R^p} b_j(y)dy} = \frac{b_j(y)}{\displaystyle\int_{R^p} b_j(y)dy} = \frac{b_j(y)}{V_j} = p_{B_j}(y). \tag{2.162}$$

The marginal density $p_{B_j}(y)$ lets us define the variance $\sigma^2_{B_j}$ of the (implicitly normalized) then-part set B_j:

$$\sigma^2_{B_j} = \int_{-\infty}^{\infty} (y - E_{B_j}(y))^2 p_{B_j}(y)dy \tag{2.163}$$

$$= \int_{-\infty}^{\infty} (y - c_j)^2 p_{B_j}(y)dy. \tag{2.164}$$

This holds since the then-part set centroid c_j is just the mean with respect to the jth then-part marginal density $p_{B_j}(y)$:

$$c_j = \frac{\displaystyle\int_{-\infty}^{\infty} y b_j(y)dy}{\displaystyle\int_{-\infty}^{\infty} b_j(y)dy} \tag{2.165}$$

$$= \int_{-\infty}^{\infty} y p_{B_j}(y)dy \tag{2.166}$$

$$= E_{B_j}(Y). \tag{2.167}$$

The following corollary uses these facts to simplify the first-order and second-order conditional statistics of a SAM system. It gives the SAM conditional variance as a convex sum of the local then-part variances plus a global dispersion term.

Corollary. (*SAM Statistics*) The scalar-valued SAM system

$$F(x) = Centroid\left(\sum_{j=1}^{m} w_j a_j(x) B_j\right)$$

has conditional mean and variance

$$E[Y|X = x] = F(x) = \sum_{j=1}^{m} p_j(x) c_j \tag{2.168}$$

$$V[Y|X = x] = \sum_{j=1}^{m} p_j(x)\sigma_{B_j}^2 + \sum_{j=1}^{m} p_j(x)(c_j - F(x))^2 \tag{2.169}$$

with convex coefficients

$$p_j(x) = \frac{w_j a_j(x) V_j}{\sum_{k=1}^{m} w_k a_k(x) V_k} \tag{2.170}$$

Proof. The conditional expectation (2.168) follows from Theorem 2.2 and (2.165)–(2.167). The conditional variance (2.169) follows from (2.159)–(2.160) and (2.161)–(2.164) as follows:

$$V[Y|X = x,\ B_j'] = V[Y|X = x,\ B_j] \tag{2.171}$$

$$= \int_{-\infty}^{\infty} (y - F(x))^2 p_{B_j'}(y|x)dy \tag{2.172}$$

$$= \int_{-\infty}^{\infty} (y - F(x))^2 p_{B_j}(y|x)dy \tag{2.173}$$

$$= \int_{-\infty}^{\infty} [(y - c_j) + (c_j - F(x))]^2 p_{B_j}(y)dy \tag{2.174}$$

$$= \int_{-\infty}^{\infty} (y - c_j)^2 p_{B_j}(y)dy + (c_j - F(x))^2 \int_{-\infty}^{\infty} p_{B_j}(y)dy$$
$$+\ 2(c_j - F(x))\left[\int_{-\infty}^{\infty} y p_{B_j}(y)dy - c_j \int_{-\infty}^{\infty} p_{B_j}(y)dy\right] \tag{2.175}$$

$$= \sigma_{B_j}^2 + (c_j - F(x))^2. \tag{2.176}$$

Then the convex-sum structure of (2.157) gives the result (2.169):

$$V[Y|X = x] = \sum_{j=1}^{m} p_j(x)V[Y|X = x,\ B'_j] \tag{2.177}$$

$$= \sum_{j=1}^{m} p_j(x)[\sigma^2_{B_j} + (c_j - F(x))^2] \tag{2.178}$$

$$= \sum_{j=1}^{m} p_j(x)\sigma^2_{B_j} + \sum_{j=1}^{m} p_j(x)(c_j - F(x))^2. \tag{2.179}$$

■

The second term in (2.179) acts as a penalty term. It is positive if and only if some jth rule fires ($a_j(x) > 0$ and thus $p_j(x) > 0$) and $c_j \neq F(x)$. So it is positive iff the fuzzy system F *interpolates* to reach the output $F(x)$. Each rule comes with its own output uncertainty $\sigma^2_{B_j}$. Interpolated outputs have more variance than just the sum of these weighted rule variances. Combining two uncertain structures gives a still less certain third structure.

The corollary allows us to compute the then-part variances $\sigma^2_{B_j}$ in advance but not the unweighted dispersion terms $(c_j - F(x))^2$ that change with each x. The convex structure of the weights implies that the partial sum

$$\sum_{k \neq j}^{m} p_k(x)$$

falls to 0 as the lone rule weight $p_j(x)$ grows to 1. The global output $F(x)$ also tends toward the jth then-part centroid c_j as $p_j(x)$ grows to 1. Then $F(x) = c_j$ when $p_j(x) = 1$. Then (2.169) reduces to $V[Y|X = x] = \sigma^2_{B_j}$. The system acts as if it has just one rule. It gives the uncertainty in the output $F(x)$ as just the uncertainty or variance in the (normalized) then-part set B_j. The system has a least uncertain output $F(x)$ if $p_j(x) = 1$ and if the jth then-part set has the least variance: $\sigma^2_{B_j} \leq \sigma^2_{B_k}$ for all k. The more the jth if-part $a_j(x) > 0$ "fires" then the more $p_j(x)$ grows and the more $F(x)$ resembles c_j.

The center of gravity or COG case gives a still simpler form of the SAM conditional variance. Here $w_1 = \ldots = w_m > 0$ and $V_1 = \ldots = V_m > 0$. The COG case also assumes that the then-part sets B_j have the same shape (since we can just as well recast them as delta pulses or point masses). Then all then-part sets B_j have the same variance: $\sigma^2_{B_j} = \sigma^2$ for all j. The common variance σ^2 factors from (2.169) and then the convex sum adds to 1:

$$V[Y|X = x] = \sigma^2 + \sum_{j=1}^{m} p_j(x)(c_j - F(x))^2. \tag{2.180}$$

There is least uncertainty in the output when $p_j(x) = 1$ holds for any j and thus when just one rule fires and the system does not interpolate between then-part centroids.

The COG case shows yet again why all fuzzy-system outputs do not have the same uncertainty status. Consider two COG fuzzy systems of m rules each. They have the same sets and rules except for one difference. The first system F_1 has then-part sets B_j with less variance than the then-part sets of F_2. The sets B_j have the same centroids c_j and unity volumes V_j. So both systems define the same mapping and have $F_1(x) = F_2(x)$ for all x and so have the same first-order statistics. But $V[F_1|X = x] < V[F_2|X = x]$ holds for each x and so $V[F_1|X] < V[F_2|X]$ holds.

This example shows that we cannot replace then-part sets B_j with rectangles or delta-pulse spikes and do so with complete impunity. A spike centered at the then-part centroid c_j may simplify the first-order computation of output values $F(x)$. But it assumes a level of expert confidence and precision that seldom holds in practice. This uncertainty view also comports with the view that larger rule patches are less certain than smaller ones and give a rougher function approximation.

We end this section with an application of Theorem 2.2 to the SAM system combination scheme of the previous section. The recursive relations (2.133)–(2.136) give the corollary in recursive form:

$$E[Y|X = x] = c_m + \sum_{j=1}^{m-1} p_j(x)[c_j - c_m] \tag{2.181}$$

$$V[Y|X = x] = \sigma^2_{B_m} + (c_m - F(x))^2 + \sum_{j=1}^{m-1} p_j(x)\left[\sigma^2_{B_j} - \sigma^2_{B_m}\right]$$

$$+ \sum_{j=1}^{m-1} p_j(x)\left[(c_j - F(x))^2 - (c_m - F(x))^2\right] \tag{2.182}$$

These laws can act as learning laws of sorts as the SAM system adds new rules.

Below we shall see that both supervised learning and optimal rule selection also depend on the gap $c_j - F(x)$ between the local and global centroids. We first review the formal theory of fuzzy function approximation ($F \approx f$) and representation ($F = f$) and then show how learning can move the fuzzy approximator F closer to the approximand f.

2.10 FUNCTION APPROXIMATION WITH ADDITIVE FUZZY SYSTEMS

Additive fuzzy systems $F\colon R^n \to R^p$ can uniformly approximate any continuous function $f\colon U \subset R^n \to R^p$ on a compact (closed and bounded) domain U [22]. This result holds for fuzzy sets of all types and for additive systems other than SAMs. The constructive proof below exploits the convex expansion of $F(x)$ in (2.31)–(2.33).

The result holds for other methods [57] of "defuzzifying" B in (2.12) that trap $F(x)$ between the same centroidal bounds. Centroidal additive outputs always lie within these bounds. The homework problems show that picking the max or supremum point of B can lie outside the centroidal bounds. Nonadditive systems need not satisfy these bounds. This is true of the max combiner in (2.52) as the homework problems also show.

A finer and finer rule-patch cover is the thrust of the uniform fuzzy approximation theorem. Figure 2.1 shows this process for "blind" approximation. The next section deals with the special case where we have complete knowledge of the approximand f. Even in the blind case we can always find a finite number of rule patches to cover the graph of f to keep the distance $|f(x) - F(x)|$ as small as we please for all x. In practice we must guess at the rules or use a neural or statistical scheme to learn them from data. We must also balance the fineness of the patch approximation with how much it costs to process and tune the rules. The combinatorial explosion of rules in Section 2.2 quickly takes its toll in high dimensions.

The uniform approximation lets us pick the error level ε in advance. Then for all $\varepsilon > 0$ we have $|f(x) - F(x)| < \varepsilon$ for all x in X. The domain X is a compact (closed and bounded) subset of R^n. X may be an interval $[u, v]$ where $u < v$ or a product of intervals $[u_1, v_1] \times \cdots \times [u_n, v_n]$. The fuzzy approximation theorem below ensures that we can in theory find m rules that satisfy the ε-criterion.

In practice we may never find the rules or we may find only some of them in some regions of the state space $X \times Y$. We can guess at the rules or ask experts for them or watch experts or real processes to try to learn the rules from sample data. These first rules initialize the learning process. We can then tune them with gradient descent or random hill climbing or other search techniques. But the rules learned from a finite stream of data may not give an F ε-close to f. Section 2.12 reviews some methods for learning the rule patches from data with both unsupervised and supervised techniques.

The history of fuzzy function approximation starts in the early 1990s and tracks the history of neural function approximation that starts in the late 1980s. Fuzzy systems have a left-to-right feedforward structure that acts much as a feedforward multilayer neural network acts. Nodes combine inputs and pass those signals to the next layer of nodes. The same learning schemes can use the same sample data to tune the neural system $N: R^n \to R^p$ or the fuzzy system $F: R^n \to R^p$. We often call such adaptive fuzzy systems "neural fuzzy systems" or "fuzzy neural systems." This wrongly suggests that all use of unsupervised clustering or supervised gradient descent depends on neural networks. Feedforward neural and fuzzy systems have similar architectures and even have the same architecture in the case of most radial basis function networks [33] and Gaussian SAMs. So it is no surprise that results in fuzzy function approximation have quickly tracked results in neural function approximation.

In 1989 Hornik and White [13] first used the Stone-Weierstrass theorem of functional analysis [39] to show uniform convergence of such neural networks. The Stone-Weierstrass theorem states that $A = C(X)$ if $C(X)$ is the sup-norm space of continuous functions on a compact and Hausdorf space X and if the set of functions $A \subset C(X)$ is a closed algebra and if A is self-adjoint and separates points and contains the constant functions. Cybenko [4] published a like result in the same year.

These neural results do not show how to build or learn real neural systems. The neurons and synapses do not correspond to fuzzy rule patches and their patch geometry. Open a neural net's black box and you find the "connectionist glop" of a massive synaptic web just as you would find in a brain. Radial basis nets come close to additive fuzzy systems with Gaussian fuzzy sets and some are the same thing. In 1990 Hartman and Keeler [12] showed that radial basis nets are universal approximators and thus extended a wide literature on Gaussian-sum approximators [33, 44]. Radial basis nets (Gaussian SAMs) also suffer from the curse of dimensionality.

Most neural approximation theorems use any number of "hidden" neurons between the n input and the p output neurons in the neural system $N: R^n \rightarrow R^p$. So far no one has shown how to choose or even interpret these neurons. The user must guess at both the number of layers of "hidden" neurons in the network and the number of these neurons in each layer. This reflects the problem of the "connectionist glop." Neural learning further complicates the problem as it tunes the distributed synaptic memory. Neural nets forget old patterns as they learn new patterns. Users cannot tell which old patterns the trained neural net forgot or to what degree it forgot or changed them. In this sense neural nets model real brains all too well.

The first fuzzy approximation theorem appeared in 1991 [21]. It showed that additive fuzzy systems can define simple functions and so can uniformly approximate bounded measurable functions. Then even bivalent expert systems can act as universal approximators if they use enough binary rules. We review this simple result below. The fuzzy approximation theorem or the FAT theorem for additive fuzzy systems appeared the next year in 1992 [22]. It was the first such theorem and it holds for fuzzy sets of all shapes.

The FAT theorem used the simple geometry of shrinking rule patches or cubes and the simple scheme of sums and products of the additive model. The proof traps the fuzzy system's output $F(x)$ between the centroids of the scaled then-part fuzzy sets as Theorems 2.1 and 2.2 show. This geometry and ease of computation soon led to new learning algorithms [5–7] and led in part to an array of low-cost software tool kits and digital VLSI chips. Wang and Mendel [54] used the Stone-Weierstrass theorem to prove the special case of the FAT theorem for continuous functions $f: R^n \rightarrow R$ on a compact domain when the additive system uses the Gaussian set functions (2.36) in the constant-volume SAM (2.44)–(2.45) and hence when the SAM coincides with a radial basis function network. Equations (2.55)–(2.60) show how the SAM model reduces to such a radial basis network. This is also the case when an additive fuzzy system coincides with the Gaussian Parzen estimator of Specht's 1991 generalized regression neural network [44].

Watkins [55, 56] extended the FAT theorem in the scalar case. He showed that an additive fuzzy system F with just two rules can *represent* any bounded function $f: R \rightarrow R$ in the sense that $F(x) = f(x)$ for all x in R. We present this clever result in the next section. The scalar function f need not be continuous. Watkins also showed that SAMs cannot represent all functions on compact spaces of higher dimension. They can of course approximate them. No finite set of fuzzy rules will give an additive F such that $F(x, y) = f(x, y)$ for all x and y if $f(x, y) = xy$

on some compact domain that contains the origin in R^2. Watkins also showed that whether a vector SAM system F can represent f often depends on whether a certain differential equation has a solution.

Other fuzzy approximation theorems have since appeared for nonadditive systems. Buckley and Hayashi [2] showed in 1993 that some abstract fuzzy control nets act as uniform approximators for continuous maps $f: R^2 \to R$ on compact domains if the approximator uses actual samples $(x, f(x))$ from the function. The result does not show how to build such a system. Dubois and Grabisch [10] used the max-min "extension principle" [9] to map if-part fuzzy sets to then-part sets in rules and then used sets of these rules to approximate *monotone* functions but not arbitrary functions on compact domains. These abstract proofs do not show how to build or learn real fuzzy systems. Indeed their complex operations can compound the high costs of computation that all fuzzy systems face when the number of rules grows.

We first review how simple additive systems can uniformly approximate bounded measurable functions. Most functions of practical interest are measurable but we here omit the technical definition [40]. The approximation result amounts to an exercise in abstract real analysis when applied to the SAM additive combiner

$$B(x) = \sum_{j=1}^{m} a_j(x) B_j \tag{2.183}$$

$$= \sum_{j=1}^{m} a_j(x) c_j \tag{2.184}$$

where the centroid value c_j replaces the then-part set B_j as the singleton set $\{c_j\}$. The m rules have the form "If $X = A_j$ then $Y = c_j$." Then combined set B has a generalized set function $b_j: X \to R$ of the additive form

$$b(x) = \sum_{j=1}^{m} c_j a_j(x). \tag{2.185}$$

We now force the set function b to act as a *simple* function $s: X \to Y$ or a function that maps all points in the input space X into just m output values $c_1, \ldots, c_m$. Now suppose the m if-part sets $A_j \subset X$ are not fuzzy and contain the inverse images of the simple-function values:

$$A_j = \{x \in X: b(x) = c_j\} = b^{-1}(c_j). \tag{2.186}$$

Then the if-part set functions have the binary form

$$a_j(x) = \begin{cases} 0 & if\ b(x) \neq c_j \\ 1 & if\ b(x) = c_j. \end{cases} \tag{2.187}$$

Then the m if-part sets partition the input space X: $A_j \cap A_k = \emptyset$ if $j \neq k$ and

$$X = \bigcup_{j=1}^{m} A_j.$$

So the set function b in (2.185) is a proper simple function. Then standard results in analysis [40] show that there is a sequence of simple functions b_n each of the form (2.185) that increases to any measurable function f. So for all $\varepsilon > 0$ there is a simple function $F_\varepsilon = b_\varepsilon$ ε-close to f on X. Boundedness of f ensures a uniform approximation.

This result shows that a large enough artificial intelligence (AI) expert system can approximate any bounded measurable function and reminds us that fuzzy rules reduce in the bivalent case to decision-tree or expert-system rules. It does not show that an additive fuzzy system with multivalued sets A_j *converges* uniformly to f. So in practice the result gives no error bound. The problem arises because triangles, trapezoids, bell curves, and other fuzzy sets need not converge uniformly to rectangles.

Uniform convergence holds for fuzzy sets if we work with continuity instead of measurability. The FAT theorem below requires that the vector map $f: X \to Y$ is continuous and that X is a compact (closed and bounded) subset of R^n. This holds in practice since no one has yet found an unbounded set of parameters in the physical universe. The FAT theorem shows that in principle an additive fuzzy system (not just a SAM) can approximate any continuous function to any degree of accuracy for arbitrary fuzzy sets. Indeed the set of additive fuzzy systems is dense [39] in the space of continuous functions on compact domains in the same sense that the rational numbers are dense in the real numbers.

Theorem 2.3 (*Fuzzy Approximation Theorem*). An additive fuzzy system $F: X \to Y$ uniformly approximates $f: X \to Y$ if X is compact and f is continuous.

Proof. Pick any small $\varepsilon > 0$. We must show that $|F(x) - f(x)| < \varepsilon$ for all $x \in X$. X is a compact subset of R^n and

$$F(x) = Centroid(B) = Centroid\left(\sum_{j=1}^{m} B'_j\right)$$

holds in an additive model.

Continuity of f on compact X gives uniform continuity. So there is a fixed distance δ such that for all x and z in X we have $|f(x) - f(z)| < \frac{\varepsilon}{4}$ if $|x - z| < \delta$. We can construct open cubes $M_1, \ldots, M_m$ that cover X and that have ordered overlap in their n coordinates so that each cube corner lies at the midpoint c_j of its neighbors M_j. Pick then-part fuzzy sets B_j centered at $f(c_j)$. Thus the centroid of B_j is $f(c_j)$.

Pick $u \in X$. Then by construction u lies in at most 2^n overlapping open cubes M_j. Pick any w in the same set of cubes. Suppose $u \in M_j$ and $w \in M_k$. Then for all $v \in M_j \cap M_k$ we have $|u - v| < \delta$ and $|v - w| < \delta$. Uniform continuity implies that $|f(u) - f(w)| \le |f(u) - f(v)| + |f(v) - f(w)| < \frac{\varepsilon}{2}$. So for cube centers c_j and c_k we have $|f(c_j) - f(c_k)| < \varepsilon/2$.

Pick $x \in X$. Then x too lies in at most 2^n open cubes with centers c_j and such that $|f(c_j) - f(x)| < \varepsilon/2$. Along the kth coordinate of the range space R^p the kth component of the additive system centroid $F(x)$ lies as in Theorem 2.2 on or between the kth components of the centroids of the then-part B_j sets (see the discussion after (2.32) and (2.33) above). So $|F(x) - f(c_j)| < \varepsilon/2$ holds since $|f(c_j) - f(c_k)| < \varepsilon/2$ holds for all $f(c_j)$. Then $|F(x) - f(x)| \le |F(x) - f(c_j)| + |f(c_j) - f(x)| < \varepsilon/2 + \varepsilon/2 = \varepsilon$. ■

The proof traps the additive output $F(x)$ between the then-part set centroids c_j. This is clear in the scalar case when $F\colon R^n \to R$ for then the centroids are well ordered from left to right as in Figure 2.4. In the vector case $F\colon R^n \to R^p$ the vector output $F(x)$ must lie within the centroidal hyperbox defined in Section 2.4:

$$F(x) \in \left[c^1_{left}, c^1_{right}\right] \times \ldots \times \left[c^p_{left}, c^p_{right}\right] \subset R^p. \tag{2.188}$$

This always holds in the SAM or product-inference case $B'_j = a_j(x)B_j$ since the fired then-part set B'_j has the same centroid as the unfired set B_j has if $a_j(x) > 0$. It need not hold in the min-clip or minimum-inference case $B'_j = \min(a_j(x), B_j)$ unless B_j is symmetric. The same is true for other t-norm-inference schemes $B'_j = T(a_j(x), B_j)$ unless B_j is symmetric. This is a further reason for not using such schemes.

Counterexamples show that the max combiner

$$B = \bigcup_{j=1}^{m} B'_j$$

does not lie within the centroidal hyperbox in (2.188). There is a trivial exception when sum = max. Recall the identity $x + y = \min(x, y) + \max(x, y)$. Then $x + y = \max(x, y)$ iff $x = 0$ or $y = 0$. So the sum and max combiners coincide iff the then-part sets B_j (or B'_j) are pairwise disjoint as discussed above in Section 2.5. The FAT theorem holds for other defuzzifiers [57] $D(B) = F(x)$ if (2.188) holds. So far only the max or supremum defuzzifier has found any use in practical systems besides the popular centroidal fuzzifier. Then we can view $F(x)$ as computing the MAP or maximum *a posteriori* outcome if we view the normalized then-part sets $B'_j(x)/V_j$ as conditional probability densities $p(y|x)$ as discussed in the previous section.

The next section shows that we can improve on the FAT theorem's approximation $F \approx f$ to get the exact representation $F = f$ in many cases if we have complete knowledge of the approximand f. This reduces the exponential rule complexity of blind approximation to linear rule complexity.

2.11 FUNCTION REPRESENTATION WITH SAMS

A fuzzy system F can use just one rule to represent a constant function $f: R^n \to R^p$. The constant map has the form $f(x) = c$ for all x. Let the centroidal fuzzy system $F: R^n \to R^p$ have just one rule or patch $A_1 \times B_1$. Let the if-part set A_1 have any positive set function $a_1(x) > 0$ for all x. Pick the then-part set B_1 so that its centroid c_1 is just the constant value c of the map f: $c_1 = c$. Then $B = a_1(x)B_1$ and the SAM Theorem gives

$$F(x) = Centroid(B) \tag{2.189}$$

$$= \frac{a_1(x)\displaystyle\int_{R^p} y b_1(y)dy}{a_1(x)\displaystyle\int_{R^p} b_1(y)dy} \tag{2.190}$$

$$= c_1 = c. \tag{2.191}$$

This simple proof reminds us that a fuzzy system defines a flat segment when just one rule fires. We use this fact in Section 2.13 to prove that optimal lone rules cover the extrema of the approximand f.

The success of representing the constant map $f: R^n \to R^p$ leads to the fundamental question of fuzzy *representation* theory: Which functions f can a SAM system F represent in the sense of $F(x) = f(x)$ for all x? Watkins [55, 56] has explored this question and found both positive and negative results.

We first present Watkins's positive result: A SAM F needs just two rules to represent any *bounded* nonconstant scalar function $f: R \to R$. This extends the FAT theorem but at the expense of building the structure of f into the two then-part sets A_1 and A_2. The result does not say that two if-part triangles or trapezoids or other simple if-part sets can represent f. And in practice we may have little or no knowledge of f or even of its turning points.

Theorem 2.4 (*Scalar Representation.*) A SAM $F: R \to R$ with just two rules of the form

"If $X = A$ then $Y = B_1$"

"If $X = not\text{-}A$ then $Y = B_2$"

can represent a bounded nonconstant function $f: R \to R$ in the sense that $F(x) = f(x)$ for all $x \in R$.

Proof. Boundedness lets us define the lower and upper bounds $\alpha = \inf f$ and $\beta = \sup f$. Center then-part sets B_1 and B_2 of any shape at α and β with finite but constant volumes (areas): $0 < V_1 = V_2 < \infty$. So $c_1 = \alpha$ and $c_2 = \beta$. B_1 and B_2 might be rectangles or triangles of unit area. Define the set function of then-part set A_1 as the ratio

$$a_1(x) = \frac{\beta - f(x)}{\beta - \alpha}. \tag{2.192}$$

Then $0 \leq a_1(x) \leq 1$ holds for all x and so A_1 is a fuzzy set. Define the second then-part set A_2 as the complement of A_1: $a_2(x) = 1 - a_1(x)$. Then (2.192) and the SAM Theorem give the result:

$$F(x) = \frac{\sum_{j=1}^{2} a_j(x) V_j c_j}{\sum_{j=1}^{2} a_j(x) V_j} \tag{2.193}$$

$$= \frac{a_1(x)\alpha + (1 - a_1(x))\beta}{a_1(x) + 1 - a_1(x)} \tag{2.194}$$

$$= a_1(x)\alpha + (1 - a_1(x))\beta \tag{2.195}$$

$$= a_1(x)(\alpha - \beta) + \beta \tag{2.196}$$

$$= \left(\frac{\beta - f(x)}{\beta - \alpha}\right)(\alpha - \beta) + \beta \tag{2.197}$$

$$= \left(\frac{f(x) - \beta}{\alpha - \beta}\right)(\alpha - \beta) + \beta \tag{2.198}$$

$$= f(x) - \beta + \beta \tag{2.199}$$

$$= f(x). \tag{2.200}$$

■

Consider the bounded function $f(x) = \sin x$. Pick the two then-part sets B_1 and B_2 as unit-area rectangles centered at -1 and 1: $c_1 = \alpha = -1$ and $c_2 = \beta = 1 = V_1 = V_2$. Then

$$a_1(x) = \frac{1 - \sin x}{1 - (-1)} = \frac{1 - \sin x}{2} \quad \text{and} \quad a_2(x) = \frac{1 + \sin x}{2}.$$

Then the SAM Theorem gives

$$F(x) = \frac{\left(\frac{1 - \sin x}{2}\right)(-1) + \left(\frac{1 + \sin x}{2}\right)(1)}{\frac{1 - \sin x}{2} + \frac{1 + \sin x}{2}}$$

$$= \frac{\sin x - 1 + 1 + \sin x}{1 - \sin x + 1 + \sin x} = \frac{2 \sin x}{2} = \sin x.$$

Representations can fail in higher dimensions. Consider the product function $f(x, y) = xy$. Watkins [55, 56] showed that $F = f$ leads to the unique separation of the form

$$F(x, y) = \frac{\alpha x + \beta y}{\frac{\beta}{x} + \frac{\alpha}{y}}.$$

Then the product function separates away from the coordinate axes. Then no SAM can represent the product function on any domain that includes any part of the coordinate axes. Watkins has further shown that twice-differentiable functions have piecewise SAM representations if and only if a system of SAM partial differential equations has a solution. There are $\frac{n(n-1)}{2}$ such simultaneous equations for a function $f: R^n \to R$. These representations can require *bipolar* set functions $a: X \to [-1, 1]$.

Fuzzy representations reduce the exponential rule complexity of blind function approximation to linear (or quadratic) complexity but at the cost of a complete knowledge of f. Future research may find algorithms that reduce the SAM rule complexity from exponential to polynomial as the system learns more of the structure of f. Other research may show us more of the boundary between the set of functions that SAMs can and cannot represent. Representation theory remains a fertile area for future research in fuzzy systems.

The next section returns to the practical problem of blind approximation of f. We know at most a few noisy input-output pairs $(x, f(x))$ and want to use these data points to move the SAM system F closer to f. This is the domain of statistical learning theory.

2.12 LEARNING IN SAMS: UNSUPERVISED CLUSTERING AND SUPERVISED GRADIENT DESCENT

A fuzzy system learns if and only if its rule patches move or change shape in the input-output product space $X \times Y$. Learning might change the centers or widths of triangle or trapezoid sets. These changing sets then change the shape or position of the Cartesian rule patches built out of them. We show in the next section with the mean-value theorem and the calculus of variations that optimal lone rules cover the extrema or bumps of the approximand. Good learning schemes tend to quickly move rule patches to these bumps and then move extra rule patches between them as the rule budget allows. Hybrid schemes use unsupervised clustering to learn the first set of fuzzy rule patches in position and number and to initialize the gradient descents of supervised learning.

Learning changes system parameters with data. Unsupervised learning amounts to blind clustering in the system product space $X \times Y$ to learn and tune fuzzy rules or the sets that compose them. Then k quantization vectors $q_j \in X \times Y$ can move in the product space to filter or approximate the distribution of incoming data pairs $(x(t),\ y(t))$ or the concatenated data points $z(t) = [x(t)|y(t)^T]$. The simplest form of such *product space clustering* [21] centers a rule patch at each data point and thus puts $k = m$ [44, 54]. In general the data greatly outnumber the rules and so $k \gg m$.

Figure 2.5 shows how the distribution of adaptive quantization vectors can track the stream of noisy data points $z(t)$ from some unknown approximand f. One learning scheme [21] just counts the numbers of quantization vectors in a patch cell to decide whether that rule "exists" or to increase or decrease that rule's weight w_j. The simplest threshold counts a rule as present if the cell contains at least one quantization vector. The higher one sets the threshold the better the scheme tends to filter noise from the data but the more likely it will miss an important rule such as a turning-point cell. This scheme uses a fixed grid of rule cells and then works with data clusters. An alternate scheme is to let the data clusters or quantization clusters pick the rule cells and hence pick the grid. The following ellipsoid scheme does this.

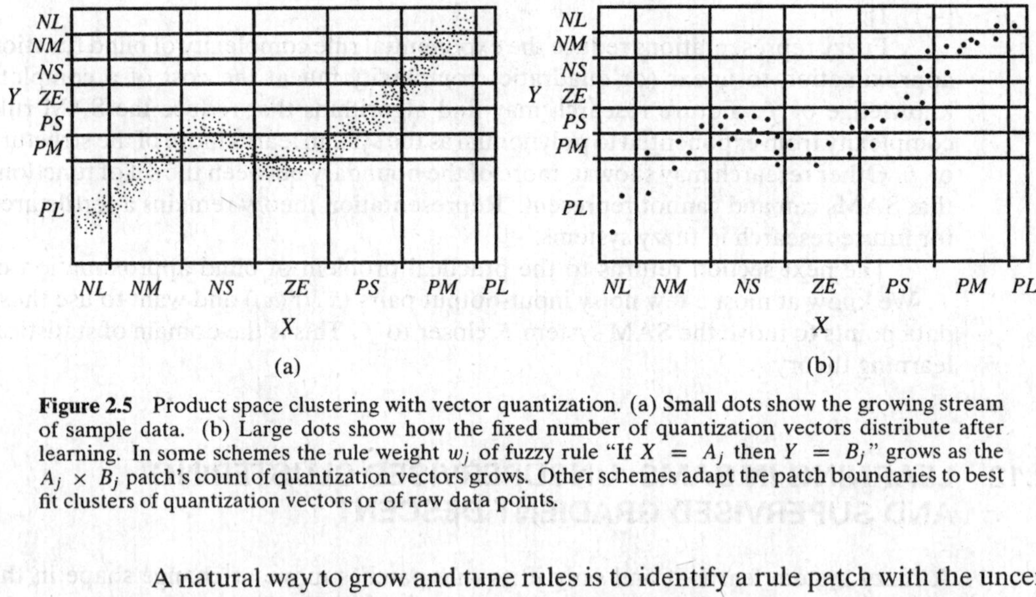

Figure 2.5 Product space clustering with vector quantization. (a) Small dots show the growing stream of sample data. (b) Large dots show how the fixed number of quantization vectors distribute after learning. In some schemes the rule weight w_j of fuzzy rule "If $X = A_j$ then $Y = B_j$" grows as the $A_j \times B_j$ patch's count of quantization vectors grows. Other schemes adapt the patch boundaries to best fit clusters of quantization vectors or of raw data points.

A natural way to grow and tune rules is to identify a rule patch with the uncertainty ellipsoid [5–7] that forms around each quantizing vector q_j from the inverse of its positive definite covariance matrix K_j. Then sparse or noisy data grows a large patch and thus a less certain rule than does denser or less noisy data. Chapter 3 looks at this learning scheme in detail. Chapter 5 applies it to the control of platoons of smart cars. Chapter 6 applies it to filtering impulsive noise and extends it to the case of unfactored joint if-part sets of ellipsoid shape. Here we briefly review the main learning laws. Unsupervised competitive learning [20, 21] can learn these ellipsoidal rules in three steps:

$$\|z(t) - q_j\| = \min(\|z(t) - q_1(t)\|, \ldots, \|z(t) - q_k(t)\|) \tag{2.201}$$

$$q_i(t+1) = \begin{cases} q_j(t) + \mu_t[z(t) - q_j(t)] & \text{if } i = j \\ q_i(t) & \text{if } i \neq j \end{cases} \tag{2.202}$$

$$K_i(t+1) = \begin{cases} K_j(t) + v_t\left[(z(t) - q_j(t))^T(z(t) - q_j(t)) - K_j(t)\right] & \text{if } i = j \\ K_i(t) & \text{if } i \neq j \end{cases} \tag{2.203}$$

for the Euclidean norm $\|z\|^2 = z_1^2 + \ldots + z_{n+p}^2$.

The first step (2.201) is the competitive step. It picks the nearest quantizing vector q_j to the incoming data vector $z(t)$ and ignores the rest. This correlation matching approximates a great deal of the competitive dynamics of nonlinear neural networks. It is just the old learning scheme of *k-means clustering* of pattern recognition theory. In neural theory it goes by the names "competitive learning" or "self-organizing mapping."

The second step updates the winning quantization or "synaptic" vector and drives it toward the centroid of the sampled data pattern class [20]. One can show that such learning locally minimizes the mean-squared error of vector quantization and does so exponentially fast.

A simple stochastic argument shows why (2.202) tends to drive quantization vectors to the centroids of the local data clusters they sample. Suppose the jth quantization vector q_j codes for the jth pattern class D_j of the product space. This is where the competitive step (2.201) takes effect. In a neural system the signal function of the jth competing neuron acts much as the set function d_j for the bivalent pattern-class set D_j:

$$d_j(z) = \begin{cases} 0 & \text{if } z \notin D_j \\ 1 & \text{if } z \in D_j. \end{cases} \tag{2.204}$$

We further assume that the pairwise disjoint pattern classes partition the product space. Then the continuous stochastic version of the discrete learning laws (2.201)–(2.202) has the form

$$\dot{q}_j = d_j(z)[z - q_j] \tag{2.205}$$

where the overdot stands for time differentiation. We can also add to the righthand side of (2.205) a zero-mean noise vector with finite covariance but we ignore it for simplicity. Learning stops at equilibrium when $\dot{q}_j = 0$. Then taking expectations of both sides of the equilibrium state gives

$$0 = E[d_j(z)[z - q_j]] \tag{2.206}$$

$$= \int_{R^n \times R^p} d_j(z)[z - q_j]p(z)dz \tag{2.207}$$

$$= \int_{D_j} [z - q_j] p(z) dz \tag{2.208}$$

$$= \int_{D_j} z p(z) dz - q_j \int_{D_j} p(z) dz \tag{2.209}$$

for some joint "occurrence" probability density function $p(z)$ that governs the statistical structure of the data clusters. Solving (2.209) for the equilibrium quantization vector $\hat{q}_j$ shows that it equals the local pattern-class centroid:

$$\hat{q}_j = \frac{\int_{D_j} z p(z) dz}{\int_{D_j} p(z) dz}. \tag{2.210}$$

We can also view this centroid as a realization of the conditional mean $E[z|D_j]$. The centroid locally minimizes the mean-squared error of vector quantization. The competitive learning law only approximates the behavior in (2.206)–(2.210). In practice many problems can arise with how well the quantization vectors spread out. The best way to avoid these problems is to pick the initial values of the quantization vectors as real sample data values z.

The third step updates the covariance matrix of the winning quantization vector. It too moves quickly toward the mean-squared optimal conditional covariance matrix. Again we initialize the quantization vector with sample data ($q_i(0) = z(i)$) to avoid skewed groupings and initialize the covariance matrix with small positive numbers on its diagonal to ensure that it is and stays positive definite. Then projection schemes can convert the ellipsoids into coordinate fuzzy sets. Supervised learning can also tune the eigenvalue parameters of the rule ellipsoids.

The sequences of learning coefficients $\{\mu_t\}$ and $\{\nu_t\}$ should decrease slowly in the sense of

$$\sum_{t=1}^{\infty} \mu_t = \infty$$

but not too slowly in the sense of

$$\sum_{t=1}^{\infty} \mu_t^2 < \infty.$$

These constraints come from the theory of stochastic approximation [21]. In practice $\mu_t \approx 1/t$. The covariance coefficients obey a like constraint as in the practical choice

$$\nu_t = 0.2 \left[1 - \frac{t}{1.2N} \right]$$

where N is the total number of data points. The supervised learning schemes below also use a like sequence $\{\mu_t\}$ of decreasing learning coefficients.

Supervised learning changes SAM parameters with error data. The error at each time t is the desired system output minus the actual SAM output: $\varepsilon_t = d_t - F(x_t)$. Unsupervised learning uses the blind data point $z(t)$ instead of the desired or labeled value d_t. The teacher or supervisor supervises the learning process by giving the desired value d_t at each training time t. Most supervised learning schemes perform stochastic gradient descent on the squared error and do so through iterated use of the chain rule of differential calculus. The literature often and wrongly refers to such supervised learning schemes as "backpropagation" learning after the scheme for supervised hill climbing in the synaptic space of mulilayer feedforward neural networks or perceptrons [21]. SAM systems involve no such "hidden" neurons and backpropagation is but one way to perform gradient descent on a multilayered network.

Supervised gradient descent can learn or tune SAM systems [26] by changing the rule weights w_j in (2.211) below. Or it can change the then-part volumes V_j or the then-part set centroids c_j or parameters of the if-part set functions a_j. The rule weight w_j enters the ratio form of the weighted SAM system

$$F(x) = \frac{\sum_{j=1}^{m} w_j a_j(x) V_j c_j}{\sum_{j=1}^{m} w_j a_j(x) V_j} \tag{2.211}$$

in the same way as does the then-part volume V_j in Theorem 2.1 or the general volume V_j' in Theorem 2.2. So both have the same learning law if we replace the nonzero weight w_j with the nonzero volume V_j or V_j':

$$w_j(t+1) = w_j(t) - \mu_t \frac{\partial E}{\partial w_j} \tag{2.212}$$

$$= w_j(t) - \mu_t \frac{\partial E}{\partial F} \frac{\partial F}{\partial w_j} \tag{2.213}$$

$$= w_j(t) + \mu_t\, \varepsilon_t \frac{p_j(x_t)}{w_j(t)}[c_j - F(x_t)] \tag{2.214}$$

for instantaneous squared error $E_t = \frac{1}{2}(d_t - F(x_t))^2$ with desired-minus-actual error $\varepsilon_t = d_t - F(x_t)$. Then the volumes change in the same way if they do not depend on the weights (which they do in most ellipsoidal learning schemes):

$$V_j(t+1) = V_j(t) - \mu_t \frac{\partial E}{\partial V_j} \tag{2.215}$$

$$= V_j(t) + \mu_t \varepsilon_t \frac{p_j(x_t)}{V_j(t)}[c_j - F(x_t)]. \tag{2.216}$$

The learning law (2.214) follows since $\partial E/\partial w_j = -\varepsilon$ and since

$$\frac{\partial F}{\partial w_j} = \frac{a_j(x)V_jc_j\sum_{i=1}^{m} w_ia_i(x)V_i - a_j(x)V_j\sum_{i=1}^{m} w_ia_i(x)V_ic_i}{\left(\sum_{i=1}^{m} w_ia_i(x)V_i\right)^2} \tag{2.217}$$

$$= \frac{w_ja_j(x)V_j}{w_j\sum_{i=1}^{m} w_ia_i(x)V_i}\left[\frac{c_j\sum_{i=1}^{m} w_ia_i(x)V_i}{\sum_{i=1}^{m} w_ia_i(x)V_i} - \frac{\sum_{i=1}^{m} w_ia_i(x)V_ic_i}{\sum_{i=1}^{m} w_ia_i(x)V_i}\right] \tag{2.218}$$

$$= \frac{p_j(x)}{w_j}[c_j - F(x)] \tag{2.219}$$

from the SAM Theorem. We again see how SAM systems depend on the gap between the local then-part centroid c_j and the global centroidal output $F(x)$.

The centroid c_j in the SAM Theorem or the conditional mean c_j' in Theorem 2.2 has the simplest learning law:

$$c_j(t+1) = c_j(t) - \mu_t\frac{\partial E}{\partial F}\frac{\partial F}{\partial c_j} \tag{2.220}$$

$$= c_j(t) + \mu_t\varepsilon_t p_j(x_t). \tag{2.221}$$

So the terms w_j, V_j, and c_j do not change when $p_j \approx 0$ and thus when the jth if-part set barely fires: $a_j(x) \approx 0$. The centroid learning law (2.221) is a convex-weighted version of the classical Widrow-Hoff LMS learning law [21].

Tuning the if-part sets involves more computation since the update law contains an extra partial derivative. Suppose if-part set function a_j is a function of l parameters: $a_j = a_j(m_j^1, \ldots, m_j^l)$. Then we can update each parameter with

$$m_j^k(t+1) = m_j^k(t) - \mu_t\frac{\partial E}{\partial F}\frac{\partial F}{\partial a_j}\frac{\partial a_j}{\partial m_j^k} \tag{2.222}$$

$$= m_j^k(t) + \mu_t\varepsilon_t\frac{p_j(x_t)}{a_j(x_t)}[c_j - F(x_t)]\frac{\partial a_j}{\partial m_j^k}. \tag{2.223}$$

Exponential if-part set functions can reduce the learning complexity. They have the form $a_j = e^{f_j(m_j^1,\ldots,m_j^l)}$ and obey

$$\frac{\partial a_j}{\partial m_j^k} = a_j\frac{\partial f_j(m_j^1, \ldots, m_j^l)}{\partial m_j^k}$$

Then the parameter update law (2.223) simplifies to

$$m_j^k(t+1) = m_j^k(t) + \mu_t \varepsilon_t p_j(x_t)[c_j - F(x_t)]\frac{\partial f_j}{\partial m_j^k}. \tag{2.224}$$

This can arise for independent exponential or Gaussian sets that have the form

$$a_j(x) = \prod_{i=1}^{n} e^{f_j^i(x_i)} = e^{\sum_{i=1}^{n} f_j^i(x_i)} = e^{f_j(x)}.$$

The exponential set function

$$a_j(x) = e^{\sum_{i=1}^{n} u_j^i(v_j^i - x_i)}$$

has parameter partial derivatives

$$\frac{\partial f_j}{\partial u_j^k} = v_j^k - x_k(t) \quad \text{and} \quad \frac{\partial f_j}{\partial v_j^k} = u_j^k.$$

This gives the exponential learning laws

$$u_j^k(t+1) = u_j^k(t) + \mu_t \varepsilon_t p_j(x)[c_j - F(x)](v_j^k - x_k) \tag{2.225}$$

$$v_j^k(t+1) = v_j^k(t) + \mu_t \varepsilon_t p_j(x)[c_j - F(x)]u_j^k \tag{2.226}$$

for vector inputs $x = (x_1, \ldots, x_n) \in R^n$.

The Gaussian set function

$$a_j(x) = e^{-\frac{1}{2}\sum_{i=1}^{n}\left(\frac{x_i - m_j^i}{\sigma_j^i}\right)^2}$$

has mean partial derivative

$$\frac{\partial f_j}{\partial m_j^k} = \frac{x_k - m_j^k}{\left(\sigma_j^k\right)^2}$$

and variance partial derivative

$$\frac{\partial f_j}{\partial \sigma_j^k} = \frac{\left(x_k - m_j^k\right)^2}{\left(\sigma_j^k\right)^3}.$$

This gives the Gaussian learning laws

$$m_j^k(t+1) = m_j^k(t) + \mu_t \varepsilon_t p_j(x)[c_j - F(x)] \frac{x_k - m_j^k}{\left(\sigma_j^k\right)^2} \tag{2.227}$$

$$\sigma_j^k(t+1) = \sigma_j^k(t) + \mu_t \varepsilon_t p_j(x)[c_j - F(x)] \frac{\left(x_k - m_j^k\right)^2}{\left(\sigma_j^k\right)^3}. \tag{2.228}$$

These Gaussian laws are the familiar learning laws of radial basis functions. And again such Gaussian set functions reduce the SAM model to Specht's [44] radial basis function network or "generalized regression neural network." The Gaussian learning laws offer a good way to cheat when tuning the much simpler (but discontinuous) triangle if-part sets found in many applications. We can use the smooth update laws (2.227) and (2.228) to update discontinuous triangles or trapezoids or other sets by viewing their centers and widths as the Gaussian means and variances.

The homework problems derive other supervised SAM learning laws. These include two strong competitors to the Gaussian SAM laws. The first is the set of Cauchy SAM learning laws

$$m_j(t+1) = m_j(t) + 2\mu_t \varepsilon_t p_j(x)[c_j - F(x)] \left(\frac{x - m_j}{d_j^2}\right) a_j(x) \tag{2.229}$$

$$d_j(t+1) = d_j(t) + 2\mu_t \varepsilon_t p_j(x)[c_j - F(x)] \left(\frac{x - m_j}{d_j}\right)^2 \frac{a_j(x)}{d_j}. \tag{2.230}$$

These laws tune the generalized Cauchy set functions $a_j\colon R \to R$ of the form

$$a_j(x) = \frac{1}{1 + \left(\dfrac{x - m_j}{d_j}\right)^2}. \tag{2.231}$$

Cauchy and Gaussian probability densities belong to the same family of alpha-stable densities that we discuss in Chapters 6 and 9. But Cauchy variables do not have finite variances or higher moments. They do have nearly the same bell-curve shape as the Gaussian bell curves. Their ratio form gives an easier set of if-part sets to compute with than do the exponentials of Gaussians.

The other set of learning laws are the sinc SAM learning laws

$$\frac{\partial a_j}{\partial m_j} = \begin{cases} \left(a_j(x) - \cos\left(\dfrac{x - m_j}{d_j}\right)\right) \dfrac{1}{x - m_j} & \text{for } x \neq m_j \\ 0 & \text{for } x = m_j \end{cases} \tag{2.232}$$

$$\frac{\partial a_j}{\partial d_j} = \left(a_j(x) - \cos\left(\frac{x - m_j}{d_j}\right)\right)\frac{1}{d_j}. \tag{2.233}$$

These laws tune the popular sinc function of signal processing:

$$a_j(x) = \frac{\sin\left(\frac{x - m_j}{d_j}\right)}{\frac{x - m_j}{d_j}}. \tag{2.234}$$

Simulations show that sinc SAMs often converge faster and more accurately than do Gaussian or Cauchy SAMs. Engineers have just begun to search the space of efficient SAM if-part sets.

2.13 OPTIMAL ADDITIVE SYSTEMS: PATCH THE BUMPS

Learning moves rule patches in the product space. But *where* should it move them? What is the best way for the fuzzy rule patches to cover the graph of the approximand f?

The best place to put a new rule is where it most reduces the approximation error e between the fuzzy system F and the approximand f. This acts as a rule-budget criterion in the face of exponential rule explosion. The goal of learning is optimality and the aim of optimality is computational efficiency. Each new rule or rule movement should decrease the error e as much as it can. It should "chop off" a bump on the error curve and try to chop off the largest error bump. There is no closed-form way to do this in general though the Remez exchange algorithm [38] can do something like it. And in practice we do not know the approximand f. After all we would not need a fuzzy system or anything else to approximate f if we knew f. We could just use the function f itself.

In practice we can often estimate the extreme or turning points of f with outlying data points or outlying data clusters. Or we might use sequential differences or iterative schemes such as Newton's method [27] or contraction maps [38] or still other techniques. This section shows that finding such extremal points $\hat{x}$ of f (such that $f'(\hat{x}) = 0$) should guide learning schemes at least to the first approximation. Many of the bumps or extrema of the error function e coincide with the extrema of the approximand f. Lone rule patches in the product state space should first cover the extrema of f or *patch the bumps* of f [25]. Extra rules can fill in between the bumps as best they can.

In the optimal case each new rule or rule movement should patch a bump of the *residual* error curve between f and the new fuzzy system F. The next chapter shows how egg-shaped or ellipsoidal rules quickly patch the bumps of known test functions f when the rule patches change with the learning laws of the previous section. The rule patches reduce most of the squared error of the function approximation as they move toward the bumps and then slowly reduce it as they hover about the bumps. This result has added force in high dimensions. Then the fuzzy map $F: R^n \rightarrow R^p$ needs on the order of k^{n+p-1} rules to cover the graph of f. The number of f extrema may well exceed the number of rules in the rule budget. Then learning reduces to the old problem of finding the zeroes of an equation or the points $\hat{x}$ of $f'(\hat{x}) = 0$ or of $\nabla f(\hat{x}) = 0$.

The local argument (2.2)–(2.11) in Section 2.2 showed that for lone rules the extrema of e are the same as the extrema of f. This holds because locally a one-rule fuzzy system gives a flat or constant output as in (2.189)–(2.191). The homework problems show that in this case the bumps in e match the bumps in f for any l^p metric. The global proof below shows why this still tends to hold in the mean-squared sense. Even the lone rule case of $F(x) = c$ involves the calculus of variations and does not tell us which bumps to cover with a rule patch. The result just says to patch a bump or extremum of f and hence of e. The problem of which bump to patch depends on knowledge of e and hence of f. We lack that knowledge in blind approximation.

The result is less helpful in the case of overlapping rules. It gives only an optimality condition that is the limiting case of supervised SAM learning laws when $c_j = F(\hat{x})$. The result also suggests that we should center the then-part sets A_j right above the extremizing input values so that just one rule fires dead-on. This matches the result in Section 2.9 that a fuzzy system minimizes its conditional variance when just one rule fires and thus when it does not interpolate. Still these results are too general to pick the shape of the if-part fuzzy sets A_j or to pick their number m. One picks these in practice by fiat or by some form of data clustering. The theory of optimal additive fuzzy systems is far from complete.

We first work with a minimal fuzzy system F of just one rule. Again we can view this lone rule as part of a multi-rule system when the rule's if-part set does not overlap with other if-part sets. The extremal result depends on the mean value theorem of calculus and Leibniz's rule of differentiating under an integral sign. The mean value theorem [27] states that if the continuous function $f: [u, v] \rightarrow R$ is differentiable on the interior (u, v) of its domain $[u, v]$, then there is at least one point $z \in (u, v)$ such that

$$f'(z) = \frac{f(v) - f(u)}{v - u} \tag{2.235}$$

for $u < v$ and where $f' = \frac{df}{dx}$. We use the special case known as Rolle's Theorem where $f(u) = f(v)$ and thus $f'(z) = 0$. We want to show in the scalar case that the z in (2.235) belongs to nonzero degree to the lone if-part set A. So we view u and v as the endpoints of the base of A. Figure 2.6 shows this proof strategy for the subinterval $[x - a, x + b]$ with non-negative constants a and b.

Leibniz's rule for differentiating an integral of the form

$$I(x) = \int_{a(x)}^{b(x)} f(w, x)dw \tag{2.236}$$

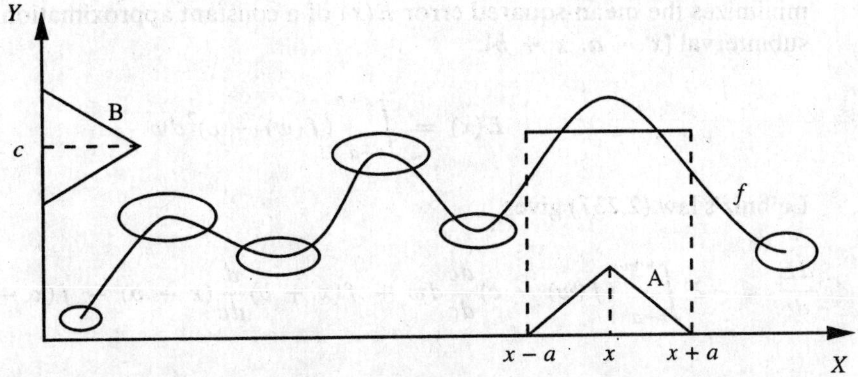

Figure 2.6 Lone optimal fuzzy rule patches cover the extrema of approximand f. A lone rule defines a flat line segment that cuts the graph of the local extremum in at least two places. The mean value theorem implies that the extremum lies between these points.

gives the derivative as

$$\frac{dI}{dx} = \int_{a(x)}^{b(x)} \frac{\partial f}{\partial x} dw + f(b(x), x)\frac{db}{dx} - f(a(x), x)\frac{da}{dx}. \tag{2.237}$$

This reduces to the more familiar rule of calculus for constant limits of integration b and a since then the last two terms equal zero.

Figure 2.6 shows that we want to slide the subinterval or base $[x - a, x + b]$ of the lone then-part set A to the best spot $\hat{x}$ in the domain interval $[u, v]$ of the smooth map f: $[u, v] \rightarrow R$. The fuzzy set A can have any shape as long as its set function obeys $a(z) > 0$ for all z in the subinterval $[x - a, x + b]$ and $a(z) = 0$ for all z outside the subinterval but in $[u, v]$. This analysis applies to all fuzzy systems with a lone rule and not just additive systems. They give the piecewise differentiable system

$$F(z) = \begin{cases} c & \text{if } z \in [x - a, x + b] \\ 0 & \text{if } z \notin [x - a, x + b] \end{cases} \tag{2.238}$$

for some constant c. This constant depends on the then-part set B in the lone rule "If $X = A$ then $Y = B$" or the lone rule patch $A \times B$. The first task is to pick the optimal set B for a fixed subinterval $[x - a, x + b]$. The harder task is to pick the optimal subinterval and hence to pick the optimal then-part set A on the entire domain $[u, v]$. We perform this analysis in the scalar case for simplicity but it holds as well in the vector case but with much more involved notation.

The centroid

$$c(x) = \frac{1}{a + b} \int_{x-a}^{x+b} f(w)dw \tag{2.239}$$

minimizes the mean-squared error $E(x)$ of a constant approximation of f over the subinterval $[x - a,\ x + b]$:

$$E(x) = \int_{x-a}^{x+b} (f(w) - c)^2 dw. \tag{2.240}$$

Leibniz's law (2.237) gives

$$\frac{dE}{dc} = -2\int_{x-a}^{x+\nu} (f(w) - c)\frac{dc}{dc}dw + f(x+b)\frac{d}{dc}(x+b) - f(x-a)\frac{d}{dc}(x-a) \tag{2.241}$$

$$= -2\int_{x-a}^{x+b} (f(w) - c)dw. \tag{2.242}$$

Then the c in (2.242) is the optimal value $\hat{c}$ iff $\frac{dE}{dc} = 0$ iff

$$\int_{x-a}^{x+b} f(w)dw = \hat{c}\int_{x-a}^{x+b} dw \tag{2.243}$$

$$= \hat{c}w|_{x-a}^{x+b} \tag{2.244}$$

$$= \hat{c}[(x+b) - (x-a)] \tag{2.245}$$

$$= \hat{c}(a+b) \tag{2.246}$$

or

$$\hat{c} = \frac{1}{a+b}\int_{x-a}^{x+b} f(w)dw \tag{2.247}$$

which confirms (2.239).

We can pick any then-part set B that has centroid $c(x)$ in (2.239). We can pick the width or volume V of B to make sure the rule patch $A \times B$ covers the extremum of the graph of f. Then the SAM Theorem and (2.189)–(2.191) show that $F_x(z) = c(x)$ for all z in the subinterval $[x - a,\ x + b]$ and $F_x(z) = 0$ for all z outside the subinterval. This defines a family of one-rule fuzzy systems $\{F_x\}$. The next task is to search the family $\{F_x\}$ for the mean-squared optimum system $F_{\hat{x}}$.

Consider first the mean value theorem applied to the centroid $c(x)$ in (2.239). Let G be the antiderivative of f so that $G'(x) = \frac{dG}{dx} = f(x)$. Then

$$c(x) = \frac{G(x+b) - G(x-a)}{(x+b) - (x-a)}. \tag{2.248}$$

So $c(x) = G'(z) = f(z)$ for some z in the subinterval $[x - a,\ x + b]$. This reminds us that the centroid cuts f over the subinterval as in Figure 2.6.

The centroid line segment $c(x)$ cuts f in just one place if $f' > 0$ or $f' < 0$ and thus if f is monotone on the subinterval $[x - a, x + b]$. Else the centroid line cuts f for two or more arguments d and e in the subinterval. So $f(d) = f(e)$ since the centroid line has zero slope. Then (2.235) reduces to Rolle's Theorem and gives $f'(z) = 0$ for some z in (d, e) and thus for some z in the open subinterval $(x - a, x + b)$. So f has an extremum in the subinterval and where $a(z) > 0$. Again we can always widen B and keep the same centroid of B to make the rule patch $A \times B$ cover this extremum graph point $(z, f(z))$. This use of the mean value theorem is the main idea in the following proof.

We will also need the centroid derivative dc/dx in the proof. It follows from Leibniz's rule (2.237):

$$\frac{d}{dx}c(x) = \frac{d}{dx}\left(\frac{1}{a+b}\int_{x-a}^{x+b} f(w)\, dw\right) \tag{2.249}$$

$$= \frac{1}{a+b}\int_{x-a}^{x+b} \frac{\partial f(w)}{\partial x} dw + \frac{f(x+b)}{a+b}\frac{d}{dx}(x+b) - \frac{f(x-a)}{a+b}\frac{d}{dx}(x-a) \tag{2.250}$$

$$= \frac{f(x+b) - f(x-a)}{a+b}. \tag{2.251}$$

We now search for the optimum system $F_{\hat{x}}$ by varying the fuzzy systems F_x in the total squared error E:

$$E(x) = \int_u^v (f(w) - F_x(w))^2 dw \tag{2.252}$$

$$= \int_u^{x-a} f^2(w)\, dw + \int_{x-a}^{x+b} (f(w) - c(w))^2 dw + \int_{x+b}^{v} f^2(w)\, dw. \tag{2.253}$$

To do this we need compute only the derivative $\frac{dE}{dx} = 0$ since each x picks out its own one-rule fuzzy system F_x We now use Leibniz's rule (2.237) to find the derivative of each integral in (2.253). This gives for the first integral

$$\frac{d}{dx}\int_u^{x-a} f^2(w)\, dw = \int_u^{x-a} \frac{\partial f^2(w)}{\partial x} dw + f^2(x-a)\frac{d}{dx}(x-a)$$

$$+ f^2(u)\frac{du}{dx} \tag{2.254}$$

$$= f^2(x-a). \tag{2.255}$$

The derivative of the third integral has the same form:

$$\frac{d}{dx}\int_{x+b}^{v} f^2(w)dw = \int_{x+b}^{v} \frac{\partial f^2(w)}{\partial x}dw + f^2(v)\frac{dv}{dx}$$

$$- f^2(x+b)\frac{d}{dx}(x+b) \tag{2.256}$$

$$= -f^2(x+b). \tag{2.257}$$

The second integral in (2.253) has a more complex derivative:

$$\frac{d}{dx}\int_{x-a}^{x+b} (f(w) - c(x))^2 dw$$

$$= -2\int_{x-a}^{x+b} (f(w) - c(w))\frac{dc}{dx}dw + (f(x+b) - c(x))^2\frac{d}{dx}(x+b)$$

$$- (f(x-a) - c(x))^2\frac{d}{dx}(x-a) \tag{2.258}$$

$$= (f(x+b) - c(x))^2 - (f(x-a) - c(x))^2. \tag{2.259}$$

This holds because (2.249)–(2.251) imply that the integral on the righthand side of (2.258) equals zero:

$$\int_{x-a}^{x+b} (f(w) - c(x))\frac{dc}{dx}dw$$

$$= \frac{dc}{dx}\int_{x-a}^{x+b} f(w)dw - c(x)\frac{dc}{dx}\int_{x-a}^{x+b} dw \tag{2.260}$$

$$= [f(x+b) - f(x-a)]\frac{\int_{x-a}^{x+b} f(w)dw}{a+b}$$

$$- c(x)\frac{[f(x+b) - f(x-a)]}{a+b}(a+b) \tag{2.261}$$

$$= [f(x+b) - f(x-a)][c(x) - c(x)] \tag{2.262}$$

$$= 0. \tag{2.263}$$

Combining these derivatives gives the derivative of the error $E(x)$ as

$$\frac{dE}{dx} = 0 \tag{2.264}$$

$$= [(f^2(\hat{x} - a) - f^2(\hat{x} + b)] + (f(\hat{x} + b) - c(\hat{x}))^2$$

$$- (f(\hat{x} - a) - c(\hat{x}))^2 \tag{2.265}$$

$$= -2c(\hat{x})[f(\hat{x} + b) - f(\hat{x} - a)]. \tag{2.266}$$

Then $c(\hat{x}) \neq 0$ implies that $f(\hat{x} + b) = f(\hat{x} - a)$. Then the mean value theorem gives a z in the open subinterval $(\hat{x} - a,\ \hat{x} + b)$ such that $f'(z) = 0$ as required. (The second derivative test for a minimum can fail at this general extremum since it has the form

$$\frac{d^2E(\hat{x})}{dx^2} = 2c(\hat{x})[f'(\hat{x} - a) - f'(\hat{x} + b)].)$$

Again we can pick B wide enough so that the fuzzy rule patch $A \times B$ contains the graph point $(z,\ f(z))$ to nonzero degree. Then the optimal lone rule patches the bump at $(z,\ f(z))$.

The special case $c(\hat{x}) = 0$ is more involved. It can hold for monotone functions and then $f(\hat{x} + b) \neq f(\hat{x} - a)$ since monotone functions have only endpoint extrema. But $c(\hat{x}) = 0$ and $f(\hat{x} + b) \neq f(\hat{x} - a)$ lead to the negative second derivative

$$\frac{d^2E(\hat{x})}{dx^2} = -\frac{2}{a + b}(f(\hat{x} + b) - f(\hat{x} - a))^2 < 0$$

and hence to a local error *maximum*. The second-derivative test fails in the case of $c(\hat{x}) = 0$ and $f(\hat{x} + b) = f(\hat{x} - a)$.

Now suppose the SAM contains m rules $(A_1,\ B_1), \ldots, (A_m B_m)$. The ith scalar if-part set A_i covers the subinterval $[z_i - d_i,\ z_i + e_i]$. Then-part set B_i has centroid c_i that equals the local centroid of f over $[z_i - d_i,\ z_i + e_i]$. The parameter vector $z = (z_1, \ldots, z_m)$ picks a SAM fuzzy system F_z. Technically this F_z is only piecewise differentiable. The best system $F_{\hat{z}}$ nulls the gradient of the error function E in the parameter space. Then we can conclude at most $F_{\hat{z}}(\hat{z}_i) = c_i$. Homework problem 2.27 shows that

$$\frac{\partial E}{\partial z_i} = -2\int_{z_i - e_i}^{z_i + e_i} [f(x) - F(w)]\frac{\partial F}{\partial z_i}dw. \tag{2.267}$$

We will work instead with the fully differentiable case for ease of analysis.

Suppose the m scalar set functions $a_j\colon R \to [0,\ 1]$ are differentiable and thus so is the scalar SAM F. This holds for many exponential set functions. Suppose each if-part set function a_j is a function of l parameters $a_j = a_j(m_j^1, \ldots, m_j^l)$ as in the previous section on supervised learning. We can think of the interval parameter z_j as one of the l set-function parameters m_j^k. We will work with just one parameter m_j for simplicity. In the learning case we used a scalar error function and we now replace it with the total or integral error function E:

$$E(m) = \int_{-\infty}^{\infty} (f(w) - F_m(w))^2 dw \tag{2.268}$$

which depends on the m-vector of parameters $m = (m_1, \ldots, m_m)$ in a double use of notation. Then the chain rule gives

$$\frac{\partial E}{\partial m_i} = \frac{\partial E}{\partial F}\frac{\partial F}{\partial m_i} = \frac{\partial E}{\partial F}\sum_{j=1}^{m}\frac{\partial F}{\partial a_j}\frac{\partial a_j}{\partial m_i} \tag{2.269}$$

$$= \frac{\partial E}{\partial F}\frac{\partial F}{\partial a_i}\frac{\partial a_i}{\partial m_i} \quad \text{since} \quad \frac{\partial a_j(m_j)}{\partial m_i} = 0 \quad \text{if} \quad i \neq j \tag{2.270}$$

$$= \left[-2\int_{-\infty}^{\infty}(f(w) - F(w))dw\right]\frac{\partial F}{\partial a_i}\frac{\partial a_i}{\partial m_i}. \tag{2.271}$$

This gives three extremal conditions and each can give $\frac{\partial E}{\partial m_i} = 0$. The optimal parameter vector $\hat{m}$ zeroes the error in the representation case when $f = F_{\hat{m}}$. The second condition $\frac{\partial a_i}{\partial m_i} = 0$ can imply that the ith if-part set function should "peak" or that it should flatten its tangent over the input value m_i or z_i. This happens in the Gaussian or Cauchy case if m_i stands for the mean of the bell curve. We can think of this happening for z_i in the case of a triangular if-part set function except of course the triangle has no derivative at its peak.

The third condition $\frac{\partial F_{\hat{m}}}{\partial a_i} = 0$ and the derivation (2.217)–(2.219) above imply the optimal result

$$F_{\hat{m}}(\hat{m}_i) = c_i \tag{2.272}$$

as claimed. This locally reduces the case of overlapping rules to the lone-rule case. It also suggests that we should take care in how much the if-part sets overlap. The optimal result (2.272) suggests that adjacent sets should have symmetric overlap as often occurs in practice. The SAM F should reduce to the ith rule when that rule fires dead-on.

2.14 FEEDBACK SAMS AND FUZZY COGNITIVE MAPS

Feedback systems take as input their own output. Most have complex equilibria that range from fixed points to chaotic attractors. This holds in part for linear systems but holds with full force for the vastly richer nonlinear systems that include feedback SAMs and other feedback fuzzy systems. Stability is the exception and not the rule. We can prove stability for few systems. We do so below for the generalized SAM system of Section 2.6 that defines the nonlinear SAM as a convex sum of linear maps. The new results extend the well-known stability results of lone linear systems. They differ in kind from the earlier qualitative results on fuzzy system stability [11, 53] based on abstract min-max systems.

There are many tradeoffs between linear and nonlinear systems and between model-based and model-free systems. Model-based linear and linearized systems may lead to tighter closed-form analyses but apply to fewer real problems. Nonlinear model-free systems apply to a wider range of real problems but their nonlinearity and model freedom often preclude formal analyses or guarantees of stability or robustness. The final word on them may well have to come from only trial and error or computer simulation.

This final section presents feedback SAMs and adaptive fuzzy cognitive maps and their hybrids. Lyapunov techniques yield sufficient conditions for stability for the simplest type of feedback SAMs. Stronger conditions may yield such results for more general SAM systems. Fuzzy cognitive maps extend the additive structure to dynamical systems where each rule feeds to all other rules. Their dynamics can range from those of feedback neural networks to more complex semantic networks of embedded SAMs or other systems. They also learn in non-neural ways to model causal changes in a sampled environment. These rich systems admit few formal properties and often we apply them as just predictive tools. Chapter 15 applies them to the swirling undersea world of sharks and dolphins and fish in a simple fuzzy virtual world.

The SAM Theorem above applied to feedforward systems. We now apply it to autonomous dynamical systems of the continuous-time form

$$\dot{x} = F(x) \tag{2.273}$$

and of the discrete form

$$x(k+1) = F(x(k)) \tag{2.274}$$

for a generalized SAM map or vector field $F : R^n \to R^n$ with the origin as a fixed point: $F(0) = 0$. These dynamical systems are free or unforced. We can add a control input column vector u and control matrix $C(k)$ of the form $C(k)u(k)$ to steer the SAM. The unforced case assumes $u = 0$ and lets us look at the stability of the equilibrium vector

$$\lim_{k \to \infty} x(k) = x_e$$

if it exists. In most cases we can take the equilibrium vector to be the origin without loss in generality: $x_e = 0$. F is the generalized additive system (2.68)–(2.77) with linear or matrix then-part maps B_j:

$$\dot{x} = \sum_{j=1}^{m} p_j(x) \; B_j \; x \tag{2.275}$$

or of discrete form

$$x(k+1) = \sum_{j=1}^{m} p_j(x(k)) \; B_j \; x(k). \tag{2.276}$$

The controlled versions of (2.275) and (2.276) can have the convex form

$$\dot{x} = \sum_{j=1}^{m} p_j(x) \left[B_j x + C_j x\right] \tag{2.277}$$

$$x(k+1) = \sum_{j=1}^{m} p_j(x(k)) \left[B_j x(k) + C_j x(k)\right] . \tag{2.278}$$

We will derive a stability result for the smooth feedback SAM (2.275) and then derive a more restrictive result for the discrete feedback SAM (2.276). We first review Lyapunov stability and apply it to simple linear systems.

The equilibrium point x_e is *stable* in the sense of Lyapunov if small changes in the initial conditions lead to only small changes in the state trajectory [21]: For all k_0 and all $\varepsilon > 0$ there is a $\delta > 0$ such that $\|x(k_0) - x_e\| < \delta$ implies $\|x(k) - x_e\| < \varepsilon$ for all $k \geq k_0$. The equilibrium point is *asymptotically stable* if it is stable and if it attracts the state trajectory: For all k_0 there is a $\delta' > 0$ such that $\|x(k_0) - x_e\| < \delta'$ implies

$$\lim_{k\to\infty} \|x(k) - x_e\| = 0.$$

The equilibrium is globally asymptotically stable if we can pick δ' to be arbitrarily large. When these results hold they hold uniformly for the autonomous systems (2.273) and (2.274).

We seek a smooth *Lyapunov function* L: $R^n \to R^n$ for the continuous feedback SAM model (2.275) that is positive definite $L(x) > 0$ when $x \neq 0$ and $L(x) = 0$ when $x = 0$ and that grows to infinity as the squared vector norm $\|x\|^2$ grows to infinity: $L(x) \to \infty$ as $x^T x \to \infty$. This holds if we take L as the quadratic form $x^T x$ or as the more general quadratic form $x^T P x$ for some n-by-n positive definite matrix P. Then standard results in Lyapunov stability theory imply that the dynamical system (2.273) or (2.275) has a stable equilibrium $x_e = 0$ if $\dot{L} \leq 0$ and a globally asymptotically stable equilibrium $x_e = 0$ if $\dot{L} < 0$ along system trajectories or for all $x \neq 0$. A discrete Lyapunov function $L(x(k))$ leads to stability for the discrete dynamical system (2.274) or (2.276) if $\Delta L \leq 0$ and to global asymptotic stability if $\Delta L < 0$ along system trajectories.

Consider first the one-rule case $m = 1$. This gives the standard continuous-time linear system

$$\dot{x} = Bx. \tag{2.279}$$

Take $L(x) = x^T P x$ as a candidate Lyapunov function for an arbitrary positive definite matrix P. The n-by-n identity matrix I is one such positive definite matrix. Then taking the total time derivative of L gives

$$\dot{L} = \sum_{i=1}^{n} \frac{\partial L}{\partial x_i}\frac{dx_i}{dt} = \nabla L \cdot \dot{x} = 2x^T P\dot{x} \tag{2.280}$$

$$= x^T P\dot{x} + \dot{x}^T Px \tag{2.281}$$

$$= x^T PBx + x^T B^T Px \tag{2.282}$$

$$= x^T \left[PB + B^T P\right] x \tag{2.283}$$

$$< 0 \quad \text{for all } x \neq 0 \tag{2.284}$$

if and only if $PB + B^T P$ is a negative definite matrix. This is a standard result for the stability of a continuous-time linear system.

Suppose B is negative definite: $x^T Bx < 0$ for all $x \neq 0$. Suppose $P = I$. Then $B + B^T$ is negative definite since then $B + B^T = 2B$. Definite matrices with real coefficients are symmetric and hence have real eigenvalues and admit the orthonormal diagonalization $B = O^T \Lambda_B O$. The diagonal matrix Λ_B lists the n negative eigenvalues $\lambda_i(B)$ along its diagonal. A positive definite matrix P has $P = O^T \Lambda_P O$ with n positive eigenvalues along the diagonal of its diagonal matrix. The product PB is negative definite since $PB = O^T \Lambda_P \Lambda_B O$ has negative eigenvalues $\rho_i \lambda_i < 0$. So the matrix $PB + B^T P$ is negative definite if B is negative definite and if P is positive definite. We can simplify the proof if $P = I$ in the Lyapunov function. Putting $\dot{x} = Bx$ in $\dot{L} = 2x^T \dot{x}$ leads to $\dot{L} = 2x^T Bx < 0$ if and only if B is negative definite.

In general the continuous-time system (2.279) is asymptotically stable if and only if all the eigenvalues λ_i of B have negative real parts and thus lie in the left half-plane of the complex s-plane. This holds for negative definite matrices since their eigenvalues are negative real numbers. The result also holds locally for nonlinear dynamical systems of the form $\dot{x} = F(x)$ if the Jacobian matrix of first partial derivatives of F has eigenvalues with negative real parts near an equilibrium point [21]. Locally the nonlinear system behaves linearly and converges exponentially quickly to the attractor or sink point x_e.

The one-rule case $m = 1$ gives the standard discrete-time linear system

$$x(k+1) = Bx(k) \tag{2.285}$$

for an n-by-n matrix B. Again take $L(x) = x^T Px$ as a candidate Lyapunov function for an arbitrary positive definite matrix P. Then (2.285) gives

$$\Delta L(k) = L(k+1) - L(k) \tag{2.286}$$

$$= x^T(k+1)\ P\ x(k+1) - x^T(k)\ P\ x(k) \tag{2.287}$$

$$= x^T(k)\ B^T PB\ x(k) - x^T(k)\ P\ x(k) \tag{2.288}$$

$$= x^T(k)\left[B^T PB - P\right] x(k) \tag{2.289}$$

$$< 0 \quad \text{for all } x \neq 0 \tag{2.290}$$

iff the matrix $B^T PB - P$ is negative definite.

This is a more restrictive constraint for global asymptotic stability than in the continuous-time case when any negative definite B gives stability. Suppose $P = I$ and $B = -cI$ for any constant c such that $0 < c < 1$. Then $B^T PB - P = c^2 I - I = (c^2 - 1)\, I$ is negative definite since the matrix has all real negative eigenvalues $c^2 - 1$. But the negative definite matrices $B = -cI$ do not lead to a negative definite $B^T PB - P$ if $c \geq 1$ and $P = I$. Note that in general $P = I$ implies $\Delta L = x^T B^T Bx - x^T x$ or $\Delta L = \|x\|^2 \left[\dfrac{x^T B^T Bx}{\|x\|^2} - 1\right]$ since $x^T x = \|x\|^2$. So $\Delta L < 0$ if $x^T B^T Bx < \|x\|^2$.

The discrete system $x(k+1) = B\, x(k)$ is asymptotically stable when all the eigenvalues of B lie in the unit circle in the complex z-plane:

$$\lim_{k\to\infty} B^k = \emptyset \quad \text{iff} \quad |\lambda_i(B)| < 1.$$

This is easy to see if B has n distinct eigenvalues $\lambda_1, \ldots, \lambda_n$. Then it has n linearly independent eigenvectors and so admits an orthonormal diagonalization $B = O^T \Lambda_B O$. Then

$$\lim_{k\to\infty} B^k = O^T \lim_{k\to\infty} \Lambda_B^k O \quad \text{and} \quad \lim_{k\to\infty} \lambda_i^k = 0 \quad \text{iff} \quad |\lambda_i| < 1.$$

The discrete model gives $x(k+1) = B^k x(0)$ for some finite initial state vector $x(0)$ that we take as non-null. So

$$\lim_{k\to\infty} x(k) = 0$$

iff all of B's eigenvalues lie inside the unit circle. They do not need to lie in the negative half-plane of the complex plane.

The bilinear transforms

$$s = 2\frac{1 - z^{-1}}{1 + z^{-1}} \tag{2.291}$$

$$z = \frac{1 + \frac{s}{2}}{1 - \frac{s}{2}} = \frac{1 + \frac{x}{2} + i\frac{y}{2}}{1 - \frac{x}{2} - i\frac{y}{2}} \tag{2.292}$$

map the left half-plane of the complex s-plane onto the unit disc of the complex z-plane and vice versa for imaginary unit $i = \sqrt{-1}$ and rectangular complex numbers $s = x + iy$ and polar complex numbers $z = re^{i\omega}$. These two transforms connect the engineering field of control theory and the continuous Laplace transform in the s-plane to the field of signal processing and the discrete z-transform in the z-plane. They also help explain the stability results for the linear systems (2.279) and (2.285).

The bilinear transform maps the imaginary axis iy in the s-plane onto the unit circle $\{z \in C: |z| = 1\}$ and vice versa when $x = 0$. It maps the left half-plane of the s-plane (when all $x < 0$) onto the interior of the unit circle in the z-plane and vice versa. It maps the right half-plane of the s-plane (when all $x > 0$) onto the exterior of the unit circle in the z-plane and vice versa. So stability in terms of eigenvalues on the left half-plane of the s-plane corresponds to stability in terms of eigenvalues inside the unit circle in the z-plane. These eigenvalues contract the system trajectories while those on the right half-plane or outside the unit circle expand them.

Qualitative stability results can differ for the continuous-time and discrete dynamical systems as we will see for the SAM stability theorems below. Consider the matrix $B = \frac{1}{2}I$. Then the continuous-time linear system $\dot{x} = \frac{1}{2}x$ is unstable since the eigenvalues $\lambda_i = \frac{1}{2}$ lie on the right half-plane of the s-plane. Each component obeys $x_i(t) = x_i(0)e^{\frac{t}{2}} \rightarrow \infty$. But the discrete linear system

$$x(k + 1) = \frac{1}{2}x(k) = \frac{1}{2^k}x(0) \rightarrow 0$$

is stable since the eigenvalues $\lambda_i = 1/2$ lie inside the unit circle in the z-plane. The reverse holds for the matrix $B = -2I$. Then the linear system $\dot{x} = -2x$ is stable since the eigenvalues $\lambda_i = -2$ lie in the left half-plane of the s-plane. The origin attracts all trajectories exponentially quickly since $x_i(t) = x_i(0)e^{-2t} \rightarrow 0$. The discrete linear system $x(k + 1) = -2x(k) = (-2)^k x(0)$ is unstable since the eigenvalues $\lambda_i = -2$ lie outside the unit circle in the z-plane. The state $x(k)$ bounces between ever larger positive and negative integer multiples of the finite initial non-null state $x(0)$.

We now apply the above Lyapunov techniques to the continuous-time feedback SAM (2.275) and the discrete-time feedback SAM (2.276). The continuous-time case gives a much stronger result. The SAM is globally asymptotically stable if all the local rule matrixes B_j are negative definite. So the nonlinear convex sum F is stable if its local linear rule systems are stable. This does not hold in the discrete case. The global convex sum F need not be stable if each rule matrix B_j is stable in the z-plane. It requires a common positive definite matrix P that makes each matrix $B_j^T P B_j^T - P$ negative definite. Each new rule lowers the odds that we will ever find such a common stabilizing matrix P.

Theorem 2.5 (*Continuous SAM Stability*). The generalized feedback SAM system

$$\dot{x} = \sum_{j=1}^{m} p_j(x)\ B_j\ x \tag{2.293}$$

with convex coefficients

$$p_j(x) = \frac{w_j(x)\ a_j(x)}{\sum_{i=1}^{m} w_i(x)\ a_i(x)} \tag{2.294}$$

is globally asymptotically stable if each then-part rule matrix B_j is negative definite.

Proof. Choose the Lyapunov function L as the quadratic form $L(t) = x^T(t)x(t)$. Then

$$\dot{L} = 2x^T \dot{x} \tag{2.295}$$

$$= 2x^T \sum_{j=1}^{m} p_j(x) B_j x \tag{2.296}$$

$$= 2 \sum_{j=1}^{m} p_j(x) x^T B_j x. \tag{2.297}$$

At all times some convex coefficient obeys $p_k(x(t)) > 0$. So $\dot{L} < 0$ if each then-part matrix B_j is negative definite. ■

The proof directly extends the stability proof of the continuous-time linear system. The feedback SAM F is nonlinear since the convex coefficients $p_j(x)$ change with each input x. The fact that at least one term obeys $p_k(x(t)) > 0$ lets us treat the convex sum of matrices as if it were a simple sum with constant coefficients. This does not hold in the discrete case below. We can also use the quadratic form $L(x) = x^T P x$ as in the proof (2.280)–(2.284) above to derive a SAM version of (2.283):

$$\dot{L} = \sum_{j=1}^{m} p_j(x) x^T \left[P B_j + B_j^T P \right] x \tag{2.298}$$

We now extend the proof (2.286)–(2.290) to prove a sufficient condition for stability of the discrete-time feedback SAM. The result extends the like result of Tanaka [50, 51] to the more general SAM framework.

Theorem 2.6 (*Discrete SAM Stability*). The generalized discrete-time feedback SAM system

$$x(k+1) = \sum_{j=1}^{m} p_j(x(k))\ B_j\ x(k) \tag{2.299}$$

with convex coefficients

$$p_j(x(k)) = \frac{w_j(x(k))\ a_j(x(k))}{\sum_{i=1}^{m} w_i(x(k))\ a_i(x(k))} \tag{2.300}$$

is globally asymptotically stable if there exists a common positive definite matrix P such that all m of the then-part matrices $B_j^T P B_j - P$ are negative definite.

Proof. Choose the Lyapunov function L as the quadratic form $L(t) = x^T(t)Px(t)$. Then

$$\Delta L(k) = L(k+1) - L(k) \tag{2.301}$$

$$= x^T(k+1)Px(k+1) - x^T(k)Px(k) \tag{2.302}$$

$$= \left[\sum_{i=1}^{m} p_i(x(k))x^T(k)B_i^T\right] P \left[\sum_{j=1}^{m} p_j(x(k))B_j x(k)\right] - x^T(k)Px(k) \tag{2.303}$$

$$= x^T \left[\left(\sum_{i=1}^{m} p_i(x)B_i^T\right) P \left(\sum_{j=1}^{m} p_j(x)B_j\right)\right] x - x^T P x \tag{2.304}$$

$$= x^T \left[\sum_{i=1}^{m} p_i(x)B_i^T P \sum_{j=1}^{m} p_j(x)B_j - P\right] x \tag{2.305}$$

$$= x^T \left[\sum_{i=1}^{m} p_i(x)B_i^T P \sum_{j=1}^{m} p_j(x)B_j - \sum_{i=1}^{m}\sum_{j=1}^{m} p_i(x)p_j(x)P\right] x \tag{2.306}$$

since $\sum_{i=1}^{m}\sum_{j=1}^{m} p_i(x)p_j(x) = \sum_{i=1}^{m} p_i(x) \sum_{j=1}^{m} p_j(x) = 1$

$$= x^T \left[\sum_{i=1}^{m} \sum_{j}^{m} p_i(x) p_j(x) \left(B_i^T P B_j - P \right) \right] x \tag{2.307}$$

$$= \sum_{j=1}^{m} p_j^2(x)\, x^T \left[B_j^T P B_j - P \right] x + \sum_{i \neq j}^{m} \sum^{m} p_i(x)\, p_j(x)\, x^T \left[B_i^T P B_j - P \right] x. \tag{2.308}$$

The first sum in (2.308) is negative since we assume all the matrices $B_j^T P B_j - P$ are negative definite. We now show that this implies that the mixed matrices $B_i^T P B_j - P$ are negative definite as well. Then $\Delta L(k) < 0$ holds along system trajectories. We expand the summand in the second sum in (2.308) for $i \neq j$:

$$B_i^T P B_j - P + B_j^T P B_i - P = \left(B_i^T P B_j + B_j^T P B_i \right) - 2P \tag{2.309}$$

$$= \left(- \left[B_i - B_j \right]^T P \left[B_i - B_j \right] + B_i^T P B_i + B_j^T P B_j \right) - 2P \tag{2.310}$$

$$= - \left[B_i - B_j \right]^T P \left[B_i - B_j \right] + \left(B_i^T P B_i - P \right) + \left(B_j^T P B_j - P \right). \tag{2.311}$$

The last two terms in (2.311) are negative definite by assumption. The result follows if we can show that the first term in (2.311) is nonpositive definite since sums of nonpositive definite matrices are negative definite if at least one of the matrices is negative definite. This holds since $A^T P A$ is non-negative definite if P is positive definite and so $-A^T P A$ is nonpositive definite. So $\Delta L(k) < 0$ holds if there is some positive definite P such that all matrices $B_j^T P B_j - P$ are negative definite. ■

The discrete feedback SAM need not be stable even if all the then-part rule matrices are stable and hence have eigenvalues in the unit circle of the z-plane. Tanaka [50, 51] gives the two-rule example with then-part "set" matrices

$$B_1 = \begin{pmatrix} 1 & -\frac{1}{2} \\ 1 & 0 \end{pmatrix} \quad \text{and} \quad B_2 = \begin{pmatrix} -1 & -\frac{1}{2} \\ 1 & 0 \end{pmatrix}.$$

Homework problem 2.28 shows that B_1 has the two eigenvalues

$$\lambda_i = \frac{1}{2} \pm i\frac{1}{2}$$

and B_2 has the two eigenvalues

$$\lambda_i = -\frac{1}{2} \pm i\frac{1}{2}.$$

All four eigenvalues lie in the unit circle and so both matrices or linear subsystems are stable. Tanaka shows that the convex sum (2.299) is unstable for trapezoidal if-part sets A_1 and A_2 and constant or unity rule weights $w_1 = w_2 > 0$. He also shows that the discrete SAM is robust against small input changes. This holds for all the feedback SAM models discussed here with convex coefficients.

The discrete SAM stability theorem gives no guidance for finding the common positive definite matrix P. Each new rule can decrease the chance of finding such a P. Exponential rule explosion compounds the search complexity. We next present a corollary that shows that the choice $P = I$ gives global asymptotic stability for stable *diagonal* matrices B_j with real coefficients. The result holds for any positive definite matrix cI with constant $c > 0$.

Corollary (*Discrete Diagonal SAM Stability*). The generalized discrete-time feedback SAM system

$$x(k+1) = \sum_{j=1}^{m} p_j(x(k))\, B_j\, x(k) \tag{2.312}$$

is globally asymptotically stable if all m then-part matrices B_j are diagonal and stable.

Proof. The diagonal matrix $B_j = \text{Diag}[b_j(1, 1), \ldots, b_j(n, n)]$ is symmetric and lists its eigenvalues along its main diagonal. So the choice $P = I$ in the discrete SAM stability theorem gives

$$B_j^T P B_j - P = B_j^2 - I. \tag{2.313}$$

The diagonal matrix $B_j^2 - I$ is negative definite iff all its eigenvalues are negative. This holds iff for each diagonal entry $b_j^2(i, i) - 1 < 0$ iff $b_j^2(i, i) < 1$ iff $|b_j(i, i)| < 1$. The last condition is just that of stability for matrix B_j. ■

Fuzzy cognitive maps or FCMs are dynamical systems that relate fuzzy sets and rules [16, 19, 21]. They are more complex than feedback SAMs and do not admit a simple Lyapunov analysis. An FCM has the topology of a directed fuzzy graph with cycles or feedback. It is a set of nodes and edges. The concept nodes C_i are fuzzy sets or even fuzzy systems. The edges e_{ij} define rules or causal flows $C_i \rightarrow C_j$ between the concept nodes. At time t the state of the FCM is the concept vector $C = (C_1, \ldots, C_n) \in [0, 1]^n$ or point in the fuzzy-cube state space. The n concepts or nodes all belong to the event C to some degree at time t. The n-by-n matrix E lists the n^2 rules or pathways in the causal web.

The simplest FCMs act as asymmetrical networks of threshold or continuous neurons and converge to limit cycles. At this level they differ from neural nets in how they learn as we discuss in the following section and in Chapter 15. More complex FCMs differ in the nonlinear structure of their concept nodes and hence in their global feedback dynamics. Researchers have applied FCMs to model a wide range of medical [47, 48] and social [3, 8, 60] problems and have put forth many schemes to grow and combine FCMs [41–43]. Here we review just their dynamical structure.

An FCM F carves the fuzzy-cube state space I^n into k many attractor regions. Some regions may contain attracting fixed points or limit cycles while other regions contain attracting chaotic or aperiodic equilibria. Any change to an FCM node or edge can change this carving of the state space. FCM inference unfolds as the FCM dynamical system moves the initial state vector $C(0)$ to an equilibrium attractor in the fuzzy cube. So the FCM digs a path or trajectory in the fuzzy cube. The path starts at the input initial condition and ends where the FCM "cools down" or equilibrates to a dynamic equilibrium. The initial state might model a drug injection or a social policy. The equilibrium can yield a "hidden pattern" in the tangle of causal rules or edges.

The goal of FCM analysis is to find a fixed FCM F of n nodes that can approximate an arbitrary dynamical system $\dot{x} = f(x)$ of m nodes where $n \geq m$ and thus where the FCM may contain extra or "hidden" nodes. No one has yet found such an approximation result. Most FCM research deals with decision aids and qualitative prediction and not control. Chapter 5 shows how a feedforward SAM may need tens of thousands of rules to approximate a chaotic attractor of only low dimension. FCMs may offer a way to approximate many such complex attractors with a *fixed* number of rules. The feedback involved may help model a real problem of hormonal or battlefield feedback but it largely prevents a closed-form analysis.

FCM dynamics depend on the dynamics of the concept nodes and causal edges. The edges e_{ij} are constant weights in the simplest case and only the nodes change in time. We first discuss neural-type nodes where the ith node C_i uses a bounded nonlinear signal function S_i to map a real causal activation value $x_i(t)$ to a number in [0, 1] or in [−1, 1]. The threshold signal function defines the simplest concept node:

$$S_i(x_i) = \begin{cases} 0 & \text{if } x_i \leq T_i \\ 1 & \text{if } x_i > T_i \end{cases} \tag{2.314}$$

for some real threshold value T_i. Such bivalent FCMs are the most common in practice and in commercial FCM software. Most systems use $T_i = 0$ and update the entire concept vector $C(t)$ in synchrony rather than just update some subset of its concept nodes in asynchrony.

Then the 2^n distinct states or events lie at the 2^n corners of the fuzzy cube I^n. The FCM inference path $C(0), C(1), C(2), \ldots$ hops from binary vertex to binary vertex in the fuzzy cube and ends in a limit cycle or a fixed point. The limit cycle starts with the first repeated state vector.

The causal activation $x_i(t)$ can have a SAM-like or neural additive activation update equation of the form

$$x_i(t_{k+1}) = \sum_{i=1}^{n} e_{ji}(t_k)\, C_j(t_k) + I_i(t_k) \tag{2.315}$$

where $e_{ji}(t_k)$ states the rule strength at time t_k of the causal link $C_j \rightarrow C_i$ and

$$C_j(t_k) = C_j(x_j(t_k)) = S_j\left(\sum_{k=1}^{n} e_{kj}(t_{k-1})\, C_k(t_{k-1}) + I_j(t_{k-1})\right).$$

The external input I_i is often zero. Continuous FCMs can map these activation values to a bounded range with the logistic signal function

$$S_i(x_i) = \frac{1}{1 + e^{-cx_i}} \tag{2.316}$$

for $c > 0$ or with some other signal function such as a Gaussian or Cauchy bell curve. The S-shaped logistic signal function gives a truly fuzzy concept node for small c and gives in effect a binary threshold function for large c. The logistic signal function also has the activation derivative $S_i' = \frac{dS_i}{dx_i} = cS_i(1 - S_i)$ that can simplify learning algorithms.

The n^2 causal edge weights e_{ij} can take on values in the fuzzy bipolar interval $[-1, 1]$. Node C_i causally increases C_j if $e_{ij} > 0$ and causally decreases C_j if $e_{ij} < 0$ and has no causal effect on C_j if $e_{ij} = 0$. Simple or trivalent FCMs have causal edge values in the set $\{-1, 0, 1\}$ and give rise to 3^{n^2} possible causal webs. Concepts are not neurons that can stimulate themselves. So most FCM models assume $e_{ii} = 0$ to capture the idea that nothing causes itself. Other models retain the self-feedback to give the FCM more approximation power. The square edge matrix E leads to the vector version of the FCM nonlinear update equation:

$$C(t_{k+1}) = S(E(t_k)\, C(t_k) + I(t_k)) \tag{2.317}$$

where $S(x) = (S_1(x_1), \ldots, S_n(x_n))$.

FCM combination acts as the simplest form of learning. Here q experts draw causal pictures or FCMs $F_1, \ldots, F_q$. Each expert has a real credibility weight w_k as in the case of combined fuzzy systems above. The experts need not draw the same concept nodes. Some of the node sets will be disjoint but most will tend to overlap. Each expert does draw all the nodes in the sense that he draws some with no causal or rule links. So we can zero-pad the q edge matrices to make them all of size n-by-n where n is the total number of distinct node concepts. We add zero rows and zero columns for all missing concepts. Then we permute the rows and columns of each matrix to make the q matrices $E_1, \ldots, E_q$ conformable for addition. Then we make the combined FCM system an additive system and weight and sum to get the final combined edge matrix E:

$$E = \sum_{k=1}^{q} w_k E_k. \tag{2.318}$$

This additive scheme works for any number of FCMs. It allows the knowedge engineer to quickly build a joint FCM from documents or questionnaires or classroom assignments. The weights w_k might come from professional rankings or test results. The lone FCM edge matrices are trivalent in most cases even though the combined matrix E is multivalent. Knowledge sources tend to give trivalent edge values more reliably than they give numerical magnitudes. Summing these trivalent matrices can lead to a final matrix E that reflects the joint statistics or "consensus" of the expert pool.

Most causal learning schemes are unsupervised since no one knows the "real" causal structure of the world. The additive combiner (2.318) is an unsupervised scheme. It does not depend on error feedback from a teacher or supervisor. Homework problem 2.31 shows how much computational effort it takes to make even a small local change to an FCM with supervised learning. The formal supervised SAM learning schemes in Section 2.12 require such error feedback and hold only for feedforward systems.

We still lack tools to force a limit cycle or chaotic attractor into FCM dynamics. The best that unsupervised schemes can do is to *suggest* such equilibria. Chapter 15 shows how to add a bipolar limit cycle to the sum in (2.318). The bipolar correlation matrix $X_k X_{k+1}^T$ forms the associative link from state $C(k)$ to state $C(k+1)$ where X_k replaces the 0s in binary state vector $C(k)$ with -1s: $X_k^T = (2C_1(k) - 1, \ldots, 2C_n(k) - 1)$. This is a crude form of Hebbian or correlation learning [21]. It need not drive the FCM to follow the desired sequence of binary vectors. It also gives a poor causal model since it says A causes B if both events occur at the same time. Such synaptic learning laws quickly lead to webs of spurious causal rules.

The *differential* Hebbian learning (DHL) law is an unsupervised causal learning law [19, 21]. The DHL law correlates the changes or signal velocities $\dot{C}_i \dot{C}_j$ and not the signals $C_i C_j$ themselves. It correlates time derivatives or differences to measure what empiricist philosopher John Stuart Mill called "concomitant variation" in his 1843 classic text *A System of Logic* on causal inference: "Whatever phenomenon varies in any manner whenever another phenomenon varies in some particular manner, is either a cause or an effect of that phenomenon, or is connected with it through some fact of causation." The term $\dot{C}_i \dot{C}_j$ models "concomitant" as product and models "variation" as time change.

Here FCM dynamics differ most clearly from the neural or synaptic dynamics of modern neural networks. A neural Hebbian law would correlate the concept signals C_i and C_j in a simple first-order model:

$$\dot{e}_{ij} = -e_{ij} + C_i C_j. \tag{2.319}$$

Then the causal rule $C_i \rightarrow C_j$ tends only to grow in time and to encode an exponentially weighted average of the paired events:

$$e_{ij}(t) = e_{ij}(0)\, e^{-t} + \int_0^t C_i(s)\, C_j(s)\, e^{s-t}\, ds. \tag{2.320}$$

The DHL law

$$\dot{e}_{ij} = -e_{ij} + \dot{C}_i \dot{C}_j \tag{2.321}$$

does not only grow in time. The edge or rule strength

$$e_{ij}(t) = e_{ij}(0)\, e^{-t} + \int_0^t \dot{C}_i(s)\, \dot{C}_j(s)\, e^{s-t}\, ds \tag{2.322}$$

can be negative or positive or zero since the integrand can be so even if the nodes C_i and C_j do not take on negative values.

The DHL law grows a positive edge between two nodes if they both increase or both decrease since then $\dot{C}_i \dot{C}_j > 0$. It grows a negative edge if they move in opposite directions since then $\dot{C}_i \dot{C}_j < 0$. We tend not to suspect a causal link between a person's arm and the light in a room. But we do suspect a link if the light turns on and off as the arm moves up and down in a like pattern of change. Time series data can grow or tune the FCM matrix E with some discrete form of the DHL law (2.321). Chapter 15 uses a discrete version that changes only when a concept changes:

$$e_{ij}(t+1) = \begin{cases} e_{ij}(t) + \mu_t[\Delta C_i(t)\Delta C_j(t) - e_{ij}(t)] & \text{if } \Delta C_i(t) \neq 0 \\ e_{ij}(t) & \text{if } \Delta C_i(t) = 0 \end{cases} \tag{2.323}$$

where $\Delta C_i(t) = C_i(t) - C_i(t-1)$. The learning coefficient μ_t decreases slowly in time to help forget old causal-edge strengths in favor of new ones:

$$\mu_t = 0.1\left[1 - \frac{t}{1.1N}\right].$$

Neural FCMs extend to SAM FCMs when each node C_i acts as a SAM map $F_i\colon R^n \to R$. The SAM map uses m_i rules to approximate the causal nonlinearity of the ith node. The simplest SAM FCM uses the scalar map $F_i\colon R \to R$. Then

$$C_i(t+1) = F_i(x_i(t)) = \sum_{k=1}^{m_i} p_k(x_i(t))c_k. \tag{2.324}$$

The SAM rules map the causal activation x_i in (2.315) to the node signal. Then the signal passes through the n edges that lead out from it. The scalar SAM node needs few rules to model most causal throughputs and admits the 2-rule representation above for known bounded functions f.

Vector SAM FCMs can model small but complex semantic networks. Node C_i acts as a SAM map $F_i\colon R^n \to R$ and replaces the scalar activation x_i with the path-weighted vector of inputs $(C_1 e_{1i}, \ldots, C_n e_{ni})$. Each node must deal with its own k^n rule complexity. These hybrid feedback systems can model parts of many systems in fine detail but they may do so at the cost of high computation and inscrutable dynamics.

REFERENCES

[1] Black, M,, "Vagueness: An Exercise in Logical Analysis," *Philosophy of Science*, vol. 4, 427–455, 1937.

[2] Buckley, J. J., and Hayashi, Y., "Fuzzy I/O Controllers as Universal Approximators," *Proceedings of the 1993 World Congress on Neural Networks (INNS WCNN–93)*, vol. 2, 92–96, July 1993.

[3] Craiger, P., "Causal Structure, Model Inferences, and Fuzzy Cognitive Maps: Help for the Behavioral Scientist," *Proceedings of the 1994 World Congress on Neural Networks (INNS WCNN–94)*, vol. I, 836–841, June 1994.

[4] Cybenko, G., "Approximation by Superpositions of a Sigmoidal Function," *Mathematics of Control, Signals, and Systems*, vol. 2, 303–314, 1989.

[5] Dickerson, J. A., and Kosko, B., "Fuzzy Function Learning with Covariance Ellipsoids," *Proceedings of the IEEE International Conference on Neural Networks (IEEE ICNN–93)*, 1162–1167, March 1993.

[6] Dickerson, J. A., and Kosko, B., "Fuzzy Function Approximation with Supervised Ellipsoidal Learning," *Proceedings of the World Congress on Neural Networks (INNS WCNN–93)*, vol. 2, 9–17, July 1993.

[7] Dickerson, J. A., and Kosko, B., "Fuzzy Function Approximation with Ellipsoidal Rules," *IEEE Transactions on Systems, Man, and Cybernetics*, vol. 26, no. 4, 542–560, August, 1996.

[8] Dickerson, J. A., and Kosko, B., "Virtual Worlds as Fuzzy Cognitive Maps," *Presence*, vol. 3, no. 2, 173–189, Spring 1994.

[9] Dubois, D., and Prade, H., *Fuzzy Sets and Systems: Theory and Applications*, Academic Press, 1980.

[10] Dubois, D., Grabisch, M., and Prade, H., "Synthesis of Real-Valued Mappings Based on Gradual Rules and Interpolative Reasoning," *Proceedings of the 13th International Joint Conference on Artificial Intelligence (IJCAI–93) Workshop on Fuzzy Logic in AI*, 29–40, September 1993.

[11] Gupta, M. M., Kiszka, J. B., and Nikiforuk, P. N., "Energetistic Stability of Fuzzy Dynamic Systems," *IEEE Transactions on Systems, Man, and Cybernetics*, vol. 15, no. 6, 783–792, November 1985.

[12] Hartman, E., Keeler, J. D., and Kowalski, J., "Layered Neural Networks with Gaussian Hidden Units as Universal Approximators," *Neural Computation*, vol. 2, 210–215, 1990.

[13] Hornik, K., Stinchcombe, M., and White, H., "Multilayer Feedforward Networks Are Universal Approximators," *Neural Networks,* vol. 2, 35–366, 1989.

[14] Kandel, A., *Fuzzy Mathematical Techniques with Applications*, Addison-Wesley, 1986.

[15] Klir, G. J., and Folger, T. A., *Fuzzy Sets, Uncertainty, and Information*, Prentice Hall, 1988.

[16] Kosko, B., "Fuzzy Cognitive Maps," *International Journal of Man-Machine Studies*, vol. 24, 65–75, January 1986.

[17] Kosko, B., "Fuzzy Knowledge Combination," *International Journal of Intelligent Systems*, vol. I, no. 4, 293–320, 1986.

[18] Kosko, B., *Foundations of Fuzzy Estimation Theory*, Ph.D. dissertation, Department of Electrical Engineering, University of California at Irvine, June 1987; Order Number 8801936, University Microfilms International, 300 N. Zeeb Road, Ann Arbor, MI 48106.

[19] Kosko, B., "Hidden Patterns in Combined and Adaptive Knowledge Networks," *International Journal of Approximate Reasoning*, vol. 2, no. 4, 377–393, 1988.

[20] Kosko, B., "Stochastic Competitive Learning," *IEEE Transactions on Neural Networks*, vol. 2, no. 5, 522–529, September 1991.

[21] Kosko, B., *Neural Networks and Fuzzy Systems: A Dynamical Systems Approach to Machine Intelligence*, Prentice Hall, 1991.

[22] Kosko, B., "Fuzzy Systems as Universal Approximators," *IEEE Transactions on Computers*, vol. 43, no. 11, 1329–1333, November 1994; an earlier version appears in the *Proceedings of the First IEEE International Conference on Fuzzy Systems (IEEE FUZZ–92)*, 1153–1162, March 1992.

[23] Kosko, B., and Isaka, S., "Fuzzy Logic," *Scientific American*, vol. 269, no. 1, 76–81, July 1993.

[24] Kosko, B., "The Probability Monopoly," *IEEE Transactions on Fuzzy Systems*, vol. 2, no. 1, 32–33, February 1994.

[25] Kosko, B., "Optimal Fuzzy Rules Cover Extrema," *International Journal of Intelligent Systems*, vol. 10, no. 2, 249–255, February 1995; an earlier version appears in the *Proceedings of the 1994 World Congress on Neural Networks (INNS WCNN–94)*.

[26] Kosko, B., "Combining Fuzzy Systems," *Proceedings of the IEEE FUZZ–95*, vol. IV, 1855–1863, March 1995.

[27] Kreysig, E., *Advanced Engineering Mathematics*, 6th ed., John Wiley & Sons, 1988.

[28] Kruse, R., "The Strong Law of Large Numbers for Fuzzy Random Variables," *Information Sciences*, 233–241, 1982.

[29] Laviolette, M., and Seaman, J. W., "The Efficacy of Fuzzy Representations of Uncertainty," *IEEE Transactions on Fuzzy Systems*, vol. 2, no. 1, 4–15, February 1994.

[30] Mamdani, E. H., and Assilian, S., "An Experiment in Linguistic Synthesis with a Fuzzy Logic Controller," *International Journal of Man-Machine Studies*, vol. 7, 1–13, 1977.

[31] Mamdani, E. H., "Application of Fuzzy Logic to Approximate Reasoning Using Linguistic Synthesis," *IEEE Transactions on Computers*, vol. C–26, no. 12, 1182-1191, December 1977.

[32] Mead, C., *Analog VLSI and Neural Systems*, Addison-Wesley, 1989.

[33] Moody, J., and Darken, C., "Fast Learning in Networks of Locally Tuned Processing Units," *Neural Computation*, vol. 1, 281–294, 1989.

[34] Nguyen, H. T., "On Random Sets and Belief Functions," *Journal of Mathematical Analysis and Applications*, vol. 65, 531–542, 1978.

[35] Pacini, P. J., and Kosko, B., "Adaptive Fuzzy System for Target Tracking," *Intelligent Systems Engineering*, vol. 1, no. 1, 3–21, Fall 1992.

[36] Papoulis, A., *Probability, Random Variables, and Stochastic Processes*, 2nd ed., McGraw-Hill, 1984.

[37] Puri, M. L., and Ralescu, D., "Fuzzy Random Variables," *Journal of Mathematical Analysis and Application*, vol. 114, 409–422, 1986.

[38] Rice, J. R., *The Approximation of Functions*, Addison-Wesley, 1964.

[39] Rudin, W., *Functional Analysis*, McGraw-Hill, 1973.

[40] Rudin, W., *Real and Complex Analysis*, 2nd ed., McGraw-Hill, 1974.

[41] Satur, R., Liu, Z., and Gahegan, M., "Multi-Layered FCMs Applied to Context Dependent Learning," *Proceedings of the IEEE FUZZ–95*, vol. II, 561–568, March 1995.

[42] Schneider, M., Shnaider, E., Kandel, A., and Chew, G., "Constructing Fuzzy Cognitive Maps," *Proceedings of the IEEE FUZZ–95*, vol. IV, 2281–2288, March 1995.

[43] Silva, P.C., "Fuzzy Cognitive Maps Over Possible Worlds," *Proceedings of the IEEE FUZZ–95*, vol. II, 555–560, March 1995.

[44] Specht, D. F., "A General Regression Neural Network," *IEEE Transactions on Neural Networks*, vol. 4, no. 4, 549–557, 1991.

[45] Sugeno, M., "An Introductory Survey of Fuzzy Control," *Information Sciences*, vol. 36, 59–83, 1985.

[46] Sugeno, M., and Kang, G. T., "Structure Identification of Fuzzy Model," *Fuzzy Sets and Systems*, vol. 28, 15–33, 1988.

[47] Taber, W. R., "Knowledge Processing with Fuzzy Cognitive Maps," *Expert Systems with Applications*, vol. 2, no. 1, 82–87, February 1991.

[48] Taber, W. R., and Siegel, M., "Estimation of Expert Weights Using Fuzzy Cognitive Maps," *Proceedings of the IEEE 1987 International Conference on Neural Networks (IEEE ICNN–87)*, vol. II, 319–326, June 1987.

[49] Takagi, T., and Sugeno, M., "Fuzzy Identification of Systems and Its Applications to Modeling and Control," *IEEE Transactions on Systems, Man, and Cybernetics*, vol. 15, 116–132, 1985.

[50] Tanaka, K., and Sugeno, M., "Stability Analysis and Design of Fuzzy Control Systems," *Fuzzy Sets and Systems*, vol. 45, no. 2, 135–156, January 24, 1992.

[51] Tanaka, K., and Sano, M., "A Robust Stabilization Problem of Fuzzy Control Systems and Its Application to Backing up Control of a Truck-Trailer," *IEEE Transactions on Fuzzy Systems*, vol. 2, no. 2, 119–134, May 1994.

[52] Terano, T., Asai, K., and Sugeno, M., *Fuzzy Systems Theory and Its Applications*, Academic Press, 1992.

[53] Tong, R. M, "Some Properties of Fuzzy Feedback Systems," *IEEE Transactions on Systems, Man, and Cybernetics*, vol. 10, no. 6, 327–331, June 1980.

[54] Wang, L., and Mendel, J. M., "Fuzzy Basis Functions, Universal Approximation, and Orthogonal Least-Squares Learning," *IEEE Transactions on Neural Networks,* vol. 3., no. 5, 807–814, September 1992.

[55] Watkins, F. A., "Fuzzy Engineering," Ph.D. dissertation, Department of Electrical Engineering, University of California at Irvine, 1994; University Microfilms International, 300 North Zeeb Road, Ann Arbor, MI 48106.

[56] Watkins, F. A., "The Representation Problem for Additive Fuzzy Systems," *Proceedings of the IEEE FUZZ–95*, vol. I, 117–122, March 1995.

[57] Yager, R. R., and Filev, D. P., "On the Issue of Defuzzification and Selection Based on a Fuzzy Set," *Fuzzy Sets and Systems*, vol. 55, 255–273, 1993.

[58] Zadeh, L. A., "Fuzzy Sets," *Information and Control,* vol. 8, 338–353, 1965.

[59] Zadeh, L. A., *Fuzzy Sets and Applications: Selected Papers by L.A. Zadeh*, R. R. Yager, S. Ovchinnikov, R. M. Tong, and H. T. Nguyen, editors, John Wiley & Sons, 1987.

[60] Zhang, W. R., Chen, S. S., and Bezdek, J. C., "Pool 2: A Generic System for Cognitive Map Development and Decision Analysis," *IEEE Transactions on Systems, Man, and Cybernetics*, vol. 19, no. 1, 31–39, January 1989.

[61] Zimmermann, H. J., *Fuzzy Set Theory and its Application*, Kluwer, 1985.

HOMEWORK PROBLEMS

2.1. Use one fuzzy rule in a SAM $F: R \longrightarrow R$ to best approximate the scalar function $f(x) = a^2 - x^2$ on the compact set $[-a, a]$ for some constant $a > 0$. The rule patch $A \times B$ is a fuzzy subset of the plane R^2. Suppose if-part set A has set function $a(x) > 0$ for all x in$(-a, a)$. Which then-part fuzzy sets B minimize the mean-squared error of the function approximation? Sketch the results. Does the rule-encoding method affect the choice of B? Does it affect the choice of A?

2.2. Find a SAM F with 2 rules that *represents* $f(x) = a^2 - x^2$ on $[-a, a]$ for some constant $a > 0$: $F(x) = f(x)$ for all x in $[-a, a]$.

2.3. Sketch the polynomial $f(x) = \frac{x^4}{4} + \frac{x^3}{3} - 3x^2 + 1$ for all x in $[-4, 4]$. Find the first set of fuzzy rules that "patch the bumps." Use triangular or bell-shaped sets. Then find a SAM F with 2 rules that represents f.

2.4. The general SAM $F: R^n \rightarrow R^p$ gives the vector output $F(x)$ as a convex sum of vector centroids c_j:

$$F(x) = \text{Centroid}(B) = \sum_{j=1}^{m} p_j(x)c_j.$$

Suppose combined output set $B \subset R^p$ is a hyperbox with interval sides $[u_1, v_1]$, ..., $[u_p, v_p]$. Suppose hyperbox B has vector centroid $c \in R^p$. Prove that c and thus $F(x)$ has the midpoint form

$$c = \left(\frac{u_1 + v_1}{2}, \ldots, \frac{u_p + v_p}{2}\right) \in R^p.$$

2.5. Consider a nonadditive fuzzy system that combines the two then-part sets $B_1 \subset R$ and $B_2 \subset R$ into the output set B with pairwise maximum: $b(y) = \max(b_1(y), b_2(y))$. Hence $B = B_1 \cup B_2$. Then the fuzzy system output $F(x)$ equals the centroid of B or $c(B)$. Let $c(B_1)$ and $c(B_2)$ be the centroids of B_1 and B_2. Then show by counterexample that the inequality

$$c(B_1) \leq c(B) \leq c(B_2)$$

need not hold. How does this affect the fuzzy approximation theorem?

2.6. Suppose an additive fuzzy system sums the two then-part sets $B_1 \subset R$ and $B_2 \subset R$: $B = B_1 + B_2$. But the system is not centroidal. Instead it defuzzifies B by picking that value $\hat{y}$ that has the largest set value: $F(x) = \hat{y}$ iff $b(\hat{y}) = \sup\{b(y) : y \in R\}$. Then show by counterexample that such mode or supremum defuzzification need not satisfy

$$c(B_1) \leq \hat{y} \leq c(B_2).$$

2.7. Suppose a 1-rule fuzzy system $F: R \rightarrow R$ approximates the scalar function f in some local region with the constant map $F(x) = c$. Suppose $f > c$ for simplicity. Define the local l^p error function e as

$$e(x) = \frac{1}{p}(f - c)^p$$

for all $p > 0$. Show that optimal lone rules "patch the bumps" of f in the local sense that the extrema of e lie at the extrema of f.

2.8. Repeat problem 2.7 for the summed absolute and squared error

$$e(x) = |f - c| + \frac{1}{2}(f - c)^2.$$

The squared error can exaggerate the effect of outliers. The absolute error is more robust and gives less weight to outliers. The summed error adds this property to the squared error.

2.9. Finite fuzzy sets A_j and B_j are points in unit hypercubes of high dimension: $A_j \in [0, 1]^u$ and $B_j \in [0, 1]^v$. A rule matrix M_j associates the *fit* (fuzzy unit) vectors $A_j = \left(a_1^j, \ldots, a_u^j\right)$ and $B_j = \left(b_1^j, \ldots, b_v^j\right)$ in some correlation (or *Hebbian*) way. Correlation *product* encoding stores the association as the outer-product matrix $M_j = A_j^T B_j : m_{ik}^j = a_i^j b_k^j$. Correlation *minimum* encoding stores the association (A_j, B_j) as the outer-product matrix $M_j = A_j^T o B_j$: $m_{ik}^j = \min\left(a_i^j, b_k^j\right)$. In neural terms these schemes define the rule matrices M_j as a fuzzy associative memory or FAM matrix that acts as a between-cube mapping M_j: $[0, 1]^u \rightarrow [0, 1]^v$. Then similar inputs $A \approx A_j$ map to similar outputs $B \approx B_j$ for the max-min composition $B = A \; o \; M_j$: $b_k = \max\left(\min\left(a_1, m_{1k}^j\right), \ldots, \min\left(a_u, m_{uk}^j\right)\right)$. Encode $A_1 = (1, .6, .2)$ and $B_1 = (0, .2, .7, 1)$, $A_2 = (.4, 1, .3)$ and $B_2 = (.1, .5, 1, .5)$, and $A_3 = (.1, .5, 1,)$ and $B_3 = (1, .8, .3, 0)$ first with correlation-production encoding and then with correlation-minimum encoding. Then pass input set A through each FAM matrix to derive the output set B where A is the fit vector (1, 0, 0), (0, 1, 0), (0, 0, 1), (1, 1, 1), (0, 0, 0), (1, .5, .3), (.3, .8, .1), or (0, .5, .5). A FAM matrix M_j displays *perfect recall* if $A_j \, oM_j = B_j$. Do the six FAM matrices display perfect recall? State and prove a necessary and sufficient condition for perfect recall for both correlation-product and correlation-minimum encoding.

2.10. An additive fuzzy system F: $[0, 1]^u \rightarrow Y$ for discrete domain space $X = (x_1, \ldots, x_u)$ and range space $Y = (y_1, \ldots, y_v)$ is a mapping that gives $F(A) \in [y_1, \ldots, y_p]$ as the discrete centroid of the summed fired then-part sets $B_j' = A \; oM_j$:

$$F(A) = \frac{\sum_{k=1}^{v} y_k b_k}{\sum_{k=1}^{v} b_k}$$

for summed output "set"

$$B = \sum_{j=1}^{m} B_j' \in [0, m]^v .$$

Construct an additive system that stores the 3 associations (A_j, B_j) in problem 2.9. Then compute $F(A)$ for each of the input sets A in problem 2.9. Use $X = \{1, 2, 3\}$ and $Y = \{1, 2, 3, 4\}$ and round off to the nearest integer.

2.11. A BIOFAM or binary-input/binary-output FAM system F maps singleton input sets $\{x_i\} \subset X$ to singleton output sets $\{y_k\} \subset Y$ for discrete domain space $X = (x_1, \ldots, x_u)$

and discrete range space $Y = (y_1, \ldots, y_v)$. The input x_i is in effect a unit binary vector $I_x^i = (0, 0, \ldots, 0, 1, 0, \ldots, 0)$ with a 1 in the ith slot and 0s elsewhere and thus the ith row of the u-by-u identity matrix. Then

$$F(x_i) = Centroid(B(x_i)) = Centroid\left(\sum_{j=1}^{m} a_j(x_i)B_j\right)$$

in the SAM case for the discrete sets or fit vectors $A_j \subset X$ and $B_j \subset Y$. Fit vector $A_j = (a_j(x_1), \ldots, a_j(x_u)) \in [0, 1]^u$ makes up the if-part of one rule and fit vector $B_j = (b_j(y_1), \ldots, b_j(y_v)) \in [0, 1]^v$ makes up the then-part of that rule. Then prove the discrete SAM Theorem:

$$F(x_i) = \frac{\sum_{j=1}^{m} a_j(x_i)\ c(B_j)\ c_j}{\sum_{j=1}^{m} a_j(x_i)\ c(B_j)}$$

for the set cardinality measure or *count* $c(B_j)$.

$$c(B_j) = \sum_{k=1}^{v} b_j(y_k)$$

and then-part set centroid c_j

$$c_j = \frac{\sum_{k=1}^{v} y_k\ b_j(y_k)}{\sum_{k=1}^{v} b_j(y_k)}.$$

The convex nature of the SAM output means that we may need to round off the fuzzy system output $F(x_i)$ to the nearest output element $y_k \in Y$.

2.12. A discrete scalar SAM $F: R \to R$ or $F: X \to Y$ stores m if-then rules with m if-part sets $A_j = \left(a_j^1, \ldots, a_j^u\right) \in [0, 1]^u$ and m then-part sets $B_j = \left(b_j^1, \ldots, b_j^v\right) \in [0, 1]^v$. Here $A_j \subset X = \{x_1, \ldots, x_u\}$ and $B_j \subset Y = \{y_1, \ldots, y_v\}$. We can write the discrete scalar version of the SAM Theorem in the form

$$F(x_i) = \frac{\sum_{j=1}^{m} a_j^i\ c(B_j)\ c_j}{\sum_{j=1}^{m} a_j^i\ c(B_j)} = \sum_{j=1}^{m} p_j(x_i)\ c_j$$

for count $c(B_j) = b_j^1 + \cdots + b_j^v$.

Construct a SAM from the three rule pairs (A_1, B_1), (A_2, B_2), (A_3, B_3) from problem 2.9. Then $X = \{x_1, x_2, x_3\}$ and $Y = \{y_1, y_2, y_3, y_4\}$. SAM input x_1 corresponds to the

FAM set input (1, 0, 0), x_2 to (0, 1, 0), and x_3 to (0, 0, 1) as they pass through the FAM rule matrices M_1, M_2, and M_3. Use $X = \{1, 2, 3\}$ and $Y = \{1, 2, 3, 4\}$. Let F map to the interval [1, 4] and round off as needed to keep the output in Y.

2.13. Prove the discrete *set* SAM Theorem:

$$F(A) = \frac{\sum_{j=1}^{m} a_j(A)\ c(B_j)\ c_j}{\sum_{j=1}^{m} a_j(A)\ c(B_j)}$$

for the set-function correlation measure $a_j(A) = c(A \cap A_j) = a^1 \wedge a_j^1 + \ldots + a^v \wedge a_j^v$. The discrete set SAM reduces to the discrete point SAM for a unit bit vector input: $c(I_X^i \cap A_j) = a_j(x_i)$. Construct a set SAM from the three rule pairs (A_1, B_1), (A_2, B_2), (A_3, B_3) from problem 2.9. Take as the set inputs A the fit vectors (1, .5, .3), (.3, .8, 1), and (0, .5, .3). Use $X = \{1, 2, 3\}$ and $Y = \{1, 2, 3, 4\}$. Let F map to the interval [1, 4] and round off as needed to keep the output in Y.

2.14. The scalar exponential fuzzy set function a_j: $R^n \rightarrow R$ has the form

$$a_j(x) = e^{\sum_{i=1}^{n} u_j^i (v_j^i - x_i)}$$

Derive the supervised learning laws for an exponential SAM:

$$u_j^k(t+1) = u_j^k(t) + \mu_t\ \varepsilon_t\ p_j(x)\ \left[c_j - F(x)\right]\ \left(v_j^k - x_k\right)$$

$$v_j^k(t+1) = v_j^k(t) + \mu_t\ \varepsilon_t\ p_j(x)\ \left[c_j - F(x)\right]\ u_j^k$$

for vector inputs $x = (x_1, \ldots, x_n) \in R^n$.

2.15. The Gaussian bell-curve set function a_j: $R^n \rightarrow R^+$ has the form

$$a_j(x) = e^{-\frac{1}{2}\sum_{i=1}^{n}\left(\frac{x_i - m_j^i}{\sigma_j^i}\right)^2}$$

Derive the supervised learning laws for a Gaussian SAM:

$$m_j^k(t+1) = m_j^k(t) + \mu_t\ \varepsilon_t\ p_j(x)\ \left[c_j - F(x)\right]\ \frac{x_k - m_j^k}{\left(\sigma_j^k\right)^2}$$

$$\sigma_j^k(t+1) = \sigma_j^k(t) + \mu_t\ \varepsilon_t\ p_j(x)\ \left[c_j - F(x)\right]\ \frac{\left(x_k - m_j^k\right)^2}{\left(\sigma_j^k\right)^3}.$$

2.16. Define the scalar *Cauchy* bell-curve set function a_j: $R \rightarrow R$ as

$$a_j(x) = \frac{1}{1 + \left(\frac{x - m_j}{d_j}\right)^2}$$

The location parameter m_j acts as the "mean" of the Cauchy bell curve. Cauchy random variables have infinite variance and do not obey laws of large numbers (though they do obey a form of the central limit theorem). The dispersion d_j acts as a variance in that it controls the width of the bell. The Cauchy bell curve with $d_j = 10$ has a wide bell over the range $[-5, 5]$ for zero location parameter. The Cauchy bell curve with $d_j = 0.01$ has almost a spiked bell over 0. Derive the supervised learning laws for a Cauchy SAM:

$$m_j(t+1) = m_j(t) + 2\,\mu_t\,\varepsilon_t\,p_j(x)\left[c_j - F(x)\right]\left(\frac{x - m_j}{d_j^2}\right) a_j(x)$$

$$d_j(t+1) = d_j(t) + 2\,\mu_t\,\varepsilon_t\,p_j(x)\left[c_j - F(x)\right]\left(\frac{x - m_j}{d_j}\right)^2 \frac{a_j(x)}{d_j}.$$

2.17. Derive the two supervised learning laws for a SAM with clipped-parabola set functions:

$$a_j(x) = \begin{cases} 1 - \left[b_j(x - m_j)\right]^2 & \textit{if} \;\; |x - m_j| < \frac{1}{b_j} \\ 0 & \textit{else.} \end{cases}$$

This involves deriving the two partial derivatives

$$\frac{\partial a_j}{\partial b_j} = -2b_j(x - m_j)^2$$

$$\frac{\partial a_j}{\partial m_j} = 2b_j^2(x - m_j)$$

for $|x - m_j| < 1/b_j$ and zero for $|x - m_j| \geq 1/b_j$.

2.18. The sinc function

$$a_j(x) = \frac{\sin\left(\frac{x - m_j}{d_j}\right)}{\frac{x - m_j}{d_j}}$$

appears in signal processing as a linear interpolator of a continuous time signal $x(t)$ that has a finite frequency band:

$$x(t) = \sum_{n=-\infty}^{\infty} x[n] \frac{\sin[\pi(t - nT)/T]}{\pi(t - nT)/T}$$

for countably many equally spaced discrete samples $x[0]$, $x[1]$, $x[-1]$, $x[2]$, $x[-2]$, . . . and for sampling period T. Sinc functions can act as fuzzy set functions even though they have a bipolar range: a_j: $R \rightarrow [-0.217, 1]$. Sinc functions often give adaptive SAMs that converge quickly and accurately to a sampled approximand. Derive the supervised learning laws to tune parameters d_j and m_j for a SAM with sinc if-part set functions. This entails deriving the partial derivatives

$$\frac{\partial a_j}{\partial m_j} = \begin{cases} \left(a_j(x) - \cos\left(\frac{x - m_j}{d_j}\right)\right) \frac{1}{x - m_j} & \text{for } x \neq m_j \\ 0 & \text{for } x = m_j \end{cases}$$

$$\frac{\partial a_j}{\partial d_j} = \left(a_j(x) - \cos\left(\frac{x - m_j}{d_j}\right)\right) \frac{1}{d_j}.$$

2.19. Prove the Cauchy-Schwartz inequality for correlation set functions:

$$\left(a_j(A)\right)^2 \leq a(A)\ a_j(A_j).$$

Restate this result in terms of covariances and variances of scalar random variables.

2.20. Prove that the additive combination of q SAM systems F_k: $R^n \rightarrow R^p$ has the SAM form

$$F(x) = \frac{\sum_{k=1}^{q} \sum_{j=1}^{m_k} w_k\ a_j^k(x)\ V_j^k\ c_j^k}{\sum_{k=1}^{q} \sum_{j=1}^{m_k} w_k\ a_j^k(x)\ V_j^k}.$$

2.21. Prove that

$$F(x) = F_{q+1}(x) + \sum_{k=1}^{q} p_k(x)\left[F_k(x) - F_{q+1}(x)\right]$$

for $q + 1$ additively combined centroidal fuzzy systems F_k: $R^n \rightarrow R^p$ and with $q + 1$ convex coefficients

$$p_k(x) = \frac{w_k\ V^k(x)}{\sum_{j=1}^{q+1} w_j\ V^j(x)}.$$

2.22. Laws of large numbers are theorems that say that the sample mean

$$S_n = \frac{1}{n} \sum_{i=1}^{n} X_i$$

converges to the population mean

$$\mu = E_X(X_i) = \int_{-\infty}^{\infty} x\ p_X(x)\ dx$$

if the infinite sequence of scalar random variables $X_1, X_2, X_3, \ldots$ consists of random variables that are all mutually independent and have the same or identical distribution. So the independent random variables have the same finite mean and the same finite variance $\sigma^2 = V_X(X) = E_X\big((X - \mu)^2\big) < \infty$. Prove the *mean-squared* law of large numbers:

$$\lim_{n\to\infty} E\big[(S_n - \mu)^2\big] = 0.$$

Then prove the *weak* law of large numbers:

$$\text{For all } \varepsilon > 0\text{:}\ \lim_{n\to\infty} \Pr\,(|S_n - \mu| \geq \varepsilon) = 0.$$

Now prove that the mean-squared law of large numbers *implies* the weak law of large numbers. Give an example of independent identically distributed random variables that need not satisfy either law.

2.23. Define the scalar conditional random variables

$$E[Y|X = x] = \int_{-\infty}^{\infty} y\ p(y|x)\ dy$$

$$V[Y|X] = E\Big[\big(Y - E\,[Y|X]\big)^2\ |X\Big]$$

$$Cov[X,\ Y|Z] = E\big[(X - E\,[X|Z])\ (Y - E[Y|Z])\ |Z\big]\ .$$

Then prove the scalar equalities

(a) $E[Y|X] = E_Y(Y)$ if X and Y are independent $[p(x,\ y) = p_X(x)\ p_Y(y)]$

(b) $E_Y(f(y)) = E_X(E[f(Y)|X])$

(c) $E_{XY}[aX + bY|Z] = aE[X|Z] + bE[Y|Z]$ for constants a and b

(d) $V_Y(Y) = E_X\big(V[Y|X]\big) + V_X(E[Y|X])$

(e) $V[Y|X] = E\big[Y^2|X\big] - \big(E[Y|X]\big)^2$

(f) $Cov[X,\ Y|Z] = E[XY|Z] - E[X|Z]E[Y|Z]$

(g) $E[f(X)g(Y)|X] = f(X)E[g(Y)|X]$.

2.24. Prove Bayes theorem for conditional probability density functions of vector random variables X and Y:

$$p(y|x) = \frac{p(x|y)\ p_Y(y)}{\displaystyle\int_{R^p} p(x|y)\ p_Y(y)\ dy}.$$

2.25. Prove that the choice of the scalar conditional mean $\phi(x) = E[Y|X = x]$ minimizes the mean-squared error of nonlinear function approximation:

$$E\big[(Y - E[Y|X])^2\big] \leq E\big[(Y - \phi)^2\big]\ .$$

2.26. Use Leibniz's rule for differentiating under an integral sign to compute the derivative $\frac{dI(x)}{dx}$ of the integral

$$I(x) = \int_x^{x^3} \frac{\sin xw}{w} dw.$$

Show that

$$\frac{dI(x)}{dx} = \frac{4 \sin x^4 - 2 \sin x^2}{x}.$$

2.27. Use Leibniz's rule to derive the gradient condition

$$\frac{\partial E}{\partial z_i} = -2 \int_{z_i - d_i}^{z_i + e_i} [f(x) - F(w)] \frac{\partial F}{\partial z_i} dw$$

where

$$E(z) = \int_u^v (f(w) - F_z(w))^2 dw$$

for parameter vector $z = (z_1, \ldots, z_m)$ and subintervals $[z_i - d_i, z_i + e_i]$.

2.28. Find the eigenvalues of the two then-part rule matrices

$$B_1 = \begin{pmatrix} 1 & -\frac{1}{2} \\ 1 & 0 \end{pmatrix} \text{ and } B_2 = \begin{pmatrix} -1 & -\frac{1}{2} \\ 1 & 0 \end{pmatrix}.$$

2.29. The discrete feedback SAM with two rules has the form

$$x(k) = \sum_{j=1}^{2} p_j(x(k)) \, B_j \, x(k).$$

Suppose the two then-part matrices are

$$B_1 = \begin{pmatrix} \frac{1}{2} & 0 \\ 0 & -\frac{3}{4} \end{pmatrix} \text{ and } B_2 = \begin{pmatrix} 0 & \frac{1}{2} \\ 0 & \frac{1}{2} \end{pmatrix}.$$

Are the matrices stable? Are they positive definite? Does there exist a common positive definite matrix P that makes the SAM system globally asymptotically stable?

2.30. Confirm that $S' = \frac{dS}{dx} = cS\,(1 - S)$ for the logistic signal function

$$S(x) = \frac{1}{1 + e^{-cx}} \quad \text{for } c > 0.$$

2.31. Suppose we want the fuzzy cognitive map (FCM) state $C(t - p)$ to lead to the desired state vector D at time t for some small time difference $p > 0$. The actual output state vector $C(t)$ will differ from D and leads to the squared error

$$E = \frac{1}{2} \sum_{k=1}^{n} (D_k - C_k(t))^2$$

and where we assume

$$C_k(t) = S_k(x_k(t)) = \frac{1}{1 + e^{-x_k(t)}}.$$

Derive the supervised gradient-descent learning law

$$e_{ij}(t+1) = e_{ij}(t) + \mu_t \, \varepsilon_j \, S_j(t) \left[1 - S_j(t)\right] C_i(t).$$

Note that no learning occurs for binary state vectors. How do the squared-error term and the learning law change if we also want $C(t + q) = G$ for some small $q > 0$?

2.32. Show that the unsupervised Hebbian learning law

$$\dot{e}_{ij} = -e_{ij} + C_i C_j$$

for bounded signals C_i and C_j has the solution

$$e_{ij}(t) = e_{ij}(0)e^{-t} + \int_0^t C_i(s) \, C_j(s) \, e^{s-t} \, ds.$$

Suppose $C_i(t) = C_j(t) = 1$ for all t. What value does the edge e_{ij} encode?

2.33. An absolute value sign complicates supervised learning in a Laplace SAM with the Laplace set function

$$a_j(x) = \exp\left\{-\left|\frac{x - m_j}{d_j}\right|\right\}.$$

Derive the supervised learning laws

$$m_j(t+1) = m_j(t) + \mu_t \, \varepsilon_t \, p_j(x)[c_j - F(x)]\frac{\text{sign}(x - m_j)}{|d_j|}$$

$$d_j(t+1) = d_j(t) + \mu_t \, \varepsilon_t \, p_j(x)[c_j - F(x)]\text{sign}(d_j)\frac{|x - m_j|}{d_j^2}.$$

A striking property of a fuzzy rule is its simple geometry as a patch or fuzzy subset in a system's state space. Users often focus on the shape of the scalar fuzzy sets that make up a rule and ignore the shape of the rule itself. But the set shapes dictate the structure and shape of the rule. This rule structure dictates how the rule helps the overall fuzzy system map inputs to outputs.

Nonfuzzy rules have the simplest shape. They define hyper-boxes in the input-output state space. The hyper-boxes are among the few rule shapes that factor into scalar sets. Simple joint Gaussian rule patches also factor into scalar Gaussian bell curves along all the input and output axes. Most rule patches do not factor into scalar fuzzy sets. So we cannot in most cases factor a vector fuzzy system that maps into a p-dimensional vector space into p many scalar-valued fuzzy systems.

An ellipsoidal rule has the shape of an egg and does not factor in general. The simple egg shape has in turn the simple algebra of a quadratic form and a positive definite matrix. The ellipsoidal rule can also emerge as the uncertainty or covariance "ball" that surrounds a statistical estimator.

Chapter 3 explores additive fuzzy systems with ellipsoidal rules. Unsupervised learning forms and tunes the first set of ellipsoidal rules through competitive learning of a statistical estimator. Then supervised learning tunes the eigenvalue structure of the ellipsoid's positive definite matrix. A histogram scheme projects the ellipsoid onto axes as a type of shadow. The shadow forms the scalar fuzzy sets. Chapter 6 omits this step and works with unfactored ellipsoidal rules. Other schemes can exploit the many computational advantages of the ellipsoidal shape.

3

ELLIPSOIDAL FUZZY SYSTEMS

Julie A. Dickerson[1] and Bart Kosko

PREVIEW[2]

A fuzzy rule can have the shape of an ellipsoid in the input-output state space of a system. Then an additive fuzzy system approximates a function by covering its graph with ellipsoidal rule patches. It averages rule patches that overlap. The best fuzzy rules cover the extrema or bumps in the function. Neural or statistical clustering systems can approximate the unknown fuzzy rules from training data. Then neural systems can both tune these rules and add rules to improve the function approximation.

We use a hybrid neural system that combines unsupervised and supervised learning to find and tune the rules in the form of ellipsoids. Unsupervised competitive learning finds the first-order and second-order statistics of clusters in the training data. The covariance matrix of each cluster gives an ellipsoid centered at the vector or centroid of the data cluster. The supervised neural system learns with gradient

[1] Electrical and Computer Engineering Department, Iowa State University, Ames, IA, 50011.

[2] The Caltrans PATH Program supported this research (Agreement 20695MB).

descent. It locally minimizes the mean-squared error of the fuzzy function approximation. In the hybrid system unsupervised learning initializes the gradient descent. The hybrid system tends to give a more accurate function approximation than does the lone unsupervised or supervised system.

We found a closed-form model for the optimal rules when only the centroids of the ellipsoids change. We used numerical techniques to find the optimal rules in the general case.

3.1 LEARNING WITH ELLIPSOID RULES

A fuzzy system is a set of fuzzy rules that maps inputs to outputs. So it defines a function f: X $\rightarrow$ Y. The rules are if-then rules of the form If "X is A then Y is B." A and B are multivalued or fuzzy sets that contain members to some degree. A is a subset of input space X. B is a subset of output space Y.

Fuzzy rules have a simple geometry in the input-output state space X $\times$ Y. They define fuzzy patches or subsets of the state space. Less certain rules are large patches. More precise rules are small patches. The rule "If X is A then Y is B" defines the fuzzy patch or Cartesian product $A \times B$. In the precise limit the rule patch $A \times B$ is a point if A and B are binary spikes.

An additive fuzzy system adds the output or then-part of sets B' fuzzy rules in Figure 3.3. It covers the graph of the function and averages patches that overlap. Chapter 2 showed how these systems can approximate a continuous function to any degree of accuracy with a finite number of fuzzy rules. The rules and their sets can have any shape. In practice the sets have simple shapes like triangles, trapezoids, or bell curves.

We propose rules that have the shape of ellipsoids. Ellipsoids arise from the covariance matrices of statistical or neural estimators. And they lead to simple algorithms to tune the rules with training data.

A fuzzy system learns or adapts when its rules change. Then the system function f: X $\rightarrow$ Y varies with time. A rule changes when its fuzzy sets change. Neural networks can change the sets with training data or error reinforcement signals.

We used two types of neural nets to change the ellipsoidal rules in an additive fuzzy system. The first neural system used unsupervised learning to find the covariance matrices of synaptic quantization vectors. A quantization vector "hops" more in regions of sparse or noisy data. Then it has a larger covariance ellipsoid and thus gives a less certain rule. The synaptic vector hops less in regions of dense or less noisy data and gives a more precise rule. The second neural system used supervised learning to tune the rules with stochastic gradient descent. Supervised learning needs an error signal. The neural net must know the function it tries to approximate or at least know some input-output samples from it. Supervised learning changes the centers and eigenvalues of the ellipsoids. This moves, shapes, and orients the ellipsoids in the state space.

The supervised learning takes far more computation than does the unsupervised learning in this algorithm. But it often gives rules that better approximate the function. At best the supervised gradient descent finds the local minima of the error surface. How well it learns depends on how well we pick the initial set of fuzzy rules.

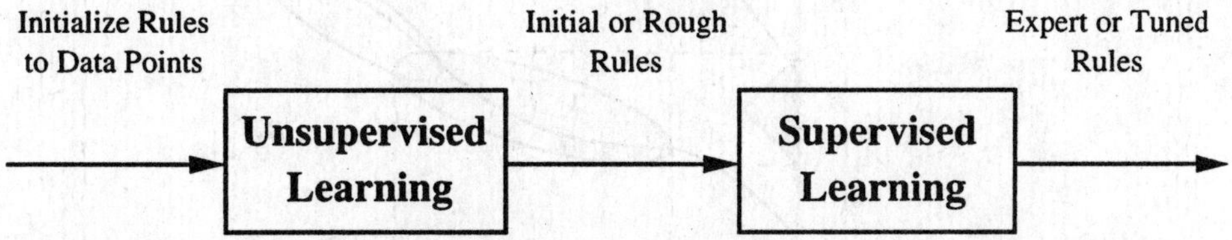

Figure 3.1 The hybrid neural system combines unsupervised and supervised neural learning to find and tune the ellipsoidal fuzzy rules.

We combined the two types of learning in a hybrid neural system that both finds and tunes the ellipsoidal rules as in Figure 3.1. Unsupervised learning picks the first set of rules based on the second-order statistics of the training data. Then supervised learning tunes these rule ellipsoids. We tested all three systems on a fifth-order polynomial. The hybrid system learned faster and had less mean-squared error than did the lone unsupervised or supervised system.

Optimal ellipsoidal rules minimize the mean-squared error of the fuzzy function approximation. We found a closed-form model of optimal ellipsoid rules when just the rule centroids change. We used the numerical Nelder-Mead algorithm [2, 22] to find the optimal rules when more rule parameters change. In simulations supervised and hybrid learning tended to converge to or near the optimal ellipsoidal rules. Optimal rules help deal with the exponential "rule explosion" as the joint dimension $n + p$ of the product space $X \times Y$ grows. Lone optimal rule patches cover the extrema of the approximand [15].

3.2 ADDITIVE FUZZY SYSTEMS

A fuzzy system approximates a function by covering its graph with fuzzy patches and averaging patches that overlap. The approximation improves as the fuzzy patches grow in number and shrink in size. The approximation improves as we add more small patches but storage and complexity costs increase. The additive fuzzy systems have a feedforward architecture that resembles the feedforward multilayer neural systems used to approximate functions [10]. Chapter 2 explores additive systems in depth.

An additive fuzzy system adds the then-parts of fired if-then rules. Other fuzzy systems combine the then-part sets with pairwise maxima. An additive fuzzy system can uniformly approximate continuous [13] or measurable [14] functions. A fuzzy system has rules of the form "If input conditions hold then output conditions hold"

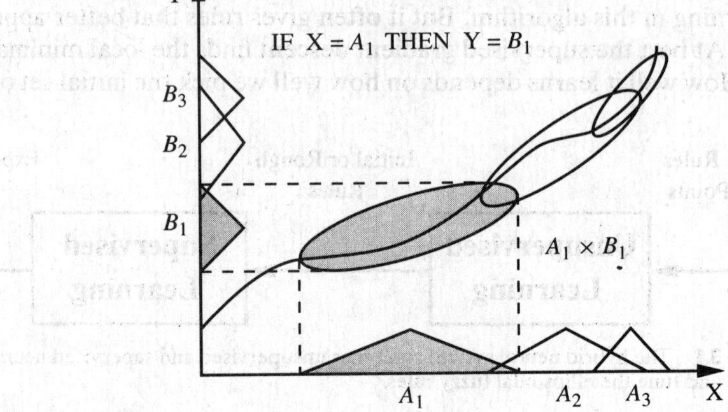

Figure 3.2 The fuzzy rule patch "If X is fuzzy set A_1 then Y is fuzzy set B_1" is the fuzzy Cartesian product $A_1 \times B_1$ in the input-output product space X × Y.

or "If X is A then Y is B" for fuzzy sets A and B. Each fuzzy rule defines a fuzzy patch or a Cartesian product $A \times B$ as shown in Figure 3.2. The fuzzy system covers the graph of a function with fuzzy patches and averages patches that overlap. Uncertain fuzzy sets give a large patch or fuzzy rule. Small or more certain fuzzy sets give small patches.

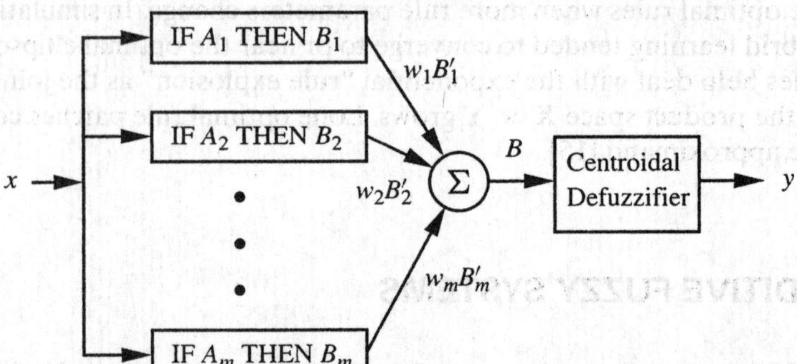

Figure 3.3 Additive fuzzy system architecture. The input x_k acts as a delta pulse (or unit bit vector) and fires each rule to some degree. The system adds the scaled output fuzzy sets. The centroid of this combined set gives the output value y_k. The system computes the conditional expectation value E[Y|X = x_k]. as a convex sum of the local centroids or centers of the then-part sets B'_j.

Additive fuzzy systems fire all rules in parallel and average the scaled then-part sets B'_j to get the output fuzzy set B as in Figure 3.3. Correlation product inference scales each then-part set B_j by the degree $a_j(x)$ that the rule "IF A_j THEN B_j" fires. Most rules fire to degree 0. Defuzzification of B gives a number or a control signal output. Centroidal defuzzification with correlation product inference [13, 19] gives the output value y or $F(x)$ given input vector $x \in R^n$:

$$y = F(x) = Centroid(B) = \frac{\int y\, b(y)\, dy}{\int b(y)\, dy} \tag{3.1}$$

$$= \frac{\sum_{j=1}^{m} w_j\, Volume(B'_j)\; Centroid(B'_j)}{\sum_{j=1}^{m} w_j\, Volume(B'_j)} \tag{3.2}$$

$$= \frac{\sum_{j=1}^{m} w_j\; V_j\; a_j(x)\; c_{y_j}}{\sum_{j=1}^{m} w_j\; V_j\; a_j(x)} \tag{3.3}$$

where V_j is the volume of the jth then-part set B_j and w_j is the weight of the jth rule (often $w_j = 1$). We can always normalize the finite volumes V_j to unity to keep some rules from dominating others. The term c_{y_j} is the centroid of the jth output set. Fit value $a_j(x)$ scales the then-part set B_j. m is the number of then-part *fuzzy sets*. In practice B is connected. It need not be. But then we could view the rule "If X is A then Y is B" as two or more rules of the form "If X is A then Y is B_1" and "If X is A then Y is B_2" where B_1 and B_2 are two of the disjoint components of B. So assume B is connected. Then the rule patch $A \times B$ is connected. Chapter 2 showed that the standard additive model (3.1) gives $F(x)$ as a convex sum of local centroids:

$$F(x) = \sum_j p_j(x)\; c_{y_j} \tag{3.4}$$

for convex coefficients

$$p_j(x) = \frac{a_j(x)\; w_j\; V_j}{\sum_{i=1}^{m} a_i(x)\; w_i\; V_i}. \tag{3.5}$$

A general additive fuzzy system is a map $F\colon R^n \to R^p$. Both in practice and in uniform approximation proofs we restrict the domain to a compact subset $U \subset R^n$ but we need not. The additive fuzzy system $F\colon R^n \to R^p$ stores m fuzzy patches $A_j \times B_j$ or rules of the form "If X is A_j then Y is B_j." Here $A_j \subset R^n$ and $B_j \subset R^p$ are multivalued or "fuzzy" sets with set functions $a_j\colon R^n \to [0,1]$ and $R^p \to [0,1]$.

3.3 ELLIPSOIDAL FUZZY RULES

3.3.1 Ellipsoids as Fuzzy Patches

A fuzzy rule patch can take the form of an ellipsoid [5]. The covariance of the pattern classes in the data can define ellipsoidal patches. The size and shape of the ellipsoid show how the inputs and outputs relate to each other in some region of the state space. We project the ellipsoid onto the input and output axes to form the fuzzy sets.

The eigenvectors and eigenvalues of a positive-definite matrix **A** define an ellipsoid in the q-dimensional input-output state space [21]. Here $q = n + p$, n is the number of inputs to the fuzzy system, and p is the number of outputs. The ellipsoid is the locus of all **z** that satisfy

$$\begin{aligned} \alpha^2 &= (\mathbf{z} - \mathbf{c})^T \mathbf{A} (\mathbf{z} - \mathbf{c}) \\ &= (\mathbf{z} - \mathbf{c})^T P \Lambda P^T (\mathbf{z} - \mathbf{c}) \end{aligned} \tag{3.6}$$

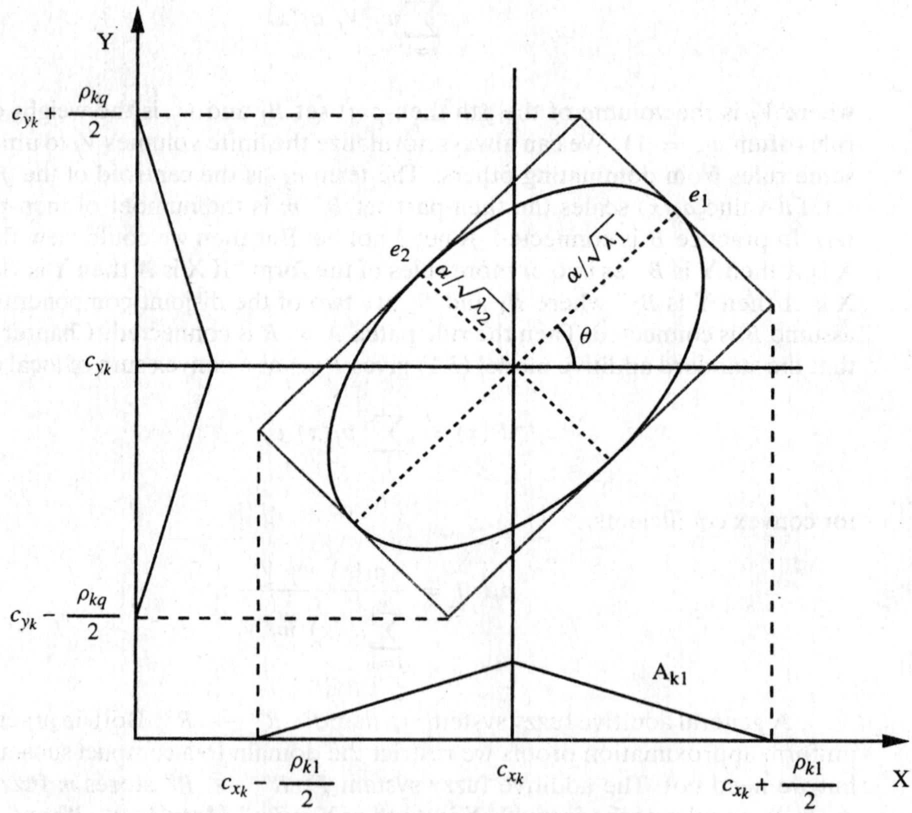

Figure 3.4 A positive definite matrix **A** defines an ellipsoid around the center **c** of the ellipsoid. The eigenvectors of **A** define the axes. The eigenvalues define the length of the axes. The ellipsoid projects onto the axes to define the input and output fuzzy sets.

where α is a positive real number and $\mathbf{c}$ is the center of the ellipsoid. Λ is a diagonal matrix of the eigenvalues $\lambda_1, \ldots, \lambda_q$ of $\mathbf{A}$. $\mathbf{P}$ is an orthogonal matrix whose columns are the unit eigenvectors $\mathbf{e}_1, \ldots, \mathbf{e}_q$ of $\mathbf{A}$. $\mathbf{P}$ rotates the coordinate system to the eigenvectors to orient the ellipsoid. The Euclidean half-lengths of the axes equal $\alpha/\sqrt{\lambda_1}, \ldots, \alpha/\sqrt{\lambda_q}$. Figure 3.4 shows the geometry of the ellipsoid in two-space. If $\mathbf{A}$ is not positive-definite the ellipsoid can concentrate on a lower dimensional hyperplane. We assume that $\mathbf{A}$ is positive definite.

To simplify the math we inscribe the ellipsoids in hyperrectangles. Then we project the hyperrectangles onto the axes of the state space to form the fuzzy sets. The kth hyperrectangle has 2^q vertices at $(\pm\alpha_k/\sqrt{\lambda_{k1}}, \ldots, \pm\alpha_k/\sqrt{\lambda_{kq}})$ in the rotated coordinate plane. The unit eigenvectors define direction cosines for each axis of the ellipse. The *direction cosine* [9] $\cos\ \gamma_{kij}$ is the angle between the jth eigenvector and the ith axis for the kth ellipsoid. The projection of the kth hyperrectangle onto the ith axis is centered at c_{ki} on the ith axis and has length ρ_{ki}:

$$\rho_{ki} = 2\alpha_k \sum_{j=1}^{q} \frac{|\cos\ \gamma_{kij}|}{\sqrt{\lambda_{kj}}}. \tag{3.7}$$

Unimodal sets can approximate the ellipsoid projections onto each axis. We use symmetric triangular sets. Figure 3.5 shows the ellipsoid patches and their triangular projections for a single-input single-output function. The ellipsoid parameters fix the

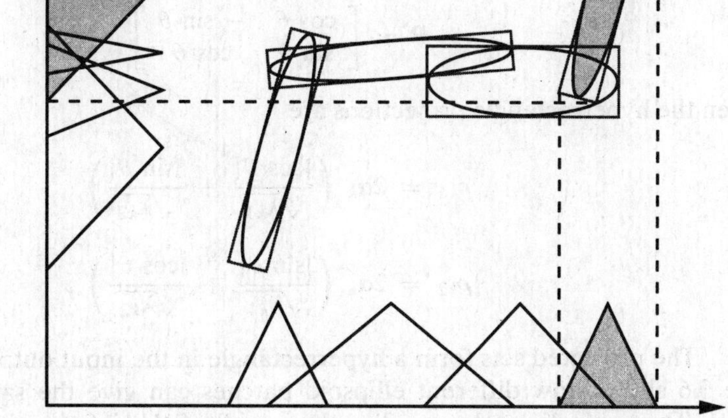

Figure 3.5 The projection of each ellipsoid onto the axes of the input-output state space defines a fuzzy set. The ellipsoid defines a fuzzy patch or rule between fuzzy subsets of inputs and outputs.

position, shape, and size of the fuzzy rules. The term ρ_{ki} defines the base of the triangular if-part fuzzy sets on the ith axis for the kth ellipsoidal rule. The volume of the triangular then-part set V_k for the kth ellipsoidal rule is

$$V_k = \frac{1}{2} \cdot 1 \cdot \rho_{kq}. \tag{3.8}$$

The fit-value degree $a_k(x)$ is

$$a_k(x) = \min_j \left(a_k^j(x) \right) \tag{3.9}$$

for multiple inputs in the if-part conjoined with AND. This gives a fuzzy rule of the form "IF x_1 is a_k^1 AND x_2 is a_k^2 AND. ... x_n is a_k^n THEN y is B_k." a_k^j is the triangular, if-part set for the kth ellipsoid's hyperrectangle projected on the jth axis:

$$a_k^j(x) = \begin{cases} 1 - \frac{2|x - c_{x_{kj}}|}{\rho_{kj}} & \text{for } |x - c_{x_{kj}}| \geq \frac{\rho_{kj}}{2} \\ 0 & \text{else.} \end{cases} \tag{3.10}$$

The orientation of the eigenvectors fixes the size of projections. The eigenvectors are *orthonormal* for a symmetric positive-definite matrix [21]. So the direction cosines for the kth ellipsoid obey

$$\begin{aligned} (\cos \gamma_{ki1})^2 + \ldots + (\cos \gamma_{kiq})^2 &= 1 \\ \cos \gamma_{ki1} \cos \gamma_{kj1} + \ldots + \cos \gamma_{kiq} \cos \gamma_{kjq} &= 0 \\ \text{for all } i, j = 1, \ldots, q, i = j. \end{aligned} \tag{3.11}$$

This gives q^2 unknowns and $q + (q-1)!$ equations. $q^2 - q - (q-1)!$ independent variables define the ellipsoid's orientation. For $q = 2$ just one number orients the ellipsoid. For a 2-D ellipsoid the rotation matrix is

$$\mathbf{P} = \begin{bmatrix} \cos\theta & -\sin\theta \\ \sin\theta & \cos\theta \end{bmatrix}. \tag{3.12}$$

Then the hyperrectangle projections are

$$\begin{aligned} \rho_{k1} &= 2\alpha_k \left(\frac{|\cos\theta|}{\sqrt{\lambda_{k1}}} + \frac{|\sin\theta|}{\sqrt{\lambda_{k2}}} \right) \\ \rho_{k2} &= 2\alpha_k \left(\frac{|\sin\theta|}{\sqrt{\lambda_{k1}}} + \frac{|\cos\theta|}{\sqrt{\lambda_{k2}}} \right). \end{aligned} \tag{3.13}$$

The projected sets form a hyperrectangle in the input output state space. Figure 3.6 shows how different ellipsoid patches can give the same fuzzy sets and rules. The projections do not directly use all of the information available in the ellipsoidal fuzzy patch such as ellipsoid volume and orientation. Performance improvements may come from using the ellipsoidal patch directly at the expense of more complicated mathematics.

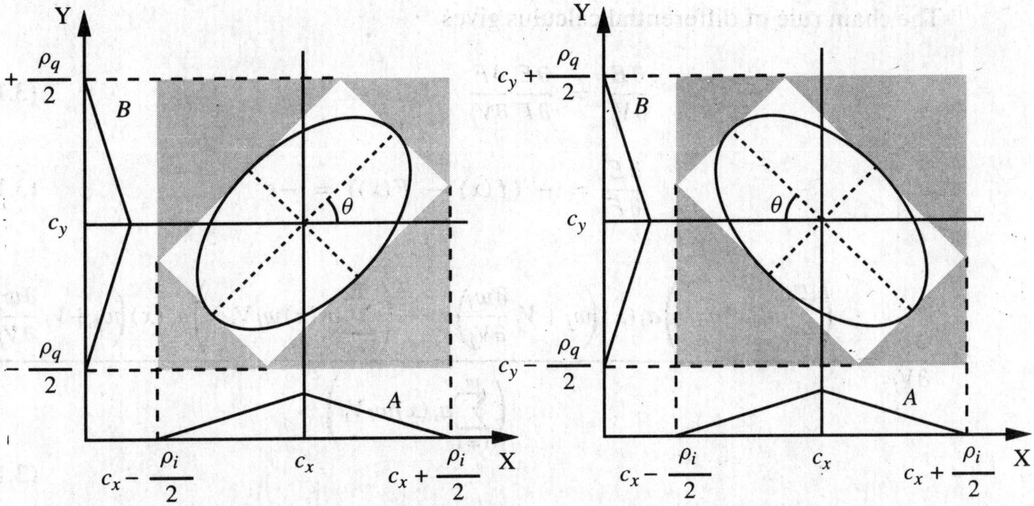

Figure 3.6 Symmetric ellipsoidal patches give the same fuzzy rules and sets. The shaded region shows the fuzzy patch in the state space when we define the fuzzy sets as ellipsoidal projections.

3.3.2 Weighting Ellipsoidal Rules

The fuzzy additive system F in (3.4) computes the global conditional mean $F(x) = E\,[Y|X = x]$ with a convex sum. The jth rule tries to make $F(x)$ look like c_{y_j} when it fires. $F(x)$ tends to c_{y_j} when the then-part volume V_j grows with respect to the other then-part volumes:

$$\lim_{V_j\to\infty} F(x) = \lim_{V_j\to\infty} \frac{\sum_{i=1}^{m} a_i(x)\; w_i\; V_i\; c_{y_i}}{\sum_{i=1}^{m} a_i(x)\; w_i\; V_i} \to c_{y_j} \quad \text{if } a_j(x) > 0. \tag{3.14}$$

The rule weights w_j weight the then-part centroids to approximate the function. To change the rule weights w_j we can change w_j itself or change the volume V_j if the weight depends on the volume: $\partial w_j/\partial V_j \neq 0$. We can learn the rule weights that minimize the mean-squared error E with a gradient algorithm as in Chapter 2:

$$V_j(t+1) = V_j(t) - \frac{\partial E}{\partial V_j} \tag{3.15}$$

for the instantaneous mean-squared error E:

$$E = \frac{1}{2}\,(f(x) - F(x))^2\,. \tag{3.16}$$

The chain rule of differential calculus gives

$$\frac{\partial E}{\partial V_j} = \frac{\partial E}{\partial F}\frac{\partial F}{\partial V_j} \tag{3.17}$$

$$\frac{\partial E}{\partial F} = -\,(f(x) - F(x)) = -\varepsilon \tag{3.18}$$

$$\frac{\partial F}{\partial V_j} = \frac{\left(\sum_{i=1}^{m} a_i(x) w_i V_i\right) a_j(x)\left(w_j + V_j \frac{\partial w_j}{\partial V_j}\right) c_{y_j} - \left(\sum_{i=1}^{m} a_i(x) w_i V_i c_{y_i}\right) a_j(x)\left(w_j + V_j \frac{\partial w_j}{\partial V_j}\right)}{\left(\sum_{i=1}^{m} a_i(x) w_i V_i\right)^2} \tag{3.19}$$

$$= \frac{a_j(x)\left(w_j + V_j \frac{\partial w_j}{\partial V_j}\right)}{\sum_{i=1}^{m} a_i(x) w_i V_i}\left[\frac{\left(\sum_{i=1}^{m} a_i(x) w_i V_i\right) c_{y_j}}{\sum_{i=1}^{m} a_i(x) w_i V_i} - \frac{\sum_{i=1}^{m} a_i(x) w_i V_i c_{y_i}}{\sum_{i=1}^{m} a_i(x) w_i V_i}\right] \tag{3.20}$$

$$= \frac{a_j(x)}{\sum_{i=1}^{m} a_i(x) w_i V_i}\left(w_j + V_j \frac{\partial w_j}{\partial V_j}\right)\left[c_{y_j} - F(x)\right] \tag{3.21}$$

$$= p_j(x)\left[c_{y_j} - F(x)\right]\left(\frac{1}{V_j} + \frac{1}{w_j}\frac{\partial w_j}{\partial V_j}\right) \tag{3.22}$$

where $p_j(x)$ is the convex coefficient in (3.5). When we combine (3.15) with the above result we get the volume update equation:

$$V_j(t+1) = V_j(t) + \varepsilon_t p_j(x_t)\left[c_{y_j} - F(x_t)\right]\left(\frac{1}{V_j} + \frac{1}{w_j}\frac{\partial w_j}{\partial V_j}\right) \tag{3.23}$$

Suppose the rule weights w_j do not depend on the volumes V_j: $\frac{\partial w_j}{\partial V_j} = 0$. Then

$$\left(\frac{1}{V_j} + \frac{1}{w_j}\frac{\partial w_j}{\partial V_j}\right) = \frac{1}{V_j} \tag{3.24}$$

$$V_j(t+1) = V_j(t) + \varepsilon \frac{p_j(x)}{V_j(t)}\left[c_{y_j} - F(x)\right] \tag{3.25}$$

If the jth rule fires a little or not at all then V_j does not change. If the jth rule fires strongly $a_j(x) \approx 1$ then learning depends strongly on how well $F(x)$ matches the then-part set centroid c_{y_j}. When V_j is large in (3.25) it tends to shut off learning since the rule is uncertain. When V_j is small its weight is large and it tries to move the output $F(x)$ toward the centroid c_{y_j}.

We can weight the then-part sets in (3.2) with an inverse volume weight:

$$w_j = \frac{1}{V_j} \tag{3.26}$$

to give rules with equal weights. For then

$$\left(\frac{1}{V_j} + \frac{1}{w_j}\frac{\partial w_j}{\partial V_j}\right) = \left(\frac{1}{V_j} - V_j\frac{1}{V_j^2}\right) = 0 \tag{3.27}$$

in (3.22). So the volume weights in (3.23) do not change. The inverse-volume weight (3.26) reduces $F(x)$ to an equal-weight or COG (center of gravity) additive fuzzy system:

$$F(x) = \frac{\sum_{j=1}^{m} a_j(x)\, c_{y_j}}{\sum_{j=1}^{m} a_j(x)} \tag{3.28}$$

We can also use an inverse square weighting scheme that gives more weight to smaller or more certain fuzzy rules:

$$w_j = \frac{1}{V_j^2} \tag{3.29}$$

$$\left(\frac{1}{V_j} + \frac{1}{w_j}\frac{\partial w_j}{\partial V_j}\right) = \left(\frac{1}{V_j} - V_j^2\frac{2}{V_j^3}\right) = -\frac{1}{V_j}. \tag{3.30}$$

The new learning increment cancels in sign that of (3.25):

$$V_j(t+1) = V_j(t) - \varepsilon\frac{p_j(x)}{V_j(t)}\left[c_{y_j} - F(x)\right] \tag{3.31}$$

This weights certain rules with a small volume more heavily than rules with a larger volume in the centroid calculation of (3.1):

$$F(x) = \frac{\sum_{j=1}^{m} \frac{a_j(x)}{V_j} c_{y_j}}{\sum_{j=1}^{m} \frac{a_j(x)}{V_j}} \tag{3.32}$$

The figures below show how learning the weights for 10 fuzzy rule patches affects the fuzzy approximation of the fifth-order polynomial $f(x) = 3x(x-1)(x-1.9)(x+0.7)(x+1.8)$. The underlying data distribution is uniform between -2.1 and 2.1. We used equation (3.51) to find the optimal mean-squared centroids c_{y_j}. The centers of the if-part of the rules are equally spaced across the x-axis. Figure 3.7 shows the case where $w_j = 1/V_j$. The weights do not change with time, as shown in (3.27). Figure 3.8 shows the case where the weights w_j are not a function of V_j. Equation (3.25) updates the weights. Figure 3.9 shows the case when $1/V_j^2$. Equation (3.31) updates the weights.

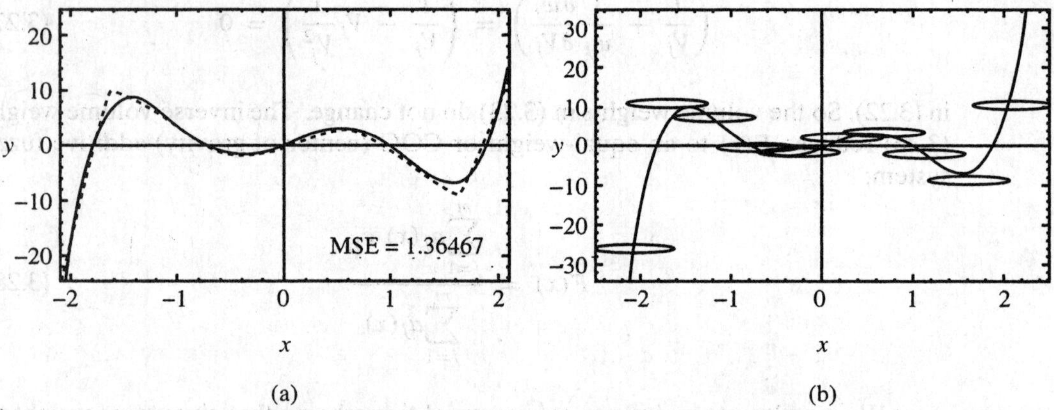

Figure 3.7 The effect of different rule weights w_k on the output function approximation when 10 rule patches approximate the function. (a) The function approximation for $w_j = 1/V_j$. Rules all have the same effect on the output calculation. (b) The rule patches for $w_j = 1/V_j$. MSE stands for mean square error. Symmetric triangular if-part sets give a piecewise-linear fuzzy system.

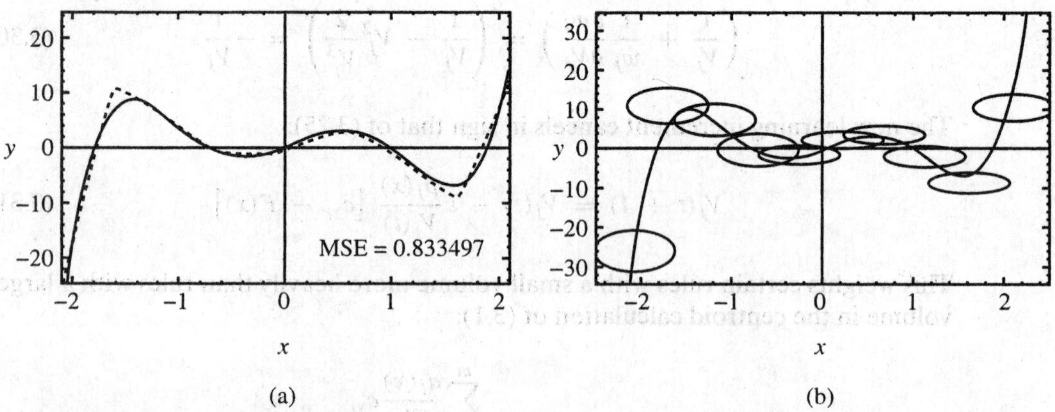

Figure 3.8 The effect of different rule weights w_k on the output function approximation when 10 rule patches approximate the function. (a) Equal weights on the rules: w_j = constant. Equation (3.25) updates the weights. (b) Rule patches after volume learning. Rules with larger volumes have more effect on the output.

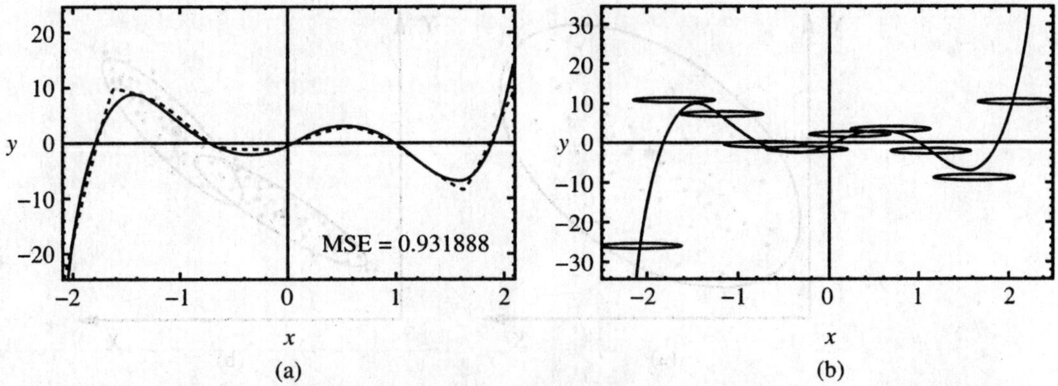

Figure 3.9 The effect of different rule weights w_k on the output function approximation when 10 rule patches approximate the function. (a) The function approximation for $w_j = 1/V_j^2$. Equation (3.31) updates the weights. (b) Weighted rule patches after volume learning. Rules with larger volumes have more effect on the output.

3.4 UNSUPERVISED COVARIANCE ELLIPSOID ESTIMATION

First-order and second-order statistics of the data can find the ellipsoidal fuzzy patches for an additive fuzzy system [5]. Unsupervised competitive learning laws estimate the covariance matrix $\mathbf{K}_k$ and centroid $\mathbf{s}_k$ of data clusters [13] as shown in Appendix 3.A at the end of this chapter. The covariance matrix $\mathbf{K}_k$ defines an ellipsoid in the input-output space for $\mathbf{A} = \mathbf{K}_k^{-1}$ in (3.6):

$$\alpha_k^2 = (\mathbf{z} - \mathbf{c}_k)^T \mathbf{K}_k^{-1} (\mathbf{z} - \mathbf{c}_k) \tag{3.33}$$

where $\mathbf{K}_k^{-1}$ is the inverse of the covariance matrix and $\mathbf{s}_k$ is the centroid of the kth pattern class.

3.4.1 Unsupervised Competitive Learning

Unsupervised competitive learning laws can learn statistics of data clusters [11, 13]. Each data cluster forms a fuzzy rule patch [13]. Adaptive vector quantization (AVQ) systems cluster quantization vectors in the input-output state space. The clustered quantizer vectors track the clusters in the incoming data. An autoassociative AVQ system combines the input $\mathbf{x}$ and the output $\mathbf{y}$ of the data to form $\mathbf{z}^T = [\mathbf{x}^T|\mathbf{y}^T]$. Each data/quantization-vector cluster forms a fuzzy rule patch in this form of *product-space clustering* [13]. Appendix 3.A reviews the unsupervised competitive learning algorithm. Other unsupervised clustering techniques such as MacQueen's adaptive k-means [16], fuzzy clustering [1] and generalized learning vector quantization [20] might also find the fuzzy rules.

The AVQ system clusters each sample with the closest centroid in Euclidean distance. Other distance metrics such as the Mahalanohbis or "city block" distance can be used depending on the data [11]. When data samples are sparse or noisy the

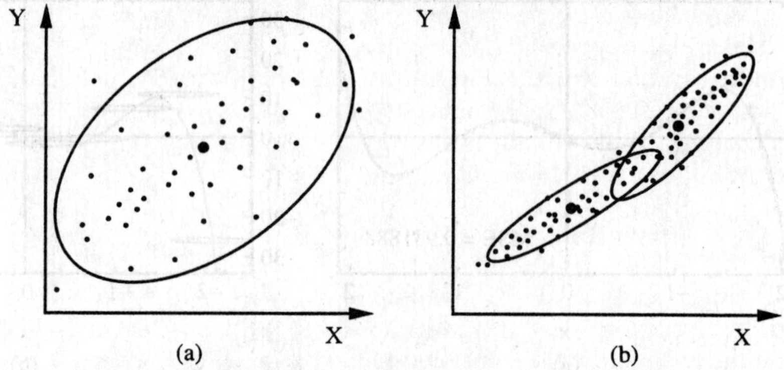

Figure 3.10 The size of the ellipsoid rule patches depends on the data. (a) When the data are sparse or noisy the ellipsoid is large. (b) When the data are dense the ellipsoid is small.

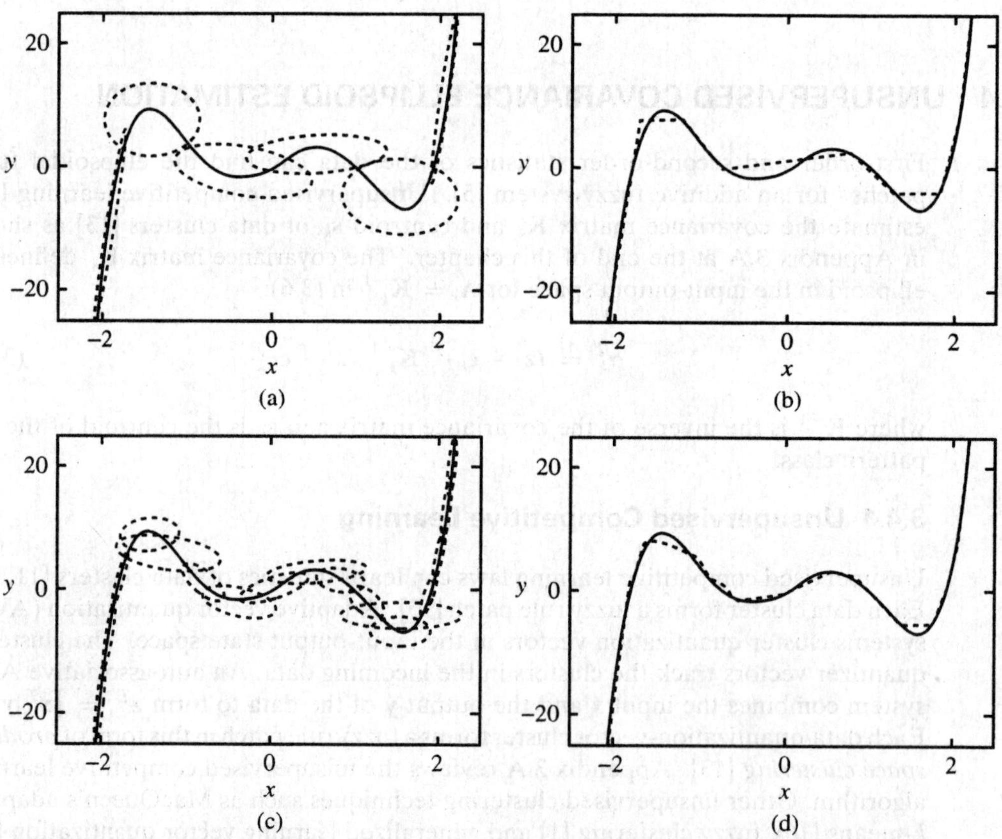

Figure 3.11 The additive fuzzy system gave a better approximation as more ellipsoids covered the graph of the fifth-order polynomial $f(x) = 3x(x-1)(x-1.9)(x+0.7)(x+1.8)$. (a) 10 ellipsoids cover the function. (b) Function approximation for 10 ellipsoids. (c) 20 ellipsoids cover the function. (d) Function approximation for 20 ellipsoids. Note that the elllipsoidal rules cover the extrema of the function in (a) and fill in between the extrema in (c).

covariance of the cluster or rule patch is large. This gives a large ellipsoid. When the data are dense the covariance of the cluster is small. This gives a smaller ellipsoid or more certain fuzzy rule. Figure 3.10 shows how the data distribution can change the size of an ellipsoid rule patch. Appendix 3.A shows that if the initial estimate of the covariance matrix $\mathbf{K}_k(0)$ is positive definite then $\mathbf{K}_k(t)$ is positive definite. So each covariance matrix forms an ellipsoid.

3.4.2 Fuzzy Rule Learning

Unsupervised ellipsoidal covariance learning finds the fuzzy rules and sets from the system input-output data [5]. The covariance estimate $\mathbf{K}_j^{-1}$ defines the jth ellipsoid in the q-dimensional input-output state space. q is the combined number of inputs n and outputs p to the fuzzy system. The projections of the ellipsoid onto the input and output axes bound the fuzzy sets.

Figure 3.11 shows the function approximation for the fifth-order polynomial $f(x) = 3x(x-1)(x-1.9)(x+0.7)(x+1.8)$ when the data distribution is uniform. The weights for each rule are unity $w_k = 1$. More patches better cover the "bumps" in the function. The size of the patches shrinks as the number of patches grows. This gives more certain or less fuzzy rules. Figure 3.12 shows the mean-squared error of the function approximation as the number of fuzzy rules grows. More patches give better MSE approximations.

The ellipsoid projections must cover the domain of the function so that the function is defined for all inputs. More overlap in the fuzzy patches smooths the function approximation. The choice of α fixes how much the ellipsoids overlap. Small α values give a step-like approximation with weights from the output set centroid cy_i. Larger α values smooth the function approximation since the ellipsoids overlap more. We used the same value for α for each of the ellipsoidal fuzzy patches in an approximation: $\alpha = \alpha_k$ for $k = 1, \ldots, r$. Optimal lone rules cover the extrema of f [15] to minimize the mean-squared error of the function approximation. The elliptical rule patches quickly move to cover these extrema in all of the learning schemes we studied.

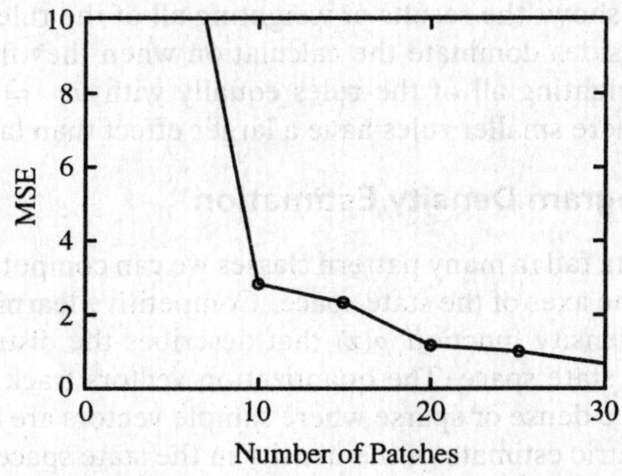

Figure 3.12 Mean-squared error of function approximation for a fifth-order polynomial falls as the number of patches increases.

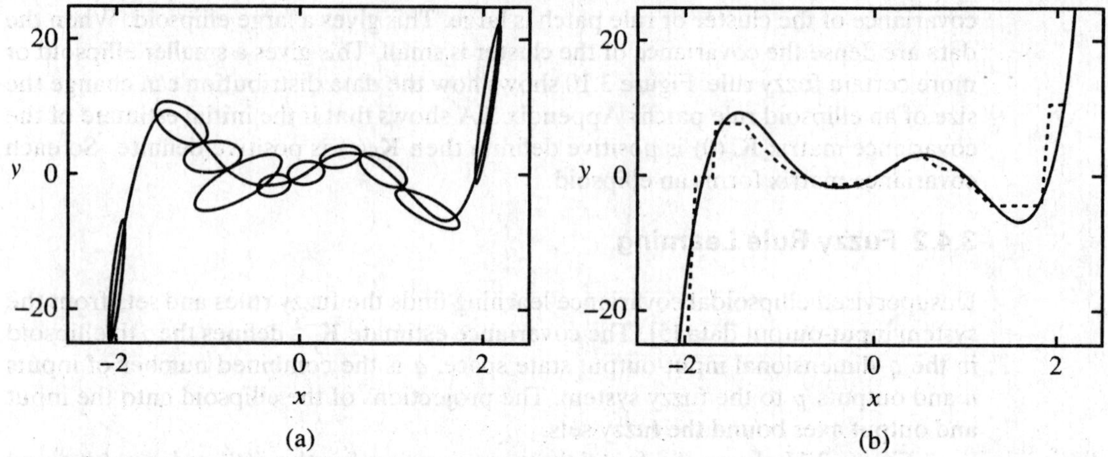

Figure 3.13 Unsupervised ellipsoid covariance learning estimates the fuzzy rule patches. (a) 10 ellipsoidal patches when the data are Gaussian. (b) Function approximation for 10 ellipsoids when the data are Gaussian.

The random distribution $p(\mathbf{z})$ of the data points $\mathbf{z}$ affects the accuracy of the approximation. Fewer or noisy data samples in a region result in larger ellipsoids since they have more variance. Smaller patches concentrate in regions of dense samples. Figure 3.13 shows the ellipsoids and the function approximation when the data are Gaussian. Most ellipsoids lie near the middle of the function. The edges of the function have fewer and larger ellipsoids since there are fewer data points there.

In practice most control systems spend their time at equilibrium and so most input-output data come from this region. AVQ systems estimate the unknown density $p(\mathbf{z})$ of the data. The quantizing vectors spread out to approximate $p(\mathbf{z})$. The rules far from equilibrium tend to be large since the data are sparse or noisy. For less frequent events the sets are larger to reflect this uncertainty. More certain rules occur where the system spends the most time. Figure 3.14a shows the ellipsoidal rules found with unsupervised learning for Gaussian distributed data. In 3.3 with unity weights ($w_k = 1$) rules with a large volume V_k have a large effect on the output. Figure 3.14a shows the results of weighting all of the rules with $w_k = 1$. The large rules on the sides dominate the calculation when they fire. Figure 3.14c shows the results of weighting all of the rules equally with $w_k = 1/V_k$. Figure 3.14d uses $w_k = V_k^2$ where smaller rules have a larger effect than larger rules have.

3.4.3 Histogram Density Estimation

When the data fall in many pattern classes we can compute the density of the projections along the axes of the state space. Competitive learning estimates the unknown probability density function $p(\mathbf{z})$ that describes the distribution of patterns in the input-output state space. The quantization vectors track the data. Quantizing vectors tend to be dense or sparse where sample vectors are dense or sparse. This gives a nonparametric estimate of the density in the state space.

Low-probability regions need enough quantizing vectors to estimate the pattern class statistics. Rare events may be as important as equilibrium events for good

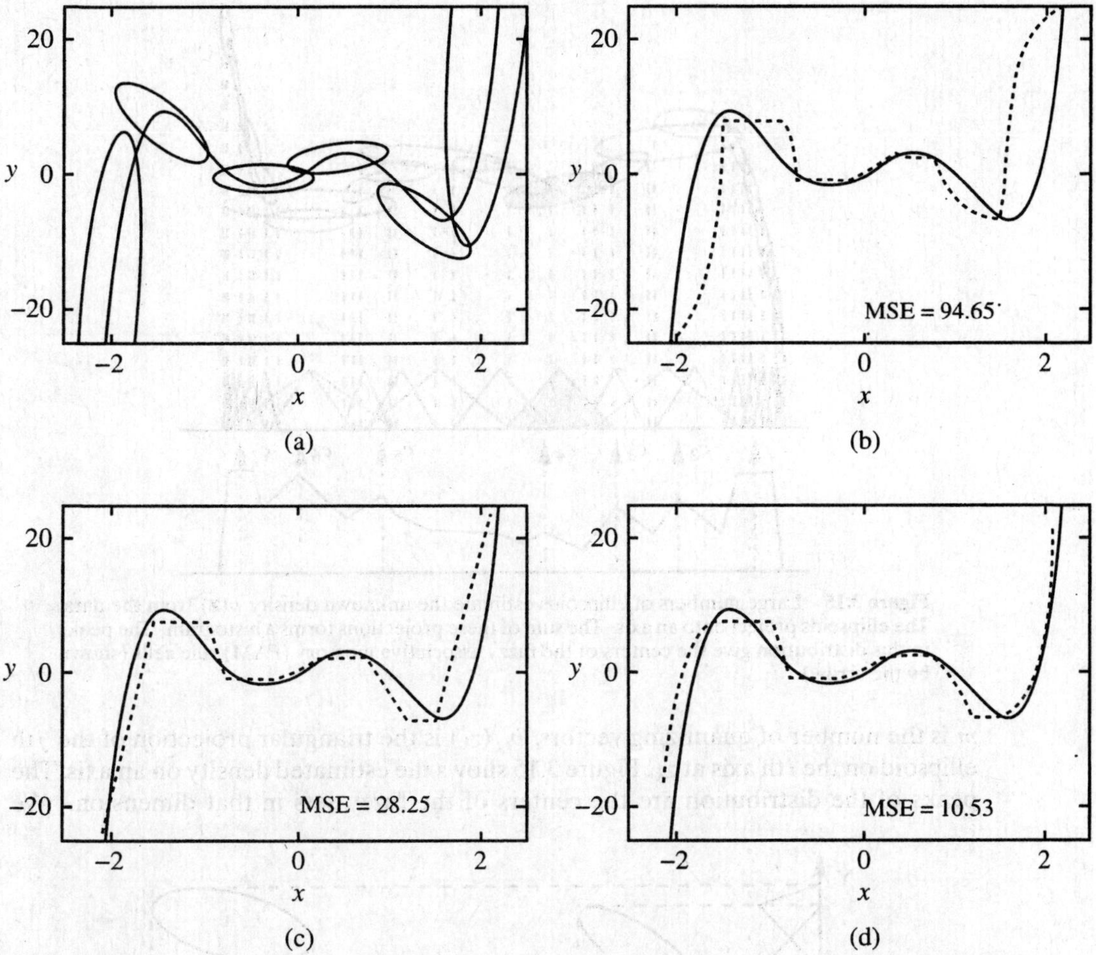

Figure 3.14 (a) The effect of different rule weights w_k on the output function approximation when six rule patches approximate the function. (b) Equal weights on the rules: $w_k = 1$. Rules with larger volumes have more effect on the output. (c) $w_k = 1/V_k$. Rules all have the same effect on the output calculation. (d) $w_k = 1/V_k^2$ gives more weight to smaller and thus more certain rules.

control or for stability. This method needs data or expert rules for these rare events. If data are not available the rules need to be added to the system by a knowledge engineer to ensure proper operation. Large numbers of fuzzy rules result if each quantizing vector counts as a rule. We can combine the quantizing vectors with histogram estimation.

Histogram estimation [7] gives a nonparametric estimate of a probability density. We sum the weighted triangular projections of the ellipsoids to form the histogram of the density on the ith axis $g_i(z_k)$:

$$g_i(z_k) = \sum_{j=1}^{m} b_{ij}(z_k) \tag{3.34}$$

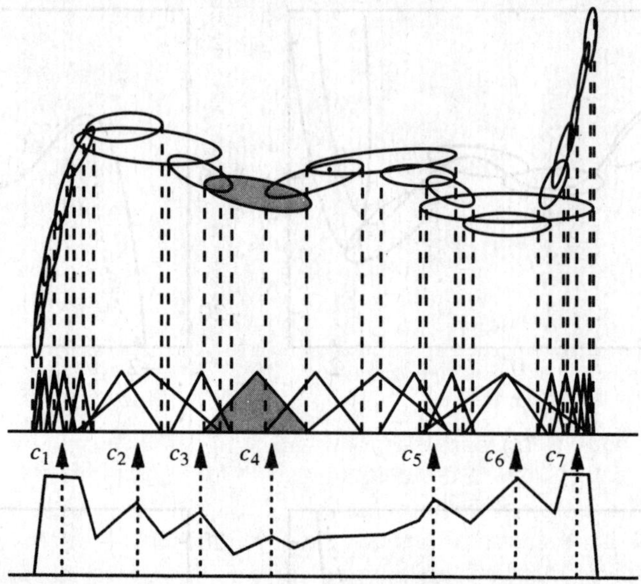

Figure 3.15 Large numbers of ellipsoids estimate the unknown density $p(\mathbf{Z})$ from the data. The ellipsoids project onto an axis. The sum of these projections forms a histogram. The peaks in this distribution give the centers of the fuzzy associative memory (FAM) rule cells (shown by the arrow).

m is the number of quantizing vectors. $b_{ij}(z_k)$ is the triangular projection of the jth ellipsoid on the ith axis at z_k. Figure 3.15 shows the estimated density on an axis. The peaks of the distribution are the centers of the fuzzy sets in that dimension. We

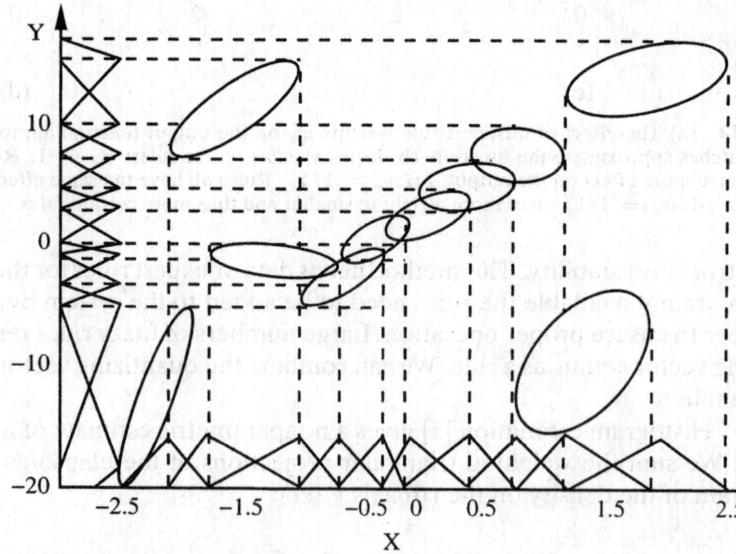

Figure 3.16 10 fuzzy rule patches found with the histogram method. 200 quantizing vectors approximated the unknown density $g(\mathbf{Z})$.

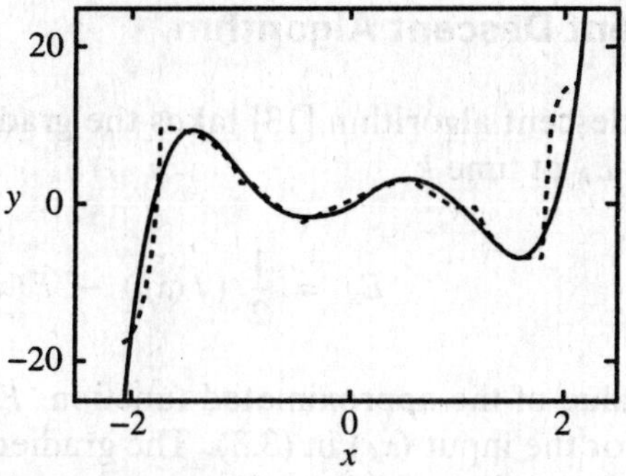

Figure 3.17 Function approximation with 10 fuzzy rule patches found with the histogram method for 200 quantizing vectors.

partition the state space along this grid and count the number of quantizing vectors in each cell. Clusters of quantizing vectors form a rule [5, 12, 13].

The ellipsoid scaling constant α changes the size of the projections on the axis. Larger values of α smooth the data. Small values of α give more resolution in areas where there are many quantizing vectors. Figure 3.16 shows the output ellipsoidal patches chosen with the histogram method. 200 quantizing vectors estimated the histogram. We used the 10 highest peaks of the histogram to give the centers of the rules. Figure 3.17 shows the function approximation.

3.5 SUPERVISED LEARNING OF FUZZY RULE PATCHES

A supervised neural system learns the ellipsoidal rules as it locally minimizes the mean-squared error of the function approximation as in Figure 3.18. The neural system learns the size and shape of the fuzzy rule patches that minimize the function error.

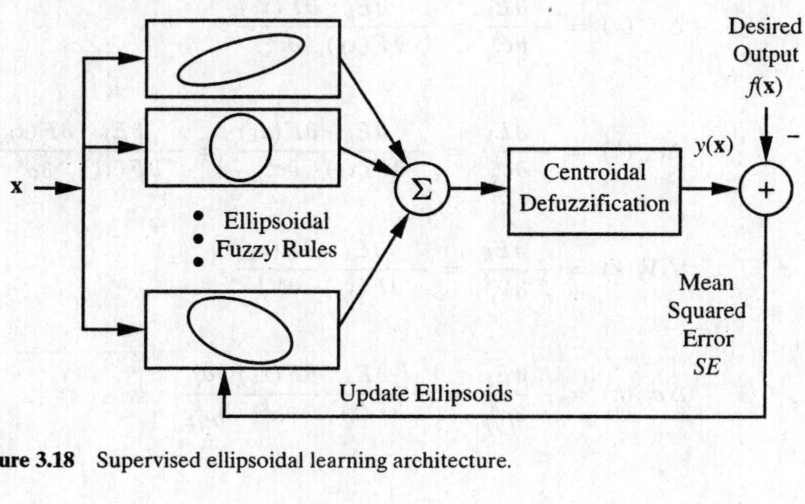

Figure 3.18 Supervised ellipsoidal learning architecture.

3.5.1 Gradient Descent Algorithm

The gradient descent algorithm [13] takes the gradient of the instantaneous mean-squared error E_k at time k

$$E_k = \frac{1}{2}\,(f(x_k) - F(x_k))^2 \tag{3.35}$$

$f(x_k)$ is the value of the approximated function. $F(x_k)$ is the output of an additive fuzzy system for the input (x_k) in (3.3). The gradient descent algorithm updates the fuzzy rule patches with the additive fuzzy system model in (3.3). Equations (3.8) and (3.10) define the volume V_i and the set functions $a_i(x)$ for the ith output fuzzy set.

The backpropagation algorithm can learn the size and shape of the projected fuzzy sets or the ellipsoid parameters [4]. The projected sets form a hyperrectangle in the input output state space. Figure 3.6 shows how symmetric ellipsoid patches give the same fuzzy sets and rules. This leaves four parameters for the ith fuzzy rule: the output set centroid c_{y_i}, the input set centroid c_{x_i}, the volume of the output set V_i, and the base of the input set ρ_{ik}.

The gradient descent algorithm uses a discrete form of gradient steepest descent:

$$\mathbf{El}_i(k+1) = \mathbf{El}_i(k) + \Omega_{ck}\nabla_{\mathbf{El}_i} E_k \tag{3.36}$$

$\mathbf{El}_i(k)$ is a concatenated vector of the ith ellipsoids' parameters. Ωc_k is a diagonal matrix of decreasing learning constants and $\nabla_{\mathbf{El}_i} E_k$ is the gradient vector of $\mathbf{El}_i(k)$. The components of $\mathbf{El}_i(k)$ are the output set centroid c_{y_i}, the input set centroid c_{x_i}, the area of the output set V_i, and the base of the input set ρ_{ik} . These variables fix the fuzzy rule patches.

Repeated applications of the chain rule of differential calculus give the ellipsoidal supervised algorithm. Four equations update the ellipsoid parameters:

$$\Delta c_{y_i}(k) = -\frac{\partial E_k}{\partial c_{y_i}^k} = -\frac{\partial E_k}{\partial F(x_k)}\frac{\partial F(x_k)}{\partial c_{y_i}^k} \tag{3.37}$$

$$\Delta c_{x_i}(k) = -\frac{\partial E_k}{\partial c_{x_i}^k} = -\frac{\partial E_k}{\partial F(x_k)}\frac{\partial F(x_k)}{\partial c_{x_i}^k} = -\frac{\partial E_k}{\partial F(x_k)}\frac{\partial F(x_k)}{\partial a_i^k}\frac{\partial a_i^k}{\partial c_{x_i}^k} \tag{3.38}$$

$$\Delta V_i(k) = -\frac{\partial E_k}{\partial V_i^k} = -\frac{\partial E_k}{\partial F(x_k)}\frac{\partial F(x_k)}{\partial V_{x_i}^k} \tag{3.39}$$

$$\Delta \rho_{ij}(k) = -\frac{\partial E_k}{\partial \rho_{ij}^k} = -\frac{\partial E_k}{\partial F(x_k)}\frac{\partial F(x_k)}{\partial a_i^k}\frac{\partial a_i^k}{\partial \rho_{ij}^k}. \tag{3.40}$$

The equations below give the partial derivatives used in (3.37)–(3.40):

$$\frac{\partial E_k}{\partial F(x_k)} = -\,(f(x_k) - F(x_k)) \tag{3.41}$$

$$\frac{\partial F(x_k)}{\partial c^k_{y_i}} = \frac{V^k_i a^k_i}{\sum\limits_{j=1}^{m} V^k_j a^k_j} \tag{3.42}$$

$$\frac{\partial F(x_k)}{\partial a^k_i} = \frac{V^k_i \sum\limits_{j=1}^{m} V^k_j a^k_j \left(c^k_{y_i} - c^k_{y_j}\right)}{\left(\sum\limits_{j=1}^{m} V^k_j a^k_j\right)^2} \tag{3.43}$$

$$\frac{\partial a^k_i}{\partial c^k_{x_{il}}} = \begin{cases} \frac{2}{\rho_{il}} & \text{if } 0 < x - c_{x_{il}} \le \frac{\rho_{il}}{2} \\ 0 & \text{else} \\ -\frac{2}{\rho_{il}} & \text{if } -\frac{\rho_{il}}{2} < x - c_{x_{il}} \le 0 \end{cases} \tag{3.44}$$

$$\frac{\partial F(x_k)}{\partial V^k_i} = \frac{a^k_i \sum\limits_{j=1}^{m} V^k_j a^k_j \left(c^k_{y_i} - c^k_{y_j}\right)}{\left(\sum\limits_{j=1}^{m} V^k_j a^k_j\right)^2} \tag{3.45}$$

$$\frac{\partial a^k_i}{\partial \rho^k_{il}} = \begin{cases} \frac{2\left|x - c_{x_{il}}\right|}{\rho^2_{il}} & \text{if } \left|x - c_{x_{il}}\right| \le \frac{\rho_{il}}{2} \\ 0 & \text{else.} \end{cases} \tag{3.46}$$

These equations are nonlinear. We used different magnitudes of learning coefficients for different parameters. This tends to improve the gradient descent algorithm [8]. Some equations contain ρ^2_{ij} in the denominator. So as the size of the sets shrinks the partial derivative becomes more sensitive to small changes in set size. We used learning coefficients of the form

$$\beta_k = \min(\zeta_k, \beta_{\max}) \tag{3.47}$$

β_k is the learning coefficient at time k[8]. $\beta_{\max}$ is the maximum coefficient size. The learning coefficient is

$$\zeta_k = \zeta \left(1 - \frac{k}{1.1N}\right) \tag{3.48}$$

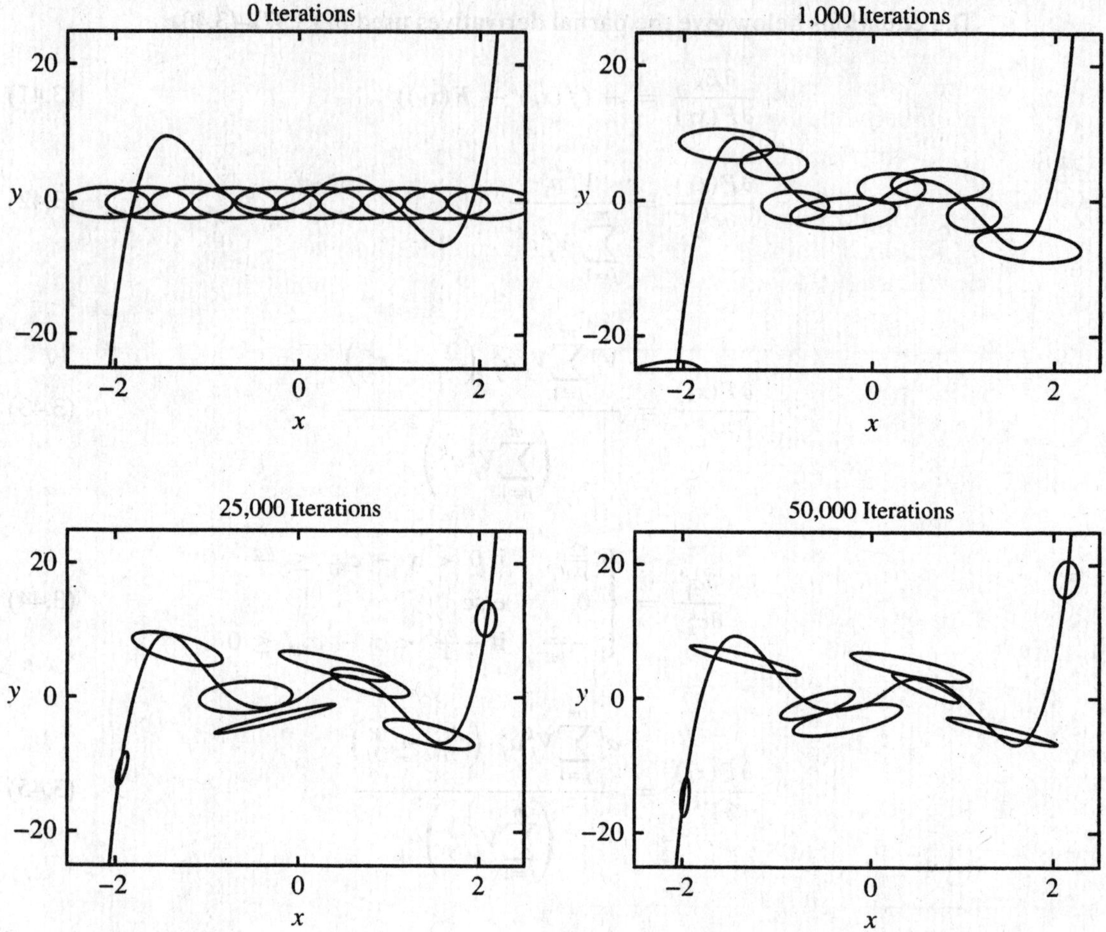

Figure 3.19 Supervised learning. Uniform initial ellipsoids cover the domain of the function. The ellipsoids quickly cover the extrema of the function. Over time the ellipsoids spread out to minimize the mean-squared error.

and ζ is the maximum learning coefficient. We kept the output set volumes V_i and the base of the triangular input sets ρ_{il} positive:

$$V_i^k = \max\left(V_i^k, \varepsilon_v\right) \tag{3.49}$$

$$\rho_{il}^k = \max\left(\rho_{il}^k, \varepsilon_\rho\right) \tag{3.50}$$

ε_v and ε_ρ are small positive constants. So each rule forms an ellipsoid.

3.5.2 Supervised Function Approximation

We initialized the ellipsoids along the x-axis. At each iteration the algorithm updated the parameter estimates of two ellipsoid fuzzy patches for the fifth-order polynomial

$f(x) = 3x(x - 1)(x - 1.9)(x + 0.7)(x + 1.8)$. The supervised algorithm found 8 partial derivatives for each patch to update 4 parameters for the single-input single-output case. The nonlinearities in the equations required small learning coefficients for stability. Figure 3.19 shows how the ellipsoids learn or move over time.

Figure 3.20 shows the ellipsoidal function approximation for this method. The ellipsoids do not cover the graph of the function as in the unsupervised case. The patch projections do cover the x-axis. So each input x belongs to at least one fuzzy set to nonzero degree and fires at least one rule. The patch's size and shape do not change if only one fuzzy rule fires for an input. Only the output set centroid can

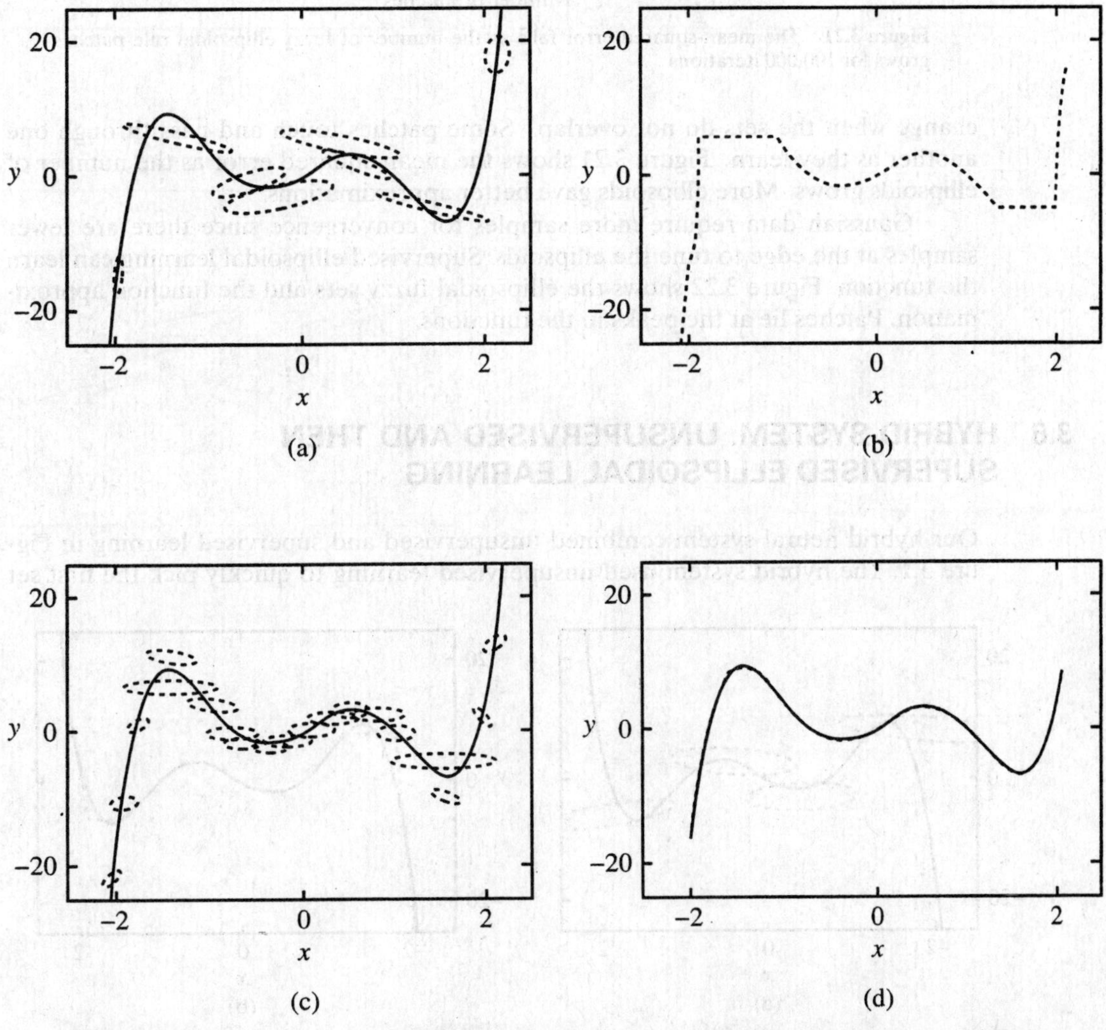

Figure 3.20 More fuzzy patches gave less mean-squared error in the function approximation. (a) Ten ellipsoidal rules found with supervised learning. (b) Function approximation for 10 rules after 50,000 iterations. (c) 20 ellipsoidal rules found with supervised learning. (d) Function approximation for 20 rules after 100,000 iterations.

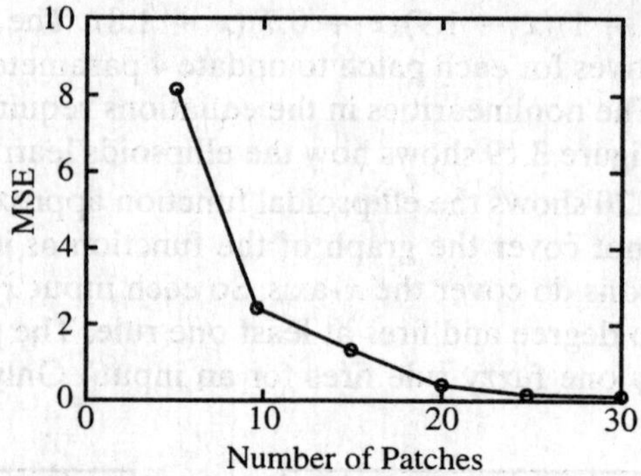

Figure 3.21 The mean-squared error falls as the number of fuzzy ellipsoidal rule patches grows for 100,000 iterations.

change when the sets do not overlap. Some patches touch and pass through one another as they learn. Figure 3.21 shows the mean-squared error as the number of ellipsoids grows. More ellipsoids gave better approximations.

Gaussian data require more samples for convergence since there are fewer samples at the edge to tune the ellipsoids. Supervised ellipsoidal learning can learn the function. Figure 3.22 shows the ellipsoidal fuzzy sets and the function approximation. Patches lie at the peaks in the functions.

3.6 HYBRID SYSTEM: UNSUPERVISED AND THEN SUPERVISED ELLIPSOIDAL LEARNING

Our hybrid neural system combined unsupervised and supervised learning in Figure 3.1. The hybrid system used unsupervised learning to quickly pick the first set

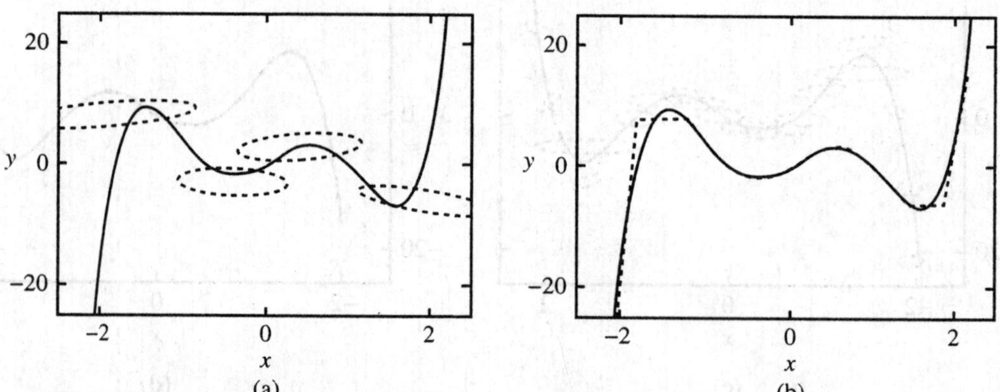

Figure 3.22 Backpropagation ellipsoidal learning of fifth-order polynomial with Gaussian data distribution. (a) 10 ellipsoidal patches, most ellipsoids are at extrema but move off the graph after 150,000 iterations. (b) Function approximation for 10 rules.

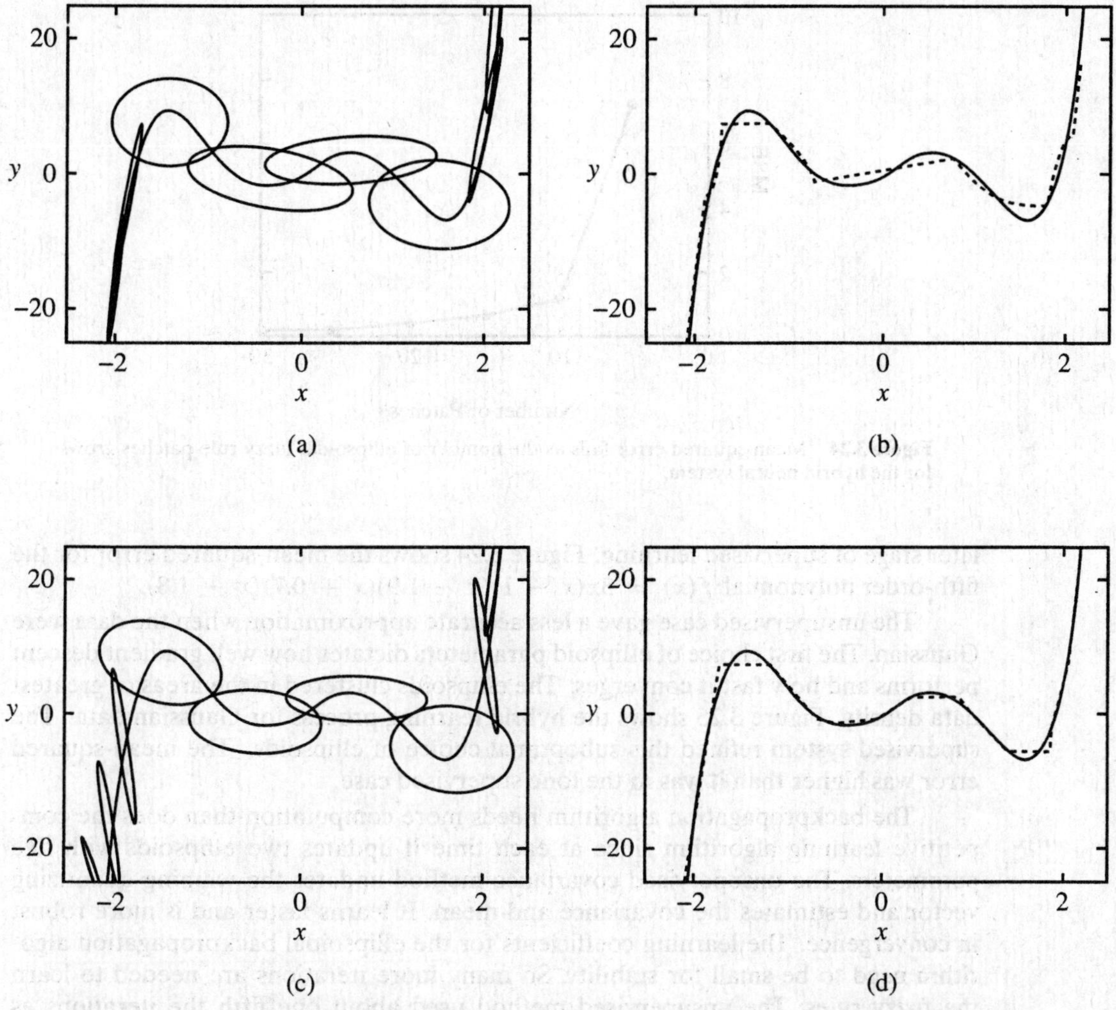

Figure 3.23 Combined ellipsoidal learning methods gave a fast function approximation. (a) 10 ellipsoidal rules found with unsupervised learning. (b) Function approximation. (c) The same 10 ellipsoidal rules tuned with supervised learning after 25,000 iterations. (d) Final function approximation.

of ellipsoidal fuzzy rules. It picked them in shape, orientation, and number. Then supervised learning tuned the rules.

The hybrid system acts as a human expert who learns to control a system. The unsupervised system acts as the expert as she first controls the system and turns her experience into a skill or a set of rough fuzzy rules. The supervised system acts as the expert as she improves her skill through trial and error.

The hybrid system performed better than did either the lone unsupervised or supervised system for the uniform case. Figure 3.23 shows the hybrid approximation for 10 ellipsoidal fuzzy rules. The first stage of unsupervised learning speeded the

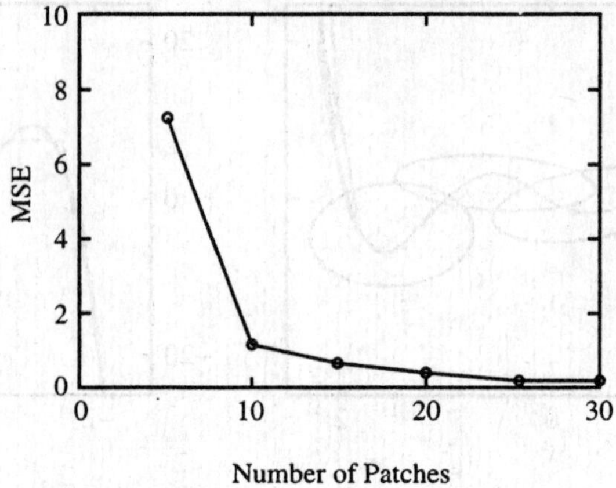

Figure 3.24 Mean-squared error falls as the number of ellipsoidal fuzzy rule patches grows for the hybrid neural system.

later stage of supervised learning. Figure 3.24 shows the mean-squared error for the fifth-order polynomial $f(x) = 3x(x - 1)(x - 1.9)(x + 0.7)(x + 1.8)$.

The unsupervised case gave a less accurate approximation when the data were Gaussian. The first choice of ellipsoid parameters dictates how well gradient descent performs and how fast it converges. The ellipsoids clustered in the areas of greatest data density. Figure 3.25 shows the hybrid learning process for Gaussian data. The supervised system refined this suboptimal choice of ellipsoids. The mean-squared error was higher than it was in the lone supervised case.

The backpropagation algorithm needs more computation than does the competitive learning algorithm since at each time it updates two ellipsoids with five parameters. The unsupervised covariance method updates the winning quantizing vector and estimates the covariance and mean. It learns faster and is more robust in convergence. The learning coefficients for the ellipsoidal backpropagation algorithm need to be small for stability. So many more iterations are needed to learn the fuzzy rules. The unsupervised method used about one fifth the iterations as did the supervised method. The hybrid system converged in about one third to one half the iterations of the supervised method. The hybrid method was stabler since the initial approximation errors tended to be smaller. The hybrid system works well when the unsupervised system gives a good estimate of the ellipsoids. The hybrid system converges to a local minimum when the first estimate does not reflect the function.

3.7 OPTIMAL FUNCTION APPROXIMATION

The three learning methods change the position and size of the fuzzy rules. In this section we compare these results with the optimal solution. Optimal ellipsoidal values minimize the mean-squared error E_f of the function approximation

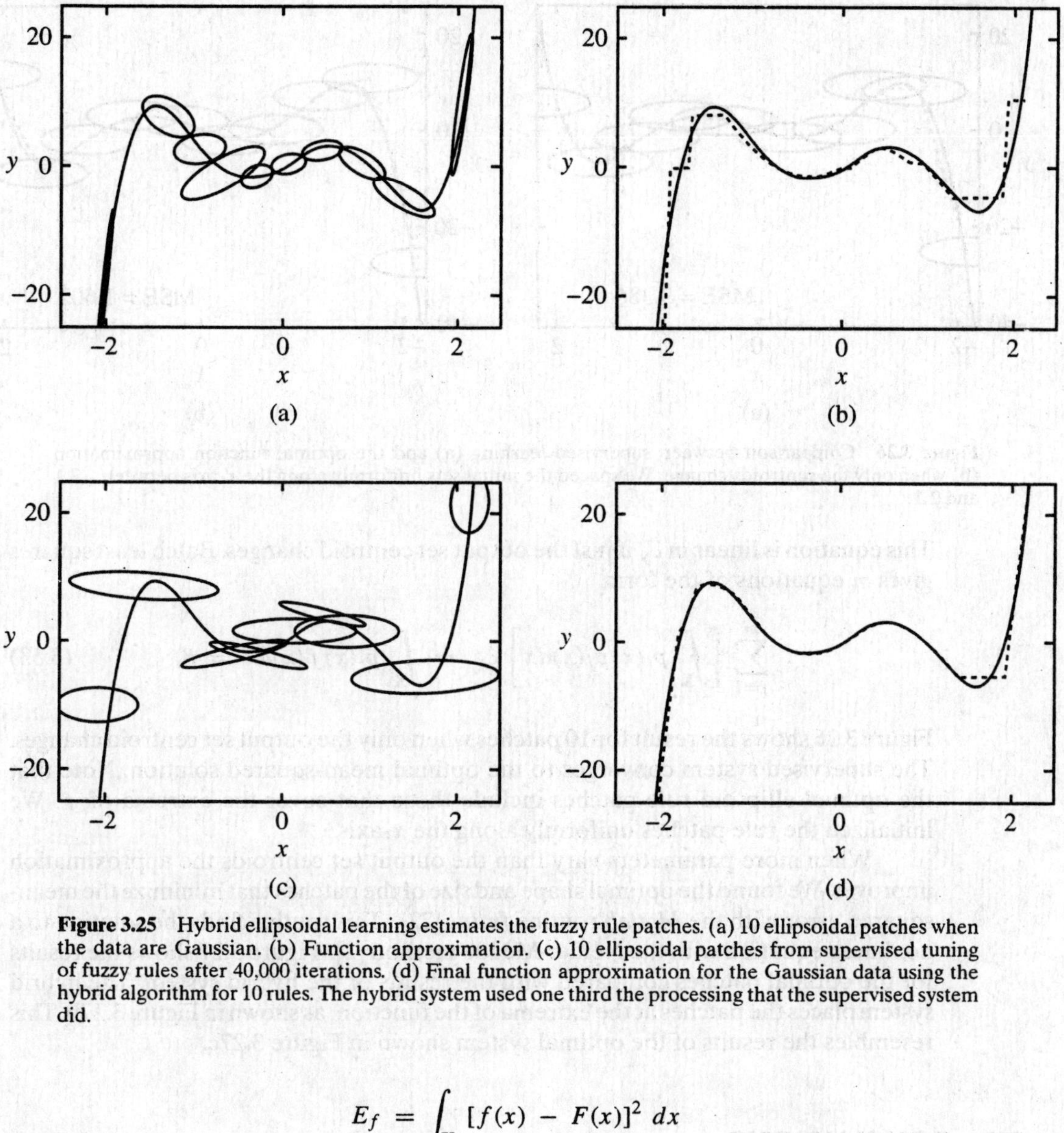

Figure 3.25 Hybrid ellipsoidal learning estimates the fuzzy rule patches. (a) 10 ellipsoidal patches when the data are Gaussian. (b) Function approximation. (c) 10 ellipsoidal patches from supervised tuning of fuzzy rules after 40,000 iterations. (d) Final function approximation for the Gaussian data using the hybrid algorithm for 10 rules. The hybrid system used one third the processing that the supervised system did.

$$\begin{aligned} E_f &= \int_{\mathbf{X}} [f(x) - F(x)]^2 \, dx \\ &= \int_{\mathbf{X}} \left[f(x) - \sum_{j=1}^{m} p_j(x) c_{y_j} \right]^2 dx \end{aligned} \tag{3.51}$$

$$p_j(x) = \frac{V_j a_j(x)}{\sum_{i=1}^{m} V_i a_i(x)}. \tag{3.52}$$

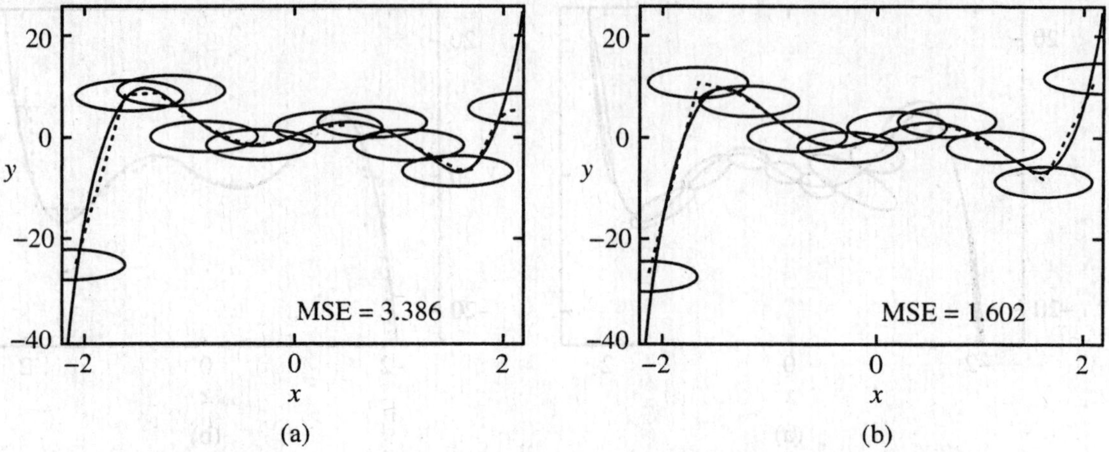

Figure 3.26 Comparison between supervised learning (a) and the optimal function approximation (b) when only the centroids change. We spaced the initial sets uniformly along the x-axis between -2.1 and 2.1.

This equation is linear in c_y if just the output set centroid changes. Batch least squares gives m equations of the form

$$\sum_{j=1}^{m} \left[\int_{\mathbf{X}} p_i(x) p_j(x) dx \right] \hat{c}_{y_j} = \int_{\mathbf{X}} p_i(x) f(x) dx = \beta_i \tag{3.53}$$

Figure 3.26 shows the result for 10 patches when only the output set centroid changes. The supervised system converges to the optimal mean-squared solution. Note that the optimal ellipsoid rule patches include those that cover the extrema of f. We initialized the rule patches uniformly along the x-axis.

When more parameters vary than the output set centroids the approximation improves. We found the optimal shape and size of the patches that minimize the mean-squared error with the *Matlab* routine *fmins* [22]. This routine finds the solution to a nonlinear equation with the Nelder-Meade method [2]. Figure 3.27 shows the results for the optimal patches compared with the results of the hybrid system. The hybrid system places the patches at the extrema of the function, as shown in Figure 3.27b. This resembles the results of the optimal system shown in Figure 3.27c.

3.8 CONCLUSIONS

Ellipsoids are a simple and natural form for fuzzy rules. Their math is well known and they arise as the covariance or uncertainty "balls" around quantizing vectors. Regions of sparse or noisy data lead to large covariance ellipsoids and thus less certain rules. Tightly clustered data lead to small ellipsoids and more certain rules. Higher-order statistics may improve estimates of fuzzy rule patch size and shape for nonlinear system data [17, 18].

Supervised learning can tune the ellipsoidal rules when we have error information. Even then the computation is heavy and someone or some system must pick the first set of ellipsoid parameters. Unsupervised learning can pick those parameters and both orient and speed the supervised gradient descent. Hybrid systems use the best of both techniques. They can learn fuzzy rules for problems of car control [3], inverse kinematics, or multimedia control and visualization [6].

Optimal ellipsoidal rules need complete knowledge of the approximand f. Learning can find some of this information as it estimates the optimal parameters. Future ellipsoidal fuzzy approximators may combine learning with closed-form models of both the shape and position of the if-part fuzzy sets.

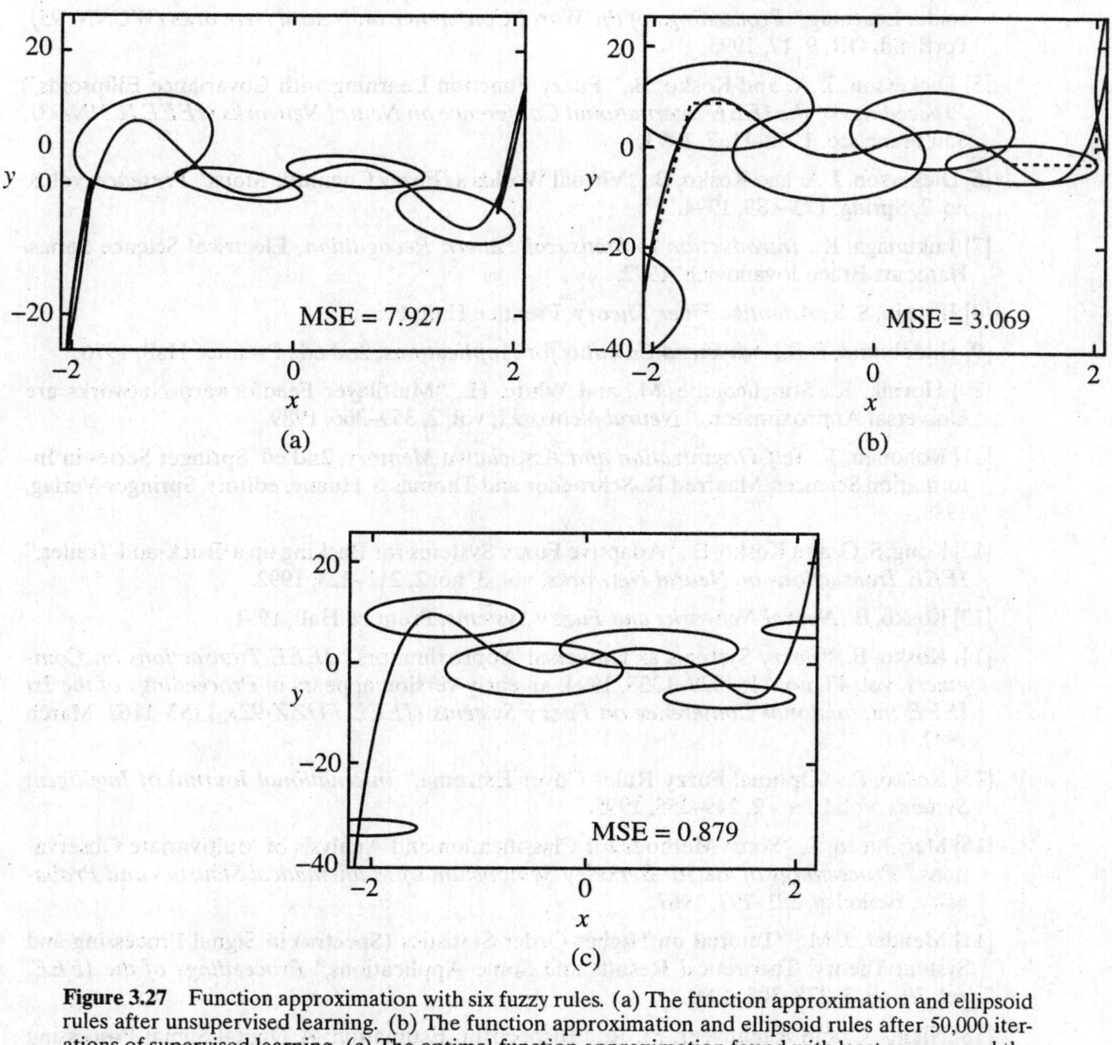

Figure 3.27 Function approximation with six fuzzy rules. (a) The function approximation and ellipsoid rules after unsupervised learning. (b) The function approximation and ellipsoid rules after 50,000 iterations of supervised learning. (c) The optimal function approximation found with least squares and the Nelder-Meade algorithm.

REFERENCES

[1] Bezdek, J. C., Tsao, E. C. and Pal, N. R., "Fuzzy Kohonen Clustering Networks," *Proceedings of the IEEE International Conference on Fuzzy Systems*, San Diego, 1035–1041, 1992.

[2] Dennis, J. E. and Woods, D. J., "New Computing Environments: Microcomputers in Large-Scale Computing," A. Wouk, editors, 116–122, SIAM, 1987.

[3] Dickerson, J. A., and Kosko, B., "Ellipsoidal Learning for Smart Car Platoons," in *Proceedings of the SPIE International Symposium on Optical Tools for Manufacturing Technologies—Applications of Fuzzy Logic Technology*, Boston, 1993.

[4] Dickerson, J. A. and Kosko, B., "Fuzzy Function Approximation with Supervised Ellipsoidal Learning," *Proceedings of the World Conference on Neural Networks (WCNN '93)*, Portland, OR, 9–17, 1993.

[5] Dickerson, J. A. and Kosko, B., "Fuzzy Function Learning with Covariance Ellipsoids," *Proceedings of the IEEE International Conference on Neural Networks (IEEE ICNN-93)*, San Francisco, 1162–1167, 1993.

[6] Dickerson, J. A. and Kosko, B., "Virtual Worlds as Fuzzy Cognitive Maps," *Presence*, vol. 3, no. 2, Spring, 173–189, 1994.

[7] Fukunaga, K., *Introduction to Statistical Pattern Recognition*, Electrical Science Series, Harcourt Brace Jovanovich, 1972.

[8] Haykin, S. S., *Adaptive Filter Theory*, Prentice Hall, 1986.

[9] Hildebrand, F. B., *Advanced Calculus for Applications*, 2nd ed., Prentice Hall, 1976.

[10] Hornik, K., Stinchcombe, M., and White, H., "Multilayer Feedforward Networks are Universal Approximators," *Neural Networks*, vol. 2, 359–366, 1989.

[11] Kohonen, T., *Self-Organization and Associative Memory*, 2nd ed. Springer Series in Information Sciences, Manfred R. Schroeder and Thomas S. Huang, editors, Springer-Verlag, 1988.

[12] Kong, S. G. and Kosko, B., "Adaptive Fuzzy Systems for Backing up a Truck-and-Trailer," *IEEE Transactions on Neural Networks*, vol. 3, no. 2, 211–223, 1992.

[13] Kosko, B., *Neural Networks and Fuzzy Systems*, Prentice Hall, 1991.

[14] Kosko, B., "Fuzzy Systems as Universal Approximators," *IEEE Transactions on Computers*, vol. 43, no. 11, 1329–1333, 1994; an early version appears in *Proceedings of the 1st IEEE International Conference on Fuzzy Systems (IEEE FUZZ-92)*, 1153–1162, March 1992.

[15] Kosko, B., "Optimal Fuzzy Rules Cover Extrema," *International Journal of Intelligent Systems*, vol. 10 no. 2, 249–255, 1995.

[16] MacQueen, J., "Some Methods for Classification and Analysis of Multivariate Observations," *Proceedings of the 5th Berkeley Symposium of Mathematical Statistics and Probability*, Berkeley, 281–297, 1967.

[17] Mendel, J. M., "Tutorial on Higher-Order Statistics (Spectra) in Signal Processing and System Theory: Theoretical Results and Some Applications," *Proceedings of the IEEE*, vol. 79, no. 3, 278–305, 1991.

[18] Nikias, C. and Raghuveer, M. R., "Bispectrum Estimation: A Digital Signal Processing Framework," *Proceedings of the IEEE*, vol. 75, 969–981, 1987.

[19] Pacini, P. J. and Kosko, B., "Adaptive Fuzzy System for Target Tracking," *Intelligent Systems Engineering*, vol. 1, no. 1, 3–21, 1992.

[20] Pal, N. R., Bezdek, J. C. and Tsao, E. C.-K., "Generalized Clustering Networks and Kohonen's Self-Organizing Scheme," *IEEE Transactions on Neural Networks*, vol. 4, no. 4, 549–557, 1993.

[21] Strang, G., *Linear Algebra and Its Applications*, 2nd ed., Academic Press, 1980.

[22] The Mathworks, *The Student Edition of MATLAB*, Prentice Hall, 1992.

APPENDIX 3.A: UNSUPERVISED COMPETITIVE LEARNING

Unsupervised competitive learning laws can learn the statistics of data clusters [13]. Adaptive vector quantization (AVQ) systems cluster quantization vectors in the input-output state space. The clusters in the quantization vectors track the clusters in the data. An autoassociative AVQ system combines the input **x** and the output **y** of the data to form $\mathbf{z}^T = [\mathbf{x}^T | \mathbf{y}^T]$. Figure 3.28 shows the geometry of an autoassociative AVQ system. Each data cluster forms a fuzzy rule patch in the form of *product space clustering* [13].

Neurons compete for activation from input patterns **z**. The r quantizing vectors $\mathbf{s}_j$ define the r columns of the synaptic connection matrix **S**. **S** connects the q input neurons in the combined input field F_Z to the r competing neurons in the output field F_W:

$$\mathbf{w} = \mathbf{z}^T \mathbf{S}. \tag{3.54}$$

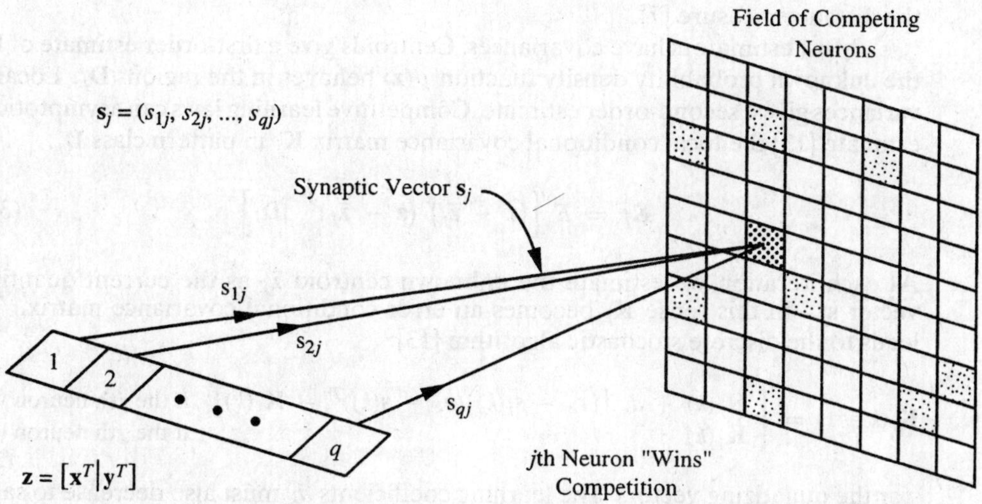

Figure 3.28 Autoassociative AVQ architecture combines the input **x** and the output **y**. Each quantizing vector competes for activation from input pattern **z**. The winning neuron learns the input pattern as its synaptic quantizing vector moves toward **z**.

The AVQ system compares the current vector random sample $\mathbf{z}(t)$ with the r columns of the synaptic connection matrix $\mathbf{S}$. The jth neuron wins if the jth quantizing vector $\mathbf{s}_j(t)$ is closest in Euclidean distance:

$$\|\mathbf{s}_j(t) - \mathbf{z}(t)\| = \min_i \|\mathbf{s}_i(t) - \mathbf{z}(t)\| . \tag{3.55}$$

The winning neuron learns the input pattern.

Competitive learning estimates the first-order statistics of the data with the stochastic difference equation

$$\mathbf{s}_j(k+1) = \begin{cases} \mathbf{s}_j(k) + \psi_k \left[\mathbf{z}_k - \mathbf{s}_j(k)\right] & \text{if the } j \text{ neuron wins} \\ \mathbf{s}_j(k) & \text{if the } j \text{ neuron loses.} \end{cases} \tag{3.56}$$

The learning coefficients ψ_k should decrease to satisfy the conditions for convergence [13]. The winning vector $\mathbf{s}_j$ learns or changes. The losing vectors $\mathbf{s}_k$ do not change and so do not forget what they have learned. We used $\psi_k = 0.2\ [1 - t_k(1.2N)]$ where N is the number of data points.

The competitive quantizing vectors converge exponentially quickly to pattern-class centroids [12, 13]. At equilibrium the average quantizing vector $E[\mathbf{s}_j]$ equals the jth centroid. Centroid estimation requires that a "signal function" $R_j(w_j)$ approximate the indicator function of the cluster $\mathbf{D}_j$ with a near binary win-lose function:

$$R_j(w_j) = \begin{cases} 1 & \text{if } \mathbf{z} \in \mathbf{D}_j \\ 0 & \text{if } \mathbf{z} \notin \mathbf{D}_j . \end{cases} \tag{3.57}$$

We scale the components of the data sample $\mathbf{z}(t)$ so all features have equal weight in the distance measure [7].

Most estimators have covariances. Centroids give a first-order estimate of how the unknown probability density function $p(\mathbf{z})$ behaves in the regions $\mathbf{D}_j$. Local covariances give a second-order estimate. Competitive learning laws can asymptotically estimate [13] the local conditional covariance matrix $\mathbf{K}_j$ in pattern class $\mathbf{D}_j$:

$$\mathbf{K}_j = E\left[\left(\mathbf{z} - \bar{\mathbf{z}}_j\right)\left(\mathbf{z} - \bar{\mathbf{z}}_j\right)^T \middle| \mathbf{D}_j\right] . \tag{3.58}$$

At each iteration we estimate the unknown centroid $\bar{\mathbf{z}}_j$ as the current quantizing vector $\mathbf{s}_j$. In this sense $\mathbf{K}_j$ becomes an error conditional covariance matrix. This leads to the discrete stochastic algorithm [13]:

$$\mathbf{K}_j(k+1) = \begin{cases} \mathbf{K}_j(k) + d_k \left[\left(\mathbf{z}_k - \mathbf{s}_j(k)\right)(\mathbf{z}_k - \mathbf{s}(k))^T - \mathbf{K}_j(k)\right] & \text{if the } j\text{th neuron wins} \\ \mathbf{K}_j(k) & \text{if the } j\text{th neuron loses} \end{cases} \tag{3.59}$$

for the quantizing vectors. The learning coefficients d_k must also decrease to satisfy the conditions for convergence [13]. We used $d_k = 0.2\ [1 - t_k/(1.2N)]$ where N is the number of data points. $\mathbf{K}_j(k+1)$ converges to the local conditional covariance matrix. The covariance estimate defines ellipsoidal patches in the state space if we set $\mathbf{A}$ in (3.6) to $\mathbf{K}_j^{-1}$.

We assume that $\mathbf{K}_j(k+1)$ is positive definite by the following logic. Assume that the initial covariance estimate $\mathbf{K}_j(0)$ is positive definite [$\mathbf{x}^T\mathbf{K}_j(0)\mathbf{x} > 0$ for all nonzero vectors $\mathbf{x}$]. If the jth neuron wins (3.59) updates the covariance matrix:

$$\mathbf{K}_j(1) = [1 - d_0]\mathbf{K}_j(0) + d_0\left(\mathbf{z}_k - \mathbf{s}_j(0)\right)\left(\mathbf{z}_k - \mathbf{s}_j(0)\right)^T \tag{3.60}$$

$0 < d_0 < 1$ is the learning coefficient in (3.60). The first term $\mathbf{A} = (1 - d_0)\mathbf{K}_j(0)$ is positive definite if $(1 - d_0)$ is positive [21]. The outer product matrix $\mathbf{B} = \left(\mathbf{z}_k - \mathbf{s}_j(0)\right)\left(\mathbf{z}_k - \mathbf{s}_j(0)\right)^T$ is a positive semidefinite matrix of rank 1 with one positive eigenvalue and the rest zero ($\mathbf{x}^T\mathbf{B}\mathbf{x} \geq 0$ for all nonzero vectors $\mathbf{x}$) [21]. A positive-definite matrix $\mathbf{A}$ plus a positive semidefinite matrix $\mathbf{B}$ form a positive definite matrix [$\mathbf{x}^T(\mathbf{A} + \mathbf{B})\,\mathbf{x} > 0$ for all non zero vectors $\mathbf{x}$]. $\mathbf{K}_j(1)$ is positive definite. If the jth neuron loses the covariance estimate does not change. The proof proceeds by induction. We initialized the covariance matrix estimate $\mathbf{K}_j(0)$ to small positive numbers on the diagonal to ensure that $\mathbf{K}_j$ is positive definite.

Steps 1 through 3 give the competitive AVQ algorithm [12]:

Competitive AVQ Algorithm

1. Initialize the quantizing vectors: $\mathbf{s}_i(0) = \mathbf{z}(i)$, $i = 1, \ldots, r$.
2. For a random data sample $\mathbf{z}(t)$, find the closest or "winning" quantizing vector $\mathbf{s}_j(t)$:

$$\left(\mathbf{s}_j(t) - \mathbf{z}(t)\right)^T \Omega\left(\mathbf{s}_j(t) - \mathbf{z}(t)\right) = \min_i\,(\mathbf{s}_i(t) - \mathbf{z}(t))^T\,\Omega\,(\mathbf{s}_i(t) - \mathbf{z}(t)) \tag{3.61}$$

where $\Omega = \begin{bmatrix} \omega_1^2 & 0 & \cdots & 0 \\ 0 & \ddots & & \vdots \\ \vdots & & \ddots & 0 \\ 0 & \cdots & 0 & \omega_q^2 \end{bmatrix}$ is a scaling matrix.

3. Update the winning quantizing vector $\mathbf{s}_j(t)$ with (3.56) and its local covariance estimate $\mathbf{K}_j(t)$ with (3.59).

We used the scaling matrix Ω:

$$\Omega = \begin{bmatrix} 1 & 0 \\ 0 & 0.15^2 \end{bmatrix}. \tag{3.62}$$

This normalized the Euclidean distance calculation in (3.61) so one variable did not dominate the choice of winners. We scaled the components of the data sample $\mathbf{z}(t)$ so all features have equal weight in the distance measure [7]. Then no feature dominates and the classification is scale invariant.

HOMEWORK PROBLEMS

3.1. The update equation for the covariance matrix (3.60) is

$$\mathbf{K}_j(k+1) = \begin{cases} \mathbf{K}_j(k) + d_k\left[(\mathbf{z}_k - \mathbf{c}_j(k))\,(\mathbf{z}_k - \mathbf{c}_j(k))^T - \mathbf{K}_j(k)\right] & \text{if the } j\text{th neuron wins} \\ \mathbf{K}_j(k) & \text{if the } j\text{th neuron loses.} \end{cases}$$

Prove that this learning law always gives at least a semipositive definite matrix $\mathbf{K}_j$. Show that if $\mathbf{K}_0$ is positive-definite then $\mathbf{K}_j$ is positive-definite.

3.2. The ellipse $x^2 + 9y^2 = 4$ gives $\mathbf{A} = \begin{bmatrix} 1 & 0 \\ 0 & 9 \end{bmatrix}$. Find the eigenvectors and eigenvalues (find $\mathbf{P}$ and Λ) and sketch the ellipse.

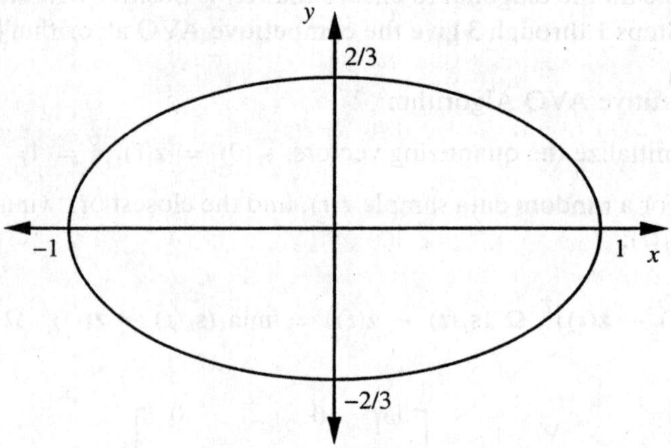

3.3. The eigenvectors orient the ellipse. Equation (3.7) shows how the eigenvectors change as the ellipsoid rotates in 2-D. Calculate new values of P in problem 2 for $\theta = 45°$ and $\theta = 135°$ and sketch the ellipses. Calculate the projections of the hyperrectangle that inscribes the ellipse with (3.8).

3.4. The equation $\lambda_1 x_1^2 + \lambda_2 x_2^2 + \lambda_3 x_3^2 = 1$ is an ellipsoid in 3-D space when all $\lambda_i > 0$. What other types of surfaces do you get when one or more of the λ_i is equal to zero (when $\mathbf{A}$ is positive semidefinite)?

3.5. Find an exact value for (3.7) when the fuzzy sets ϕ are triangles.

$$\phi_i = \begin{cases} \frac{a - c_{x_i} + x}{a} & \text{if } c_{x_i} - a \le x \le c_{x_i} \\ \frac{a + c_{x_i} - x}{a} & \text{if } c_{x_i} \le x \le c_{x_i} + a \\ 0 & \text{elsewhere.} \end{cases}$$

Each set extends to the center of the adjacent set, as shown in Figure H.1. This gives a piecewise linear approximation for $F(x)$. The result is a tri-diagonal matrix with a unique solution for the output set centroids.

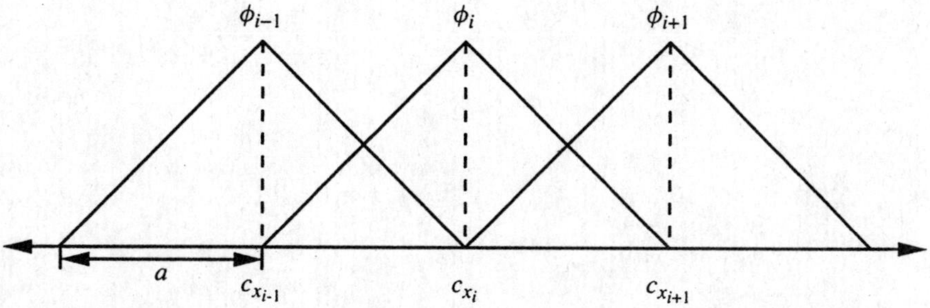

Figure 3.29 Each set reaches to the center of the adajacent set. All triangles have equal size.

SOFTWARE PROBLEMS

1. Look at the effects of different alphas on the function approximation. (Use *eplot* and data file)
2. Use an unsupervised learning program to learn the rules. Try the program for different numbers of rules or different functions. Compare the resulting function approximations for different scaling factors and ellipsoid sizes.
3. Use supervised learning to change the output set centroids.

PART III

FUZZY CONTROL AND CHAOS

Cars grow smarter by degrees each year. A car may contain over 100 microprocessors and connect these in complex sensor and data networks. Some cars have on-board navigation systems and emergency braking systems. Future cars will pass bits back and forth from their own chips to the sensors and chips in the road and in nearby cars. Head-up displays will show route options on smart windshields. Drivers will have access to tireless autopilots that depend on machine vision and on the wireless signals from the 24 orbiting Navstar Global Positioning Satellites.

Cars can also be smart in how they form groups or convoys on crowded freeways. Cars can travel in single-lane platoons just as bytes travel in packets in computer networks. Each smart car can act as a platoon leader or follower and must obey enter and exit protocols when changing lanes. Highspeed platoons can in theory increase traffic flow four-fold or five-fold but at the cost of complex nonlinear and often nonstationary dynamics. Platoons may also lead to large crashes and pile-ups and thus involve complex legal liability.

Chapter 4 uses additive fuzzy systems to control the throttle and braking systems of small smart-car platoons. The rules came from common sense, simulation experiments, and tested math models. Real tests on California Interstate–15 refined the rules and learning schemes. Highspeed platoon dynamics remains a rich nonlinear testing ground for both model-based and model-free system approximators.

4

FUZZY CONTROL FOR PLATOONS OF SMART CARS

Julie A. Dickerson[1], Hyun Mun Kim, and Bart Kosko

PREVIEW

An additive fuzzy system can control the throttle of cars in single-lane platoons. The system uses fuzzy controllers for velocity control, gap control, car following, and braking. Fuzzy controllers create, maintain, and divide platoons on the highway. Each car's controller uses data from its car and the car in front of it. Cars drop back during platoon maneuvers to avoid the "slinky effect" of tightly coupled platoons. When the lead car reaches its goal the follower cars return to the proper platoon spacing.

The controllers maintain the platoons up and down hills. Some car and engine types need their own fuzzy rules and sets. The hybrid neural fuzzy system in Chapter 3 learns the fuzzy rules and sets from input-output data. We compute new fuzzy rules and sets for a truck velocity controller. The learned system controls the velocity of the truck with no overshoot or slow response.

We tested the fuzzy gap controller first with a car model and then with a real car on highway I–15 in California. Each car's controller used data only from sensors on the car. We used this controller to drive the "smart" car on the highway in a two-car platoon. We also designed a fuzzy brake controller and tested it in computer simulations.

---

[1] The Caltrans PATH Program supported this research (Agreement 20695 MB). VORAD Incorporated provided the car and radar sensor for the gap controller test.

4.1 PLATOONS OF SMART CARS

Traffic clogs highways around the world. Platoons of cars can increase the flow and mean speed on freeways [7]. Platoons are high-speed groups of smart cars in single lanes on freeways of the future. Electronic links tie the cars together. Computer control speeds response times to road hazards so cars can travel more safely on their own or in groups.

A platoon is a group of cars with a lead car and one or more follower cars that travel in the same lane. The lead car plans the course for the platoon. It picks the velocity and car spacing and picks which maneuvers to perform. Platoons use four maneuvers: merge, split, velocity change, and lane change [4]. A merge combines two platoons into one. A split splits one platoon into two. A lane change moves a single car into an adjacent lane. Combinations help cars move through traffic.

Standard control systems use an input-output math model of the car and its environment. Fuzzy systems do not use an input-output math model or exact car parameters. A fuzzy system is a set of fuzzy rules that maps inputs to outputs [9]. The fuzzy platoon controller uses rules that act like the skills of a human driver. The rules have the form "If input conditions hold to some degree then output conditions hold to some degree" or "If X is **A** then Y is **B**" for fuzzy sets **A** and **B**. Each fuzzy rule defines a fuzzy patch or a Cartesian product $\mathbf{A} \times \mathbf{B}$ in the input-output state space $\mathbf{X} \times \mathbf{Y}$. To approximate the function the fuzzy system covers its graph with fuzzy patches and averages patches that overlap [8].

The fuzzy platoon controller (FPC) is a distributed control system for future freeways that drives a car in or out of a platoon. The FPC includes an integrated maneuver controller (IMC) for course selection and an individual vehicle controller (IVC) for throttle, brake, and steering control as in Figure 4.1.

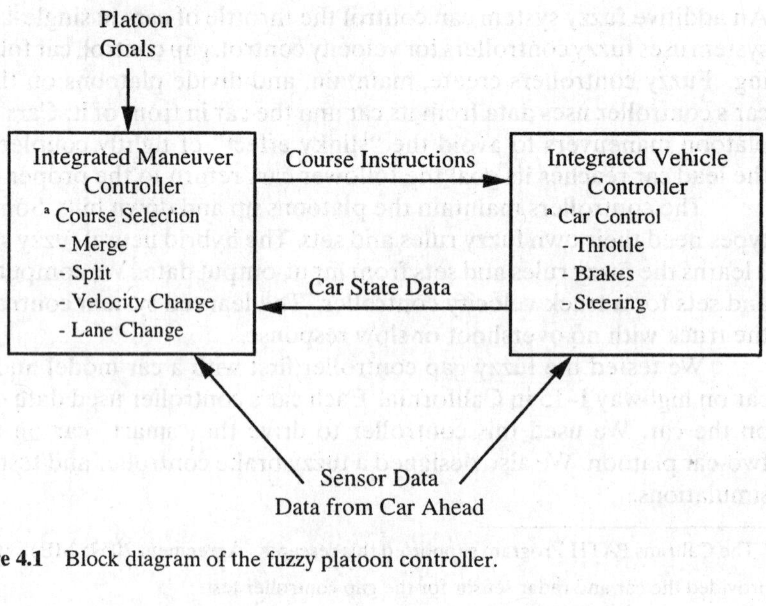

Figure 4.1 Block diagram of the fuzzy platoon controller.

We designed a fuzzy throttle controller (FTC) for velocity and gap control in smart platoons [1]. The controller uses the throttle only and has three subsystems for control of velocity and gap distance as in Figure 4.2. The FTC gets information from its own sensors, the car ahead, and from the platoon goals. Simulations show the platoon as it changes velocity, merges, and splits. We simulated only cars but the fuzzy system can control other vehicles such as trucks if it uses other fuzzy rules and sets.

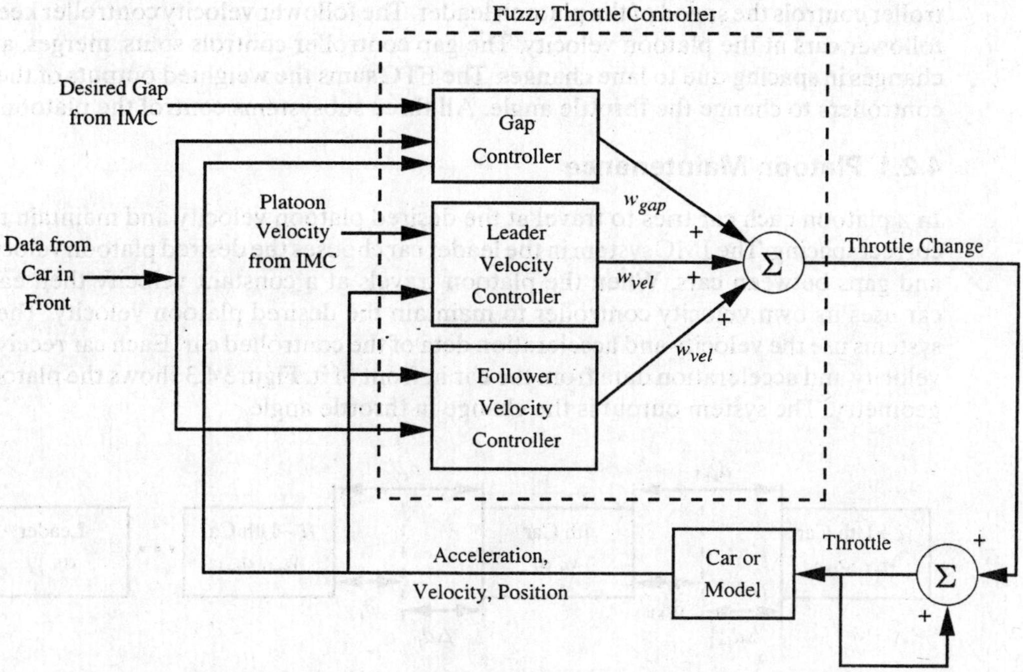

Figure 4.2 Additive fuzzy throttle controller structure.

Platoons travel at high speeds and need precise longitudinal control for safety. We used a hybrid neural-fuzzy system that finds the ellipsoidal rule patches with unsupervised and supervised learning. The hybrid system uses unsupervised learning to pick quickly the first set of ellipsoidal fuzzy rules. Then supervised learning tunes the rules using gradient descent. Each rule defines a fuzzy subset or connected region of state space [2] and thus relates throttle response, acceleration, and velocity. Chapter 3 describes the hybrid ellipsoidal learning system. We used ellipsoidal learning to find the fuzzy rules and sets for a truck velocity controller.

We tested the fuzzy gap controller on the highway. The controller got information from its own sensor and from the platoon goals. The controlled car followed the lead car as it changed speed and went over hills. The system performed smoothly in all cases. We also designed a fuzzy brake controller and tested it in computer simulations.

4.2 FUZZY THROTTLE CONTROLLER

No one knows the nonlinear dynamics of driving a high-speed car on a complex freeway. The true math models are too complex. Fuzzy systems give a model-free estimate of a nonlinear control function.

In our case the fuzzy throttle controller (FTC) performs all the longitudinal maneuvers for the platoon. Figure 4.2 shows the fuzzy subsystems for velocity and gap control that change the throttle angle for cars in the platoon. The velocity controller controls the speed of the platoon leader. The follower velocity controller keeps follower cars at the platoon velocity. The gap controller controls splits, merges, and changes in spacing due to lane changes. The FTC sums the weighted outputs of these controllers to change the throttle angle. All three subsystems control the platoon.

4.2.1 Platoon Maintenance

In a platoon each car tries to travel at the desired platoon velocity and maintain the correct spacing. The IMC system in the leader car chooses the desired platoon velocity and gaps between cars. When the platoon travels at a constant velocity then each car uses its own velocity controller to maintain the desired platoon velocity. These systems use the velocity and acceleration data of the controlled car. Each car receives velocity and acceleration data from the car in front of it. Figure 4.3 shows the platoon geometry. The system output is the change in throttle angle.

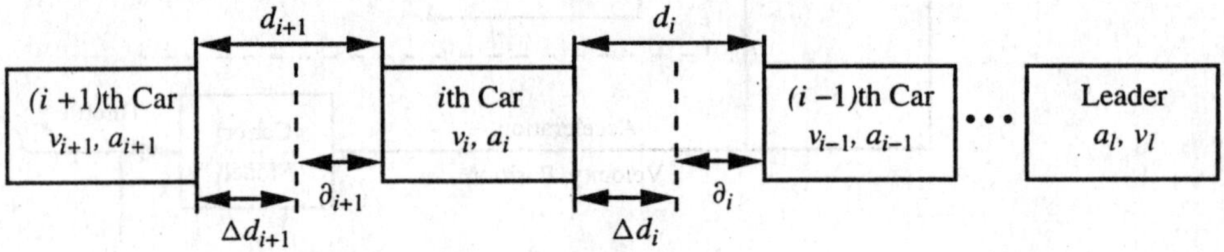

Figure 4.3 Geometry of single-lane car platoon.

The inputs to the velocity controller for the ith car in the platoon are

$$\Delta v_i(t) = v_{\text{platoon}} - v_i(t) \tag{4.1}$$

$$a_i(t) = a_i(t). \tag{4.2}$$

The output is the change in throttle angle $\partial_{\text{throttle}}$. The fuzzy rules for this case take the form IF ($a_i(t)$ is ZE) AND ($\Delta v_i(t)$ is MN) THEN ($\partial_{\text{throttle}}$ is MN).

Figure 4.4 shows the fuzzy sets for the velocity control variables. The sets near equilibrium are smaller to give finer control [9]. Figure 4.5 shows the fuzzy rules we used for the velocity control variables. The fuzzy rules and the sets define a control surface for the velocity controller (Figure 4.6). This surface shows the change in throttle angle for different combinations of Δv and a. The fuzzy system encodes "expert" knowledge for velocity control. It limits the size of accelerations and decelerations for different size changes in velocity. A human driver follows a unique acceleration

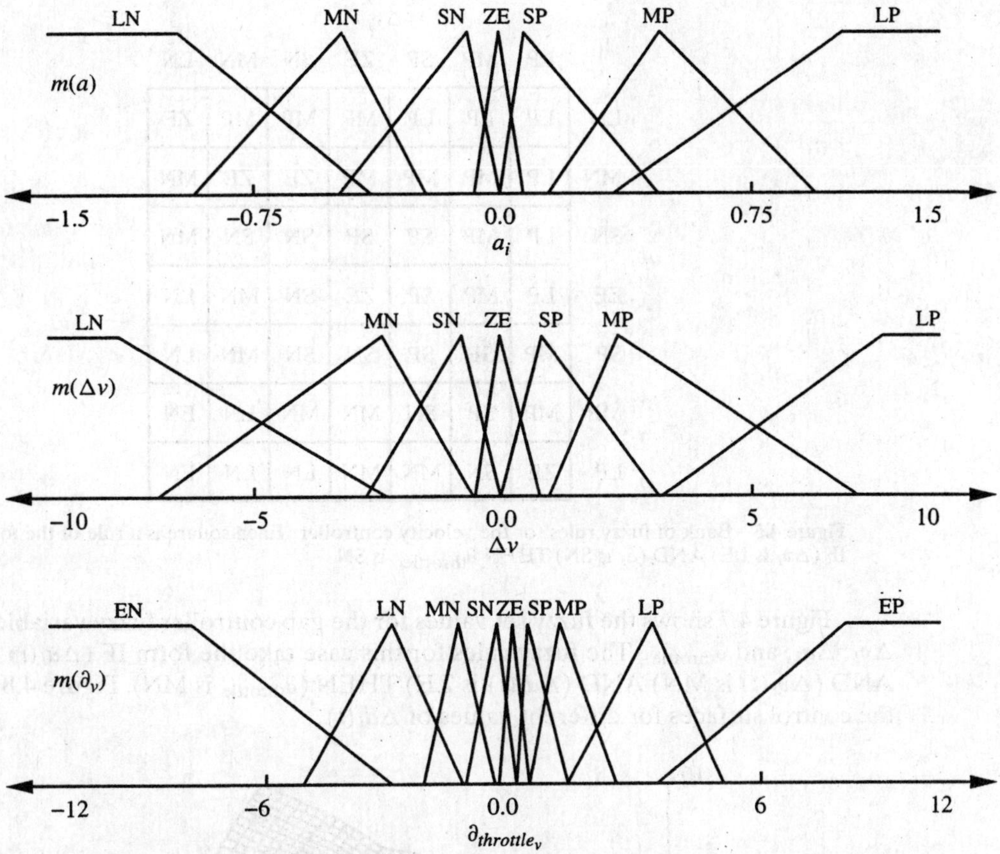

Figure 4.4 Fuzzy sets for the velocity controller. The fuzzy sets near SMALL velocity and SMALL acceleration errors are narrower to give more precise control.

profile in each case. If you wish to speed up a little bit and your acceleration is small then do not apply more gas. If you want to make a large change in speed then use a higher rate of acceleration.

The gap controller corrects the distance error when it is too large. The gap controller for platoon followers uses the differences in acceleration and velocity between cars and the distance error to achieve or maintain a constant spacing. The distance error $\Delta d_i(t)$ is the difference between the desired gap between the cars and the actual gap (Figure 4.3). A range-finding system on each car in the platoon measures the distance between the cars. The inputs for the gap controller in the ith car are

$$\Delta d_i(t) = d_{\text{desired}} - d_i(t) \tag{4.3}$$

$$\Delta v_i(t) = v_{i-1}(t) - v_i(t) \tag{4.4}$$

$$\Delta a_i(t) = a_i(t) - a_{i-1}(t). \tag{4.5}$$

a_i \ Δv_i	LP	MP	SP	ZE	SN	MN	LN
LN	EP	LP	LP	MP	MP	MP	ZE
MN	LP	MP	MP	MP	ZE	ZE	MN
SN	LP	MP	SP	SP	SN	SN	MN
ZE	LP	MP	SP	ZE	SN	MN	LN
SP	MP	SP	SP	SN	SN	MN	LN
MP	MP	SP	SN	MN	MN	LN	EN
LP	ZE	SN	MN	MN	LN	LN	EN

Figure 4.5 Bank of fuzzy rules for the velocity controller. Each square is a rule of the form IF (Δv_i is LE) AND (a_i is SN) THEN $\partial_{\text{throttle}_v}$ is SN.

Figure 4.7 shows the fuzzy set values for the gap controller fuzzy variables Δd_i, Δv_i, Δa_i, and $\partial_{\text{throttle}}$. The fuzzy rules for this case take the form IF ($\Delta a_i(t)$ is ZE) AND ($\Delta v_i(t)$ is MN) AND ($\Delta d_i(t)$ is ZE) THEN ($\partial_{\text{throttle}}$ is MN). Figure 4.8 shows the control surfaces for different values of $\Delta d_i(t)$.

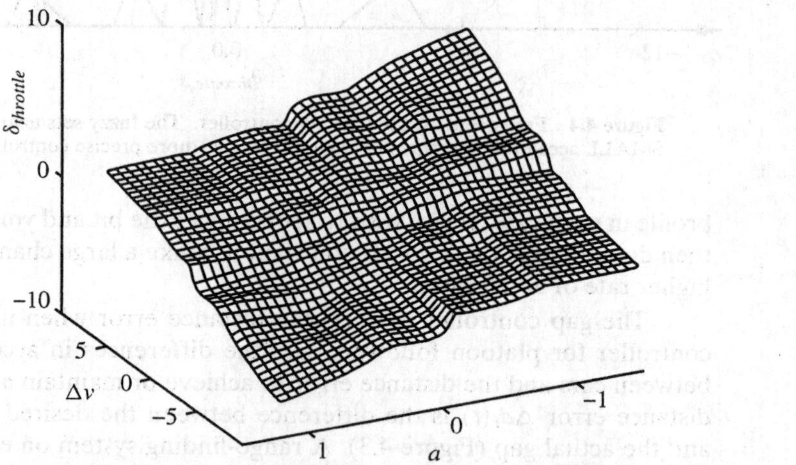

Figure 4.6 Controller surface for the velocity controller. It shows how the system output σ_{throttle} changes with respect to Δv_i and a_i.

The system weights and sums the outputs of the gap and velocity controllers for followers as in Figure 4.2. The weights depend on the error in distance. If the error is large then the gap controller restores the desired spacing. If the error is small then the velocity controller maintains a constant velocity. As the distance error grows the weight on the output of the gap controller grows linearly while the weight on the

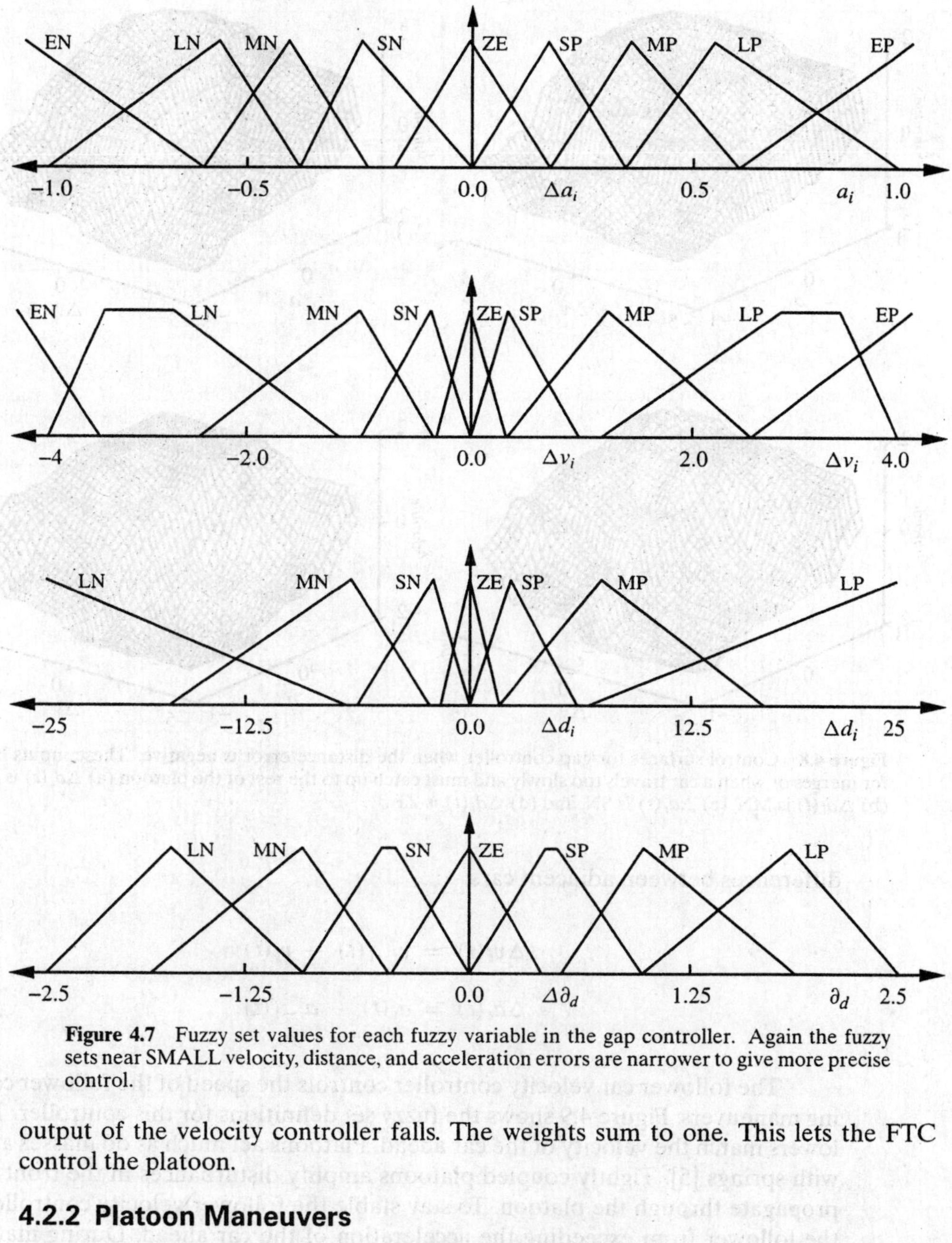

Figure 4.7 Fuzzy set values for each fuzzy variable in the gap controller. Again the fuzzy sets near SMALL velocity, distance, and acceleration errors are narrower to give more precise control.

output of the velocity controller falls. The weights sum to one. This lets the FTC control the platoon.

4.2.2 Platoon Maneuvers

The FTC performs merges, splits, and velocity changes. The maneuvering car uses the velocity controller for velocity changes and the gap controller for splits and merges that involve changes in car spacing. It does not use both at once. The lead car in a merge or split maneuver uses the gap controller to move to the desired spot. The inputs to the follower car velocity controller change to the velocity and acceleration

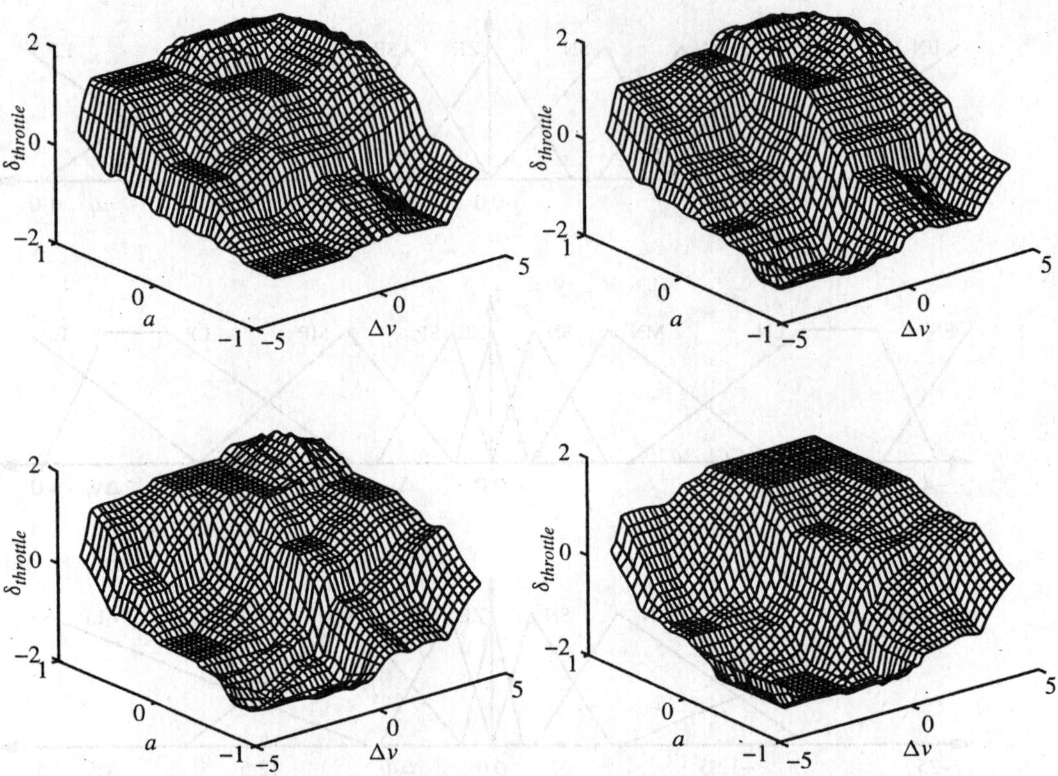

Figure 4.8 Control surfaces for gap controller when the distance error is negative. These inputs hold for merges or when a car travels too slowly and must catch up to the rest of the platoon (a) $\Delta d_i(t)$ is LN, (b) $\Delta d_i(t)$ is MN, (c) $\Delta d_i(t)$ is SN, and (d) $\Delta d_i(t)$ is ZE.

differences between adjacent cars:

$$\Delta v_i(t) = v_{i-1}(t) - v_i(t) \tag{4.6}$$

$$\Delta a_i(t) = a_i(t) - a_{i-1}(t). \tag{4.7}$$

The follower car velocity controller controls the speed of the follower cars during maneuvers. Figure 4.9 shows the fuzzy set definitions for this controller. The followers match the velocity of the car ahead. Platoons act much as do masses attached with springs [5]. Tightly coupled platoons amplify disturbances in the front as they propagate through the platoon. To stay stable the follower velocity controller stops the follower from exceeding the acceleration of the car ahead. During maneuvers we let the gap distance between follower cars increase as the speed changes [5]. This helps prevent much of the "slinky" effect seen in real and simulated platoons. The extra degree of freedom in distance removes much of the oscillation [5]. When the platoon completes the maneuver it redistributes itself to the proper spacing with the gap controller.

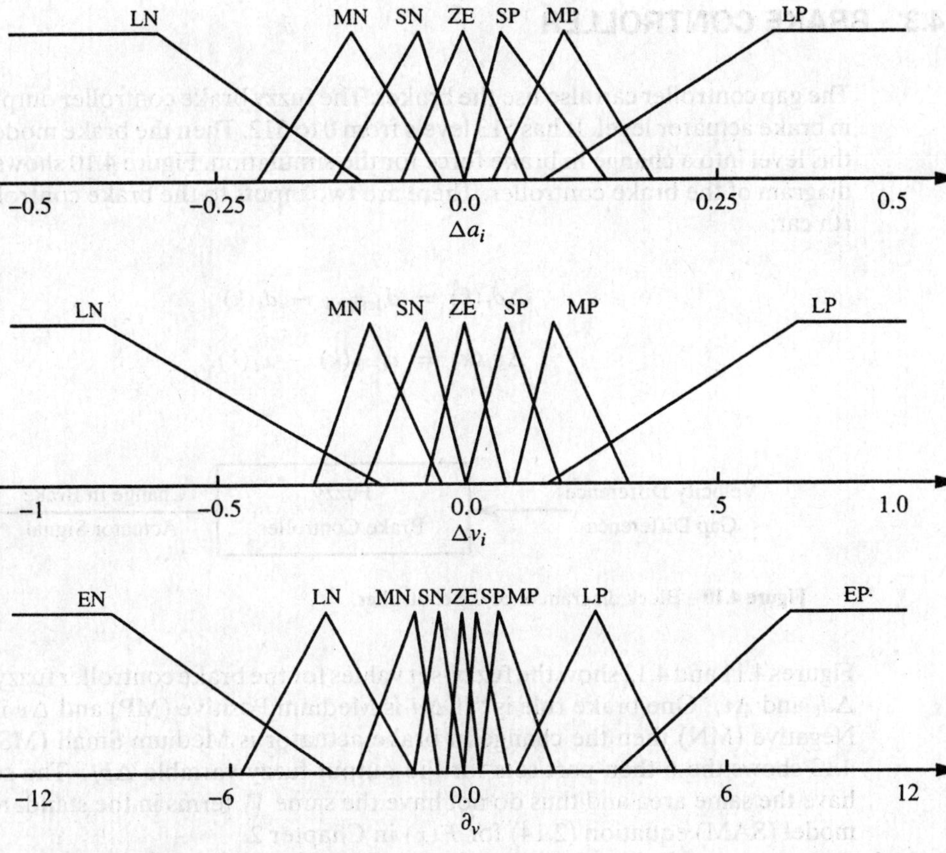

Figure 4.9 Fuzzy sets for the vehicle follower controller.

When a platoon merges or splits the gap controller in the leader moves the car to the desired spacing. When the platoon completes the maneuver the velocity controller in the leader maintains a constant velocity. There are three design rules for merges:

1. IF distance error is Large Negative (LN) THEN accelerate to a catch-up velocity.

2. IF distance error is Medium Negative (MN) THEN decelerate slowly with drag.

3. IF distance error is Small Negative (SN) THEN decelerate faster until you reach desired distance error and velocity difference.

Figure 4.7 lists the fuzzy sets for the gap controller. Figure 4.8 shows the control surfaces for the gap controller for distance errors with fuzzy sets LN, MN, SN, and ZE. Splits use similar rules but they reverse acceleration and deceleration.

4.3 BRAKE CONTROLLER

The gap controller can also use the brakes. The fuzzy brake controller outputs change in brake actuator level. It has 513 levels from 0 to 512. Then the brake model converts this level into a change in brake force for the simulation. Figure 4.10 shows the block diagram of the brake controller. There are two inputs to the brake controller for the ith car:

$$\Delta d_t(k) = d_{\text{desired}} - d_i(k) \tag{4.8}$$

$$\Delta v_i(k) = v_{i-1}(k) - v_i(k). \tag{4.9}$$

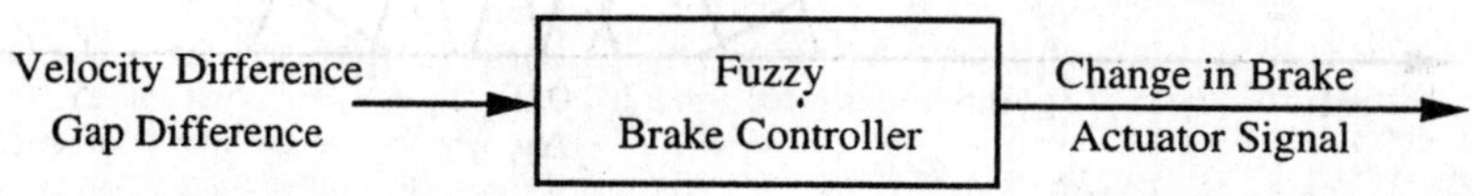

Figure 4.10 Block diagram of brake controller.

Figures 4.11 and 4.12 show the fuzzy-set values for the brake controller fuzzy variables Δd_i and Δv_i. One brake rule is "If Δd is Medium Positive (MP) and Δv is Medium Negative (MN) then the change in brake actuator is Medium Small (MS)." Figure 4.13 shows the 6 then-part sets for the output fuzzy variable Δb_i. The sets do not have the same area and thus do not have the same V_j terms in the standard additive model (SAM) equation (2.14) for $F(x)$ in Chapter 2.

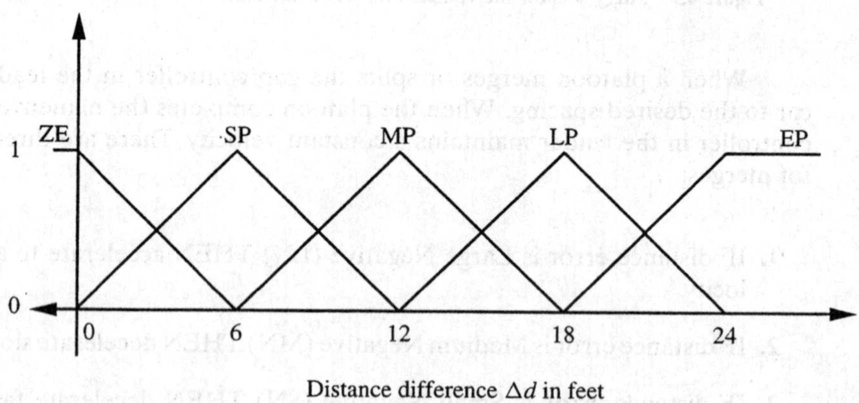

Figure 4.11 If-part fuzzy-set functions for distance difference input to the brake controller.

The brake controller has $5 \times 5 = 25$ fuzzy rules. Figure 4.14 shows the fuzzy rules for the brake controller. Nine fuzzy sets quantize the fuzzy variables Δd_i and Δv_i.

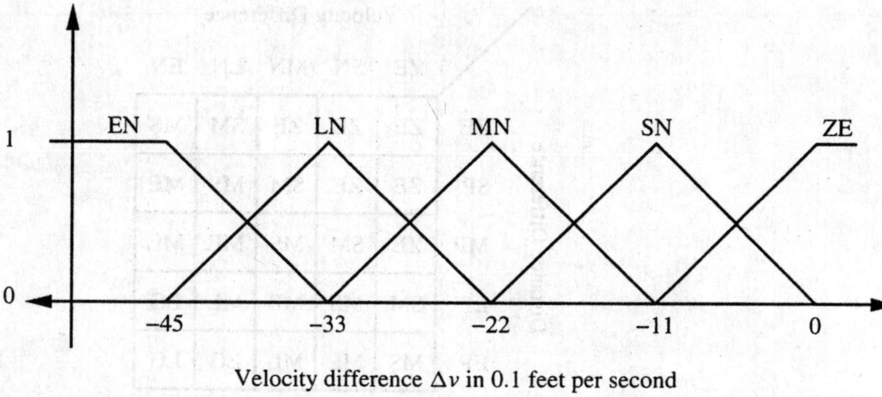

Figure 4.12 If-part fuzzy-set functions for velocity difference input to the brake controller.

EP: Extreme Positive
LP: Large Positive
MP: Medium Positive
SP: Small Positive
ZE: Zero
SN: Small Negative
MN: Medium Negative
LN: Large Negative
EN: Extreme Negative.

Figure 4.15 shows the control surface of the brake controller. The brakes are on only when the distance difference is positive (when the follower car is too close) and the relative velocity is negative (when the follower car goes faster than the lead car).

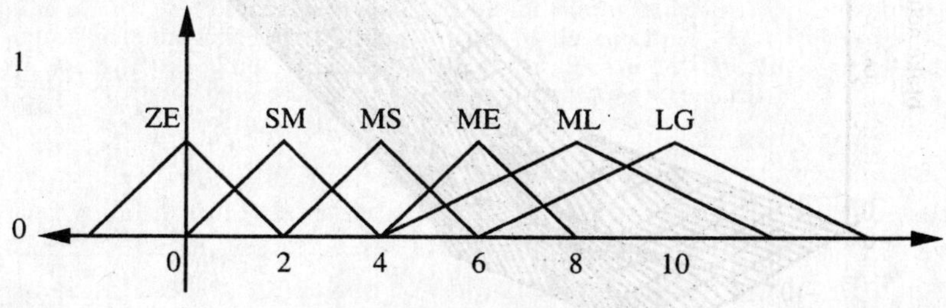

Figure 4.13 Then-part fuzzy sets for the brake actuator change. The standard additive fuzzy system F uses just the centroids c_j and the areas or volumes V_j of the 6 then-part sets when it computes the output $F(x)$.

We combined the throttle and brake outputs using a logic switch that transitions between the brake and the throttle. Figure 4.16 shows how the system used the brake

Distance Difference \ Velocity Difference	ZE	SN	MN	LN	EN
ZE	ZE	ZE	ZE	SM	MS
SP	ZE	ZE	SM	MS	ME
MP	ZE	SM	MS	ME	ML
LP	SM	MS	ME	ML	LG
EP	MS	ME	ML	LG	LG

Figure 4.14 Fuzzy rules for the brake controller. The if-part fuzzy sets are Δv and Δd. The then-part sets give the change Δb in the brake actuator signal.

and the throttle for different distance errors and velocity differences. The brake region shows when the brake is on and the throttle is off. The brake comes on only when the car is closer than the desired distance and the follower car goes faster than the car ahead.

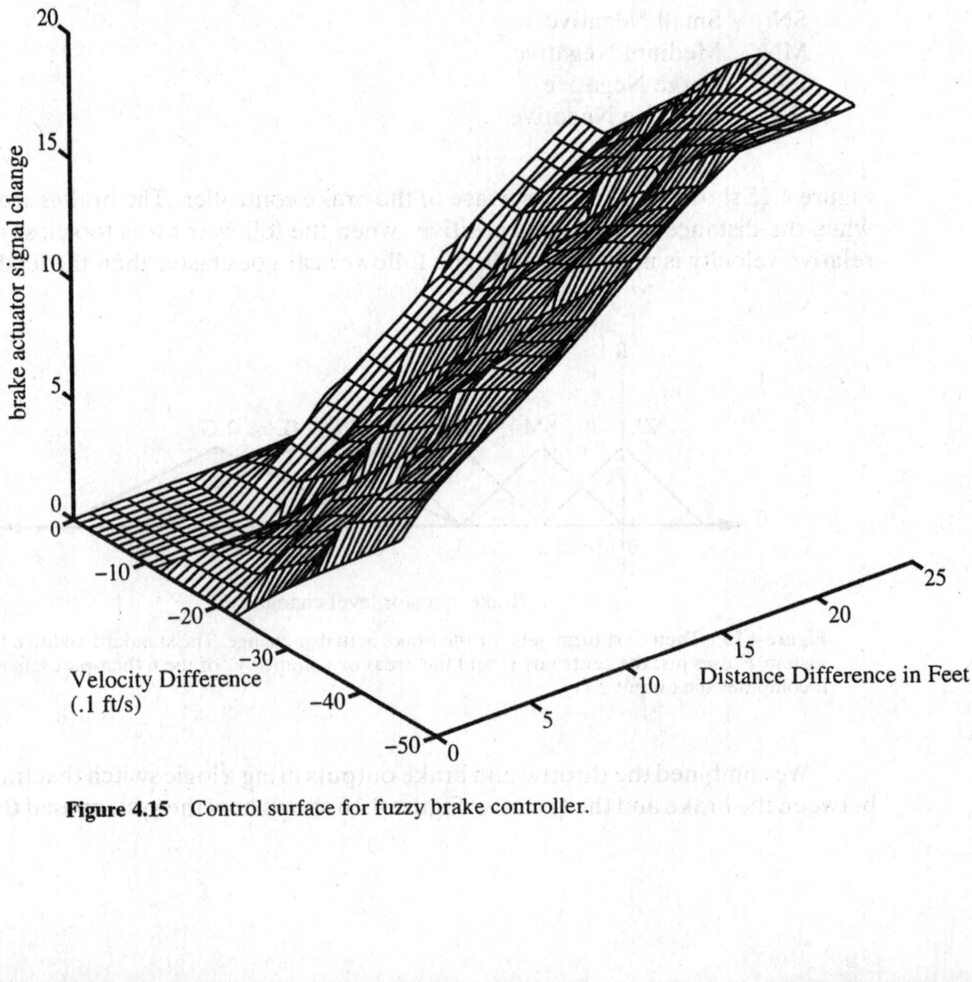

Figure 4.15 Control surface for fuzzy brake controller.

A "neutral region" [5] can help avoid frequent transitions between the throttle and brake fuzzy systems. The brake control signal does not change and the throttle is off when the inputs are in the neutral region. Figure 4.16 shows the neutral region for the fuzzy brake controller. The neutral region covers small values of the inputs Δv and Δd. The neutral region is that part of the control surface in Figure 4.15 that equals zero.

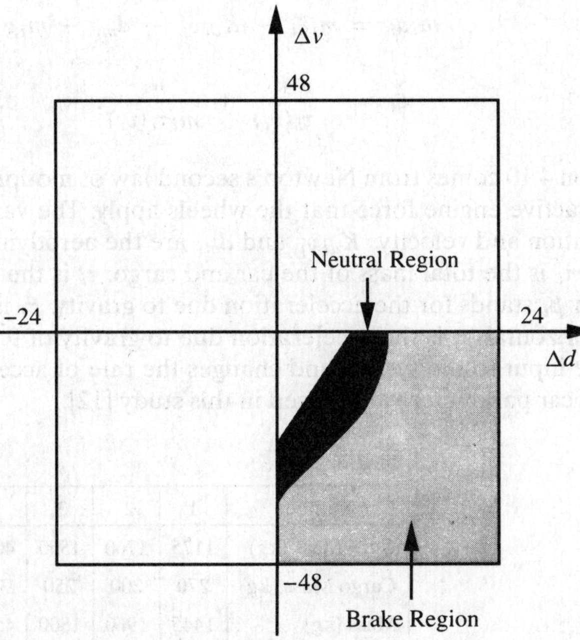

Figure 4.16 Domain of look-up table for the fuzzy gap controller.

The hardware on the test car had limited memory and used only integer operations. It stored the fuzzy gap controller as a look-up table based on the fuzzy sets and rules for throttle and brake control. Figure 4.16 shows the domain for the look-up table. The rectangle shows the control region for the gap controller look-up table. The white region shows the throttle system. The shaded region shows the brake system and the black region shows the neutral region. The horizontal axis gives the distance difference in feet (ft). The vertical axis gives the velocity difference in 1/10 ft/s. We thresholded all inputs outside these limits to the maximum or minimum values of the domain. The fourth quadrant shows the brake look-up table. One fuzzy rule for this case has the form IF ($\Delta v_i(k)$ is MN) AND ($\Delta d_i(k)$ is MP) THEN (∂_{brake} is MS).

The look-up table for the throttle system had $3 \times 49 \times 97 = 14{,}259$ entries. The distance difference varied from -24 ft to 24 ft in 1-foot units and had 49 entries. The velocity difference had 97 entries as it varied from -4.8 ft/s to 4.8 ft/s in 1/10 ft/s units. The look-up table for the brake system had $25 \times 49 = 1{,}225$ entries since it applied only in the fourth quadrant.

4.4 FTC SIMULATION RESULTS

4.4.1 Car Model

We tested the FTC with the dynamic car model in [11]. That study assumed a flat straight horizontal road with no perturbing forces such as wind or hills. In our study we added hills. We used the second-order car model [11]:

$$m_i a_i = m_i \xi_i - K_{d_i} v_i^2 - d_{m_i} - m_i g \sin \beta_i \tag{4.10}$$

$$\dot{\xi}_i = -\frac{\xi_i}{\tau_i(v_i)} + \frac{\mu_i}{m_i \tau_i(v_i)}. \tag{4.11}$$

Equation 4.10 comes from Newton's second law of motion $F = ma$. The term $m_i \xi_i$ is the tractive engine force that the wheels apply. The variables a_i and v_i stand for acceleration and velocity. $K_{di} v_i^2$ and d_{mi} are the aerodynamic and mechanical drag forces. m_i is the total mass of the car and cargo. τ_i is the engine time lag. The term $m_i g \sin \beta_i$ stands for the acceleration due to gravity. β_i is the inclination of the hill from horizontal. g is the acceleration due to gravity of 9.8 m/s^2. The throttle angle u_i is the input to the system and changes the rate of acceleration or jerk. Table 4.1 lists the car parameter values used in this study [12].

Car Number	1	2	3	4
Curb Mass (kg)	1175	1760	1550	4000
Cargo Mass (kg)	270	200	250	325
Mass (kg)	1445	1960	1800	4500
τ (seconds)	0.2	0.25	0.2	0.3
K_d (kg/m)	0.44	0.49	0.52	0.55
d_m (N)	352	392	408	390

TABLE 4.1: Car Parameters for Platoon Simulations [12, 13]

4.4.2 Brake Models

We modeled the brake to test the fuzzy brake controller. The Ford-tuned car model in Figure 4.27 needs an input brake torque. The brake controller gives as output a bit value to the brake actuator.

We modeled the brake using brake data from VORAD Incorporated. The test car was a Lincoln Town Car. This data gave the deceleration based on the brake actuator bit input value and the car velocity. Newton's second law $F = ma$ gave the total force that equals the input bit value for the mass m of 5,460 pounds.

The nonlinear validated longitudinal car model gave a total drag force Y in pounds:

$$Y = 50 + 0.819 \times V + .0192 \times V^2 \tag{4.12}$$

V is the car velocity in miles per hour (mph). Subtracting the drag force from the total force gives the brake force that corresponds to the brake actuator input bit value:

$$F_b = F - Y - D \tag{4.13}$$

F_b is the brake force and D is the force from an external disturbance such as a hill. We interpolated between data points to get the curve in Figure 4.17. It shows the brake force with respect to the actuator bit using $F = ma$ and (4.12) with brake data from the Lincoln Town Car.

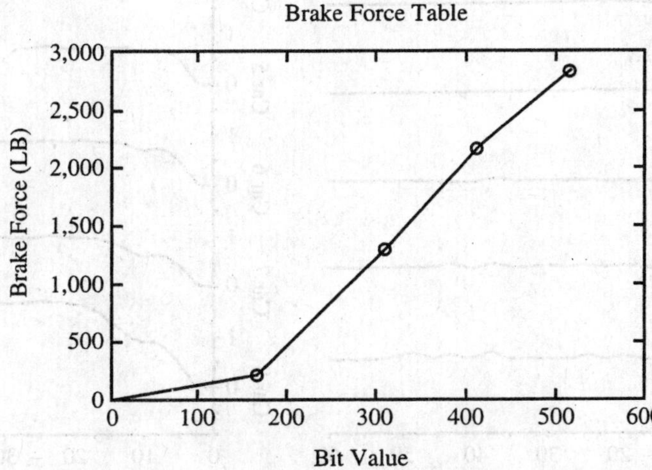

Figure 4.17 Brake force as a function of the actuator bit value.

4.4.3 Fuzzy Controller Results

We simulated platoon velocity changes, splits, and merges with the FTC. The cars used the parameters in Table 4.1 (numbers 1–3). The FTC used the same fuzzy rules and sets for all three cars. Unlike the math-model controllers in [11, 12] the fuzzy controller did not need exact knowledge of the car parameters. The simulation assumed sensor measurement and communication delays of 0.1 seconds for the data from the car in front and 0.03 seconds for the controlled car itself.

Figure 4.18a shows the velocity curves for the cars as an 8-car platoon that increases its velocity from 28 m/s to 32 m/s. Figure 4.18b shows the gap errors. As the platoon speeds up the follower cars drop back. When the leader reaches the new speed the followers move back into position one by one. The gap controller positions each car. At the end of 55 seconds the entire platoon is at the new velocity with the proper spacing restored.

Figure 4.19 shows the results of merging a four-car platoon with a leading two-car platoon. These figures show the distance and velocity profiles for the merging platoon. The two-car platoon in the front moves at the same velocity during the

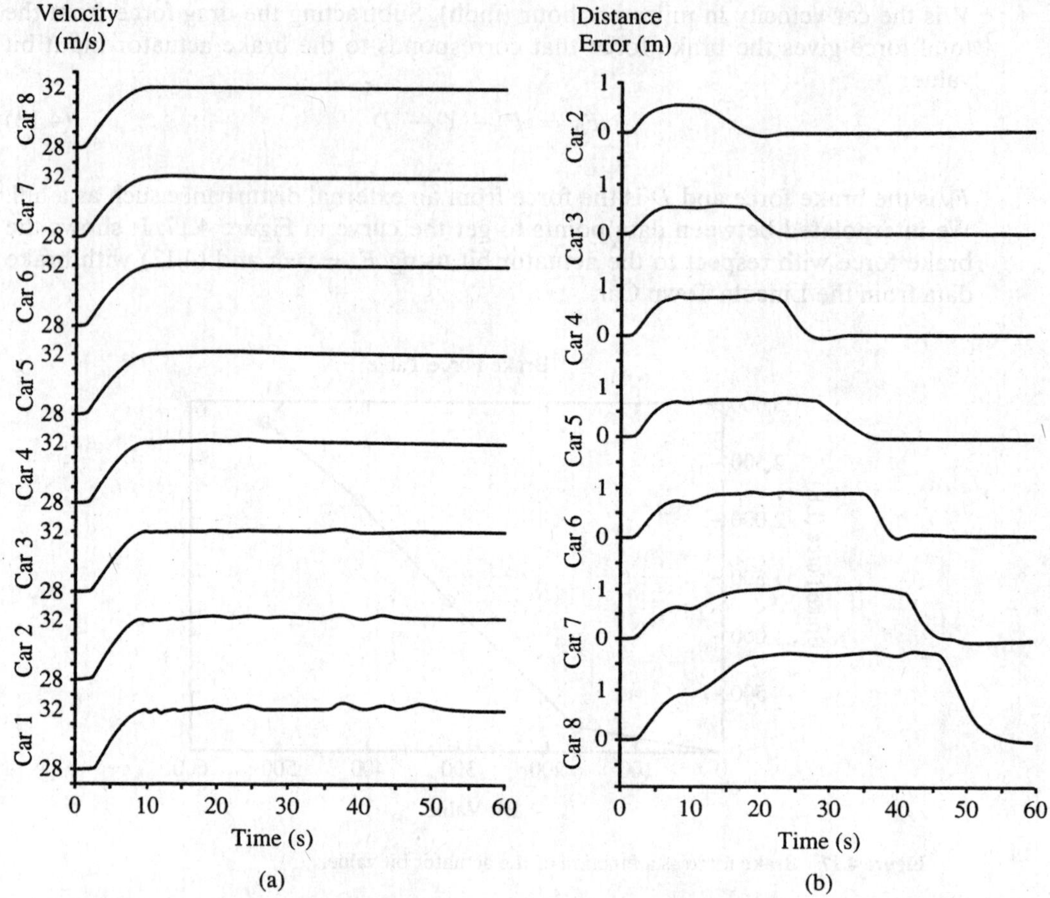

Figure 4.18 Velocity and gap profiles for a platoon changing velocity from 28 to 32 m/s. (a) The velocity for each car in the platoon. (b) The distance error for each car in the platoon. As the platoon accelerates then each car drops back from the car in front of it.

entire maneuver. The following platoon accelerates to a constant velocity faster than the lead platoon. When the mergers reach a medium negative distance away from the car ahead then they slow down with drag. The following platoon makes fine changes to nudge itself into position. The gradual accelerations of the merging platoon leader allow follower cars to keep pace. At the end of the merge the cars regain their desired gap one car at a time. The maneuver takes 30 seconds.

Figure 4.20 shows the platoon as it drives over hills. The platoon stays together even as it drives up and down hills with grades of up to 10%. But long steep downhill grades cause the cars to go too fast. As the cars go down hills they creep up on the cars in front. The following distance grows as the cars go up hills with an 8% grade (Figure 4.20a). The FTC has only throttle control. So the cars accelerate down steep hills even when no throttle input occurs. Figure 4.20b shows how the platoon velocity increases due to acceleration from gravity. The platoon system needs brake control on steep hills to keep its cars from going too fast.

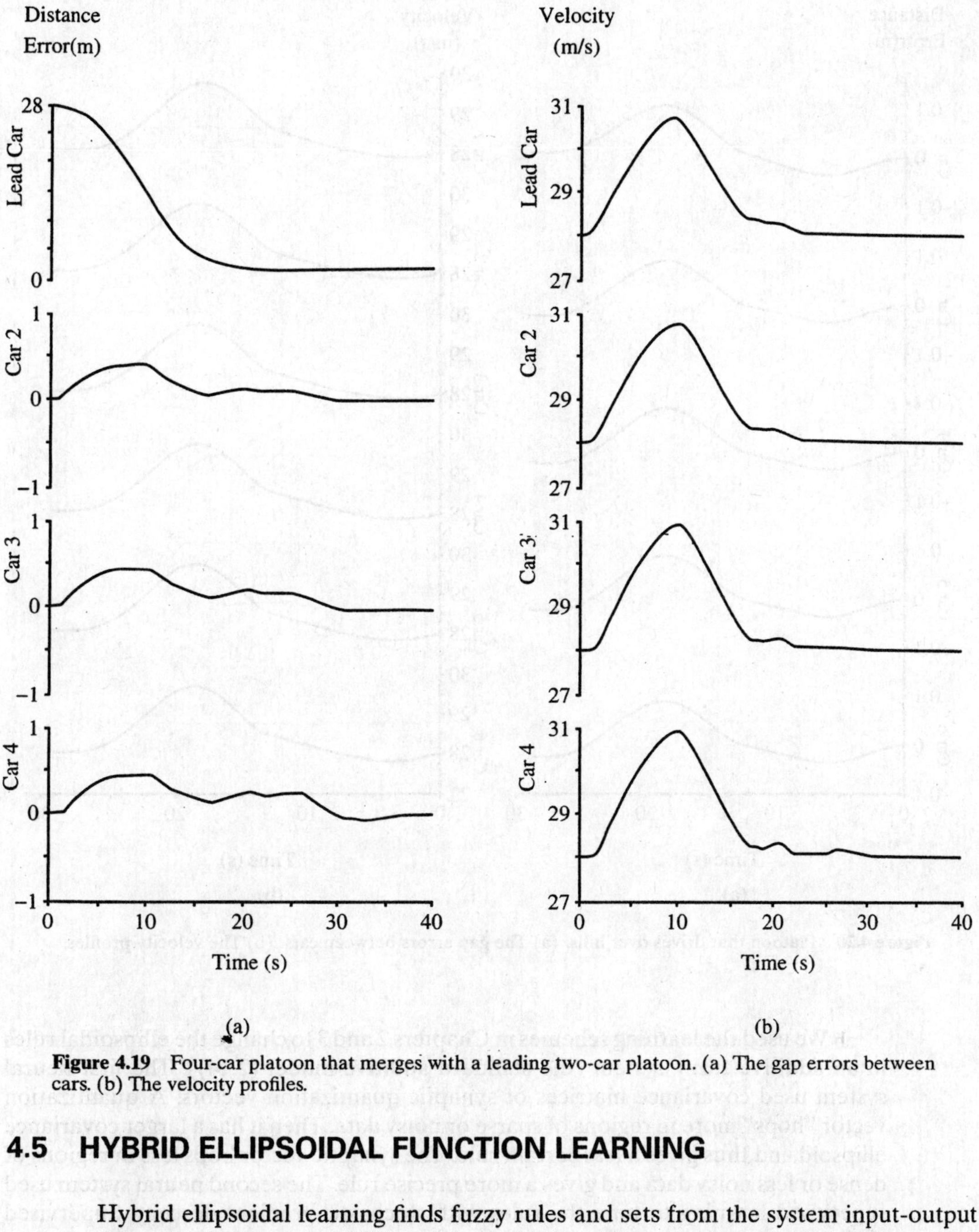

Figure 4.19 Four-car platoon that merges with a leading two-car platoon. (a) The gap errors between cars. (b) The velocity profiles.

4.5 HYBRID ELLIPSOIDAL FUNCTION LEARNING

Hybrid ellipsoidal learning finds fuzzy rules and sets from the system input-output data. We designed the FTC system in Section 4.3 by hand. Rules and sets approximated the desired system behavior. A fuzzy system learns or adapts when its rules change. Then the system function $f\colon X \rightarrow Y$ varies with time. Neural networks can change the sets with training data or error reinforcement signals.

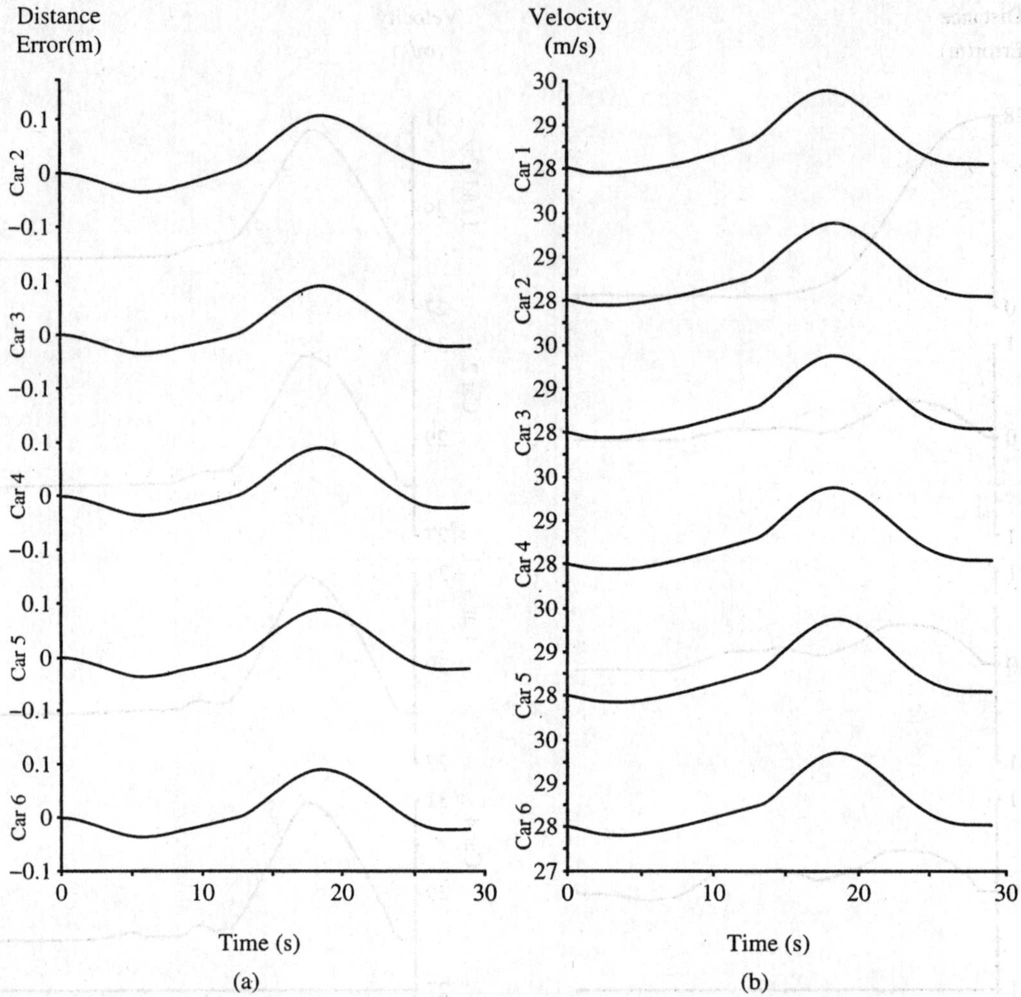

Figure 4.20 Platoon that drives over hills. (a) The gap errors between cars. (b) The velocity profiles.

We used the learning schemes in Chapters 2 and 3 to change the ellipsoidal rules in an additive fuzzy system (the standard additive model (2.14)). The first neural system used covariance matrices of synaptic quantization vectors. A quantization vector "hops" more in regions of sparse or noisy data. Then it has a larger covariance ellipsoid and thus gives a less certain rule. The synaptic vector hops less in regions of dense or less noisy data and gives a more precise rule. The second neural system used supervised learning to tune the rules with stochastic gradient descent. Supervised learning needs an error signal. The neural net must know the function it tries to approximate or at least know some input-output samples from it. Supervised learning changes the centers and eigenvalues of the ellipsoids. This moves, shapes, and orients the ellipsoids in the state space.

Supervised learning takes far more computation than does unsupervised learning. But it gives rules that better approximate the function [2]. At best the supervised gradient descent finds the local minima of the error surface. How well it learns depends on how well we pick the first set of fuzzy rules as discussed in Chapter 3.

We combined the two types of learning in a hybrid neural system (Figure 4.21) that both finds and tunes the ellipsoidal rules [2]: Unsupervised learning picks the first set of rules based on the second-order statistics of the training data. Then supervised learning tunes these rule ellipsoids. We tested the hybrid system with the car model given in Section 4.4.1. Unsupervised learning picks the first set of rules based on the second-order statistics of the training data. Then supervised learning tunes these rule ellipsoids. We tested the hybrid system with the model given in Section 4.4.1.

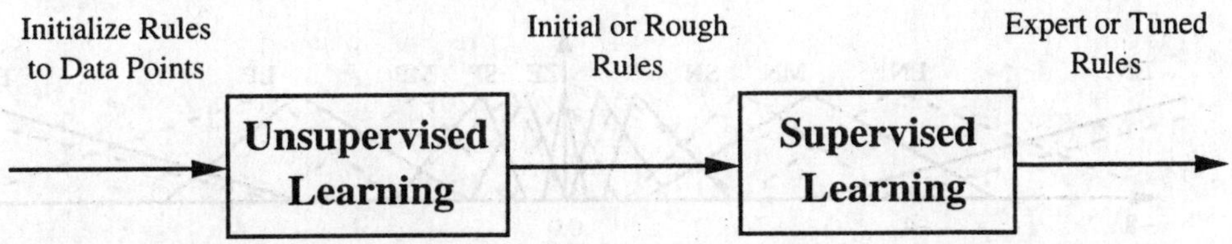

Figure 4.21 The hybrid neural system combines unsupervised and supervised neural learning to find and tune the ellipsoidal fuzzy rules.

4.6 CONTROL SURFACE LEARNING

The leader velocity controller learned its rules from the car model in [10]. The leader velocity controller in [1] generated 7,500 training samples in 200 trajectories for a sport utility vehicle. The training vectors ($\Delta\mathbf{a}$, $\Delta\mathbf{v}$, $\Delta\theta$) defined points in the three-dimensional input-output space. Unsupervised ellipsoid covariance learning clustered the data and computed its local statistics. The AVQ system had 450 synaptic vectors or local pattern classes. The sum of the ellipsoid projections onto each axis of the state space gave a histogram of the density of the pattern classes. We chose seven sets in each of the input dimensions. The center of each fuzzy rule set matched a peak in the histogram.

We then partitioned the state space into a grid of possible rule patches. We found the rules by counting the number of synaptic vectors in each cell. Clusters of synaptic vectors in fuzzy rule cells defined the rules [9]. Figure 4.22 shows the fuzzy sets after unsupervised ellipsoidal learning. The sets for the truck are larger and coarser than those for the smaller and lighter cars.

Figure 6.2 shows the results of changing the platoon velocity for the truck using the hand-tuned and unsupervised fuzzy sets. The controller found with unsupervised learning had no overshoot when the car sped up.

Next we used supervised learning to tune the rules. We optimized the rules to minimize mean-squared error for the training data. There were 49 rules. The supervised system took 30,000 iterations to refine the data. Figure 4.24 shows the control surfaces for the unsupervised and hybrid systems.

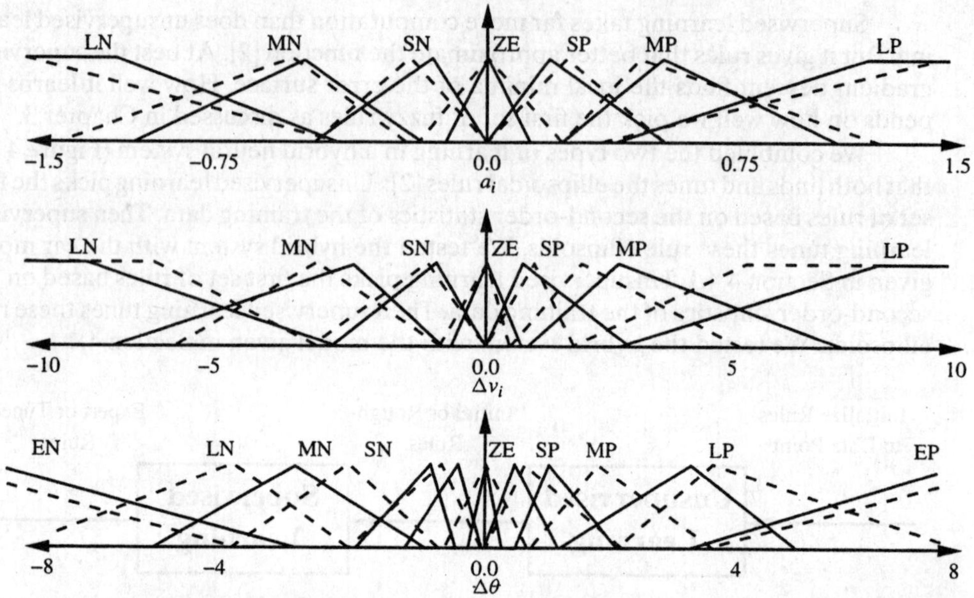

Figure 4.22 Comparison of the fuzzy sets for the velocity controller. The solid lines show the fuzzy sets from the hand-designed FTC in Section 4.2. The dashed lines show the fuzzy sets found with unsupervised ellipsoidal learning.

Figure 4.25 shows the results of changing the platoon velocity for the truck using the unsupervised and hybrid controllers. Both controllers had no overshoot when the car sped up. The hybrid controller accelerated faster than did the unsupervised learning system. The hybrid controller had no overshoot at the desired speed. The hybrid controller performed better when the car slowed down.

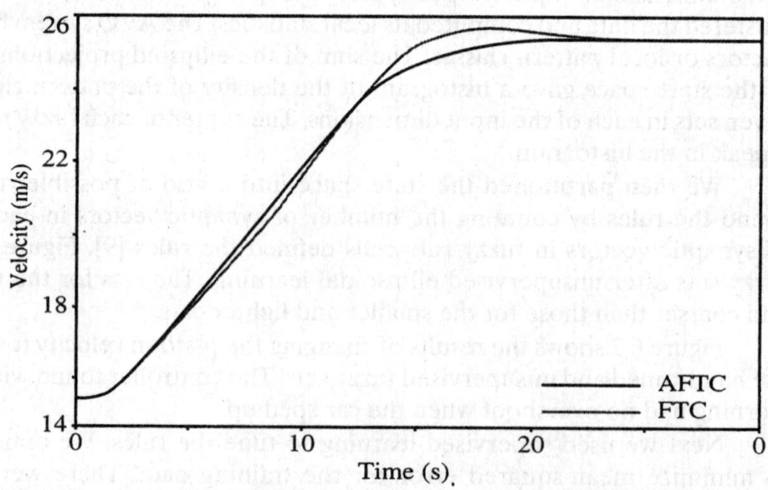

Figure 4.23 Comparison of the FTC from Section 4.2 and the adaptive controller for sport utility vehicles. Ellipsoid covariance learning found the fuzzy rules and sets.

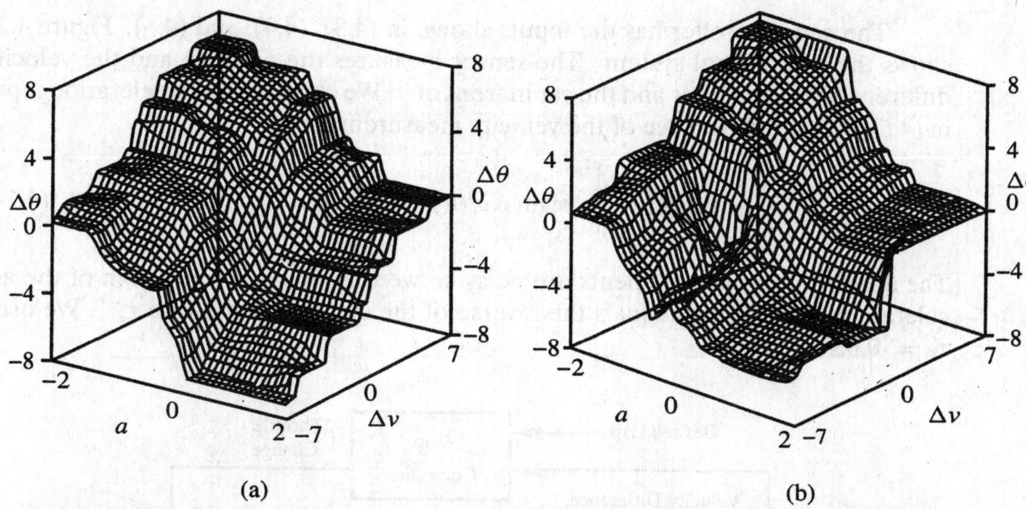

Figure 4.24 Ellipsoids estimate the control surface for a car velocity controller. (a) Control surface after unsupervised learning. (b) Control surface for the hybrid controller after supervised learning tuned the fuzzy sets.

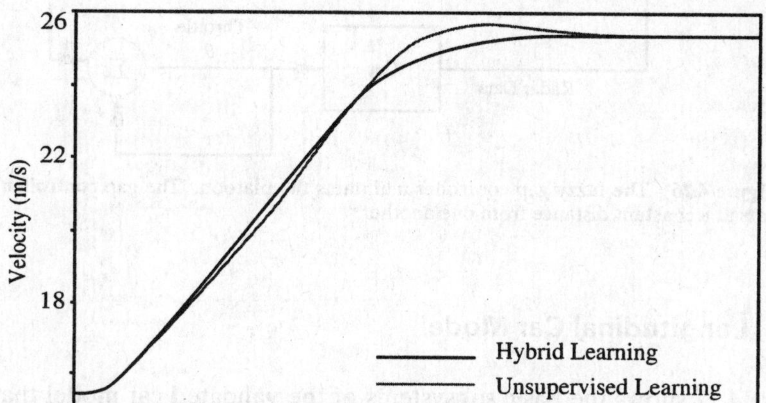

Figure 4.25 Comparison of the fuzzy throttle controller found with unsupervised ellipsoidal learning and the hybrid adaptive controller for sport utility vehicles. The hybrid accelerates faster with no overshoot.

4.7 GAP CONTROL TEST

We tested the gap controller on highway I–15 in Escondido, California. First we tested small platoons with a realistic car model. Then we put the controller in a real car from VORAD Incorporated. In this test the follower cars got data from only their sensors.

The gap controller has the inputs shown in (4.3), (4.4), and (4.5). Figure 4.26 shows the gap control system. The sensor measures the distance and the velocity difference between a car and the car in front of it. We estimate the acceleration input in (4.5) with the difference of the velocity measurements:

$$\Delta a_i(t_k) = \text{sign}(\Delta v_i(t_k) - \Delta v_i(t_{k-1}))c_a. \tag{4.14}$$

The acceleration measurements are noisy so we use only the scaled sign of the acceleration. The term c_a equals the inverse of the system update rate τ_s^{-1}. We used $\tau_s = 0.05$ seconds.

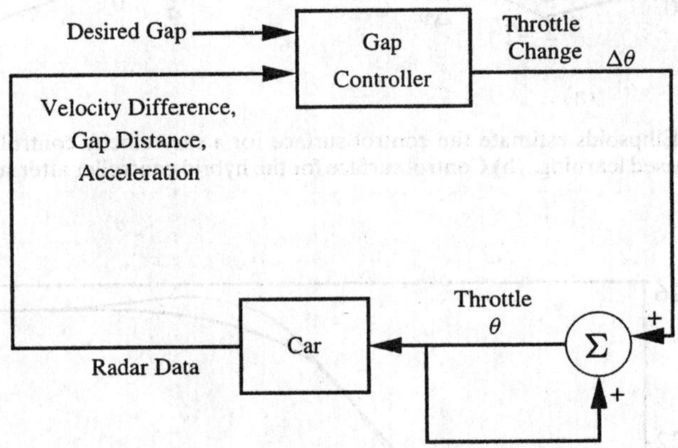

Figure 4.26 The fuzzy gap controller maintains the platoon. The gap controller keeps the cars at a constant distance from one another.

4.7.1 Longitudinal Car Model

Figure 4.27 shows the basic subsystems of the validated car model that we used to test our controller. The inputs to this model are the throttle angle and the brake torque. The outputs are the speed and acceleration of the car. This study looks only at throttle control. We set the brake torque to zero. The system has a mechanical delay of 0.25 seconds.

Each block of the model is a car subsystem. The output of the engine subsystem is the engine torque. The engine torque is a nonlinear function of the air-fuel ratio, the exhaust gas recirculation, the cylinder total mass charge, the spark advance, the engine speed, the drivetrain load, and the throttle angle [6].

The transmission transfers engine torque to the drivetrain. How the transmission performs depends on the car speed and the engine condition. The transmission is automatic with hydraulic torque coupling and four forward gears. For a given gear state the transmission torque is a linear function of the engine torque. The gear state is a nonlinear function of the throttle angle, engine speed, and car speed [6].

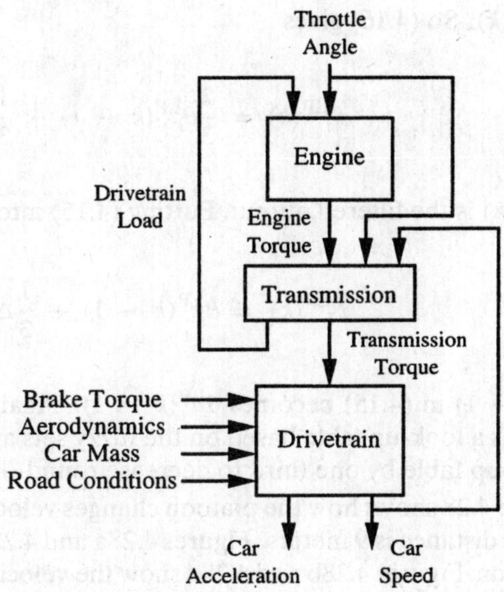

Figure 4.27 Longitudinal car model [6].

The drivetrain transfers engine and brake torque into the car dynamics. The outputs are the speed and acceleration of the car. The outputs are a nonlinear function of engine torque, road conditions, aerodynamic drag, and car mass [6].

We assume a pulse-doppler radar system. The radar measures the distance and the velocity difference between the computer-controlled car and the car ahead. The radar has a measurement delay of 0.05 seconds. The radar locks on a target on a car ahead. When the system acquires a new target on a turn it may lock onto a target in another lane.

4.7.2 Simulation Results

We simulated small platoons over a range of velocity changes. Our goal was a real test on the freeway. So we restricted the memory size and complexity of the controller. We stored the fuzzy controller as a look-up table based on the fuzzy sets and rules.

The output of the throttle controller in the ith car is the change in the throttle angle $\Delta\theta_i(k)$. The input to the car is $\theta_i(k)$

$$\theta_i(k) = \theta_i(k-1) + \Delta\theta_i(k) \tag{4.15}$$

where $\theta_i(k-1)$ is the prior input to the car. A low-pass filter $H_{\text{LP}}(z)$

$$H_{\text{LP}}(z) = \frac{\frac{1}{3}}{1 - \frac{2}{3}z^{-1}} \tag{4.16}$$

smooths $\theta_i(k)$. So (4.16) gives

$$\theta_i^{\text{LP}}(k) = \frac{2}{3}\theta_i^{\text{LP}}(k - 1) + \frac{1}{3}\theta_i(k) \tag{4.17}$$

where $\theta_i^{\text{LP}}(k)$ is the filtered output. Putting (4.15) into (4.17) gives

$$\theta_i^{\text{LP}}(k) = \theta_i^{\text{LP}}(k - 1) + \frac{1}{3}\Delta\theta_i(k) \tag{4.18}$$

since $\theta_i(k - 1)$ in (4.15) becomes $\theta_i^{\text{LP}}(k - 1)$. Again we stored the fuzzy throttle controller as a look-up table based on the fuzzy sets and rules. We scaled the output of the look-up table by one third to decrease round-off errors.

Figure 4.28 shows how the platoon changes velocity as the terrain changes. The desired gap distance is 9 meters. Figures 4.28a and 4.28c show the gaps between cars in the platoon. Figures 4.28b and 4.28d show the velocity profiles of the follower cars.

4.7.3 Test Results

We tested the gap controller on highway I–15 in Escondido, California. We put our controller in a Lincoln Town Car from VORAD Incorporated. The follower car got data from the radar system on the front of the car.

The radar measured the distance and the velocity difference between the computer-controlled car and the car ahead. The radar tracked the car gap as the platoon accelerated. Figure 4.29a shows the follower car gap as the platoon accelerated. Figure 4.29b shows the closing rate between the cars. Figure 4.29c shows the throttle value as the car accelerated. The desired gap was 125 feet. The follower car dropped back because the initial gap was too short.

The platoon went up and down hills in the second test. The desired gap was again 125 feet. The follower car dropped back as the platoon started up the hill. Figure 4.30(a) shows the gap distance as the platoon went up a hill. The follower dropped back and then moved to the right gap. Figure 4.30(b) shows the closing rate between the cars. The spike at 15 seconds occurred when the radar sensor briefly lost the lead car in the platoon. The follower car maintained a constant throttle until the sensor detected a new target. Figure 4.30(c) shows the throttle value as the car went up the hill.

4.7.4 Gap Controller with Throttle and Brake

The gap controller must use the brakes to slow the car if the engine torque is not sufficient. We simulated cases where braking can avoid a collision as when the platoon moves up and down hills or when the platoon slows down.

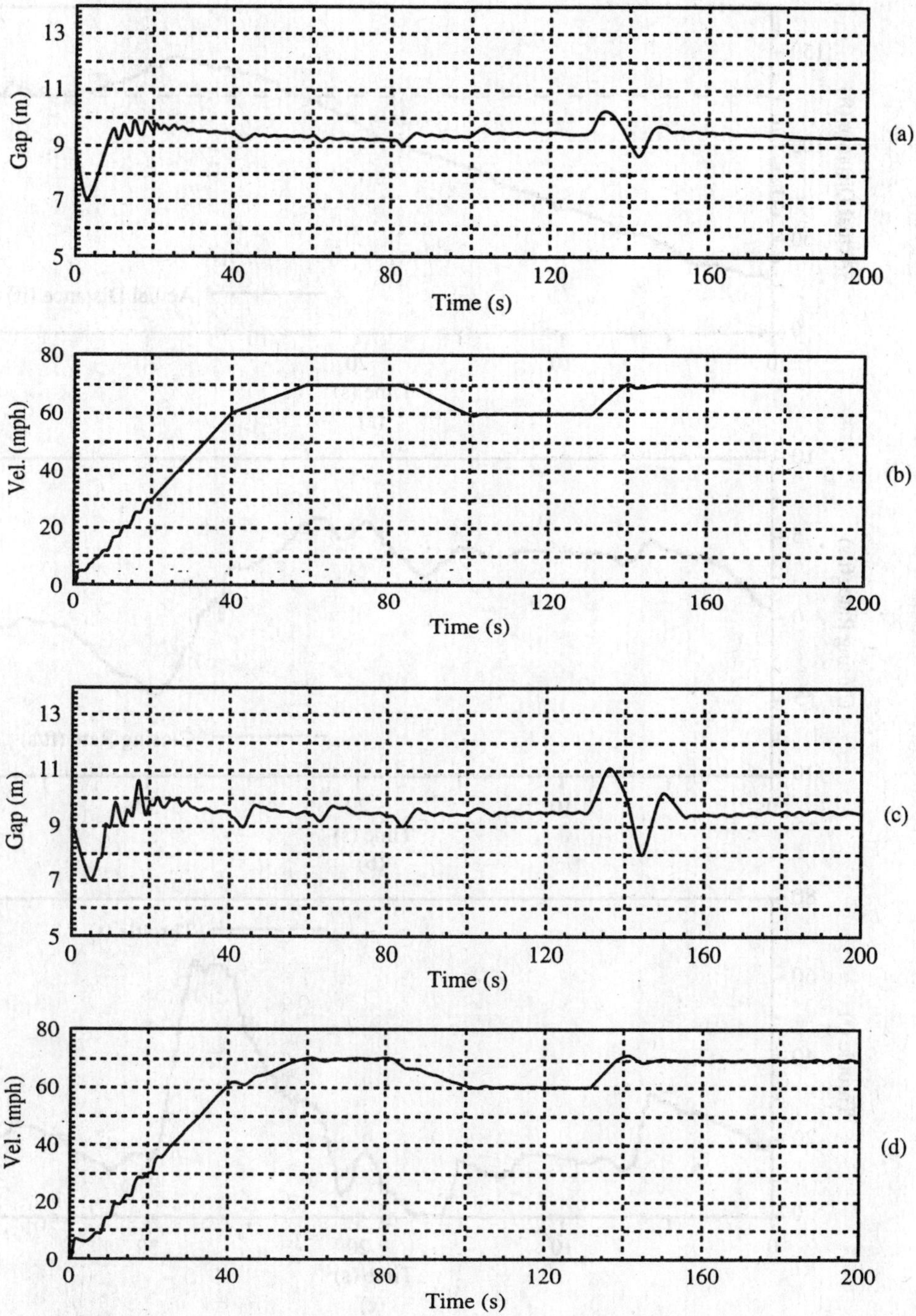

Figure 4.28 Simulation results for follower cars with a single sensor and no communication between cars: (a) and (c) show the gaps in front of the second and third follower cars. (b) and (d) show the speeds of the second and third follower cars.

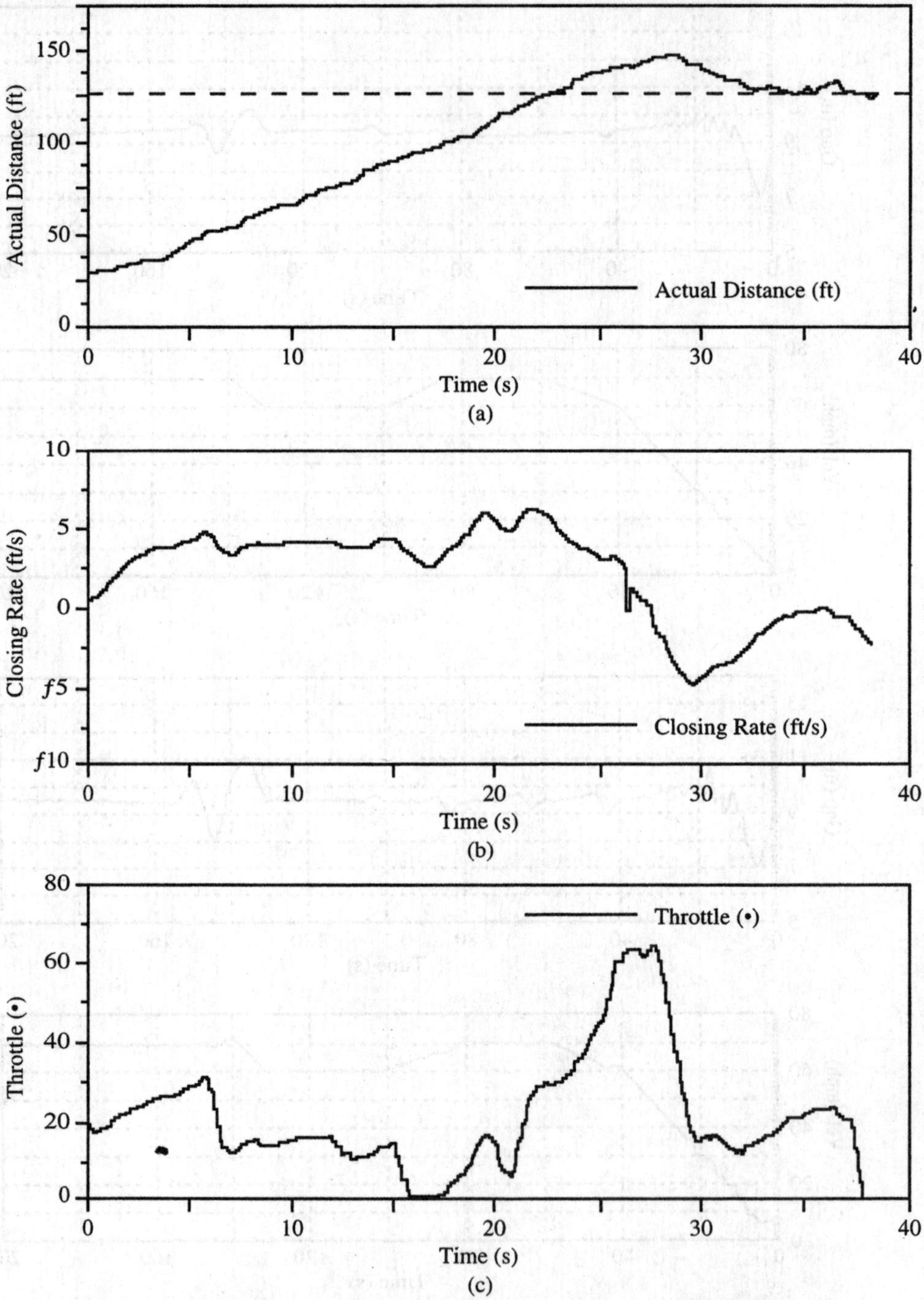

Figure 4.29 Follower car data as a two-car platoon accelerates on the highway. (a) The distance between the lead and follower cars. (b) Closing rate of the velocity differences between the cars. (c) Throttle values for the follower cars.

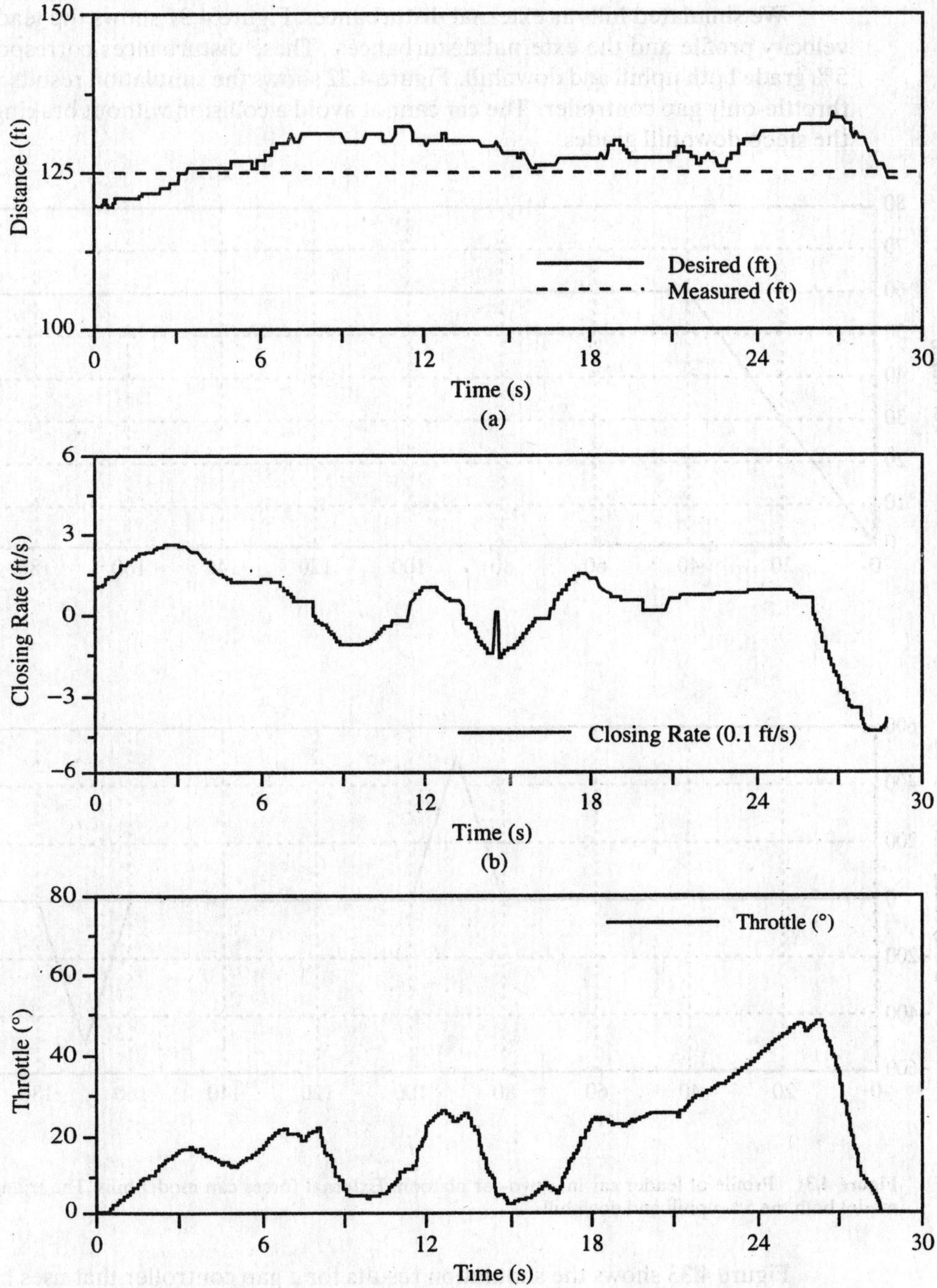

Figure 4.30 Follower car data as a two-car platoon climbs a hill. (a) The distance between the lead and follower cars. (b) Closing rate of the velocity differences between the cars. The spike at 15 seconds shows where the radar sensor briefly lost the lead car in the platoon. (c) Throttle values for the follower cars.

We simulated hills as external disturbances. Figure 4.31 shows the leader car's velocity profile and the external disturbances. These disturbances correspond to a 5% grade both uphill and downhill. Figure 4.32 shows the simulation results with the throttle-only gap controller. The car cannot avoid a collision without braking due to the steep downhill grades.

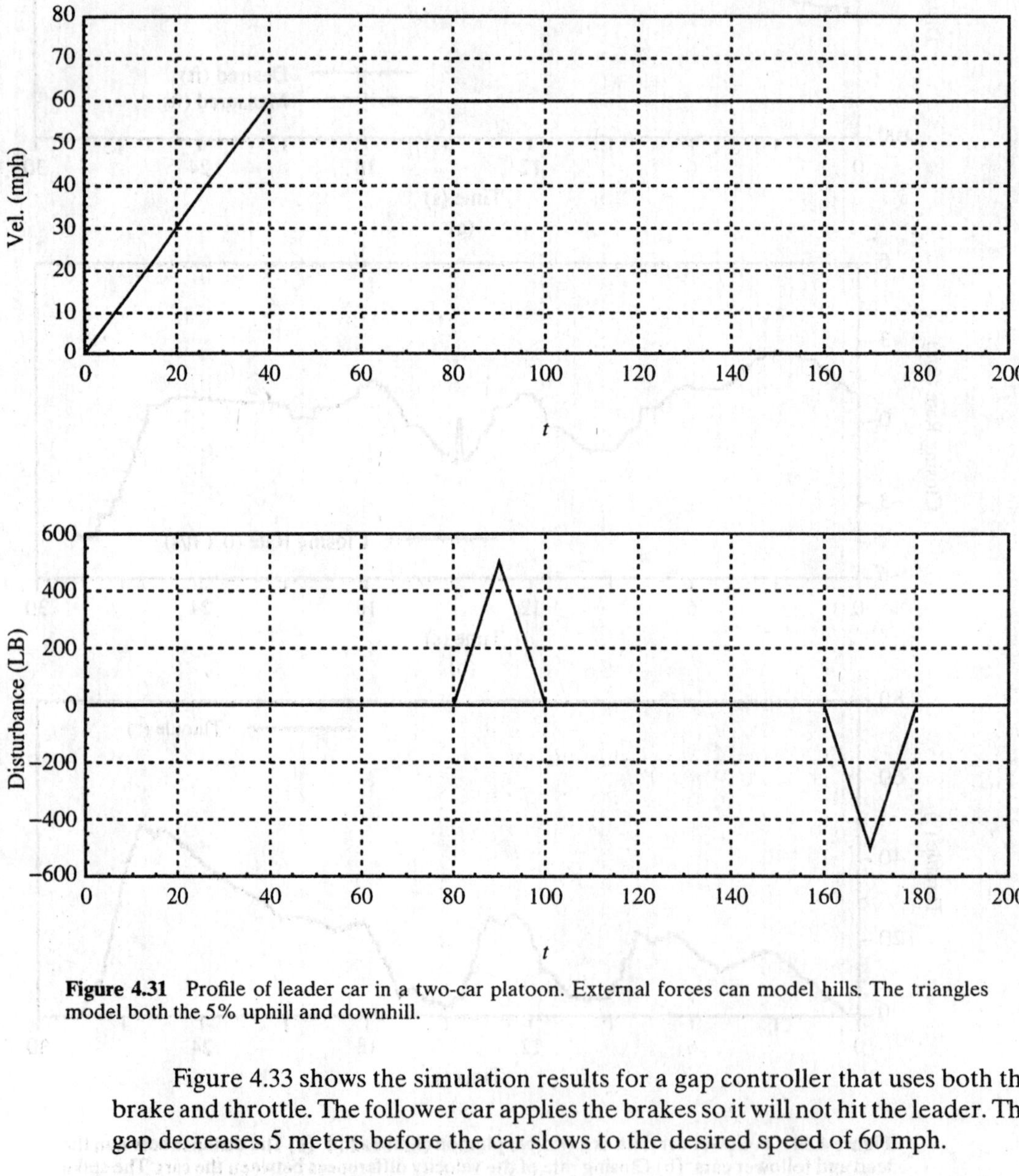

Figure 4.31 Profile of leader car in a two-car platoon. External forces can model hills. The triangles model both the 5% uphill and downhill.

Figure 4.33 shows the simulation results for a gap controller that uses both the brake and throttle. The follower car applies the brakes so it will not hit the leader. The gap decreases 5 meters before the car slows to the desired speed of 60 mph.

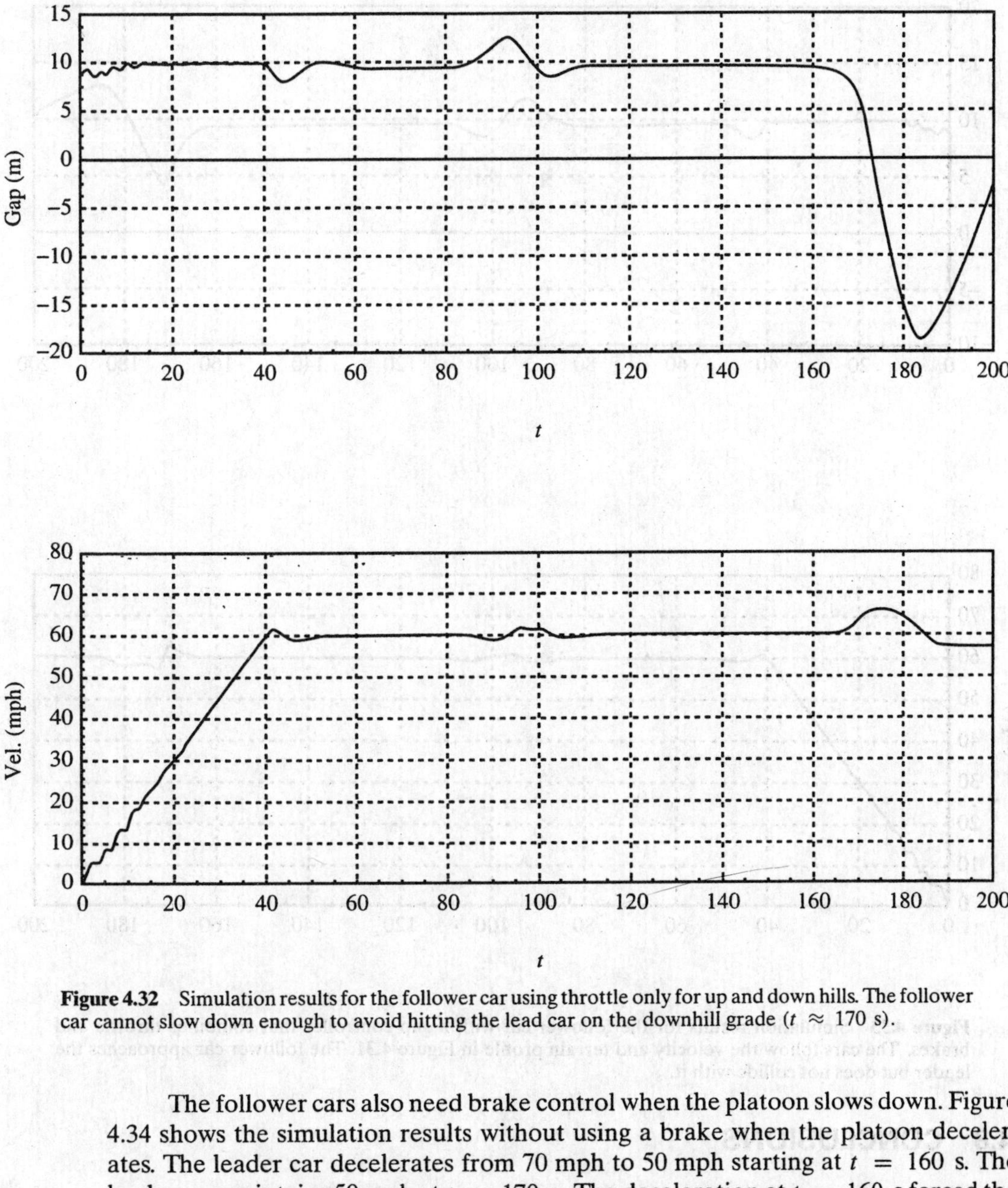

Figure 4.32 Simulation results for the follower car using throttle only for up and down hills. The follower car cannot slow down enough to avoid hitting the lead car on the downhill grade ($t \approx 170$ s).

The follower cars also need brake control when the platoon slows down. Figure 4.34 shows the simulation results without using a brake when the platoon decelerates. The leader car decelerates from 70 mph to 50 mph starting at $t = 160$ s. The leader car maintains 50 mph at $t = 170$ s. The deceleration at $t = 160$ s forced the follower car to brake to avoid a collision. Engine torque cannot slow the car down enough. Figure 4.35 shows the simulation results of the combined throttle and brake system. The combined system avoids the collision and does not oscillate between the throttle and brake.

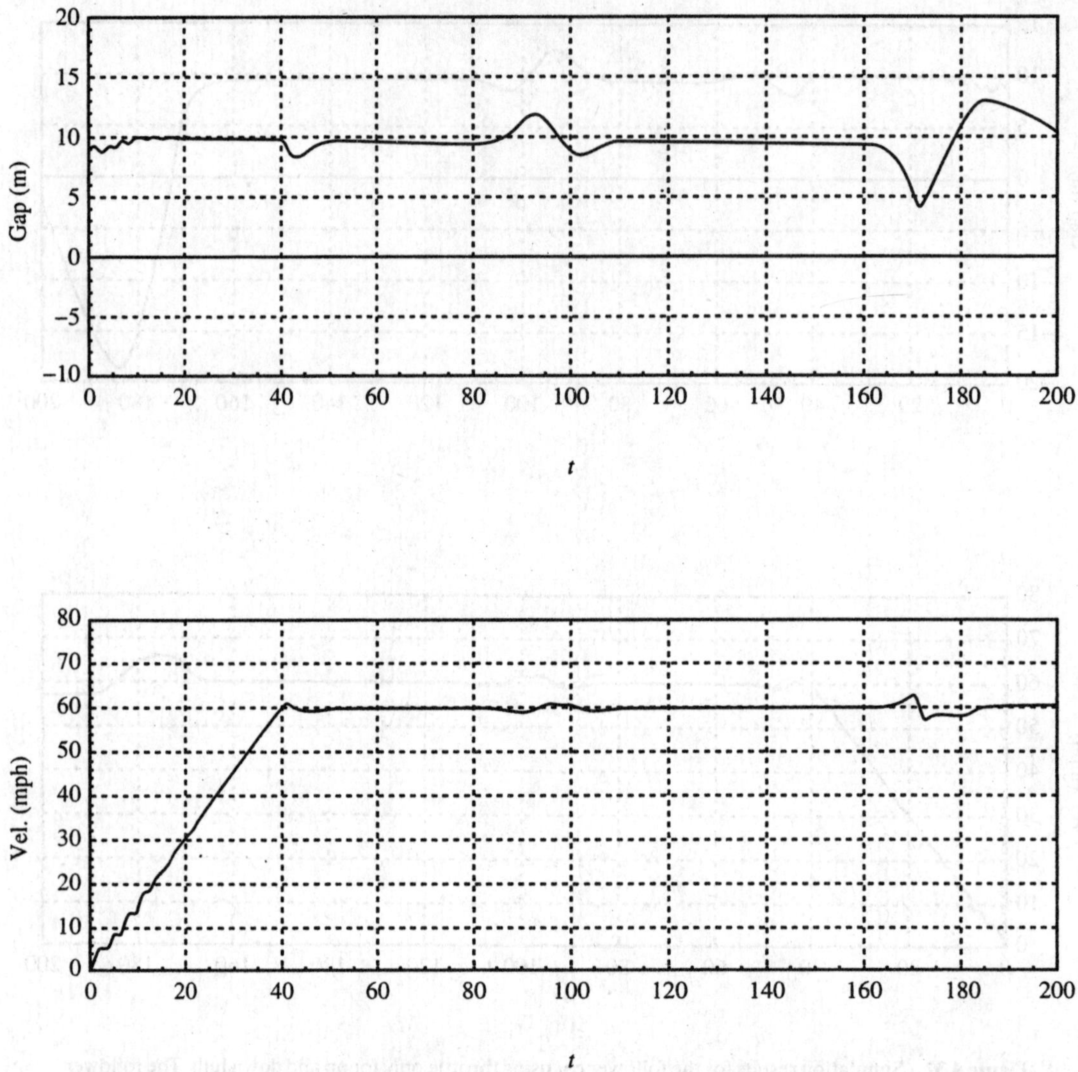

Figure 4.33 Simulation results for the follower car with a gap controller that combines throttle and brakes. The cars follow the velocity and terrain profile in Figure 4.31. The follower car approaches the leader but does not collide with it.

4.8 CONCLUSIONS

We designed and tested a fuzzy throttle controller for a car platoon. The next phase of the fuzzy platoon leader (FPL) would add our fuzzy brake controller and perhaps steering controller and lane change so the platoon can maneuver on the highway.

This controller worked well for coupled systems in which a series of objects must track and predict the object in front of them. Networks of these controllers could control the rate of message or car traffic flow through electronic and physical

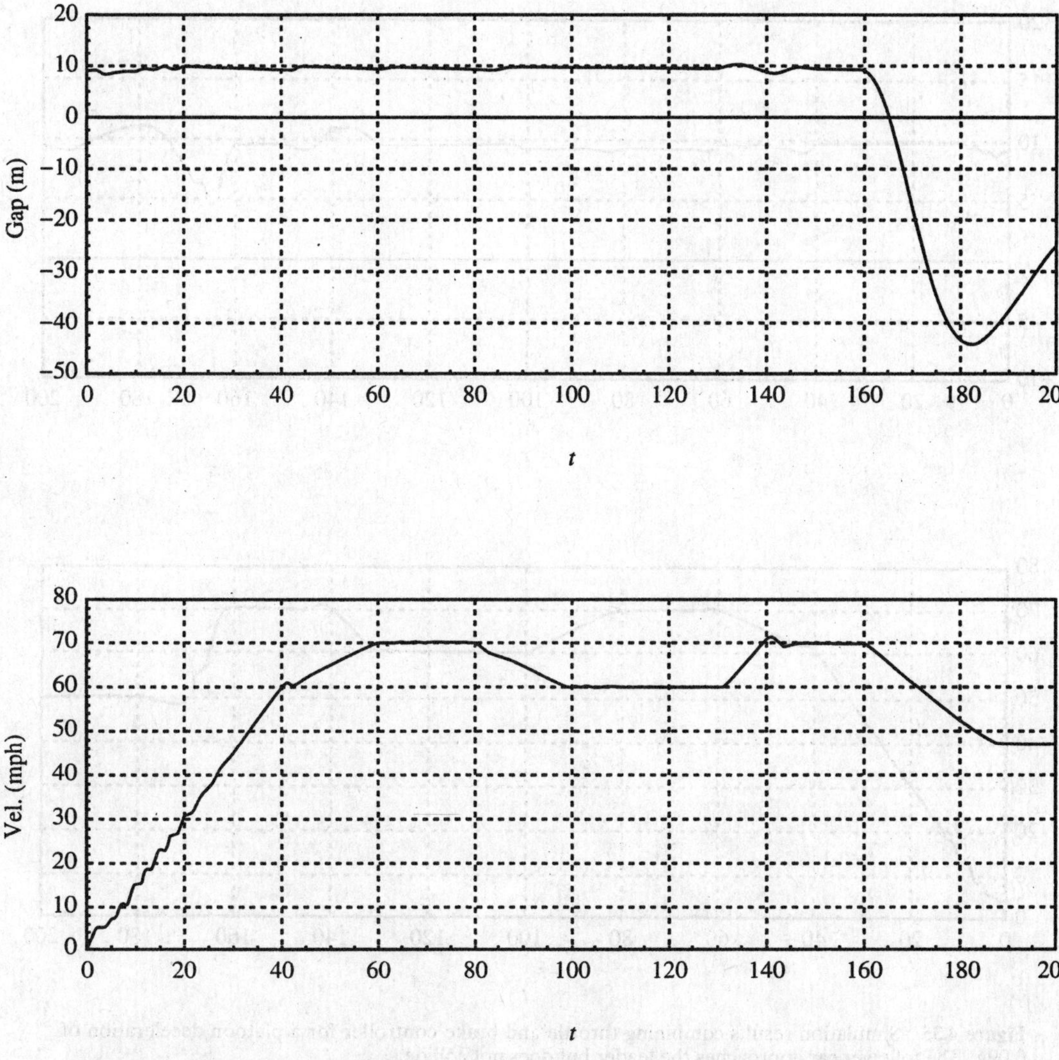

Figure 4.34 Simulation results using throttle only for a lead car deceleration of $0.09g$. The follower car collides with the leader at $t \approx 165$ s.

intersections. The coupled systems can differ. The distributed structure of the fuzzy controller could apply to factory assembly lines or to robotic limb control.

Unsupervised ellipsoidal learning tuned the fuzzy rules and sets for cars of different sizes and engine types. This gives a new way to find a fuzzy system using only data from a human driver or other controller. Ellipsoidal learning works for any control system when input-output data are available as in the control of biological or economic processes. Future research might compare how well supervised and unsupervised learning find the optimal ellipsoids and rules for fuzzy systems. On-line adaptive fuzzy control systems with ellipsoidal learning can adapt the system over time as engine parameters change.

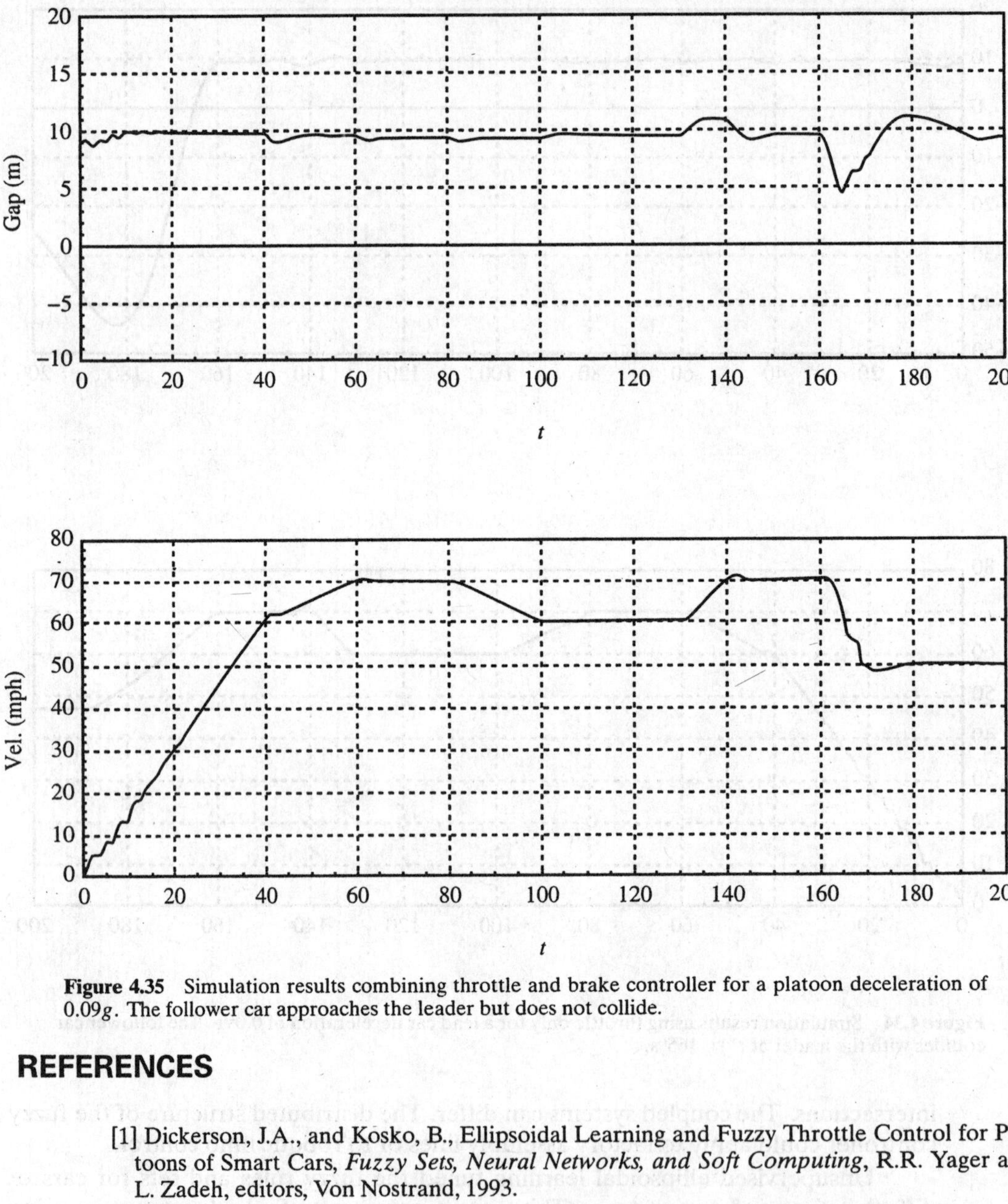

Figure 4.35 Simulation results combining throttle and brake controller for a platoon deceleration of $0.09g$. The follower car approaches the leader but does not collide.

REFERENCES

[1] Dickerson, J.A., and Kosko, B., Ellipsoidal Learning and Fuzzy Throttle Control for Platoons of Smart Cars, *Fuzzy Sets, Neural Networks, and Soft Computing*, R.R. Yager and L. Zadeh, editors, Von Nostrand, 1993.

[2] Dickerson, J.A., and Kosko, B., "Fuzzy Function Approximation with Supervised Ellipsoidal Learning," *World Congress on Neural Networks (WCNN '93)*, vol. II, 9–17, 1993.

[3] Dickerson, J.A., Kim, H.M., Kosko, B., "Hybrid Ellipsoidal Learning and Fuzzy Control for Platoons of Smart Cars," *Proceedings of the Third International Conference on Industrial Fuzzy Control and Intelligent Systems (IFIS '93)*, December 1993.

[4] Hsu, A., Eskafi, F., Sachs, S., and Varaiya, P., "The Design of Platoon Maneuver Protocols for IVHS," *PATH Research Report*, UCB-ITS-PRR–91–6, April 20, 1991.

[5] Ioannou, P., and Chien, C.C., "Autonomous Intelligent Cruise Control," *Southern California Center for Advanced Transportation Technologies, University of Southern California, Technical Report* 92–05–01, May 21, 1992.

[6] Ioannou, P., and Xu, T., "Throttle and Brake Control for Vehicle Following," *32nd Control Decision Conference (32nd CDC)*, 1993.

[7] Karaaslan, U., Varaiya, P., and Walrand, J., "Two Proposals to Improve Freeway Traffic Flow," *PATH Research Report*, UCB-ITS-PRR–90–6, *Institute of Transportation Studies, U.C. Berkeley*, 91–2, 1991.

[8] Kosko, B., "Fuzzy Systems as Universal Approximators," *IEEE Transactions on Computers*, vol. 42, no. 11, 1329–1333, November 1994; an early version appears in *Proceedings of the 1st IEEE International Conference on Fuzzy Systems (FUZZ-IEEE FUZZ92)*, 1153–1162, March 1992.

[9] Kosko, B., *Neural Networks and Fuzzy Systems*, Prentice Hall, Englewood Cliffs, 1991.

[10] Kuo, B.C., *Automatic Control Systems*, 4th ed., Prentice Hall, 1982.

[11] Sheikholesam, S., and Desoer, C.A., "Longitudinal Control of a Platoon of Vehicles with No Communication of Lead Vehicle Information: A System Level Study," *PATH Technical Memorandum*, 91–2, 1991.

[12] Sheikholesam, S., and Desoer, C.A., "Combined Longitudinal and Lateral Control of a Platoon of Vehicles: A System Level Study," *PATH Technical Memorandum*, 91–3, 1991.

[13] "Sport-utility Vehicles," *Consumer Reports*, vol. 57, No. 11, 729–735, November 1992.

HOMEWORK PROBLEMS

4.1. The equation of motion for an unforced linear system is

$$m\,\ddot{x} + c\,\dot{x} + k\,x = 0.$$

The solution to this equation corresponds to free-damped vibration. If we assume a solution of the form $x = e^{st}$ then we get the characteristic equation:

$$s^2 + \frac{c}{m}s + \frac{k}{m} = 0.$$

The general solution takes the form $x = Ae^{s_1 t} + Be^{s_2 t}$.

(a) Find the roots $s_1,\ s_2$ of the characteristic equation.

(b) If $s_1 = s_2$ then the system is *critically damped*. For what values of c, m, and k does this condition hold?

(c) If s_1 and s_2 are imaginary then the system is *underdamped*. How must c, m, and k be related to one another?

(d) Use *Euler's equation* $e^{\pm i\omega t} = \cos\,\omega t \pm i\,\sin\,\omega t$ to find the solution in the form of an exponentially damped sinusoid. What is the frequency of the sinusoidal term?

4.2. A simple model for a platoon with constant gaps is a series of masses attached with springs with dampers due to friction and air resistance:

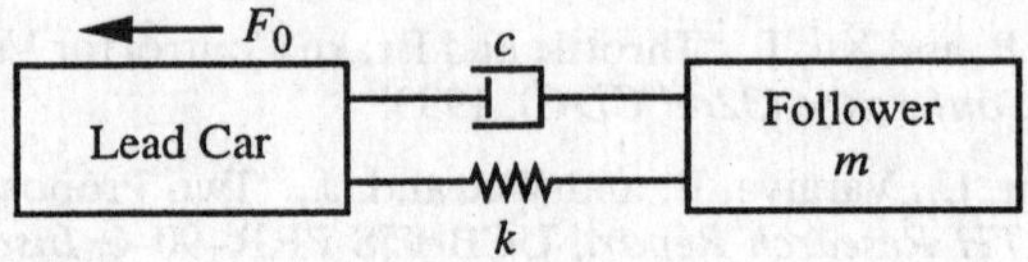

The equation of motion for the follower car in an inertial reference frame is

$$m\ddot{x} + c\dot{x} + kx = F_0.$$

The solution to this equation has the form

$$x(t) = F_0 + Xe^{-\zeta\omega_n t} \sin\left(\sqrt{1-\zeta^2}\omega_n t\right).$$

$\omega_n = \sqrt{k/m}$ is the natural frequency of undamped oscillations. $\zeta = \frac{c}{2m\omega_n}$ is the damping factor.

(a) Plot the displacement of the follower car for different values of k, m, and c.

(b) Compare the results for a heavy car ($m = 4{,}500$ kg) and for a light car ($m = 1{,}500$ kg). How large does the gap between the cars need to be to stop the follower car from hitting the lead car in each case?

4.3. Solve for the *jerk* ($\frac{da}{dt}$). Take the time derivative of (4.10) and substitute in (4.11) for $\frac{d\xi}{dt}$. A change in the throttle value changes the acceleration for the car.

(a) Use the car parameters in Table 4.1 to find the jerk when the throttle is set to 45°. Multiply the throttle value times 2,000/85 to scale the input range.

(b) If we change the throttle by 10° what is the change in acceleration for each type of car in Table 4.1?

4.4. We can model the brake force F_b as a negative force in equation (4.10). Solve for the change in acceleration when the system has brakes. Set the throttle to zero. How much brake force does it take to change the acceleration of each car in Table 4.1 by -0.5 m/s^2?

4.5. Plot the magnitude of the transfer function of the low-pass filter $\left|H_{LP}(e^{j\omega})\right|$ in (4.16):

$$H_{LP}(z) = \frac{\frac{1}{3}}{1 - \frac{2}{3}z^{-1}}$$

Show that it is a low-pass filter.

4.6. Design some rules to control the brake in a smart car. What inputs would you use to stop the brake and throttle from being on at the same time?

A simple differential equation can define the chaotic dynamics of a system that changes with time. The equation defines a path in some state space of perhaps high dimension. The path traces out an aperiodic or chaotic attractor in some region of that state space and has continuous and broadband Fourier spectra. Any two initial points lead to paths that diverge exponentially fast no matter how close the two points lie to each other.

Two types of fuzzy systems can model chaotic fuzzy systems. The first type of system is feedforward. It can model the system by covering part of the path with rule patches. The rule patches must either move or grow in number as the path moves. The feedforward system must either learn or change its rules fast enough to track the path's movement or add more rule patches to cover new sections of the system's path in the state space. The former approach builds the system's dynamics into the rule structure. The latter approach can require a growing budget of rules or an endless sequence of feedforward systems. This amounts to a new level of rule explosion.

The second type of system is a feedback fuzzy system. Such a system takes its output as its own input. It uses a fixed number of rules to directly approximate the differential equation. The feedback fuzzy system still suffers from exponential rule explosion at higher dimension as does the feedforward system. But it need not suffer from the feedforward system's second rule explosion within a given dimension. The savings comes at the risk of dynamic instability and at a far greater cost in tuning the fixed rules. Feedback fuzzy approximation remains an open and active area of research.

Chapter 5 uses a standard additive model feedforward system to model chaotic attractors of low dimension. A splitting scheme recursively partitions the state space into finer and finer rule patches and so guides the second level of rule explosion. A fine rule cover of even the simplest chaotic attractor may require tens of thousands of rules. The chaotic Lorenz attractor requires over 32,000 rules for each of these coordinates. The partitioning scheme maintains full control of the approximation process but at the high price of exponential rule growth. Future splitting schemes may trade some of this control for slower rule growth.

5

FUZZY CHAOS AND RECURSIVE PARTITIONING

Hong Liang Hiew

Chi Ping Tsang

PREVIEW

An additive fuzzy system can uniformly approximate a continuous function by covering its graph with rule patches and averaging patches that overlap. We develop a recursive algorithm that learns these patches. It recursively partitions the input-output product space until the new fuzzy system has an approximation error that falls below a preset threshold. A lower threshold improves the approximation.

We apply the partitioning algorithm to both static functions and dynamical systems with and without chaotic attractors. Rule explosion leads to thousands of rules to cover even low-dimension attractors. Fuzzy chaos emerges as the partitioning splits increase and the rule patches shrink in size and grow in number. Fine approximation chaos in the Lorenz and Rossler systems requires many thousands of rules. This suggests the need for feedback fuzzy systems that use a fixed and small number of rules.

5.1 FUZZY SYSTEMS AS RULE COVERINGS: APPROXIMATING CHAOS

A fuzzy system is a set of fuzzy if-then rules that maps inputs to outputs. Each rule defines a fuzzy subset of the input-output state space $X \times Y$. So each fuzzy system defines a unique cover or partition of the state space into rule patches and vice versa. This fuzzy mosaic changes when we add or delete rules or when we change them through learning.

Recursive partitioning cuts the state space into finer and finer fuzzy rule patches. Sample data drives each choice of partition in a form of product space clustering [5] as discussed in Section 2.12 of Chapter 2. We present a new algorithm for fuzzy recursive partitioning. The partitioning criterion depends on the local and global mean-square error of the function approximation.

Recursive partitioning also offers a way to approximate dynamical systems with feedforward fuzzy systems. An autonomous dynamical system with n variables $x_1, \ldots, x_n$ has the form

$$\dot{x} = f(x)$$

where the overdot denotes a time derivative. The system starts in state x_0 and evolves in accord with the differential equation. The trajectory ends at or near an equilibrium attractor: a fixed point, limit cycle, or aperiodic (chaotic) attractor. The attractors partition the state space into attractor regions. But approximation accuracy comes at the cost of exponential rule explosion.

Additive fuzzy systems can model a dynamical system f by modeling the n functions $f_1, \ldots, f_n$. This takes n feedforward additive fuzzy systems. The recursive partition tracks the system trajectories with finer and finer rule patches. But this leads to rule explosion. It takes thousands of rules to finely approximate a chaotic butterfly in state spaces of even low dimension. The best rule patches should cover the extrema or turning points in the functions $f_1, \ldots, f_n$ [5]. Our system does not guarantee that the final rules lie at extrema but our software simulations did show that the final patches tended toward these optimal rules.

The next section states the additive fuzzy model and the criteria for recursive partitioning. We first test the new method on the fifth-degree polynomial [1] in Chapter 3. The model uses triangular fuzzy sets and allows us to vary at will their overlap along an axis. The next sections apply the technique to a 3-D surface and to dynamical systems with and without chaotic attractors. The tradeoff between approximation accuracy and rule tractability becomes more acute as we move to systems in higher dimensions and to nonlinear systems that vary with time. We did not use the fixed-rule feedback systems of Chapter 15.

5.2 FUZZY SYSTEM DESIGN

This section presents the design of the fuzzy system. It consists of two parts. Section 5.2.1 deals with the structure of the system and how it performs fuzzy inferences. Sections 5.2.2–5.2.5 describe our recursive partitioning algorithm for adaptive rule construction.

5.2.1 Fuzzy System Architecture

We use a simple additive fuzzy system architecture [3]. The additive fuzzy system (Figure 5.1) consists of a set of fuzzy if-then rules and averages rules that overlap. An n-dimensional vector x acts as input to the system. We use triangular set functions to represent the fuzzy sets because they are simple to compute. Simulations showed that the shape of the set function did not matter much. This was especially true in cases where rules overlap a great deal.

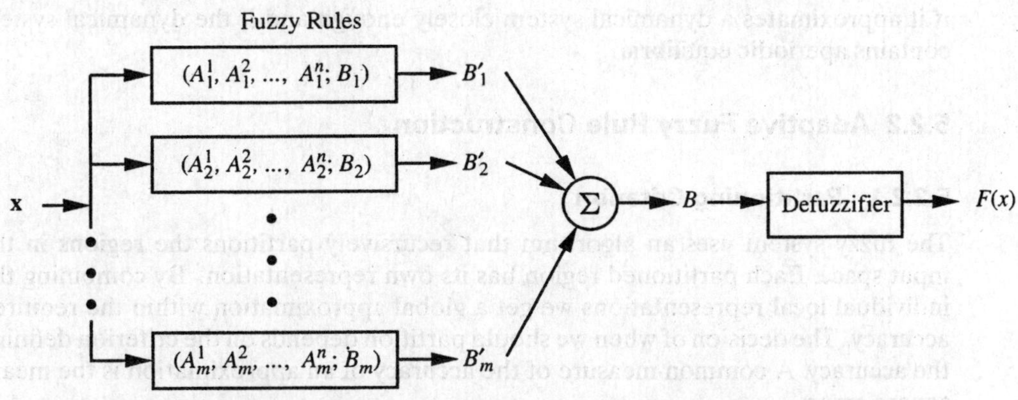

Figure 5.1 The architecture of the additive fuzzy system. The input is an n-dimensional vector $x = (x_1, x_2, \ldots, x_n)$ and the output is a scalar value $y = F(x)$ in the then-part domain.

The system uses the min if-part combiner and the centroid defuzzification scheme in [3] to obtain an output from an input. The approximation Theorem in Chapter 2 and [4] still holds if the then-part sets B'_j are symmetric. An n-dimensional input vector $(x_1, x_2, \ldots, x_n)$ and the jth fuzzy rule of the form $(A_j^1, A_j^2, \ldots, A_j^n; B_j)$ produce a scaled version of the original then-part fuzzy set B_j:

$$B'_j = B_j \min(a_j^1(x_1), a_j^2(x_2), \ldots, a_j^n(x_n)) \tag{5.1}$$

where a: $X \to [0, 1]$ is the set function of the fuzzy set $A \subset X$.

A system with m rules gives m clipped then-part fuzzy sets B'_i as in Figure 5.1. We get the final defuzzified output value $F(x)$ from

$$B = \sum_{j=1}^{m} B'_j \tag{5.2}$$

$$F(x) = \frac{\int_{-\infty}^{\infty} y\, b(y)\, dy}{\int_{-\infty}^{\infty} b(y)\, dy} = \frac{\sum_{j=1}^{m} a_j(x) V_j c_j}{\sum_{j=1}^{m} a_j(x) V_j} \tag{5.3}$$

for the SAM or standard additive model in Chapter 2. The fuzzy system $F: R^n \rightarrow R$ stores m rules or Cartesian patches $A_1 \times B_1, \ldots, A_m \times B_m$. Each if-part fuzzy set $A_j \subset R^n$ has joint set function $a_j: R^n \rightarrow [0, 1]$ that factors into n scalar set functions $a_j^i: R \rightarrow [0, 1]$. The total "set" B has set function $b: R \rightarrow [0, m]$. The m scalar then-part sets $B_1, \ldots, B_m$ have set functions $b_j: R \rightarrow [0, 1]$. Each then-part set B_j has area or volume V_j and centroid c_j.

These scalar SAM systems suffer from rule explosion or the "curse of dimensionality" on the order of k^{2n-1} rules for a SAM $F: R^n \rightarrow R^n$ that approximates a scalar approximand $f: R^n \rightarrow R$ on a compact domain. The SAM F exhibits chaos if it approximates a dynamical system closely enough and if the dynamical system contains aperiodic equilibria.

5.2.2 Adaptive Fuzzy Rule Construction

5.2.2.1 Partitioning Criterion

The fuzzy system uses an algorithm that recursively partitions the regions in the input space. Each partitioned region has its own representation. By combining the individual local representations we get a global approximation within the required accuracy. The decision of when we should partition depends on the criterion defining the accuracy. A common measure of the accuracy of an approximation is the mean-square error:

$$\text{MSE}_{\text{global}} = \frac{\sum_{i=1}^{n} (d_i - F(\mathbf{x}_i))^2}{n} \tag{5.4}$$

where

n = the number of samples

$\mathbf{x}_i$ = the input vector of a sample

d_i = the desired output of $\mathbf{x}_i$

$F(\mathbf{x}_i)$ = the approximated output value given $\mathbf{x}_i$.

We partition regions if the MSE in (5.4) exceeds a threshold. But this MSE is a global measure and does not tell us which region we need to partition. To that end we can use the local mean-square error:

$$\text{MSE}_{\text{local}}(s) = \frac{\sum_{i=1}^{m} (d_i - F_s(\mathbf{x}_i))^2}{m} \tag{5.5}$$

where

m = the number of samples in region s

$\mathbf{x}_i$ = the input vector of a sample in s

d_i = the desired output of $\mathbf{x}_i$

F_s = the fuzzy rule for region s (see Section 5.2.2.3)

$F_s(\mathbf{x}_i)$ = defuzzified output value obtained from invoking F_s with $\mathbf{x}_i$.

This local mean-square error correlates strongly with the global one from (5.4). If the MSE_{local} of all regions exceeds a threshold then the MSE_{global} tends to exceed the threshold as well. So if constructing a single rule for the current region produces an MSE_{local} greater than the threshold then we should partition the region. By partitioning a region we construct more accurate rules for that region and thus decrease the MSE_{local}. The correlation between MSE_{local} and MSE_{global} means this decreases the MSE_{global} as well. When the MSE_{local}'s for all regions are below the threshold then the MSE_{global} tends to be below the threshold.

We therefore base the partitioning criterion on the MSE_{local}.

```
function Partition ? (input-space s)
begin
    y_s = Construct-Rule(s)
    MSE = local mean square error for s if F_s represents s
    if (MSE < threshold) then
        return FALSE
    else
        return TRUE
end
```

5.2.2.2 Recursive Partitioning of the Product Space

The algorithm tests an existing region and if it meets the partitioning criterion the algorithm splits the region. The structure of the algorithm is as follows:

```
procedure Recursive-Partition (input-space s, dimension d)
begin
    if Partition ? (s) then
        begin
            partition s along dimension d into subregion 1 and
                subregion 2
            Recursive-Partition(subregion 1,
                        d + 1 modulo number of dimensions in s)
            Recursive-Partition(subregion 2,
                        d + 1 modulo number of dimensions in s)
        end
    else
        begin
            r = Construct-Rule(s)
            attach region s to r
            add r to the fuzzy system
        end
end
```

At first we consider the first dimension of the input space. If the current region we consider meets the partitioning criterion (Section 5.2.2.1) then we divide it in half along the first dimension. We now have two regions and we move on to the second dimension. For each of the two regions if they satisfy the partitioning criterion again then we divide the region in half along the second dimension. We repeat this with each iteration moving to the next dimension. When we reach the last dimension of the input space then we move back to the first dimension for the next iteration. If at any time a region fails the partitioning criterion then we use the method described in Section 5.2.2.3 to define one rule for that region.

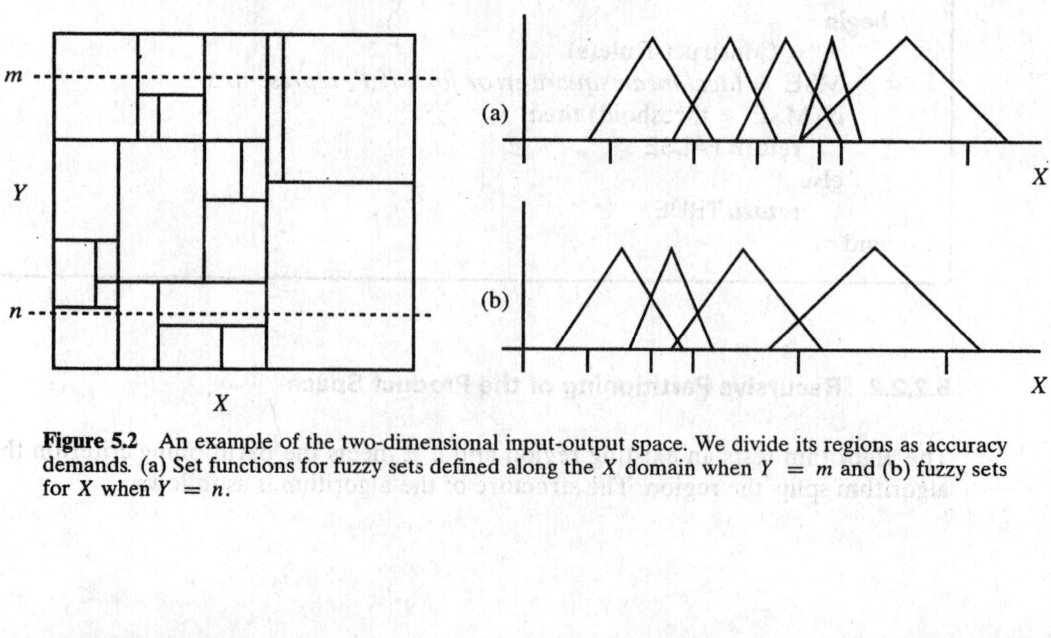

Figure 5.2 An example of the two-dimensional input-output space. We divide its regions as accuracy demands. (a) Set functions for fuzzy sets defined along the X domain when $Y = m$ and (b) fuzzy sets for X when $Y = n$.

Figure 5.2 shows the regions that may result from the partitioning. Note that we do not partition regions into two equal halves each time. All samples may fall into only one of those halves. We would have no samples with which to process the other half. Instead we partition the input space region at the average input value along the dimension we are partitioning.

5.2.2.3 Defining a Fuzzy Rule Patch

We first find a region in the input space that needs no further partitioning. Then we construct a rule for that region and define a patch that covers the samples in the region. The algorithm is as follows:

```
function Construct-Rule (input-space s)
begin
    n = number of dimensions in s
    for d = 1 to n
        begin
        range = range of s along dimension d
        extended_range = range with its maximum and minimum
                extended slightly according to the overlap rate
        A_d = fuzzy set with symmetric triangular set function
                with range extended_range
        end
    output_range = range of s along dimension d
    extended_output_range = output_range with its
            maximum and minimum
            extended slightly according to the overlap rate
    B = fuzzy set with symmetric triangular set function
            with range extended_output_range
    return rule (A^1, A^2, . . ., A^n; B)
end
```

Figure 5.3 shows the fuzzy rule construction. For each dimension of the input space region we define a fuzzy set whose set function covers the region along that dimension. The width of the set function extends according to the overlap rate. For the then-part of the rule we find the range of desired output values from samples within the region and define the set function of the then-part fuzzy set over this range. We extend this set function the same way we extended the if-part fuzzy sets. The resulting rule is a hyperrectangular patch that covers the samples in the region concerned.

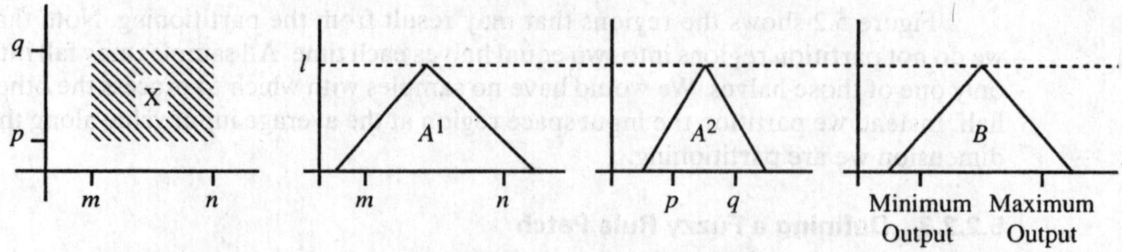

Figure 5.3 To define a rule for region X of the input space shown in (a) we take the range of values along each dimension of the region (b and c) as well as the range of desired output values (d) to form a rule (A^1, A^2; B).

5.2.2.4 Adapting to New Sample Data

Ours is a piecewise approximation system. Each piece of the system deals only with information local to its sphere of influence. Each rule patch represents only the area it covers. We leave the interpolation between patches to the system's global inference mechanism. That is the spirit we have tried to maintain in constructing the adaptive algorithm. The result is a form of product space clustering. The symmetric triangular if-part sets give a piecewise-linear fuzzy system.

All partitions carry an associated set of sample data. We first partition the input space. Then we partition its associated set of samples. A fuzzy rule patch overlaps and influences more than one partition but *only one partition defines it.* So when presented with a new sample then all we have to do is to find the partition where the new sample's input falls into and modify the rule for that partition.

```
procedure Insert-New-Sample (sample d)
begin
    r = the rule whose attached region the input vector of d falls
    s = the region attached to rule r
    delete r from the fuzzy system
    Recursive-Partition (s, 0)
end
```

The fuzzy system grows as we iteratively insert sample data points into it. This is a direct consequence of using the sample space localization property in defining the rules.

5.2.2.5 Fine Tuning the Fuzzy System

The algorithm above constructed the fuzzy system based on local mean-square error. There is no guarantee that when the recursive partitioning finishes the global mean square error will be below the threshold we require (see the next section for a discussion). We may have to fine tune the fuzzy system. This involves looking at the set of samples within each partition. If the fuzzy system gives a high mean square error for that set of samples then we must subdivide the partition. The algorithm is as follows:

```
procedure Fine Tune
begin
    repeat
        for all s ∈ partitioned regions in the input space
        begin
            if (MSE_global(s) > threshold) and
                (s has more than 1 sample)
            then
            begin
                delete from the fuzzy system the rule defined by s
                partition s along dimension d into
                    subregion 1 and subregion 2
                r1 = Construct-Rule (subregion 1)
                attach subregion 1 to r1
                r2 = Construct-Rule (subregion 2)
                attach subregion 2 to r2
                add r1 and r2 to the fuzzy system
            end
        end
    until (MSE_global ≤ threshold or no changes were made
        in the last iteration)
end
```

We get the global mean-square error of a region s by invoking the *whole* fuzzy system with points in s:

$$\text{MSE}_{\text{global}}(s) = \frac{\sum_{i=1}^{m}(d_i - F_{\text{global}}(\mathbf{x}_i))^2}{m} \tag{5.6}$$

where

m = the number of samples in region s

$\mathbf{x}_i$ = the input vector of a sample in s

d_i = the desired output of $\mathbf{x}_i$

$F_{\text{global}}(\mathbf{x}_i)$ = defuzzified output value obtained from invoking the fuzzy system with $\mathbf{x}_i$.

The global mean-square error for the whole input space (5.4) is also equivalent to

$$\text{MSE}_{\text{global}} = \frac{\sum_{s \in \text{partitions}} \eta_s \text{MSE}_{\text{global}}(s)}{\sum_{s \in \text{partitions}} \eta_s} \tag{5.7}$$

where η_s = the number of samples in partition s.

Fine tuning thus involves going through all the partitions to determine if we need to partition any of them further. This time the partitioning criterion depends on the global (rather than local) mean-square error for that partition (5.6). We know that the $\text{MSE}_{\text{local}}$ for each partition is already below the threshold. Further partitioning cannot increase $\text{MSE}_{\text{local}}$ for any subpartition. So overlapping rules from other partitions must affect any partition with an $\text{MSE}_{\text{global}}$ more than the threshold. To find and modify all the rules affecting this partition would be too inefficient. Instead we split the local partition. This will increase the strength of the local rule by bringing the sample closer to the center of the rule patch. The local rule is more accurate than outside rules so this will decrease $\text{MSE}_{\text{local}}$ for the two new subpartitions. But there is no guarantee that splitting a partition will not increase the $\text{MSE}_{\text{local}}$ of the neighboring partitions.

The fine tuning procedure may need to sweep more than once through the partitions. This is because a partition might need to be split more than once to achieve the required $\text{MSE}_{\text{global}}$. And changing the fuzzy rule patch for one partition affects the $\text{MSE}_{\text{global}}$ of its surrounding partitions (see Figure 5.4). We may need to split these surrounding partitions as well and that would then affect more partitions. So a local change can have a ripple effect. We have to keep fine tuning all partitions until we need no more changes.

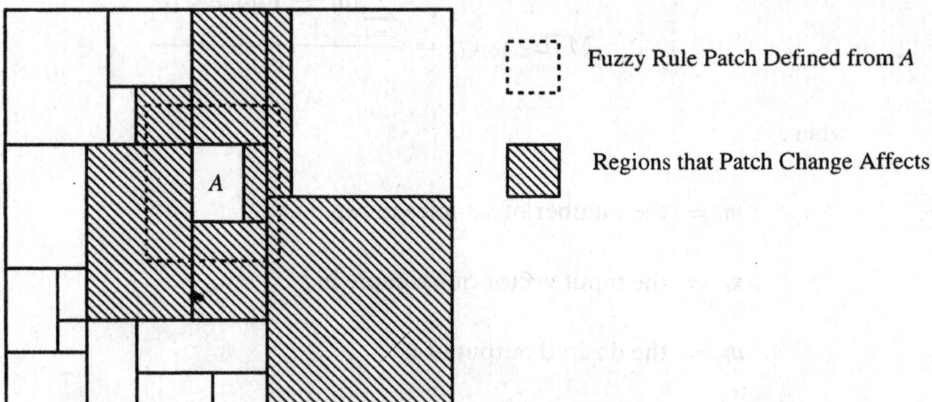

Figure 5.4 A change to the rule in region A affects not only the approximation of region A but also affects the approximation of the neighboring regions.

5.2.3 Fixed Parameters

The recursive partitioning algorithm does not vary all the parameters in the fuzzy system. The most obvious fixed factor is the overlap rate between rule patches. Much of the power of fuzzy approximation lies in the overlap between rule patches. Heavy overlap can greatly decrease the required number of rules. But heavy overlap can also decrease the accuracy of the representation. It is not clear how the overlap between two rule patches should change to improve accuracy if we do not have full knowledge of the approximand. Our algorithm holds constant the overlap rate between all patches along each dimension. We define the overlap rate (along a given dimension) as the percentage that two patches of equal length overlap each other along that dimension (see Figure 5.5). An optimal or improved overlap rate depends on the number of rules and that depends in turn on the required accuracy. Low accuracy requires less rules and each rule patch is larger. The same overlap rate means larger patches share a larger region than do smaller patches. Whether this leads to improved or reduced accuracy depends on the approximand.

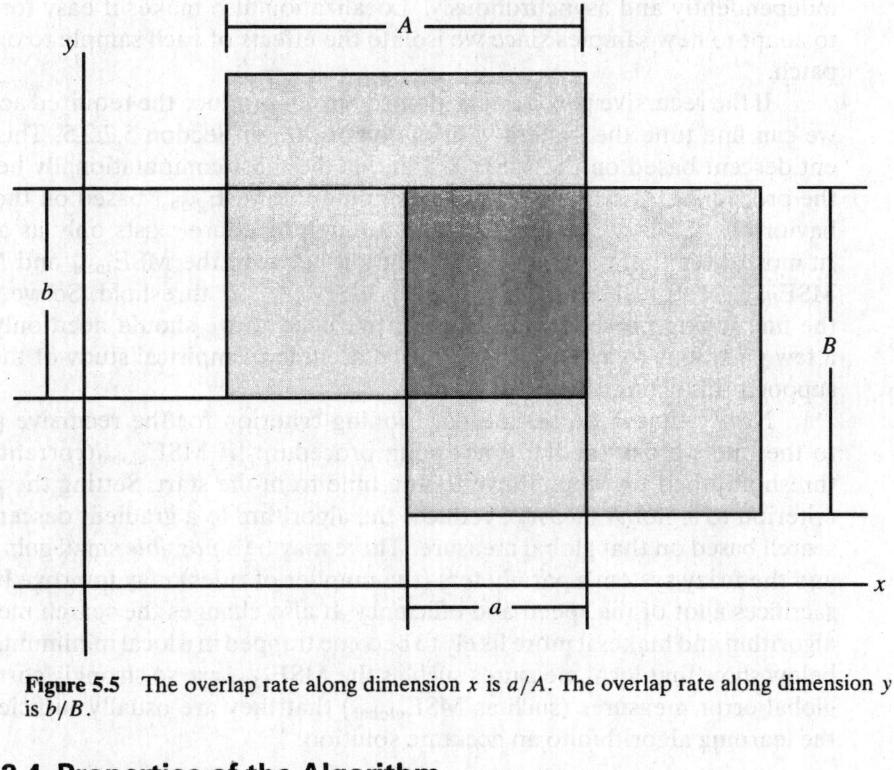

Figure 5.5 The overlap rate along dimension x is a/A. The overlap rate along dimension y is b/B.

5.2.4 Properties of the Algorithm

The MSE_{global} (5.4) for the whole input space defines the accuracy of the trained fuzzy system. The learning algorithm seeks to minimize this MSE_{global}.

5.2.4.1 Partitioning Criterion Based on Local Information

The recursive partitioning algorithm (without fine tuning) depends on the partitioning criterion. The algorithm uses a sample to construct one and only one rule. If the partitioning criterion fails for all partitions then we know that the MSE_{local} for each partition is less than the threshold we require. This does not guarantee that we have the required accuracy. The MSE_{global} for the whole input space may still exceed that threshold. Or we could have the reversed situation where the MSE_{global} is less than the threshold but the MSE_{local} for some partitions is still above the threshold. Then the fuzzy system has already achieved the required accuracy but it still creates more rules.

We allow that the above situations may arise because we can depend on a correlation between MSE_{local} and MSE_{global}. The strong correlation between MSE_{local} and MSE_{global} ensures that in most cases the final system will be at least near the required accuracy and the number of rules it creates will be close to optimal.

The main benefit of localization is that it improves speed and efficiency. At any time the algorithm requires only information about the parameter it modifies and data within the subregion it considers. We can recursively partition each subpartition independently and asynchronously. Localization also makes it easy for the system to adapt to new samples since we isolate the effects of each sample to only one rule patch.

If the recursive partitioning alone does not produce the required accuracy then we can fine tune the system with the algorithm in Section 5.2.2.5. This is a gradient descent based on the MSE_{global} and is the most computationally heavy part of the procedure. It requires that we compute the MSE_{global} based on the global behavior of the fuzzy system. This fine tuning procedure exists only as a safeguard. In most cases there is a strong correlation between the MSE_{local} and MSE_{global}—MSE_{local} < threshold usually implies MSE_{global} < threshold. So we rarely need the fine tuning procedure. Even when we need to we should need only to modify a few partitions to achieve the required accuracy. Empirical study of the algorithm supports this claim.

Notice that if we set the partitioning criterion for the recursive partitioning to the one we use for the fine tuning procedure [if MSE_{global}(current region) > threshold] then we would have to fine tune from the start. Setting the partitioning criterion to a global measure reduces the algorithm to a gradient descent statistical search based on that global measure. There may be a *possible* small gain in accuracy and the fuzzy system's parameters (the number of rules) may improve but doing so sacrifices a lot of the speed and efficiency. It also changes the search method of the algorithm and makes it more likely to become trapped in a local minimum. The results below show that local measures such as the MSE_{local} are so strongly correlated with global error measures (such as MSE_{global}) that they are usually sufficient to guide the learning algorithm to an accurate solution.

5.2.4.2 Direction of Search

Recursive partitioning is a search technique. It varies the fuzzy system's parameters (the number of rule patches and their sizes and locations) in an attempt to find an optimal or good combination. This method differs from that of neural networks in that we tightly control the search in recursive partitioning. In statistical searches the system parameters fluctuate as the search wanders through the parameter space. The direction in which the parameters change is not important as long as it improves global performance. This is not the case in recursive partitioning. The number of rule patches always increases. The sizes of the patches always decrease. And their locations always move toward the center of the subpartitions. The algorithm is closer to a form of construction than a method of search. In most cases we can safely assume that we will increase the approximation accuracy if we increase the number of rules and shrink their patch sizes and move them closer to the training samples. We can make these assumptions without information about the samples.

Having greater control over the search direction again translates to efficiency in the algorithm. We can proceed very quickly down through the partitions confident that our modifications will give a better approximation. We depend on the partitioning criterion to stop the algorithm before it goes too far. The partitioning criterion stops the algorithm when the fuzzy system is close to an optimal or near-optimal state.

5.3 APPROXIMATING FUNCTIONS

We first test the algorithm described above by trying to approximate some simple functions. These approximand functions have the form

$$f(x) = y \tag{5.8}$$

where x is a vector in the domain of f and y is a polynomial based on x. The input to the fuzzy system is the vector x and the output is the value $f(x)$.

The recursive partitioning algorithm has two fixed parameters: the overlap rate and the required mean-square error (mse). The mse achieved refers to the global mean-square error of the final trained system.

5.3.1 Approximation of a Two-Dimensional Curve

We used the algorithm to approximate a function with the *fifth-order polynomial* of Chapter 3 and [1]:

$$f(x) = 3x(x - 1)(x - 1.9)(x + 0.7)(x + 1.8). \tag{5.9}$$

We replaced the approximand function $f\colon R \to R$ with 100 sample points uniformly distributed in the interval $[-2, 2]$.

Figure 5.6 shows the effect of different parameters on the accuracy of the fuzzy approximator. A lower mean-square error led the algorithm to produce an approximation with lower final mean-square error. The figure also highlights the effect of fine tuning. In most cases the algorithm constructs an accurate approximation without having to fine tune. Fine tuning occurs when the initial partitioning based on MSE_{local} does not achieve the required accuracy. The figure shows that this happens when the overlap rate is high. A high overlap rate implies that MSE_{local} is not a good indication of MSE_{global}. We base the initial partitioning on MSE_{local} and so the algorithm stops even though MSE_{global} has not reached the required level. Fine tuning corrects this partially by further partitioning the input space. Even when this happens the fuzzy systems do not always achieve the required mean-square error. That holds when the overlap rate is too high for the algorithm to generate a proper approximation.

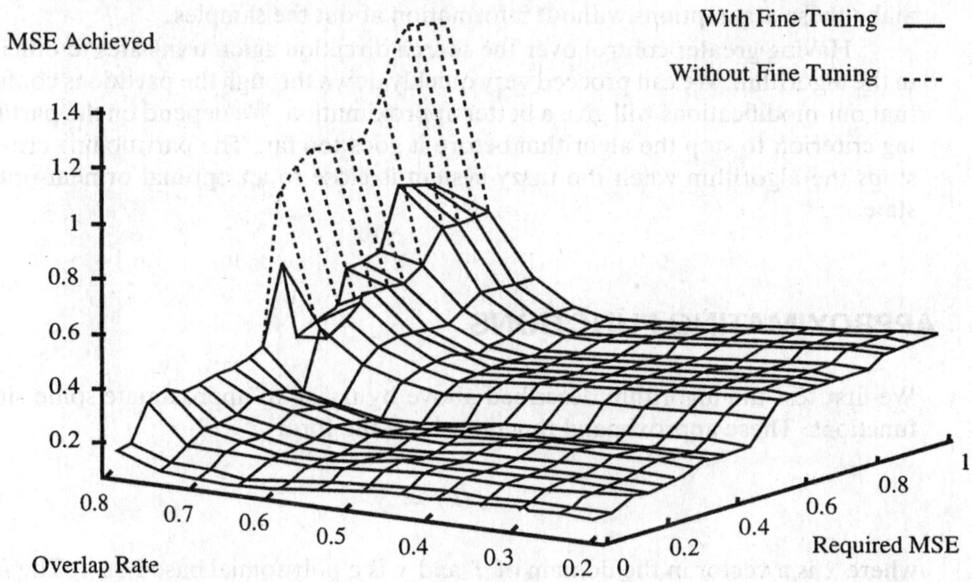

Figure 5.6 Approximating the fifth degree polynomial. Different parameters in the algorithm produce different accuracy in the fuzzy systems. This diagram displays the relationship between the required mean-square error, the fixed overlap rate, and the final global mean-square error. We compared the algorithm with and without fine tuning.

Figure 5.7 shows that the number of rules created falls exponentially as the mean-square error grows. The overlap rate has no effect on the number of rules created. This is because the overlap rate does not affect local mean-square error (5.5) and thus the partitioning criterion.

Figure 5.8 shows the performance of the trained fuzzy systems. Setting a lower required mean-square error created more rules and improved the approximation accuracy. Notice that in the rule patch plots (Figures b, d, and f) the patches do not always cover all of the range $F(x)$ but they do cover the whole domain X. Recursive partitioning occurs only in the domain or input space of the function.

Figure 5.9 shows how the overlap rate picks the size of the new rule patches. The patches are still centered at the same points as the patches in Figure 5.8d. For the approximation with 20% overlap (a and b) the approximated line is very close to being stepwise since it involves little averaging. But 80% overlap (c and d) is too much averaging and the accuracy decreases.

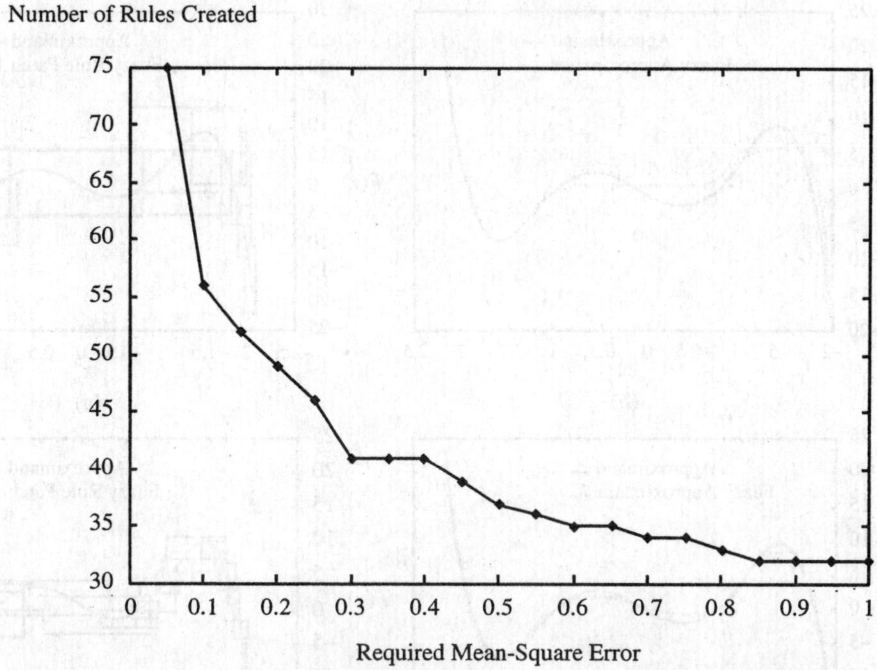

Figure 5.7 Rule count for the polynomial approximation. The algorithm creates more rules as the required accuracy grows. The overlap rate is 60%.

5.3.2 Approximation of a Three-Dimensional Surface

We next approximate the *sine-cosine wave* surface:

$$f(x, y) = \sin(x)\cos(y). \tag{5.10}$$

We replaced $f(x, y)$ with 900 sample points evenly distributed in a 30×30 grid covering the area $(-3, 3) \times (-3, 3)$. Figure 5.10 shows the surface.

Figures 5.11–5.17 show the results of the approximations. They agree with the observations we made in Section 5.3.1. The approximations get better as the number of rules grow (Figure 5.13). Very small overlaps give a steplike approximator. Very heavy overlaps lose many details of the surface (Figure 5.17). Figures 5.14–5.16 show how the algorithm partitioned the input space into 3-D patches.

5.4 APPROXIMATING DYNAMICAL SYSTEMS AND CHAOS

The above method stems from earlier work on modeling dynamical systems and chaos [2]. The objective was to find an adaptive algorithm that can capture nonlinearity easily and accurately. Recurrent neural networks define a general class of dynamical systems [7] and may work well for this purpose. The problem lies in the learning algorithm for such networks [8, 10–11]. They often try to satisfy many constraints

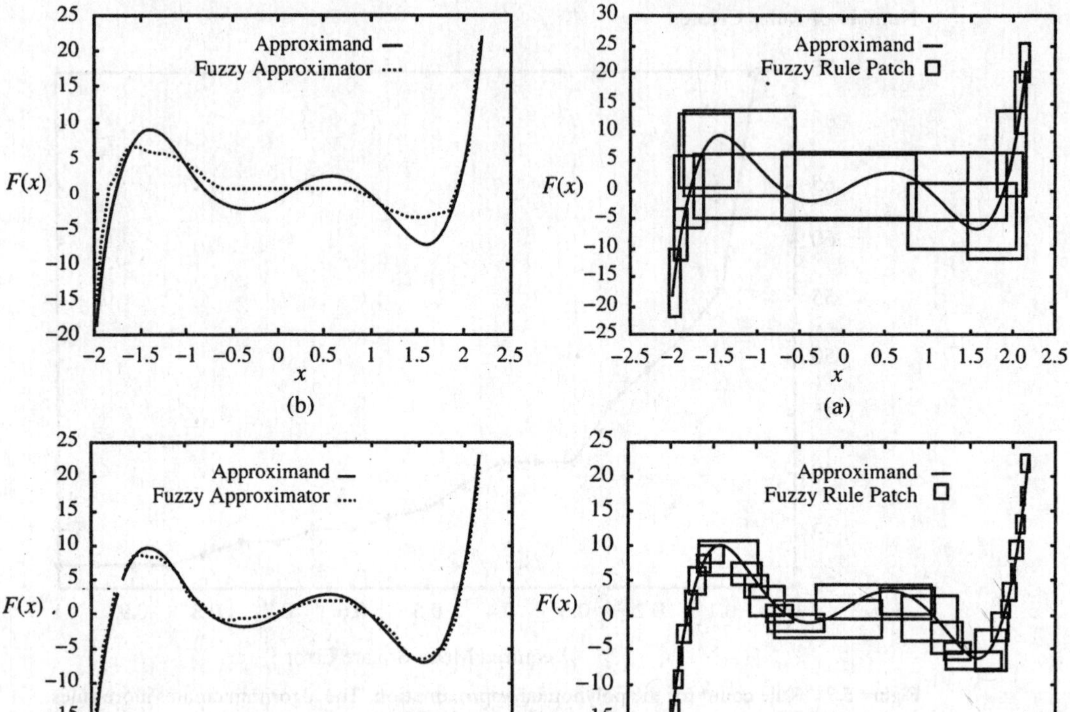

Figure 5.8 The approximation of the 5th-degree polynomial improves as the number of rules grows. The column of figures on the left shows the comparison between the original function and approximations of the trained fuzzy system. The column on the right shows how the rule patches cover the original function. The overlap rate is 60%. (a) Approximation with 12 rules. The required mse (mean square error) is set at 7 and the achieved mse is 3.4260. (b) The original function and the 12 rule patches. (c) Approximation with 32 rules (required mse = 1, mse achieved = 0.3452). (d) The function with the 32 rule patches. (e) Approximation with 76 rules (required mse = 0.05, mse achieved = 0.1090). (f) The function with the 75 rule patches.

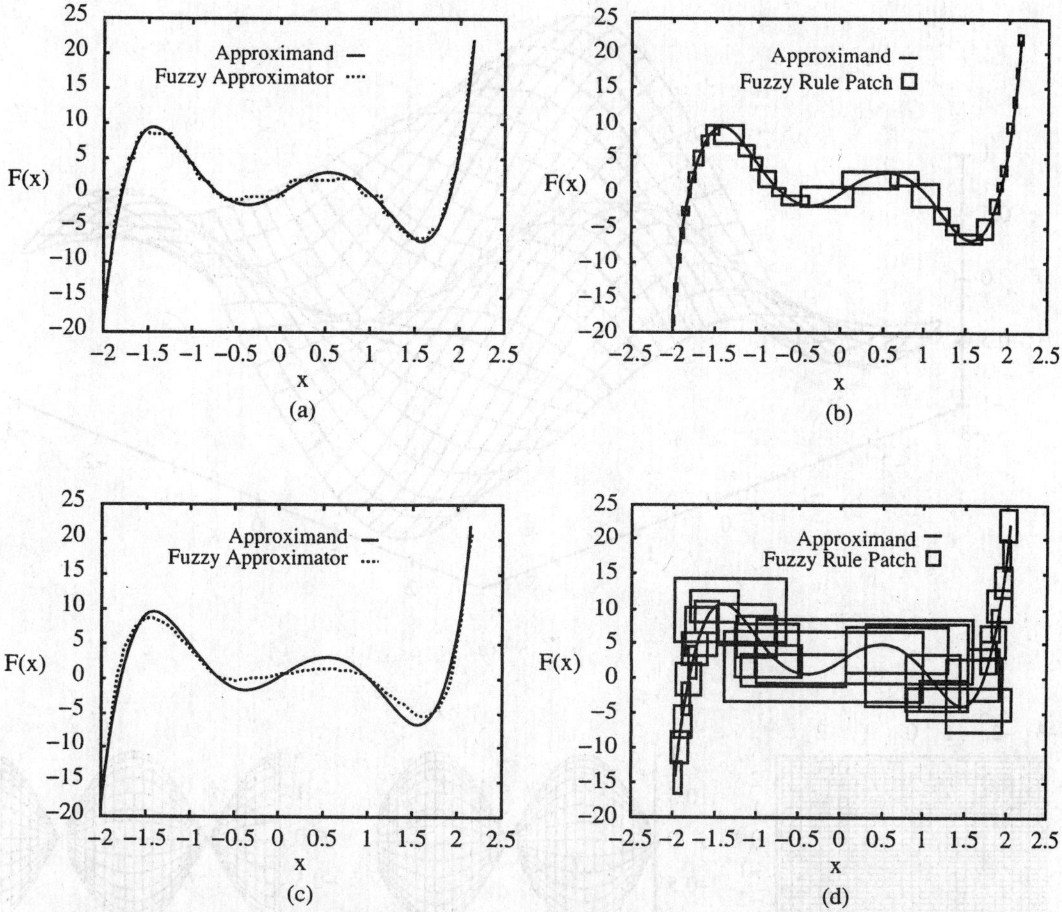

Figure 5.9 The effects of setting different overlap rates for the algorithm. The column of figures on the left shows the comparison between the original function and the new fuzzy approximators. The column on the right shows how the rule patches cover the approximand. The required mean-square error is 1 and there are 32 rules. (a) Approximation with 20% overlap. (b) The function and rule patches with 20% overlap. (c) Approximation with 80% overlap. (d) The function and rule patches with 80% overlap. We shut off the fine-tuning so that the number of new rules is the same for both cases. Compare also with Figures 5.8c and d where we set the overlap rate at 60%.

at the same time. The learning slows even with an optimal learning rate. Nonlinear behavior (especially chaos) further slows the learning algorithm as it searches through a highly irregular error surface. The error surface defines the approximation error as a function of the network's parameters. An irregular error surface has a problem property. Similar parameters can give very different approximation errors. Supervised gradient descent fails on such surfaces. Gradient descent assumes that the whole error surface slopes more or less toward a global minimum. Then it shifts the parameters toward that minimum. A highly irregular surface hides the global minimum and many good local minima.

An attractive feature of fuzzy systems is how they can grow rules from training samples. The recursive partitioning algorithm breaks the samples into clusters simple

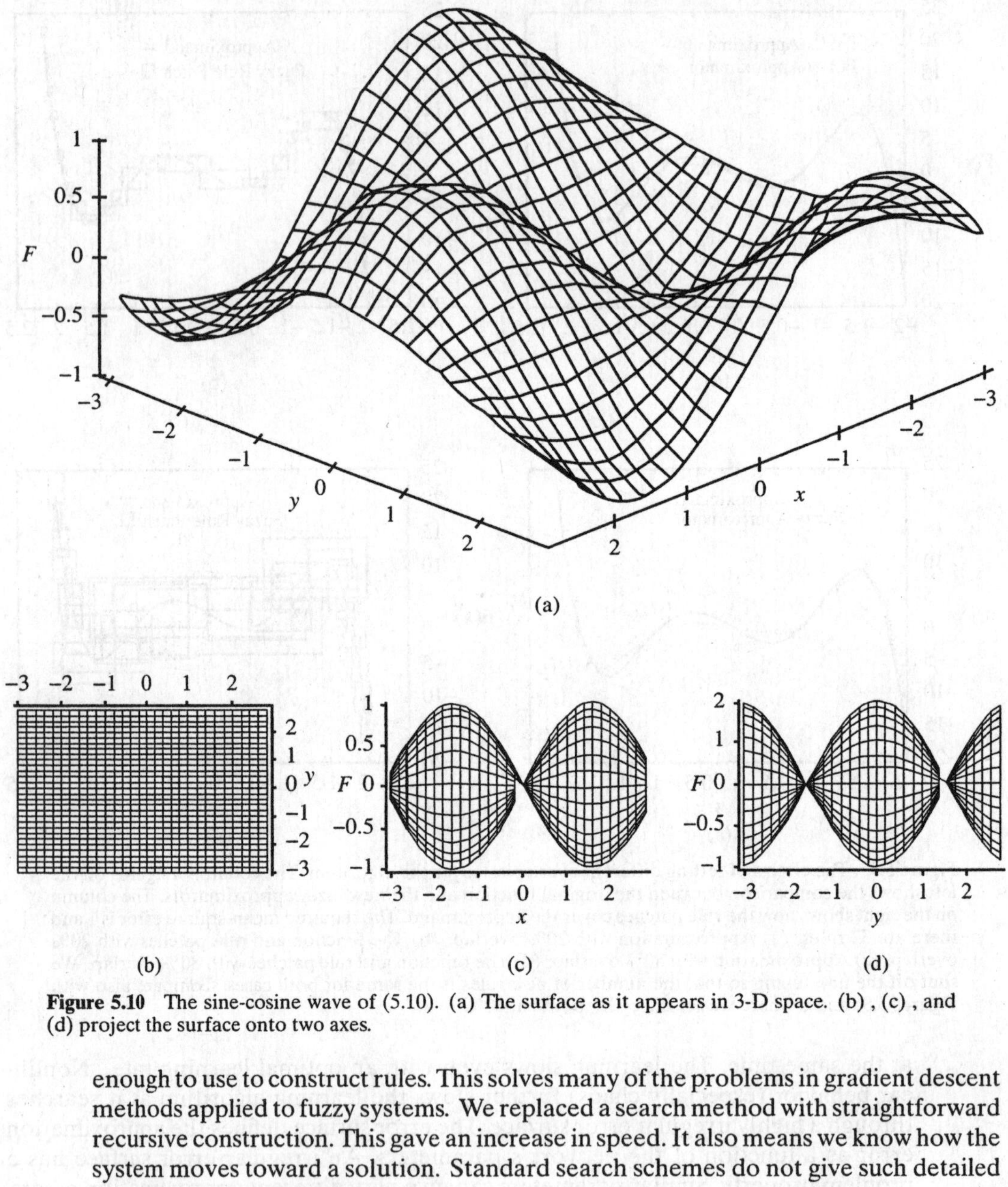

Figure 5.10 The sine-cosine wave of (5.10). (a) The surface as it appears in 3-D space. (b) , (c) , and (d) project the surface onto two axes.

enough to use to construct rules. This solves many of the problems in gradient descent methods applied to fuzzy systems. We replaced a search method with straightforward recursive construction. This gave an increase in speed. It also means we know how the system moves toward a solution. Standard search schemes do not give such detailed control over convergence.

5.4.1 A Dynamical System as a Combination of Functions

This section sets up basic notation and terms. We also explain how we can apply the above fuzzy function approximation scheme to dynamical systems.

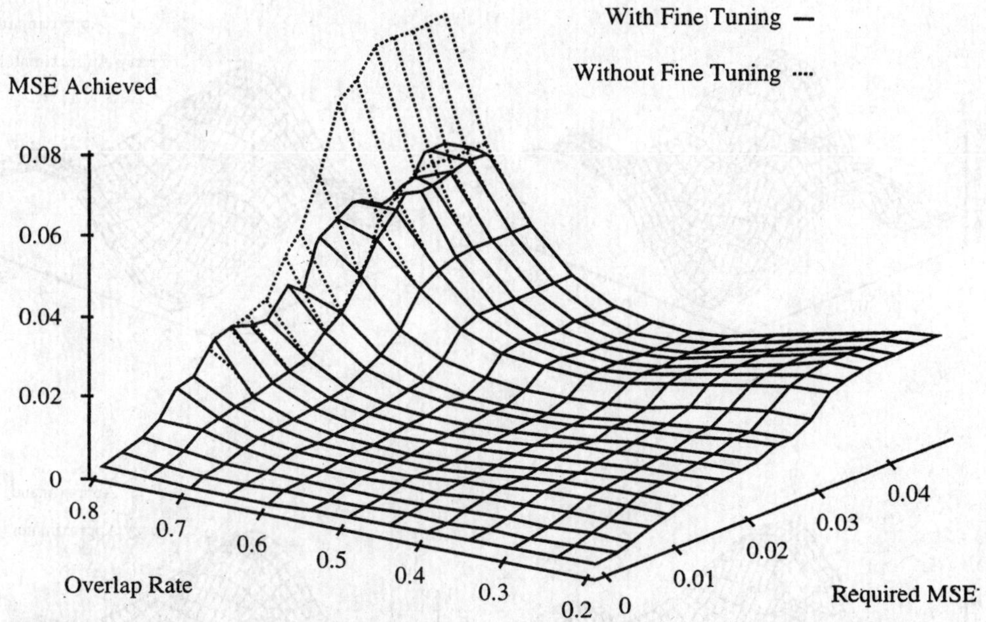

Figure 5.11 Error of sine-cosine approximation. Different parameters in the algorithm give different accuracy in the fuzzy approximators. The figure shows the relationship among the number of rules created, the overlap rate, and the final global mean-square error.

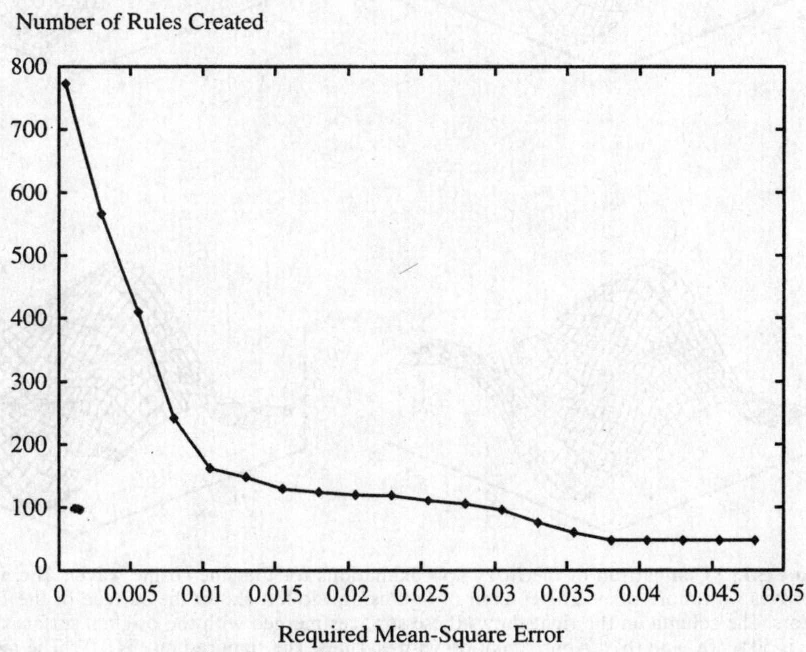

Figure 5.12 Rule count of sine-cosine approximations. The fuzzy approximators used exponentially more rules as they increased in accuracy.

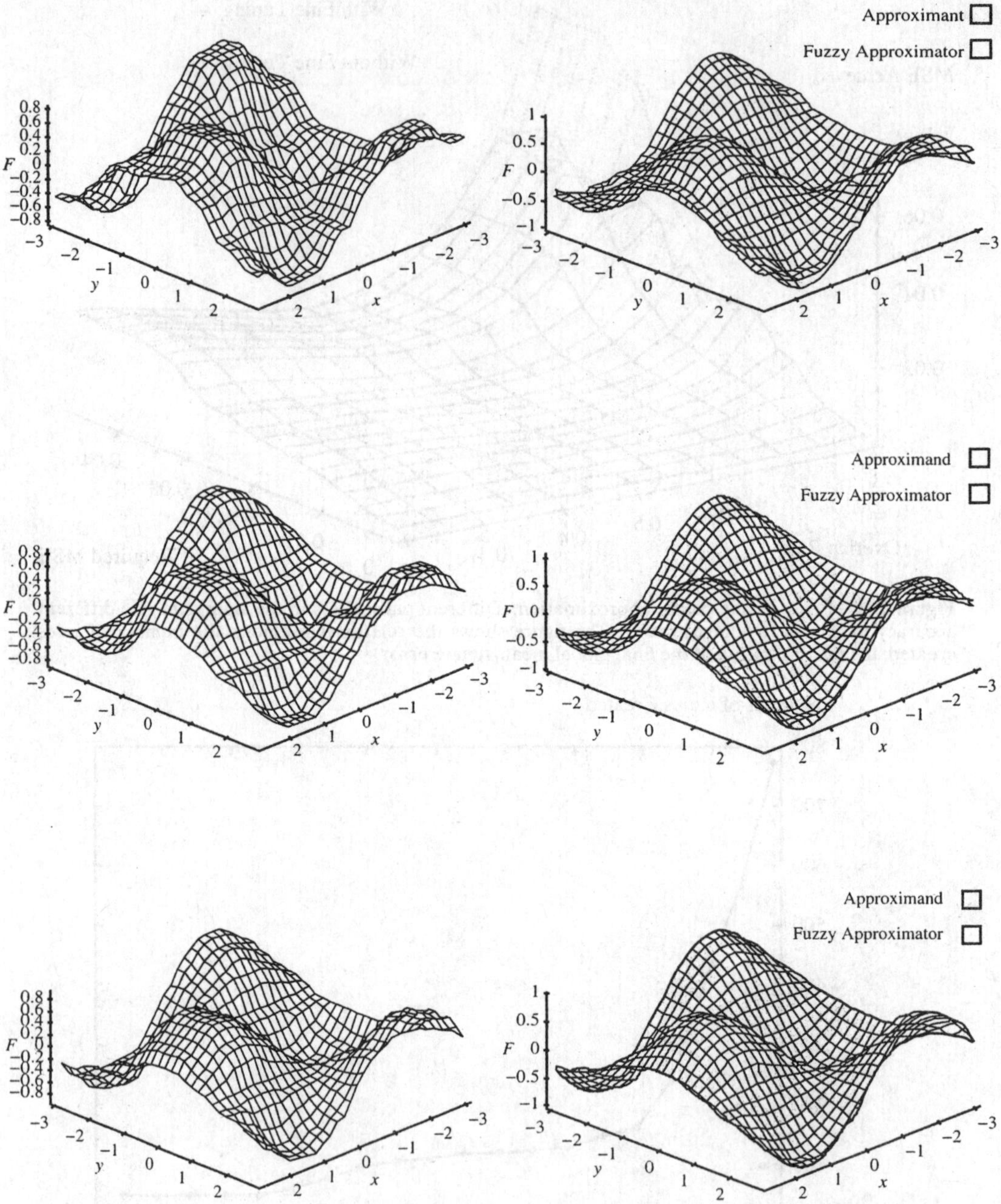

Figure 5.13 Comparison of the fuzzy approximations for the sine-cosine wave. The approximation improves with more rules. The column of figures on the left shows the surface of the fuzzy approximators. The column on the right shows these surfaces merged with the original surfaces. The overlap rate is 50%. (a) and (b) : Approximation with 96 rules. The required mse is 0.03. The mse achieved is 0.0094. (c) and (d) : Approximation with 162 rules (required mse = 0.01, mse achieved = 0.0043). (e) and (f) Approximation with 314 rules (required mse = 0.007, mse achieved = 0.0028).

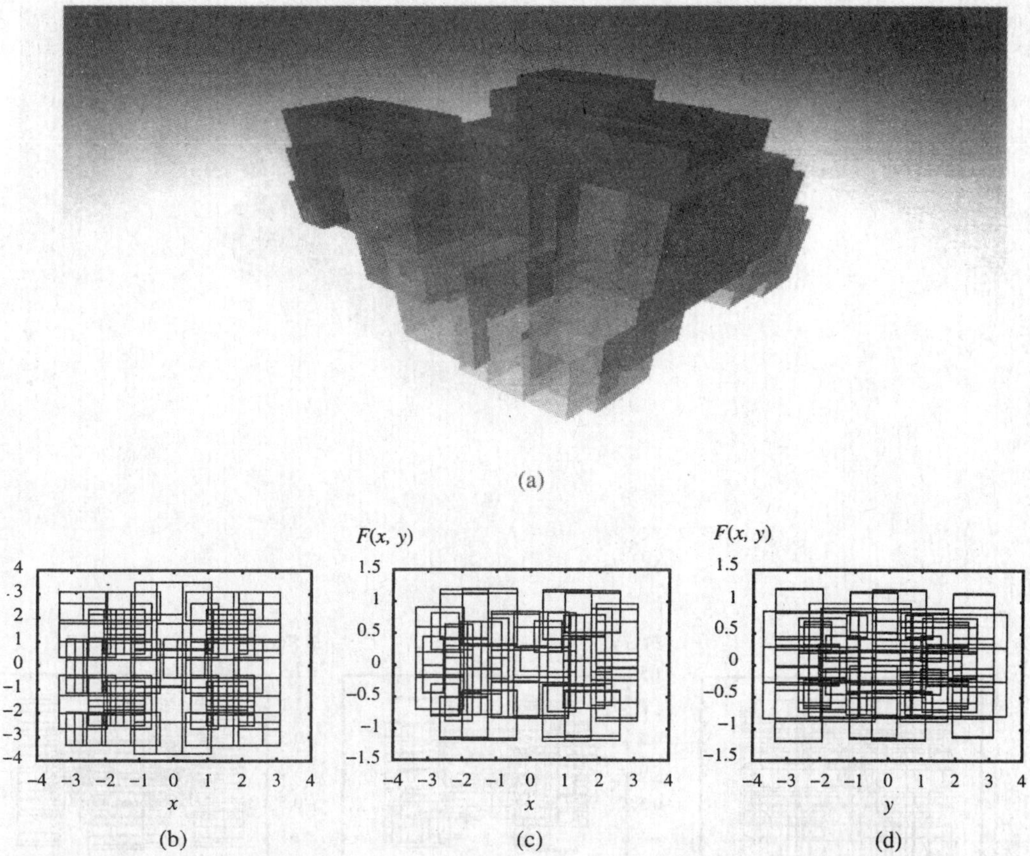

Figure 5.14 Rule-patch coverings of the sine-cosine wave surface. The required mean-square error for this approximation is 0.03. The overlap rate is 50%. The algorithm created 96 rules. (a) The 96 rule patches in 3-D: The patches are translucent boxes. (b) , (c) , and (d) The projections of the rules patches on three dimensions. Compare with Figures 5.15 and 5.16.

A dynamical system consists of a set of *state variables* x_i that change in time. Each variable stands for a feature of the system. The state of a system depends on the values the variables x_i adopt. Differential equations define how the system changes in time. An autonomous vector dynamical system with n variables has the form:

$$\dot{x} = f(x) \tag{5.11}$$

where

$$x = \text{an } n\text{-dimensional vector}$$

$$\dot{x} = \text{the time derivative of } x$$

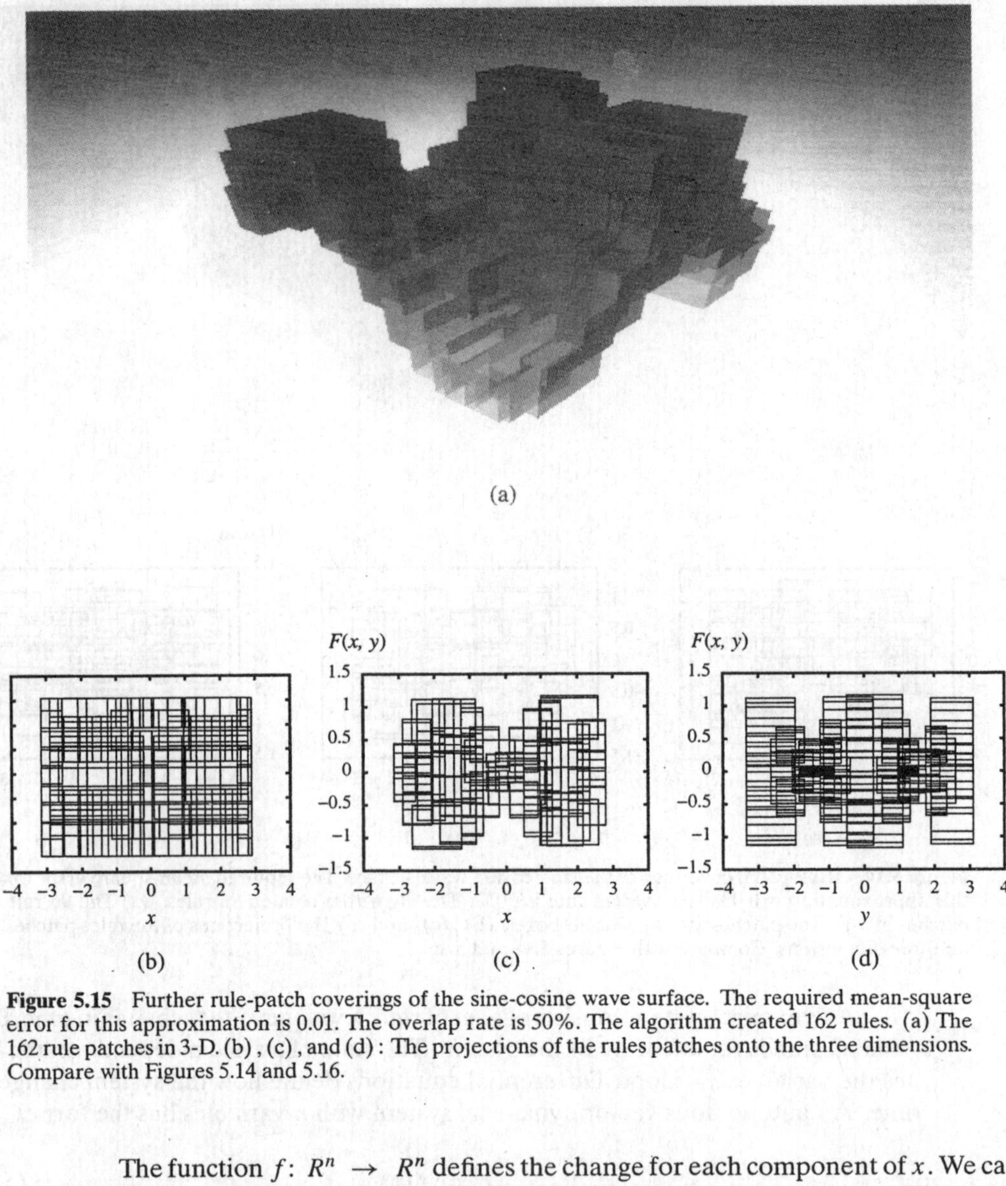

Figure 5.15 Further rule-patch coverings of the sine-cosine wave surface. The required mean-square error for this approximation is 0.01. The overlap rate is 50%. The algorithm created 162 rules. (a) The 162 rule patches in 3-D. (b), (c), and (d): The projections of the rules patches onto the three dimensions. Compare with Figures 5.14 and 5.16.

The function $f: R^n \rightarrow R^n$ defines the change for each component of x. We can represent the dynamical system geometrically in an n-dimensional space. Then each dimension represents a component of x. This space is a *phase space*. Each point in the phase space defines a unique state of the system. We define the dynamical system for all parts of the phase space. An initial point $x(t_0)$ and the solution $x(t)$ to (5.11) form a curve in phase space called a *trajectory* of the system. It traces the evolution of a dynamical system from an initial state. Generating a set of trajectories from different initial points in the phase space gives a more complete *flow diagram*. A flow diagram shows how the dynamical system behaves in different states.

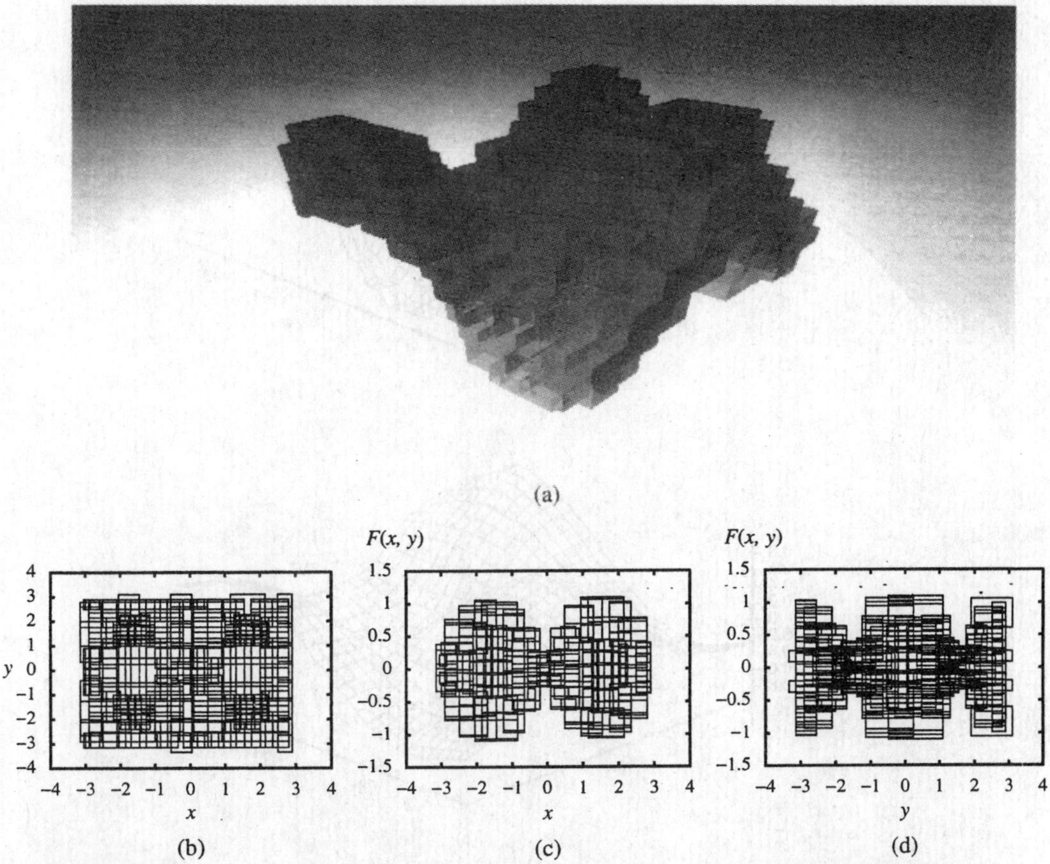

Figure 5.16 Finest rule-patch coverings of the sine-cosine wave surface. The required mean-square error for this approximation is 0.007. The overlap rate is 50%. The algorithm created 314 rules. (a) The 314 rule patches in 3-D. (b) , (c) and (d) : The projections of the rule patches on the three dimensions. Compare with Figures 5.14 and 5.15.

We approximate the function f in (5.11) to model the dynamical system. We view f as a combination of its component functions. If x is an n-dimensional vector then we decompose f into n functions of the form

$$\dot{x}_i = f_i(x) \tag{5.12}$$

where $\dot{x}_i$ is the ith component of the vector $\dot{x}$.

Then a combination of n fuzzy systems can approximate the function f. This maintains the multiple-input single-output framework we used above. The phase space acts as the input space of the functions f_i. The recursive partitioning scheme attempts to define a fuzzy rule patch covering for the whole phase space. We follow the convention in simulations of looking at the dynamical systems through sample trajectories in their phase space.

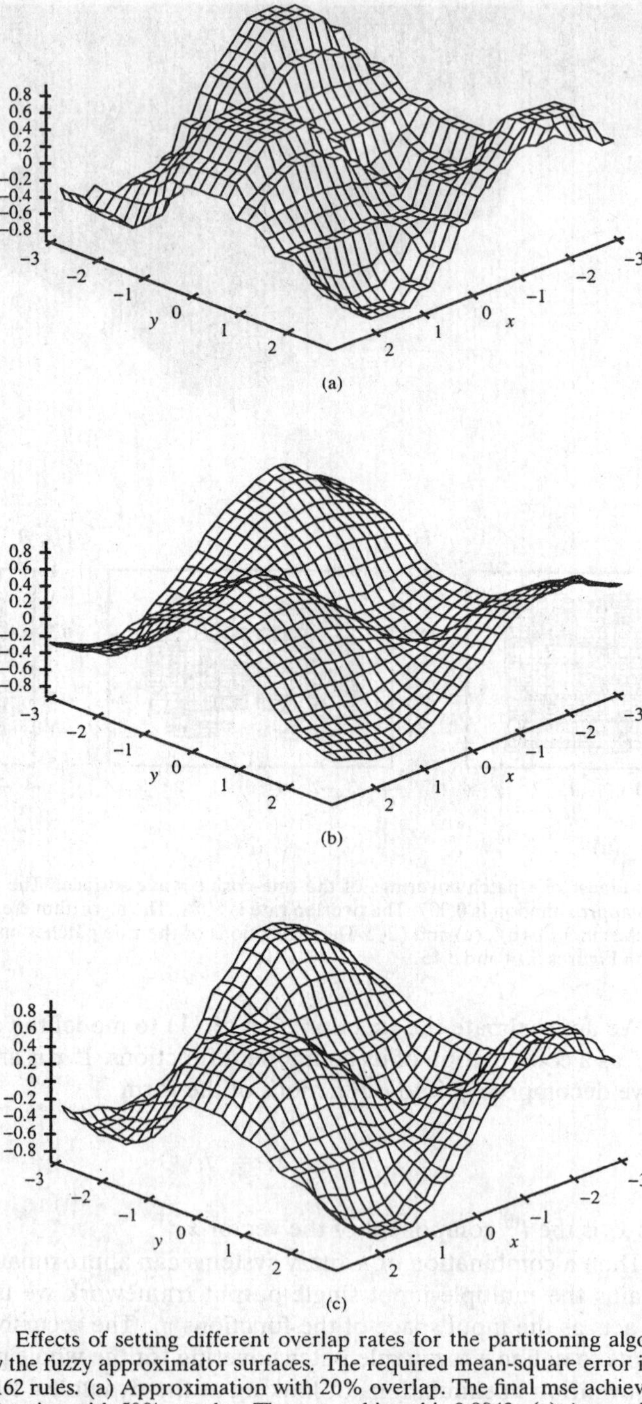

Figure 5.17 Effects of setting different overlap rates for the partitioning algorithm. The figures show the fuzzy approximator surfaces. The required mean-square error is 0.01. Each system has 162 rules. (a) Approximation with 20% overlap. The final mse achieved is 0.0067. (b) Approximation with 50% overlap. The mse achieved is 0.0043. (c) Approximation with 80% overlap. The mse achieved is 0.0133. We shut off the fine tuning so that the number of new rules was the same for all three cases.

5.4.2 Special Properties of Chaos

Chaotic systems may behave almost linearly in some parts of the phase space and behave highly nonlinearly in other parts. A few rules can approximate most of the phase space that is locally linear. Problems arise when dealing with highly nonlinear regions. Local approximation schemes such as fuzzy systems can work well in these cases. Local approximation takes advantage of the fact that we model different regions of the phase space differently. It creates more than one rule to model a region only if the region needs it.

We have thus far worked with continuous functions. We cannot be sure that a fuzzy system will approximate a discontinuous function to any arbitrary accuracy (unless it is measurable and bounded on a compact domain [3–4]). Some discontinuous functions have a level of approximation accuracy that no fuzzy system can reach. Figure 5.18 shows such an example. No number of rules can approximate the function in Figure 5.18b to an arbitrary accuracy. Demand drives recursive partitioning. It creates more and more rules as x_1 approaches a. Enough sample points could cause it to grow an unreasonably large number of rules for such a small region. We have to limit the many levels of partitioning.

Discontinuity can change the role of partitioning. So far we have partitioned a region to improve our approximation of that region. We can be sure that if we grow enough partitions (and set the right overlap rate) then we can achieve any level of accuracy. But we may never be able to approximate the discontinuous parts accurately enough. Then partitioning makes these regions as small as possible. Dynamical systems may have regions in its phase space that we can never model accurately enough. The recursive partitioning algorithm will reduce these regions to as small a size as possible. The accuracy will fall within the desired level except in these small regions.

5.4.3 The Van der Pol Oscillator

We first tested our fuzzy partitioning scheme on a nonlinear nonchaotic system called the Van der Pol oscillator [12]:

$$\dot{x} = -y$$

$$\dot{y} = x + y\mu(1 - x^2). \tag{5.13}$$

Figure 5.19a shows a trajectory grown from the initial conditions $x = 0$ and $y = 2$. Figure 5.19b shows a more complete flow diagram. A limit cycle attracts all trajectories in this system.

Figures 5.19 and 5.20 also show how the recursively grown fuzzy systems behaved. Figure 5.19 shows how the algorithm approximated one 5,000-point trajectory. The system needed over 2,000 rules to approximate both $\dot{x}$ and $\dot{y}$. This is a very large number of rules compared to the few hundred required for the simulations in Section 5.3. Growing a trajectory requires much higher accuracy. We form a new point based on previous points. An error at any stage compounds and amplifies in

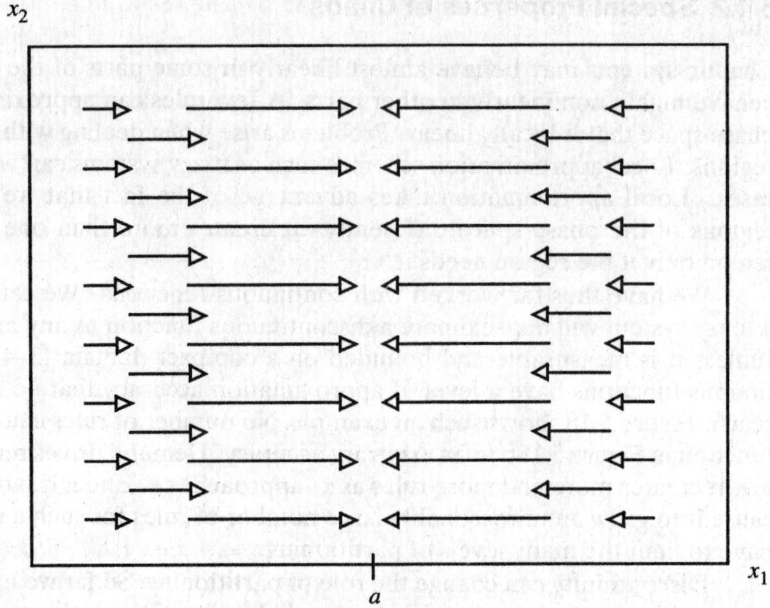

(a)

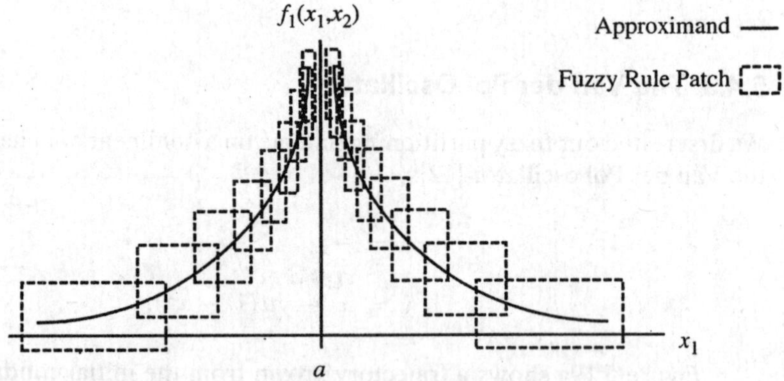

(b)

Figure 5.18 The effects of discontinuity in the approximand. (a) The phase space of a dynamical system and a few sample trajectories. The length of the lines gives the magnitude of change. The magnitude of change grows exponentially as x_1 moves toward a. (b) One of the approximand functions governing the system in (a). The discontinuity ensures that there will never be enough fuzzy rules to approximate the region around $x_1 = a$ to arbitrary accuracy. The recursive partitioning algorithm grows ever more rules as x_1 moves toward a.

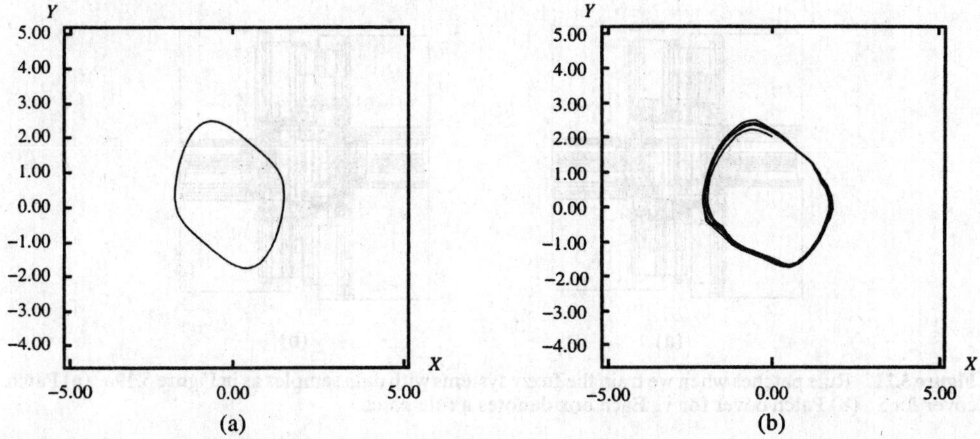

Figure 5.19 The Van der Pol Oscillator. (a) A sample trajectory of 5000 points computed using numerical solution to equations 5.13 with $\mu = 0.5$ starting from point (0,2). We grew the fuzzy system from these points. The fuzzy systems produced the 5000-point trajectory in (b). The overlap rate is 55%. The fuzzy approximator has 2,519 rules for $\dot{x}$ and 2,317 rules for $\dot{y}$.

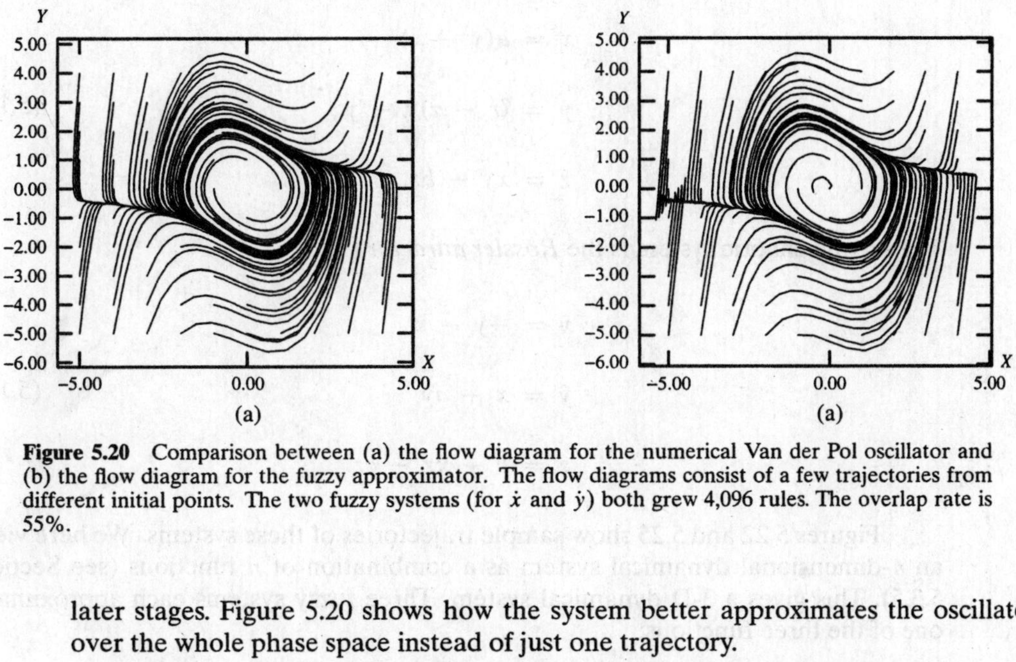

Figure 5.20 Comparison between (a) the flow diagram for the numerical Van der Pol oscillator and (b) the flow diagram for the fuzzy approximator. The flow diagrams consist of a few trajectories from different initial points. The two fuzzy systems (for $\dot{x}$ and $\dot{y}$) both grew 4,096 rules. The overlap rate is 55%.

later stages. Figure 5.20 shows how the system better approximates the oscillator over the whole phase space instead of just one trajectory.

Figure 5.21 shows how recursive partitioning covered the phase space with rule patches. We can see that the patches concentrate on regions in which there are sample points. Where there are no samples a big rule patch generalizes the behavior. The trained fuzzy approximator is three dimensional. So the rule patches will be three dimensional as well. The patches cover the whole phase space.

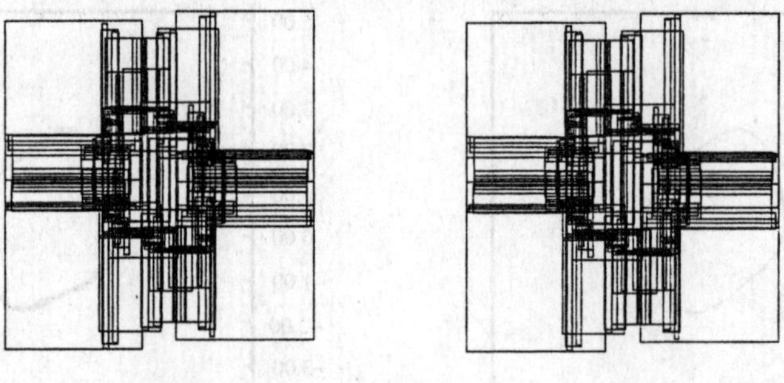

Figure 5.21 Rule patches when we train the fuzzy systems with data samples as in Figure 5.19a. (a) Patch cover for $\dot{x}$. (b) Patch cover for $\dot{y}$. Each box denotes a rule patch.

5.4.4 The Lorenz and Rossler Attractors

We next present the results from approximating two chaotic dynamical systems. The first approximand is the *Lorenz attractor* [6]:

$$\begin{aligned} \dot{x} &= a(y - x) \\ \dot{y} &= (r - z)x - y \\ \dot{z} &= xy - bz. \end{aligned} \tag{5.14}$$

The second chaotic system is the *Rossler attractor* [9]:

$$\begin{aligned} \dot{x} &= -y - x \\ \dot{y} &= x + ay \\ \dot{z} &= b + xz - cz. \end{aligned} \tag{5.15}$$

Figures 5.22 and 5.25 show sample trajectories of these systems. We here view an n-dimensional dynamical system as a combination of n functions (see Section 5.3.5). This gives a 3-D dynamical system. Three fuzzy systems each approximate one of the three functions:

$$\begin{aligned} f_1(x,\ y,\ z) &= \dot{x} \\ f_2(x,\ y,\ z) &= \dot{y} \\ f_3(x,\ y,\ z) &= \dot{z}. \end{aligned} \tag{5.16}$$

Comparing Figures 5.22 and 5.25 with 5.23 and 5.26 shows that recursive partitioning algorithm can generate accurate models for the samples from the chaotic systems. Figures 5.24 and 5.27 show how the method partitioned the phase spaces. The functions are 4-D and so the rule patches are 4-D as well. The functions range over the entire phase space. The figures show the 4-D patches projected onto two dimensions.

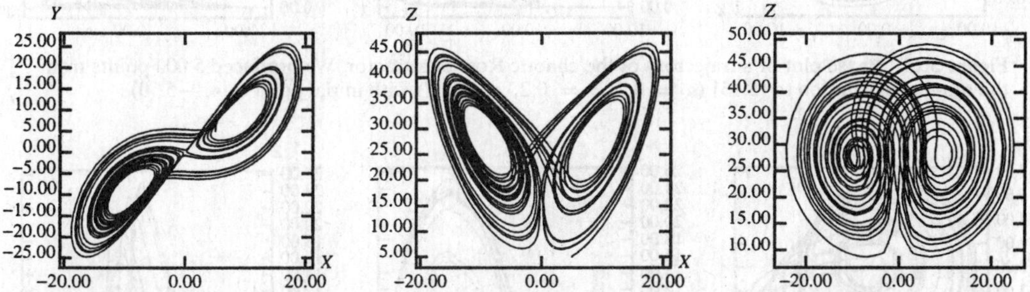

Figure 5.22 Phase plot of a trajectory of the chaotic Lorenz attractor. We produced 3,000 points from the numerical solution to (5.14) ($a = 10$, $r = 28$, $b = 8/3$) with initial point (5, 5, 5).

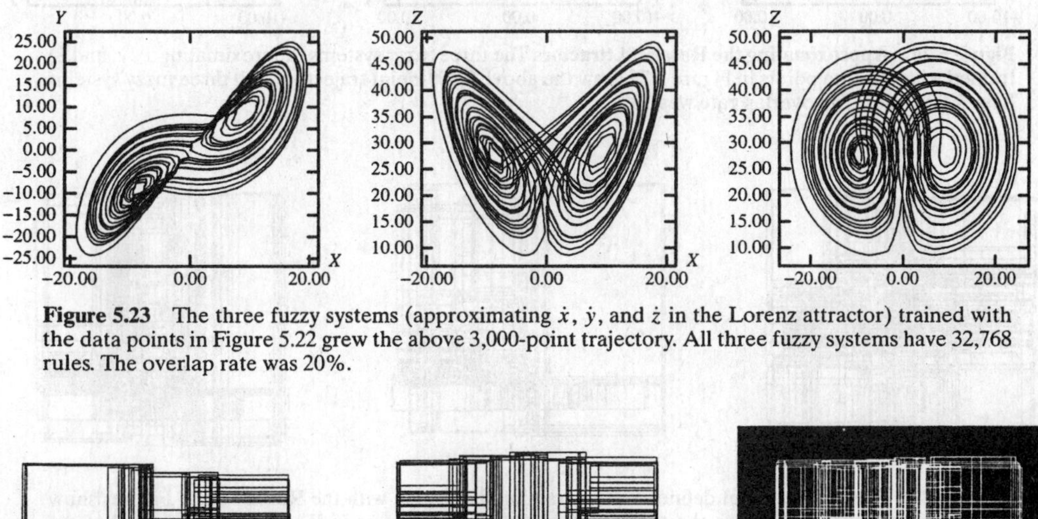

Figure 5.23 The three fuzzy systems (approximating $\dot{x}$, $\dot{y}$, and $\dot{z}$ in the Lorenz attractor) trained with the data points in Figure 5.22 grew the above 3,000-point trajectory. All three fuzzy systems have 32,768 rules. The overlap rate was 20%.

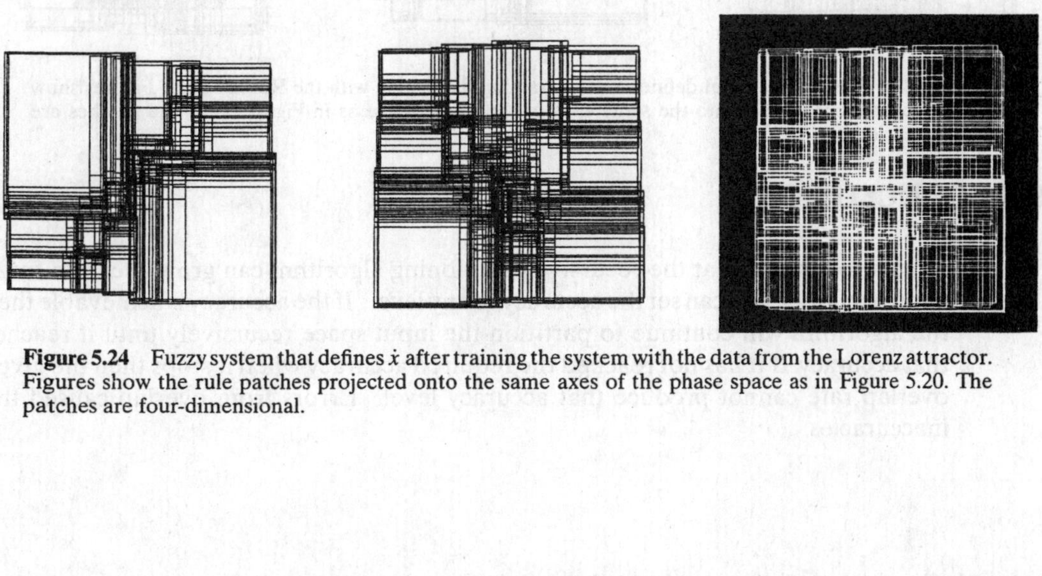

Figure 5.24 Fuzzy system that defines $\dot{x}$ after training the system with the data from the Lorenz attractor. Figures show the rule patches projected onto the same axes of the phase space as in Figure 5.20. The patches are four-dimensional.

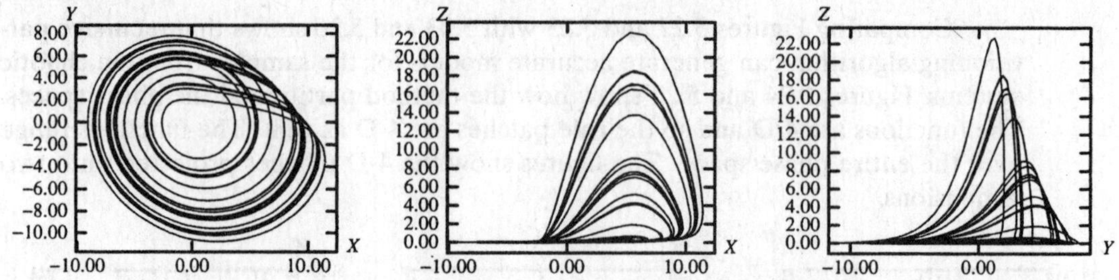

Figure 5.25 Phase plot of a trajectory of the chaotic Rossler attractor. We produced 5,000 points from the numerical solution to (5.15) ($a = 0.2$, $b = 0.2$, $c = 5.7$) with initial point $(-4, -5, 0)$.

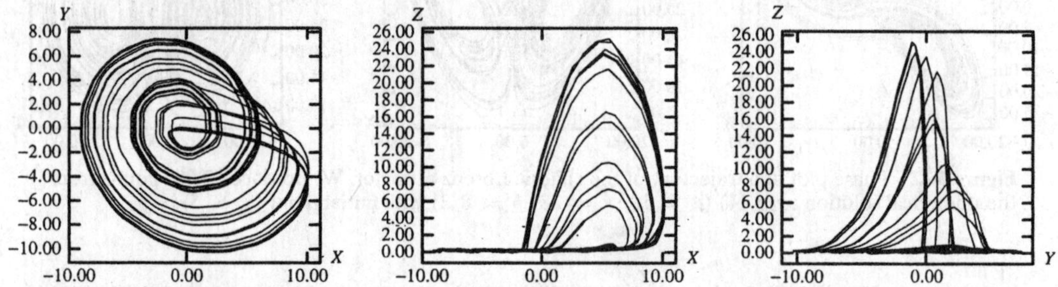

Figure 5.26 **Approximating the Rossler Attractors** The three fuzzy systems (approximating $\dot{x}$, $\dot{y}$, and $\dot{z}$) trained with the data points in Figure 5.25 grew the above 5,000-point trajectory. All three fuzzy systems have 1,024 rules. The overlap rate was 30%.

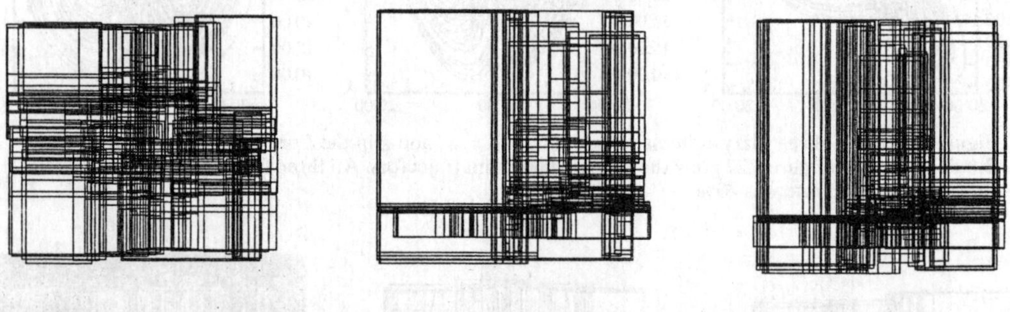

Figure 5.27 Fuzzy system that defines $\dot{x}$ after training the system with the Rossler data. Figures show the rule patches projected onto the same axes of the phase space as in Figure 5.20. The patches are four-dimensional.

5.5 CONCLUSIONS

Simulations show that the recursive partitioning algorithm can grow accurate fuzzy approximators. We can set the accuracy to any level. If the accuracy is achievable then the algorithm will continue to partition the input space recursively until it reaches that accuracy. If it has not reached the required accuracy when it stops then the given overlap rate cannot produce that accuracy level. Errors from overlap caused the inaccuracies.

The main feature of the algorithm is that it constructs rules only on demand. The partitioning criterion decides at each stage of the process if the system needs new rules. This grows rules only when the system needs them. Setting different criteria for new rules affects the structure of the algorithm. It can change the algorithm from a global gradient descent search method to a strictly local construction method or to something in between.

The overlap rate is one weakness of the algorithm. It is central to the fuzzy system and affects the approximation accuracy. Future research might explore ways to change it based on local information to ensure an increase in accuracy.

Rule explosion is the biggest problem. Feedforward fuzzy systems need tens of thousands of rules to finely approximate each of the many attractors of a nonlinear dynamical system. Optimal rules [5] can help somewhat as they try to cover the turning points of the attractors. But they too may increase in number to prohibitive levels. The feedback fuzzy systems of Chapters 2 and 15 may offer a more efficient way to approximate dynamical systems with a fixed and small number of rules. An open research problem is to find tractable learning laws that control and tune such systems.

REFERENCES

[1] Dickerson, J. A., & Kosko, B., "Fuzzy Function Approximation with Supervised Ellipsoidal Learning," *Proceedings of the INNS World Congress on Neural Networks*, vol. 2, 9–17, July 1993.

[2] Hiew, H. L., & Tsang, C. P., "An Adaptive Fuzzy System For Modelling Chaos," *Proceedings of the 1st International Conference on Fuzzy Theory and Technology*, 1992.

[3] Kosko, B., *Neural Networks and Fuzzy Systems: A Dynamical Approach to Machine Intelligence*, Prentice Hall, 1991.

[4] Kosko, B., "Fuzzy Systems As Universal Approximators," *Proceedings of the 1st IEEE International Conference on Fuzzy Systems*, 1153–1162, 1992; *IEEE Transactions on Computers*, vol. 43, no. 11, 1329–1333, November 1994.

[5] Kosko, B., "Optimal Fuzzy Rules Cover Extrema," *International Journal of Intelligent Systems*, vol. 10, no. 2, 249–255, February 1995.

[6] Lorenz, E., "Deterministic Nonperiodic Flow," *Journal of Atmospheric Science*, 20, 130, 1963.

[7] Pearlmutter, B. A., "Dynamic Recurrent Neural Networks," Technical Report CMU-CS–90–196, School of Computer Science, Carnegie Mellon University, 1990.

[8] Pearlmutter, B. A., "Learning State Space Trajectories in Recurrent Neural Networks," *Neural Computation*, 1, 263–269, 1989.

[9] Rossler, O. E., "An Equation for Continuous Chaos," *Physical Review Letters*, 57A, 397 – 398, 1976.

[10] Sato, M., "A Real Time Learning Algorithm for Recurrent Analog Neural Networks," *Biological Cybernetics*, 62, 237–241, 1990.

[11] Sato, M., "A Learning Algorithm to Teach Spatiotemporal Patterns to Recurrent Neural Networks," *Biological Cybernetics*, 62, 259–263, 1990.

[12] Van der Pol, B., "Forced Oscillations in a Circuit with Nonlinear Resistance (Receptance with Reactive Triode)," *London, Edinburg, and Dublin Phil. Mag.*, 3, 65–80, 1927.

HOMEWORK PROBLEMS

5.1. Consider the function

$$f(x) = \sin \pi x.$$

The following is a set of sample input/output pairs representing the above function:

(−1, 0), (−0.8, −0.588), (−0.6, −0.951), (−0.4, −0.951), (−0.2, −0.588), (0, 0), (0.2, 0.588), (0.4, 0.951), (0.6, 0.951), (0.8, 0.588), (1, 0).

Use (5.4) to compute the global mean-square error for the following approximations:

(a) (−1, 0), (−0.8, 0), (−0.6, 0), (−0.4, 0), (−0.2, 0), (0, 0), (0.2, 0), (0.4, 0), (0.6, 0), (0.8, 0), (1, 0)

(b) (−1, −0.294), (−0.8, −0.588), (−0.6, −0.951), (−0.4, −0.770), (−0.2, −0.588), (0, 0.294), (0.2, 0.588), (0.4, 0.951), (0.6, 0.770), (0.8, 0.588), (1, 0)

(c) (−1, −0.441), (−0.8, −0.770), (−0.6, −0.770), (−0.4, −0.476), (−0.2, −0.294), (0, 0.112), (0.2, 0.770), (0.4, 0.770), (0.6, 0.476), (0.8, 0.294), (1, 0.294)

(d) (−2, 0.294), (−1.8, 0.524), (−1.6, 0.951), (−1.4, 0.770), (−1.2, 0.411), (−1, 0.411), (−0.8, −0.770), (−0.6, −0.770), (−0.4, −0.476), (−0.2, −0.294), (0, 0.112), (0.2, 0.770), (0.4, 0.770), (0.6, 0.476), (0.8, 0.294), (1, 0.294), (1.2, −0.770), (1.4, −0.770), (1.6, −0.476), (1.8, −0.294), (2, −0.294)

(e) (−2, 0.294), (−1.8, 0.524), (−1.6, 0.939), (−1.4, 0.779), (−1.2, 0.331), (−1, −0.411), (−0.8, −0.770), (−0.6, −0.770), (−0.4, −0.476), (−0.2, −0.277), (0, 0.112), (0.2, 0.770), (0.4, 0.752), (0.6, 0.476), (0.8, 0.310), (1, 0.157), (1.2, −0.472), (1.4, −0.729), (1.6, −0.476), (1.8, −0.301), (2, −0.294).

5.2. Consider the recursive partitioning algorithm in Section 5.2.2. For simplicity let the partitioning criterion have the form

```
function Partition ? (input-space s)
begin
    if (output values among samples in s are identical) then
        return FALSE
    else
        return TRUE
end
```

All partitionings should produce two subregions of equal size.

Consider a step function

$$f(x) = \begin{cases} 0 & \text{if } x \leq 0 \\ 1 & \text{otherwise.} \end{cases}$$

Below is a set of sample input/output pairs for the pulse function:

(−5, 0), (−4, 0), (−3, 0), (−2, 0), (−1, 0), (1, 1), (2, 1), (3, 1), (4, 1), (5, 1).

(a) Plot the samples in the function's input/output space.

(b) Recursive partition the input space $x = [-5, 5]$.

(c) Recursive partition the input space $x = [-8.5, 8.5]$.

(d) Recursive partition the input space $x = [-8.5, 13.5]$.

5.3. Consider the function

$$f(x) = e^{-\frac{x^2}{2}} \cos(5x^2).$$

Below is a plot of the function:

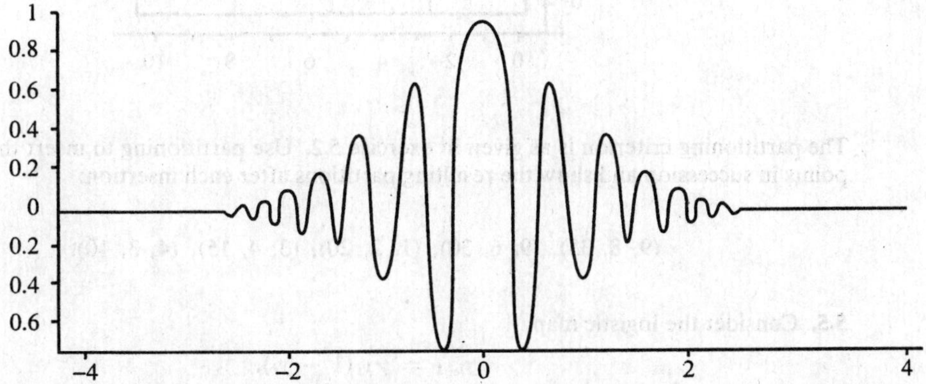

Use the recursive partitioning algorithm described in this chapter and the following partitioning criterion:

```
function Partition ? (input-space s)
begin
    if (output values among samples in s are different to
            within a threshold δ) then
        return FALSE
    else
        return TRUE
end
```

Draw partitioning lines on the diagram to show how the input space $X = [-4, 4]$ might recursively partition when

(a) $\delta = 0.4$

(b) $\delta = 0.8$

(c) $\delta = 1$

(d) $\delta = 2$.

5.4. Consider a partitioned input space with its sample data points of the form

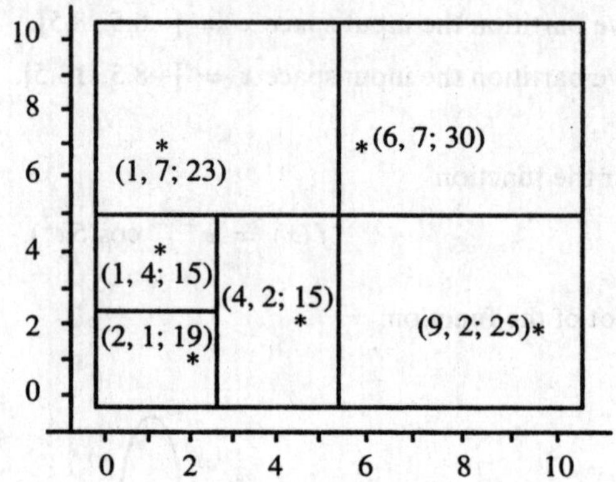

The partitioning criterion is as given in exercise 5.2. Use partitioning to insert the following points in succession and show the resulting partitions after each insertion:

(9, 8, 35), (9, 6, 30), (1, 2, 20), (3, 4, 15), (4, 3, 10).

5.5. Consider the logistic map:

$$y_{i+1} = \gamma y_i(1 - y_i).$$

(a) Let $\gamma = 3.1$. Plot the graphs of i versus y_i for $i = 0, 1, \ldots, 30$ with the following initial conditions: $y_0 = 0.2$ and $y_0 = 0.21$.

(b) Repeat part (a) with $\gamma = 3.6$.

(c) Let $y_0 = 0.21$. Plot the graphs of i versus y_i for $i = 0, 1, \ldots, 30$ with $\gamma = 2.9$, $\gamma = 3.1$, and $\gamma = 3.6$.

5.6. Consider the discrete Henon map:

$$x_{i+1} = 1 - \alpha x_i^2 + y_i$$

$$y_{i+1} = \beta x_i.$$

Let $\alpha = 1.4$ and $\beta = 0.3$.

(a) Plot the graphs of i versus x_i and i versus y_i for $i = 0, 1, \ldots, 30$ with the initial conditions $x_0 = 0$ and $y_0 = 0$.

(b) Use the same initial conditions in part a, plot the graph x_i versus y_i. Do the same for different initial conditions. Compare the graphs.

5.7. Rewrite the following maps and differential equations as a combination of functions so that you can model them using the normal fuzzy approximation framework. Plot the functions by projecting them down onto two dimensions.

(a)

$$\dot{x}_1 = -x_2 - x_3$$

$$\dot{x}_2 = -x_1 - \alpha x_2$$

$$\dot{x}_3 = \beta + x_1 x_3 - \gamma x_3$$

(b)

$$\dot{x}_1 = -x_2$$

$$\dot{x}_2 = x_1 - x_2$$

(c)

$$x_{i+1} = 1 - \alpha x_i^2 + y_i$$

$$y_{i+1} = \beta x_i$$

PART IV

FUZZY SIGNAL PROCESSING

Flashing lightning is an impulsive process. It can appear as impulsive noise on a radar screen or radio channel. Impulsive noise can also take the form of how arctic ice cracks or how new phone calls interrupt a phone or data line or how cosmic rays and particles bombard our bodies.

Most scientists and engineers model noise with a Gaussian bell curve. A Gaussian bell curve has a simple closed form that depends on just the mean and variance. The central limit theorem also yields a Gaussian bell curve as the limiting case of certain sums of independent statistics.

Other bell curves can give a better model of impulsive noise. The alpha-stable family of bell curves includes the Gaussian bell curve as the lone special case that has finite variance. Yet all independent alpha-stable statistics give rise to a central limit theorem. So the many central-limit claims put forth for a Gaussian bell curve argue for only some alpha-stable bell curve. And even ratios of Gaussian statistics give a Cauchy statistic. The Cauchy bell curve is the archetype closed-form alpha-stable bell curve with infinite variance and infinite higher-order moments.

Engineers have made only modest progress in filtering and prediction in the presence of alpha-stable noise. This stems in part from simple lack of knowledge of stable processes. It stems also from the less tractable math of alpha-stable processes. They do not have covariances. They have instead only a complex and asymmetric measure of covariation.

Chapter 6 uses the black-box power of additive fuzzy systems to filter impulsive noise from a signal and to predict signals in its presence. The approach extends the common standard additive model in two ways. It presents a new measure of covariation that acts like a covariance and so grows ellipsoidal fuzzy rules as in Chapter 3. It also works only with a metric-based joint if-part set function. This differs from the usual approach of factoring the joint set functions into scalar set functions. Chapter 9 also uses additive fuzzy systems to detect signals in alpha-stable impulsive noise. The results suggest that fuzzy systems may work well in the filtering or control of other impulsive processes that might preclude standard closed-form approaches.

6

FUZZY FILTERS FOR IMPULSIVE NOISE

Hyun Mun Kim and Bart Kosko

PREVIEW

Additive fuzzy systems can filter impulsive noise from signals. Alpha-stable statistics model the impulsiveness as a parameterized family of probability density functions or unit-area bell curves. The bell-curve parameter α ranges through the interval (0, 2] and gives the Gaussian bell curve when $\alpha = 2$ and gives the Cauchy bell curve when $\alpha = 1$. The impulsiveness grows as α falls and the bell curves have thicker tails. Only the Gaussian statistics have finite variances or finite higher moments.

An additive fuzzy system can learn ellipsoidal fuzzy rule patches from a new pseudo-covariation matrix or measure of alpha-stable covariation. Mahalanobis distance gives a joint set function for the learned if-part fuzzy sets of the if-then rules. The joint set function preserves input correlations that factored set functions ignore.

Competitive learning tunes the local means and pseudo-covariations of the alpha-stable statistics and thus tunes the fuzzy rules. Then the covariation rules can both predict nonlinear signals in impulsive noise and filter the impulsive noise in time-series data. The fuzzy system filtered such noise better than did a benchmark radial basis neural network.

6.1 FILTERING IMPULSIVE NOISE

Impulsive noise is not Gaussian. The bell-curve density of impulsive noise has thicker tails than a Gaussian bell curve has. The thicker tails give rise to more frequent bursts of noise. The Cauchy density $f(x) = \frac{1}{\pi(1+x^2)}$ has this property and so do all *alpha-stable* probability densities [13] for index parameter α in $0 < \alpha < 2$. The Cauchy density is the special case when $\alpha = 1$. The thicker polynomial tails give an infinite variance and give infinite higher-order moments. The *lower*-order fractional moments are still finite.

The Gaussian density is the special case of an alpha-stable density when $\alpha = 2$. It is unique in this family because it has exponential tails and has finite variance and higher-order moments. Figure 6.1 shows how alpha-stable noise grows more impulsive as the index α falls from 2 to 1.

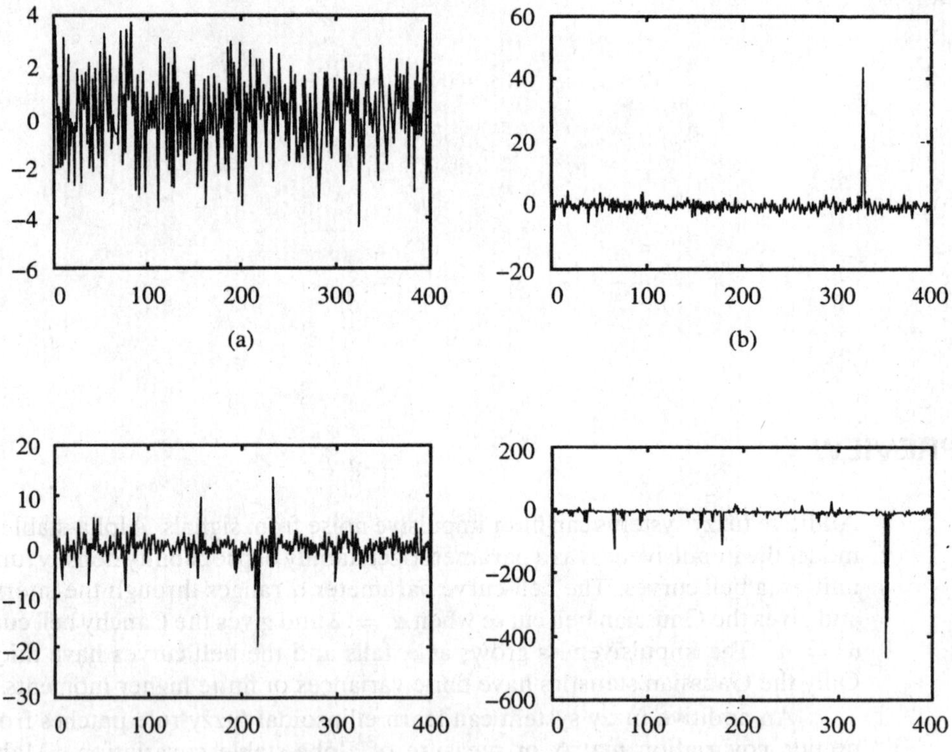

Figure 6.1 Impulsive noise modeled as independent samples from alpha-stable bell-curve densities. The noise magnitudes are not to scale. A noise density with a thicker tail has more frequent bursts of high energy. (a) The classical Gaussian case when $\alpha = 2$. Then the noise has finite variance. (b) More impulsive noise when $\alpha = 1.9$ and the noise has infinite variance. (c) More impulsiveness when $\alpha = 1.5$. (d) Still more impulsiveness in the Cauchy case when $\alpha = 1$.

The Gaussian bell curve is by far the most common bell curve in science and engineering. It fits some data well and depends on just the first and second moments of a process. This gives simple closed-form solutions to many problems. Limiting sums of independent random variables converge to a standard Gaussian variable if the variables have finite variance. The key term e^{-x^2} is in effect invariant under linear transforms and appears in the structure of Brownian motion. The Gaussian bell curve is the maximum-entropy density when we know just the first two (finite) moments of a process [12]. And it minimizes the "uncertainty principle" of physics and signal processing [6].

But locally a Gaussian curve looks like a Cauchy curve or other alpha-stable bell curve. Both have infinitely long tails and so are not "realistic." The tails have nonzero mass and so both can emit or explain a noise impulse of any magnitude. And both involve a central limit theorem: Limiting sums of independent alpha-stable variables converge to an alpha-stable random variable [13]. The real difference is that researchers first found and applied the Gaussian curve and that alpha-stable curves with $\alpha < 2$ better model impulsive noise as found in telephone lines, underwater acoustics, low-frequency atmospheric signals, and fluctuations in gravitational fields and stock prices and in many other processes. Alpha-stable models seem to work well [13] when the signal data contain outliers. We seek the best alpha-stable fit to a measured noise pattern. In general this best-fit value α^* will not pick out the Gaussian case and so $\alpha^* < 2$ will hold.

The main problem with alpha-stable models is their math complexity. Alpha-stable statistics do not have covariance matrices. Joint alpha-stable variables do have a measure of *covariation* that acts much like a covariance. But the covariation is not symmetric.

Below we define a pseudo-covariation measure that is symmetric and then use competitive learning to update it. This new measure gives a positive definite matrix and that gives a fuzzy ellipsoidal [4–5] rule as in Chapter 3. We also use a new way to project the rule ellipsoids onto joint multivariable fuzzy sets. This helps preserve correlations in the input data. We apply this method to predicting and filtering signals awash in impulsive noise.

6.2 ALPHA-STABLE NOISE: COVARIATION VERSUS COVARIANCE

This section reviews the family of alpha-stable probability densities and the measure of covariation that relates alpha-stable statistics. We then define a symmetric pseudo-covariation measure and show that it acts in the limit much as the asymmetric covariation measure acts in the limit.

6.2.1 Stable Distribution and Its Properties

An alpha-stable probability density f has the characteristic function [1–2, 13] φ:

$$\varphi(t) = \exp[iat - \gamma|t|^{\alpha}(1 + i\beta \text{sign}(t) \tan \frac{\alpha\pi}{2})] \qquad \text{for} \qquad \alpha \neq 1 \tag{6.1}$$

and

$$\varphi(t) = \exp[iat - \gamma |t| (1 + 2i\beta \ln|t| \operatorname{sign}(t)/\pi)] \quad \text{for} \quad \alpha = 1 \tag{6.2}$$

where

$$\operatorname{sign}(t) = \begin{cases} 1 & \text{if } t > 0 \\ 0 & \text{if } t = 0 \\ -1 & \text{if } t < 0 \end{cases} \tag{6.3}$$

and $i = \sqrt{-1}$, $0 < \alpha \le 2$, $-1 \le \beta \le 1$, and $\gamma > 0$. The α is the characteristic exponent parameter. An alpha-stable density with $\alpha < 2$ has finite moments only of order less than α [13]. The variance of an alpha-stable density distribution with $\alpha < 2$ does not exist. a is a location parameter. The location parameter a is the mean of the density when $\alpha > 1$. β is a skewness parameter. The density is symmetric about a when $\beta = 0$. The dispersion parameter γ acts like a variance. γ is half of the Gaussian variance when $\alpha = 2$. Figure 6.2 shows a set of symmetric alpha-stable bell curves.

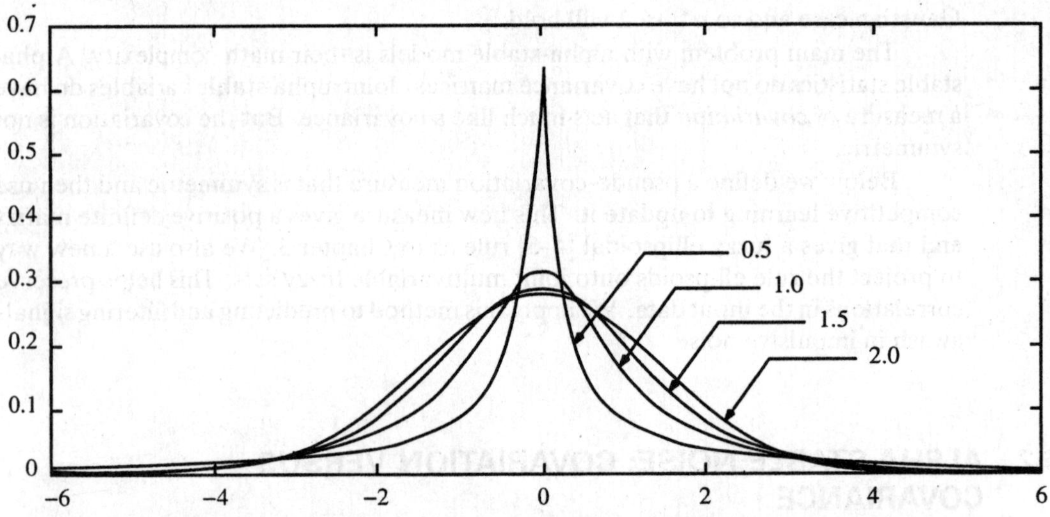

Figure 6.2 Alpha-stable probability densities. Only the Gaussian bell curve ($\alpha = 2$) has exponential tails and finite variation.

The alpha-stable characteristic function in (6.1) reduces to

$$\varphi(t) = e^{iat - \gamma |t|^2} \tag{6.4}$$

if $\alpha = 2$ and $\beta = 0$. (6.4) is the characteristic function of the Gaussian density with mean a and variance 2γ. The characteristic function in (6.2) reduces to

$$\varphi(t) = e^{-\gamma |t|} \tag{6.5}$$

if $a = 0, \alpha = 1$, and $\beta = 0$. The exponential (6.5) is the characteristic function of the Cauchy density function $f(x) = \frac{\gamma}{\pi(\gamma^2+x^2)}$. The characteristic function of (6.1) and (6.2) leads to the following theorem [13].

Theorem 6.1. Suppose $X_1, X_2, \ldots, X_n$ are independent random variables and have an alpha-stable distribution with the same parameters (α, β). Then all linear combinations of the form

$$\sum_{i=1}^{n} a_i X_i$$

are alpha stable with the parameter

$$\left(\alpha, \frac{\sum_{i=1}^{n} |a_i|^\alpha \operatorname{sign}(a_i)}{\sum_{i=1}^{n} |a_i|^\alpha} \beta \right)$$

if $\alpha \neq 1$.

6.2.2 Covariation and Pseudo-covariation

Local means and covariances define ellipsoidal fuzzy rules in Chapter 3 and in [4] and [5]. But alpha-stable random variables do not have a covariance if $\alpha < 2$. They do have a measure of covariation.

Let X and Y be symmetric alpha-stable random variables with $1 < \alpha \leq 2$, zero location parameters a_x and a_y, and dispersions γ_x and γ_y. The *covariation* of X with Y is [13]

$$(X, Y)_\alpha = \frac{E(X\operatorname{sign}(Y))}{E(|Y|)} \gamma_y. \tag{6.6}$$

Then

$$(X, X)_\alpha = \gamma_x \quad \text{and} \quad (Y, Y)_\alpha = \gamma_y. \tag{6.7}$$

So the covariations are not symmetric. The dispersion γ is also hard to estimate. Shao and Nikias [13] use three methods to estimate γ. These methods are computationally heavy and do not favor competitive learning. We can replace the dispersion γ with lower-order fractional moments from Theorem 6.2 [13].

Theorem 6.2. Let X be a symmetric alpha-stable random variable with dispersion γ and location parameter $a = 0$. Then

$$E(|X|^p) = C(p, \alpha)\gamma^{\frac{p}{\alpha}} \quad \text{for} \quad 0 < p < \alpha \tag{6.8}$$

where $C(p, \alpha)$ is a constant that depends only on α and p.

Then

$$\gamma = C_1 E(|X|^p) \quad \text{as} \quad p \to \alpha \tag{6.9}$$

if $C_1 = 1/C(p, \alpha)$.

Now define a *pseudo*-covariation that is symmetric. The pseudo-covariation of X with Y is

$$(X, Y)_{\alpha,\text{pseudo}} = E(\text{sign}(XY)|X|^{\frac{p}{2}}|Y|^{\frac{p}{2}}) \quad \text{for} \quad 0 < p < \alpha. \tag{6.10}$$

Then (6.9) implies

$$(X, X)_\alpha = C_1(X, X)_{\alpha,\text{pseudo}} \quad \text{as} \quad p \to \alpha. \tag{6.11}$$

We use lower-order fractional moments instead of dispersion for competitive learning. We use the same power for each random variable to make the pseudo-covariation symmetric. Note that the pseudo-covariation equals the covariance when $p = 2$. Later we use $p = 1$ to learn fuzzy rules that can filter Cauchy-level impulsive noise.

We can define a pseudo-covariation matrix C_X for $X = (X_1, \ldots, X_n)^T$ that acts as a symmetric and positive definite covariance matrix:

$$C_X = \begin{bmatrix} (X_1, X_1)_{\alpha,\text{pseudo}} & \cdots & (X_1, X_n)_{\alpha,\text{pseudo}} \\ (X_2, X_1)_{\alpha,\text{pseudo}} & \cdots & (X_2, X_n)_{\alpha,\text{pseudo}} \\ \cdots & \cdots & \cdots \\ (X_n, X_1)_{\alpha,\text{pseudo}} & \cdots & (X_n, X_n)_{\alpha,\text{pseudo}} \end{bmatrix}. \tag{6.12}$$

Appendix 6.A presents the competitive learning system [8] to estimate the local covariation matrix. Then the local mean and covariation matrix define hyperellipsoids in the input–output space. We use Mahalanobis distance to produce joint if-part fuzzy sets from the ellipsoidal covariation rules. Section 6.6 tests how well these covariation rules filter alpha-stable impulsive noise.

6.3 ADDITIVE FUZZY SYSTEMS

This section reviews afresh the standard additive model (SAM) of the fuzzy system $F: R^n \to R$ from Chapter 2 that we use to predict and filter nonlinear signals in alpha-stable noise. The pseudo-covariation measure of (6.10)–(6.11) acts as standard covariance and yields ellipsoidal fuzzy rules. We use a new scheme based on Mahalanobis distance to project the ellipsoidal rule onto joint if-part fuzzy sets that maintain the correlation structure among the inputs. Other projection schemes factor the input sets and ignore the correlation.

6.3.1 Additive Fuzzy Systems

A fuzzy system $F: R^n \rightarrow R$ stores m rules of the word form "If $X = A_j$ then $Y = B_j$" or the patch form $A_j \times B_j \subset X \times Y = R^n \times R$. The if-part fuzzy sets $A_j \subset R^n$ and then-part fuzzy sets $B_j \subset R$ have set functions $a_j: R^n \rightarrow [0, 1]$ and $b_j: R \rightarrow [0, 1]$. The system can use the joint set function a_j or some factored form such as $a_j(x) = a_j^1(x_1) \cdots a_j^n(x_n)$ or $a_j(x) = \min(a_j^1(x_1), \ldots, a_j^n(x_n))$ or any other conjunctive form for input vector $x = (x_1, \ldots, x_n) \in R^n$.

An additive fuzzy system [8–9] sums the "fired" then part sets B_j':

$$B = \sum_{j=1}^{m} B_j' = \sum_{j=1}^{m} a_j(x) B_j. \tag{6.13}$$

Figure 6.3 shows the parallel fire-and-sum structure of the SAM system. These systems can uniformly approximate any continuous (or bounded measurable) function f on a compact domain [9].

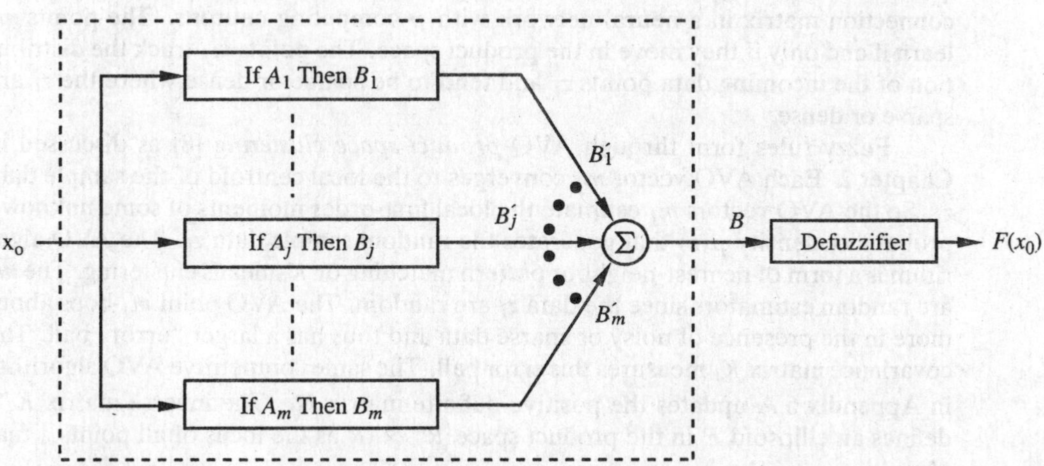

Figure 6.3 Architecture of an additive fuzzy system $F: R^n \rightarrow R^p$ with m rules. Each input $x_0 \in R^n$ enters the system F as a numerical vector. At the set level x_0 acts as a delta pulse $\delta(x - x_0)$ that combs the if-part fuzzy sets A_j and gives the m set values $a_j(x_0)$: $a_j(x_0) = \int_{R^n} \delta(x - x_0) a_j(x) dx$. The set values "fire" the then-part fuzzy sets B_j to give B_j'. A standard additive model (SAM) scales each B_j with $a_j(x)$. The system then sums the B_j' sets to give B. The system output $F(x_0)$ is the centroid of B.

The scaling choice $B'_j = a_j(x)B_j$ gives a *standard additive model* or SAM. Taking the centroid of B in (6.13) gives [8]–[11] the SAM ratio of Chapter 2:

$$F(x) = \frac{\sum_{j=1}^{m} a_j(x)V_jc_j}{\sum_{j=1}^{m} a_j(x)V_j}. \tag{6.14}$$

V_j is the nonzero volume or area of then-part set B_j. c_j is the centroid of B_j or its center of mass.

The SAM theorem (6.14) implies that the fuzzy structure of the then-part sets B_j does not matter. The ratio depends on just the volume V_j and location c_j of the then-part sets B_j. Our SAM has then-part sets of the same area and so the volume terms V_j cancel from (6.14). We need to pick only the scalar centers c_j. The projection method in the next section picks the multivariable if-part sets a_j.

6.3.2 Covariation Rules and Mahalanobis Sets: Product Space Clustering and Projection

Fuzzy rule patches $A_j \times B_j$ can take the form of ellipsoids [4–5] as in Chapter 3. Input–output data (x_i, y_i) drive the competitive learning process in Appendix 6.A. We form the concatenated vector $z_i^T = [x_i^T, y_i]$ in the product space $R^n \times R$. Then we loose p quantization vectors m_j on the same product space. These vectors learn or adapt as they code for the local sample statistics and give rise to an *adaptive vector quantization* (AVQ). The AVQ vectors m_j act as the synaptic fan-in columns of a connection matrix in a neural network with p competing neurons. The points m_j learn if and only if they move in the product space. The points m_j track the distribution of the incoming data points z_i and tend to be sparse or dense where the z_i are sparse or dense.

Fuzzy rules form through AVQ *product space clustering* [8] as discussed in Chapter 2. Each AVQ vector m_j converges to the local centroid of the sample data z_i. So the AVQ vectors m_j estimate the local first-order moments of some unknown probability density $p(z)$ that generates the random sample data z_i. The AVQ algorithm is a form of nearest-neighbor pattern matching or K-means clustering. The m_j are random estimators since the data z_i are random. The AVQ point m_j hops about more in the presence of noisy or sparse data and thus has a larger "error" ball. The covariance matrix K_j measures this error ball. The same competitive AVQ algorithm in Appendix 6.A updates the positive-definite matrix K_j. The inverse matrix K_j^{-1} defines an ellipsoid E in the product space $R^n \times R$ as the locus of all points z that obey

$$e^2 = (z - m_j)^T K_j^{-1}(z - m_j) \tag{6.15}$$

for centering constant $e > 0$. Then the jth ellipsoid defines the jth rule in the additive fuzzy system $F\colon R^n \to R$. We use the pseudo-covariation matrix in (6.10) with $p = 1$ to grow ellipsoidal *covariation rules*.

Fuzzy sets arise when the ellipsoids project down onto the axes. The projection is not unique because an ellipsoid $E \subset R^n \times R$ does not factor into the Cartesian product of $n + 1$ sets $E_1 \times \cdots \times E_n \times E_{n+1}$. The SAM theorem (6.14) lets us replace the then-part set E_{n+1} (or B_j) with just an area or volume V_j and a centroid c_j. This presents a problem since noisy rules will have larger V_j values than smaller and more certain rule patches have. Then the SAM ratio (6.14) would give more weight to the least certain rules. Chapter 3 deals with this by weighting each rule in (6.13) and (6.14) with a weight w_j that varies with the inverse of V_j or with its square. We deal with it by normalizing all then-part sets to have the same area or volume. Then $V_1 = \cdots = V_m > 0$. This ignores some of the statistical structure of E but simplifies the SAM ratio (6.14) for computing $F(x)$. It also finds each then-part set B_j as just the centroid c_j of the projection of the jth ellipsoid onto the output axis R.

Chapter 3 projects the ellipsoid E onto the n input axes to give triangular fuzzy sets. Figure 6.4 shows how a 2-D ellipsoid projects onto the x_1 and x_2 axes to give two triangles. The two vector points u and v lie the same distance from the center of the ellipsoid. So their components belong to the same degree to the fuzzy triangular sets A^1 and A^2: $a^1(u_1) = a^1(v_1)$ and $a^2(u_2) = a^2(v_2)$. We lose the correlation structure among the input components with a factored set function $a(u) = \min(a^1(u_1), a^2(u_2)) = a(v)$. The same holds for pairwise multiplication or other pairwise conjunctive operators.

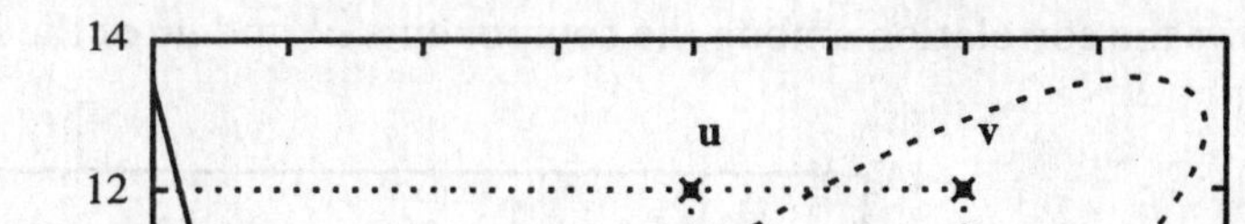

Figure 6.4 Triangular fuzzy sets from ellipsoidal projection onto input axes. The vectors u and v have components that belong to the same degree to the scalar fuzzy sets A^1 and A^2. Such projection ignores correlation among the input components.

We propose a metrical set function $a(x)$ that preserves the correlations among the input components and that exploits the second-order statistics of the sample data z. The form of the ellipsoids in (6.15) suggests the squared *Mahalanobis distance*

$$d(x)^2 = (x - m_j^i)^T (K_j^i)^{-1} (x - m_j^i) \tag{6.16}$$

as the basis for a similarity set function $a(x)$. The n-vector m_j^i in (6.16) stands for the first n components of the quantization vector m_j in the input–output product space $R^n \times R$ with covariance submatrix K_j^i. $d(x)$ is the covariance-scaled distance from input vector x to local mean m_j^i.

We define the joint if-part set function a_j as the Mahalanobis similarity measure

$$a_j(x) = \begin{cases} 1 - \frac{d(x)}{q} & \text{if } d(x) \le q \\ 0 & \text{if } d(x) > q \end{cases} \tag{6.17}$$

for some normalizing factor $q > 0$. We used $q = 2$. Figure 6.5 shows ellipsoidal contours of equal Mahalanobis distance. Then $d(u) > d(v)$ and the vector components maintain their correlation structure. So $a(u) < a(v)$ as the figure makes clear: u has less set value than v has in the concentric interior ellipsoids. Figure 6.6 shows the fuzzy structure of the joint if-part Mahalanobis set function a. The cone structure shows the correlation among the components a^1 and a^2 of the if-part set function a.

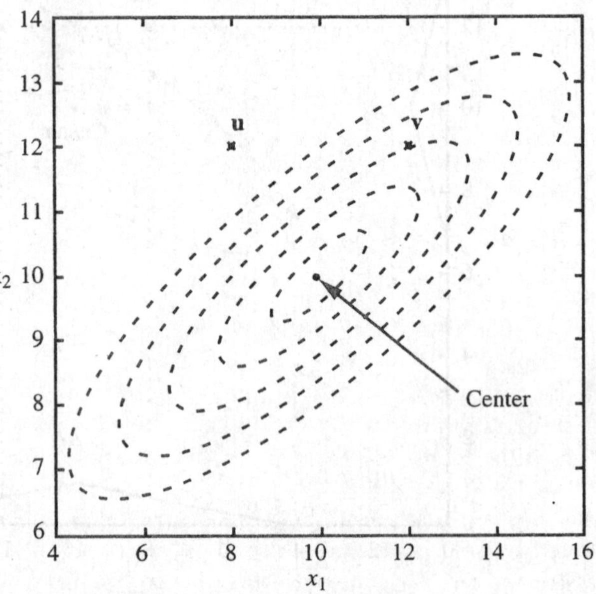

Figure 6.5 Ellipsoidal contours of equal Mahalanobis distance. The point v is closer to the center in Mahalanobis distance than point u is. Both points have the same Euclidean distance to the center when the covariance matrix K is the identity matrix I.

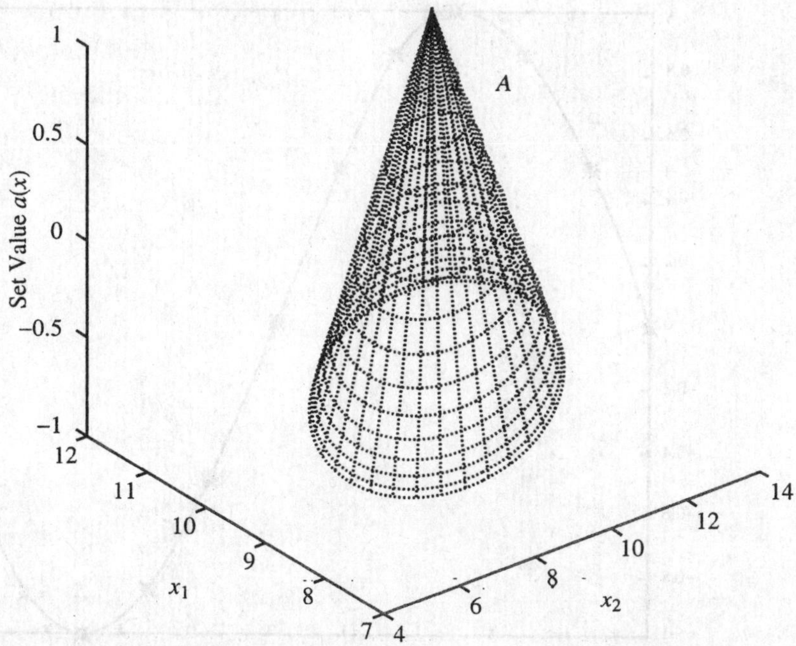

Figure 6.6 Joint input fuzzy set A that arises from the ellipsoidal Mahalanobis isometric contours in Figure 6.5. The conical fuzzy-set patch A does not factor into a Cartesian product of two uncorrelated fuzzy sets A^1 and A^2.

6.4 FUZZY PREDICTION WITH COVARIATION RULES

This section shows how a SAM fuzzy system $F: R^n \to R$ with pseudo-covariation rules from Section 6.3 predicts a nonlinear function in the presence of impulsive alpha-stable noise. The test function was a sine wave sampled at 12 points for each period as in Figure 6.7.

The goal was to predict the next sample value $y(k)$ based on the past two sample differences $y(k-1) - y(k-2)$ and $y(k-2) - y(k-3)$. Additive measurement noise $n(k)$ corrupted each input to the fuzzy system. The noise $n(k)$ came from pseudo-random numbers for the alpha-stable values $\alpha = 2, \alpha = 1.9, \alpha = 1.5$, and $\alpha = 1$.

The noise $n(k)$ had finite variance for only the Gaussian case $\alpha = 2$. So we used the finite unbiased sample variance

$$\hat{\sigma}^2 = \frac{1}{n-1} \sum_{i=1}^{n} (x_i - \bar{x})^2 \tag{6.18}$$

with sample mean

$$\bar{x} = \frac{1}{n} \sum_{i=1}^{n} x_i .$$

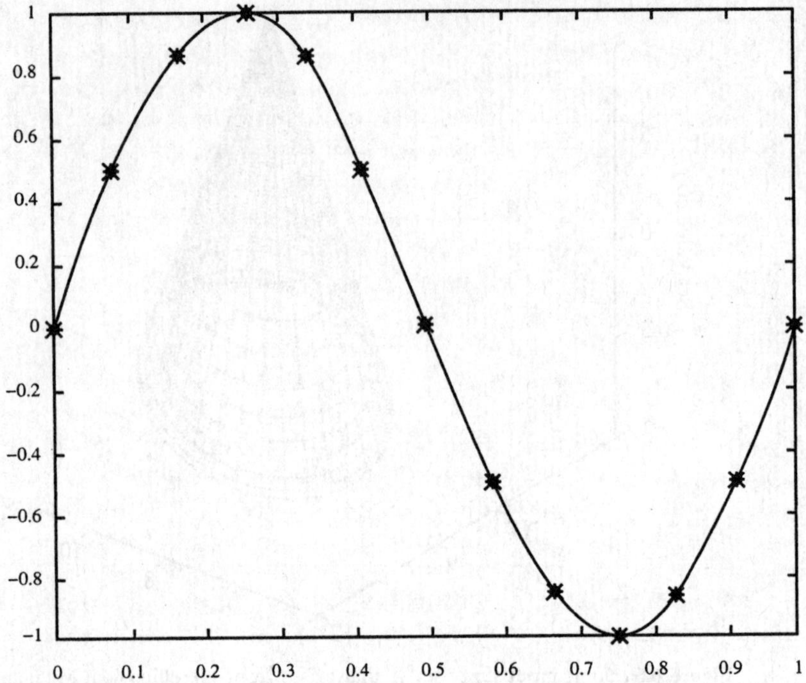

Figure 6.7 Sampling points of sinusoidal function.

The fuzzy system F gave the predicted value

$$\hat{y}(k) = F(x) = F(x_1(k), x_2(k)) \tag{6.19}$$

for input components

$$x_1(k) = y(k-2) - y(k-3) + n(k) \text{ and}$$

$$x_2(k) = y(k-1) - y(k-2) + n(k).$$

We created 1,000 data pairs and used the first 500 as training data and the second 500 as test data. We used $p = 2$ in the pseudo-covariation measure of (6.10) to simplify the competitive learning process. There were 12 clusters since there were 12 sample points for each period. Figure 6.8 shows the resulting ellipsoidal covariation rules for the Gaussian case. Figure 6.9 shows the projected Mahalanobis sets.

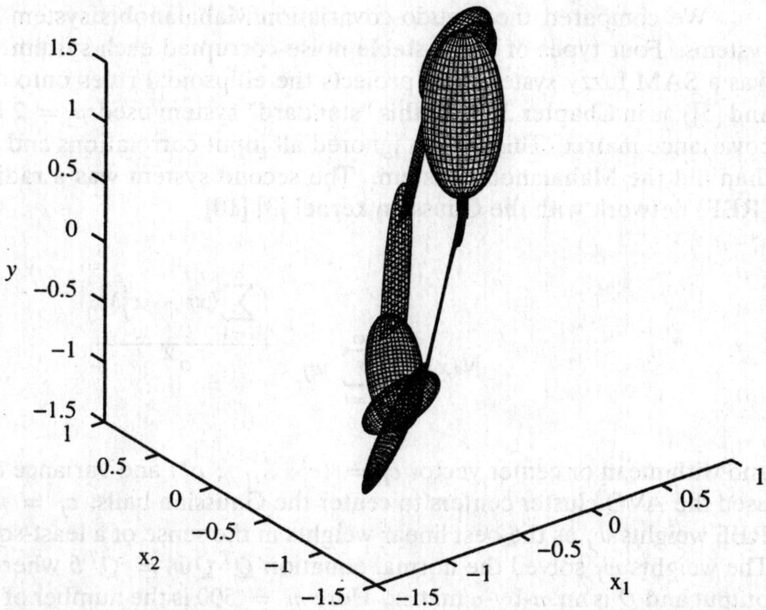

Figure 6.8 Ellipsoidal pseudo-covariation rules for the noisy fuzzy predictor. The noise in this case was additive Gaussian noise.

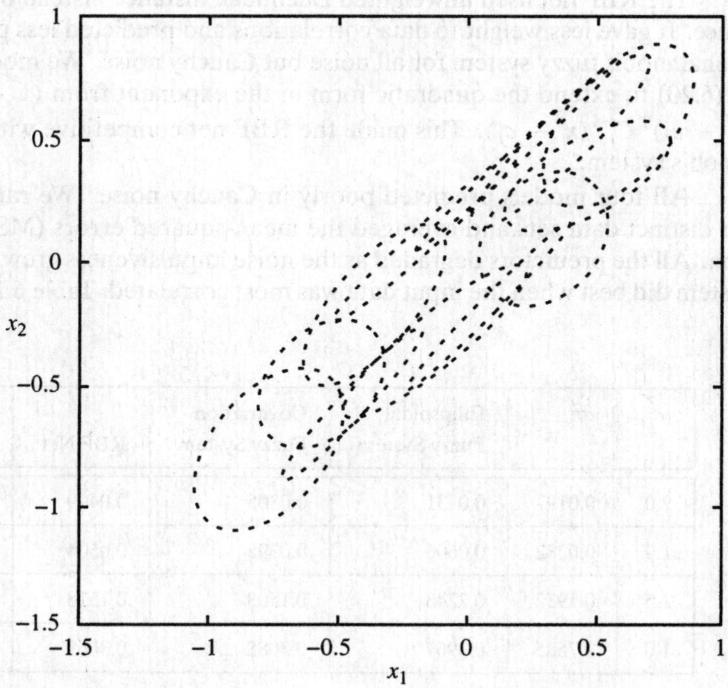

Figure 6.9 Mahalanobis joint fuzzy sets of the covariation rules in Figure 6.8.

We compared the pseudo-covariation/Mahalanobis system with three other systems. Four types of alpha-stable noise corrupted each system. The first system was a SAM fuzzy system that projects the ellipsoidal rules onto axis triangles ([4] and [5]) as in Chapter 3. Thus this "standard" system used $p = 2$ in (6.10) to give a covariance matrix. This system ignored all input correlations and did much poorer than did the Mahalanobis system. The second system was a radial basis function (RBF) network with the Gaussian kernel [3] [10]

$$N(x) = \sum_{j=1}^{m} w_j \, e^{-\frac{\sum_{i=1}^{n}(x_i - c_j^i)^2}{\sigma^2}} \tag{6.20}$$

and with mean or center vector $c_j = (c_j^1, \ldots, c_j^n)$ and variance or spread σ^2. We used the AVQ cluster centers to center the Gaussian balls: $c_j = m_j$. We chose the RBF weights w_j as the best linear weights in the sense of a least-squares data fit [3]. The weights w_j solved the normal equation $Q^T Q\bar{w} = Q^T b$ where b is the desired output and Q is an m-by-n matrix. Here $m = 500$ is the number of training data and $n = 12$ is the number of weights. $Q^T Q$ is invertible if the columns of Q are linearly independent. Then $\bar{w} = (Q^T Q)^{-1} Q^T b$. This greatly increased the computational load of tuning the RBF net but increased its predictive accuracy.

The RBF net used unweighted Euclidean distance instead of Mahalanobis distance. It gave less weight to data correlations and predicted less poorly than did the Mahalanobis fuzzy system for all noise but Cauchy noise. We modified the RBF net in (6.20) to extend the quadratic form in the exponent from $(x - c_j)^T (x - c_j)$ to $(x - c_j)^T C_j^{-1} (x - c_j)$. This made the RBF net competitive with the fuzzy Mahalanobis system.

All four models predicted poorly in Cauchy noise. We ran the test 10 times for distinct data sets and averaged the mean-squared errors (MSEs) of the prediction. All the predictors degraded as the noise impulsiveness grew. The Mahalanobis system did best when the input data was most correlated. Table 6.1 shows the results.

α	$\hat{\sigma}^2$	Ellipsoidal Fuzzy System	Covariation Fuzzy System	RBF Net	Mahalanobis RBF Net
2.0	0.0197	0.0731	0.0305	0.0499	0.0308
1.9	0.0282	0.0606	0.0395	0.0504	0.0313
1.5	0.1932	0.2248	0.1503	0.1518	0.1357
1.0	2.7585	0.3967	0.3088	0.3069	0.3191

TABLE 6.1: Mean-squared error of prediction for the pseudo-covariation/Mahalanobis, standard ellipsoidal, and radial basis function (RBF) predictors.

6.5 FUZZY FILTERS WITH COVARIATION RULES

This section tests the fuzzy Mahalanobis system as a filter or noise canceler. The test signal is the standard [4–5] fifth-degree polynomial from Chapters 3 and 5:

$$f(x) = 3x(x - 1)(x - 1.9)(x + 0.7)(x + 1.8). \tag{6.21}$$

The same four types of additive noise $n(k)$ in Section 6.4 corrupted the polynomial signal as Figure 6.10 shows. The goal was to cancel the overall noise in the sense of mean-squared error.

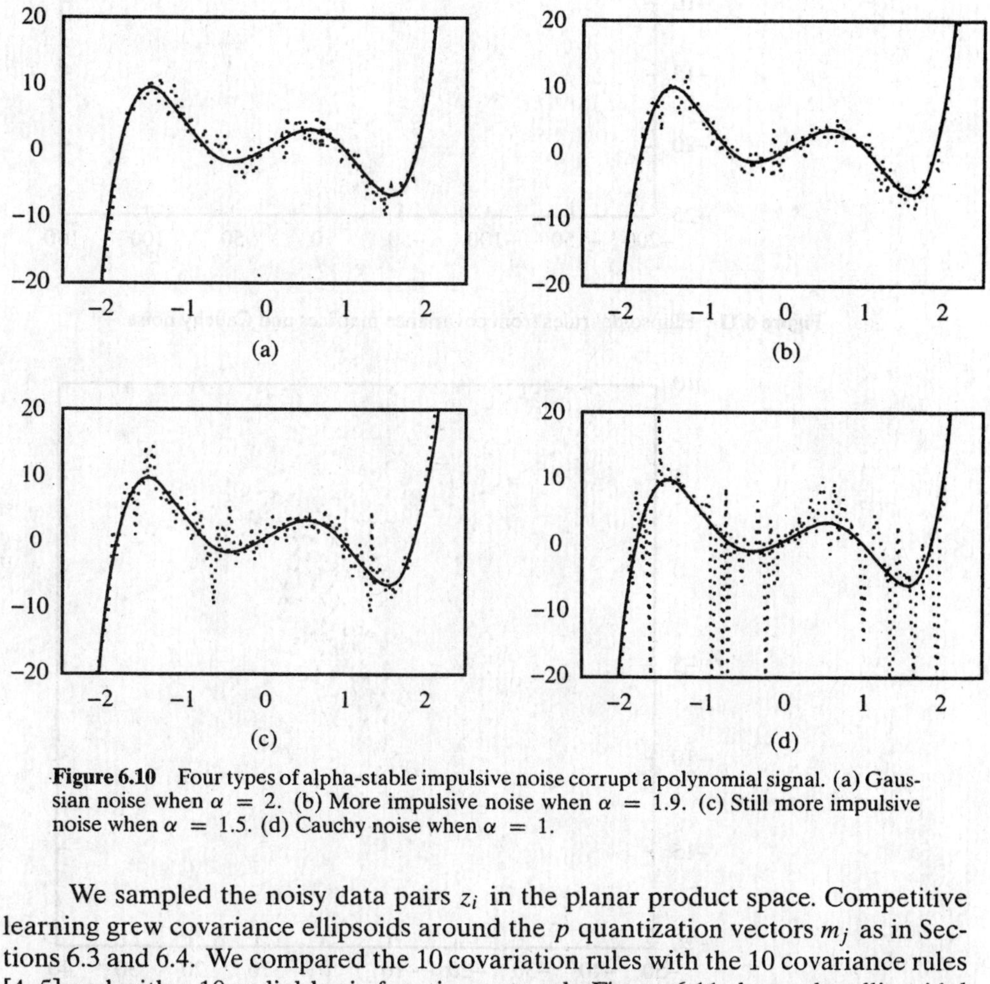

Figure 6.10 Four types of alpha-stable impulsive noise corrupt a polynomial signal. (a) Gaussian noise when $\alpha = 2$. (b) More impulsive noise when $\alpha = 1.9$. (c) Still more impulsive noise when $\alpha = 1.5$. (d) Cauchy noise when $\alpha = 1$.

We sampled the noisy data pairs z_i in the planar product space. Competitive learning grew covariance ellipsoids around the p quantization vectors m_j as in Sections 6.3 and 6.4. We compared the 10 covariation rules with the 10 covariance rules [4–5] and with a 10-radial-basis-function network. Figure 6.11 shows the ellipsoidal rules from the covariance matrices. These rule patches cover the signal polynomial in Figure 6.13. Figure 6.12 shows the ellipsoidal rules from the covariation matrices. These rule patches cover the same signal polynomial in Figure 6.14. The impulsive

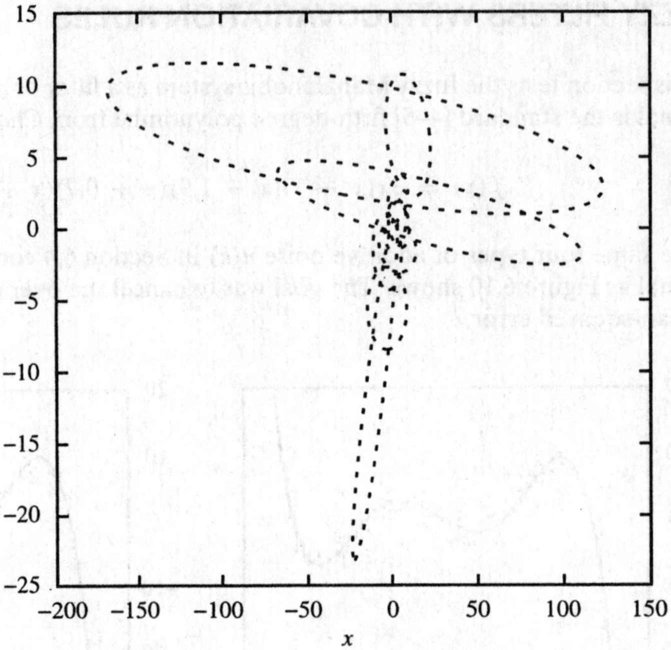

Figure 6.11 Ellipsoidal rules from covariance matrices and Cauchy noise.

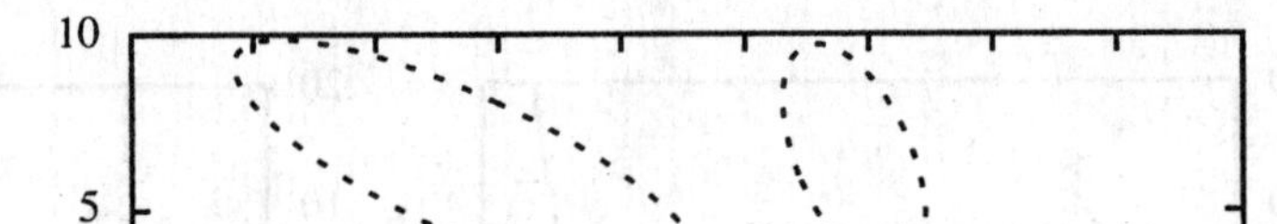

Figure 6.12 Ellipsoidal rules from pseudo-covariation matrices and Cauchy noise.

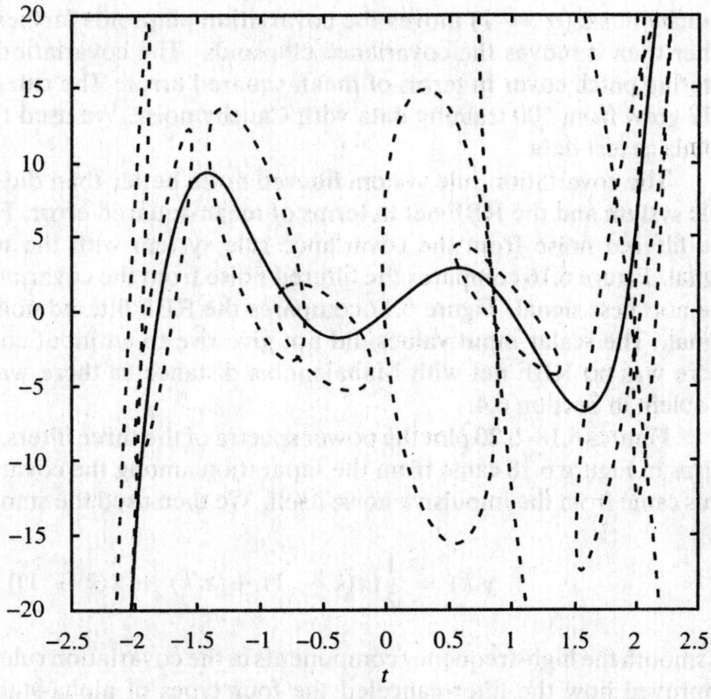

Figure 6.13 Patch cover of covariance rules.

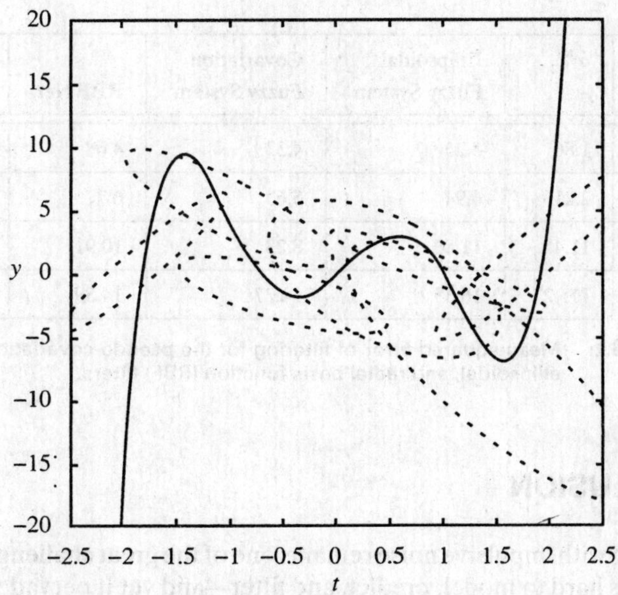

Figure 6.14 Patch cover of pseudo-covariation rules.

Cauchy noise ($\alpha = 1$) moves the covariation ellipsoids farther apart from one another than it moves the covariance ellipsoids. The covariation rules gave a better filtering patch cover in terms of mean-squared error. The rules in Figures 6.11 and 6.12 grew from 500 training data with Cauchy noise. We used the other 500 sample points as test data.

The covariation rule system filtered noise better than did both the covariance rule system and the RBF net in terms of mean-squared error. Figure 6.15 compares the filtered noise from the covariance rule system with the noiseless polynomial signal. Figure 6.16 compares the filtered noise from the covariation rule system with the noiseless signal. Figure 6.17 compares the RBF filtered noise with the noiseless signal. The scalar input values did not give rise to an input covariance matrix. So there was no RBF net with Mahalanobis distance as there was for the prediction problem in Section 6.4.

Figures 6.18–6.20 plot the power spectra of the three filters. The high-frequency terms in Figure 6.18 came from the separation among the covariation ellipsoids and thus came from the impulsive noise itself. We then used the smoothing-by-3s system

$$y(k) = \frac{1}{3}(x(k-1) + x(k) + x(k+1)) \tag{6.22}$$

to smooth the high-frequency components in the covariation rule system. This greatly improved how the filter canceled the four types of alpha-stable noise. Table 6.2 compares the four filters in terms of mean-squared error.

α	$\hat{\sigma}^2$	Ellipsoidal Fuzzy System	Covariation Fuzzy System	RBF Net	Smoothed Covariation Fuzzy System
2.0	1.86	5.25	4.33	4.68	4.24
1.9	2.21	6.94	5.63	6.77	5.13
1.5	11.45	11.88	8.23	10.91	6.12
1.0	775.27	16.35	14.27	14.59	11.85

TABLE 6.2: Mean-squared error of filtering for the pseudo-covariation/Mahalanobis, standard ellipsoidal, and radial basis function (RBF) filters.

6.6 CONCLUSION

Dealing with impulsive noise remains one of the great challenges of modern engineering. It is hard to model, predict, and filter—and yet it pervades the world. Fuzzy systems omit the modeling step that burdens most standard systems. They also dispense with the old working fiction that real noise is Gaussian noise of finite variance. Adaptive fuzzy systems can grow and tune their rules from any source of signals or noise.

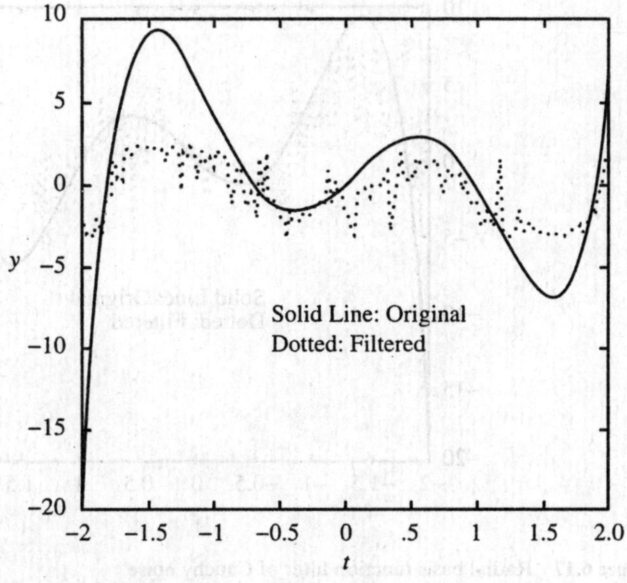

Figure 6.15 Covariance fuzzy filter of Cauchy noise.

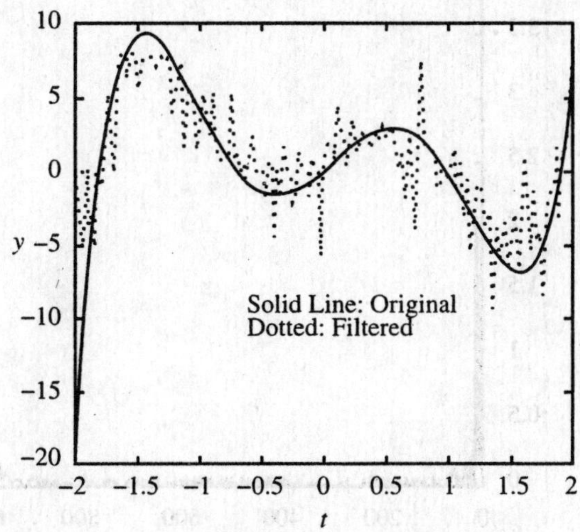

Figure 6.16 Pseudo-covariation fuzzy filter of Cauchy noise.

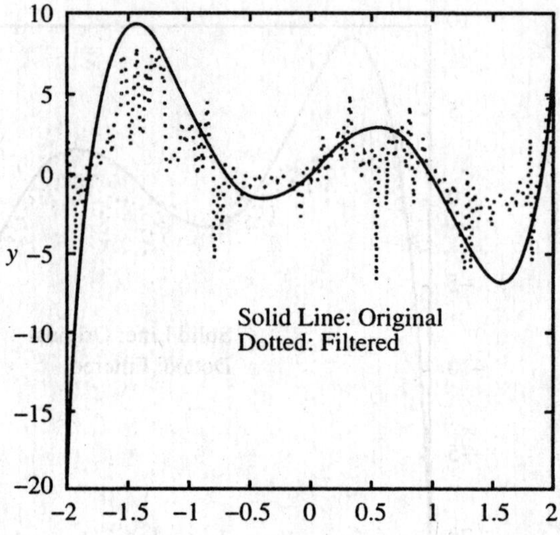

Figure 6.17 Radial basis function filter of Cauchy noise.

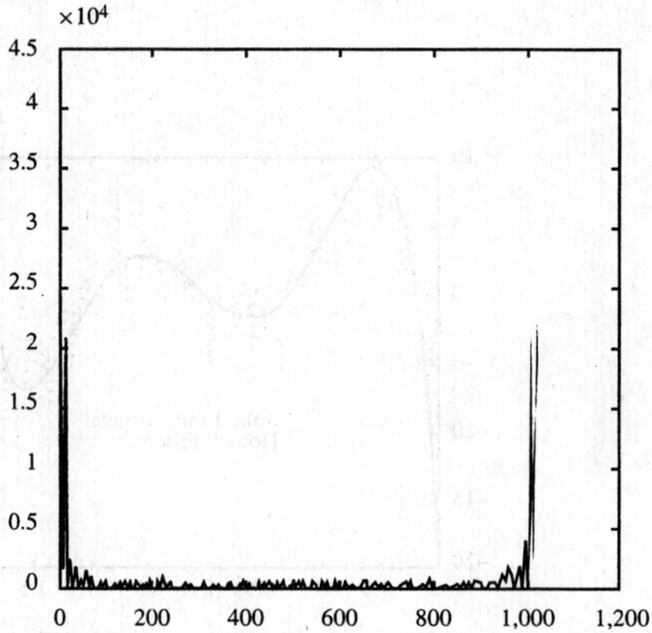

Figure 6.18 Power spectrum of covariance-filtered signal.

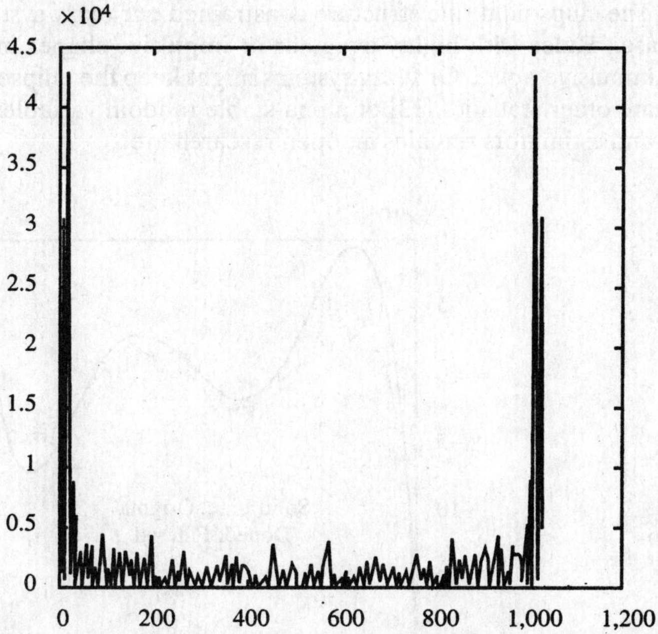

Figure 6.19 Power spectrum of pseudo-covariation-filtered signal.

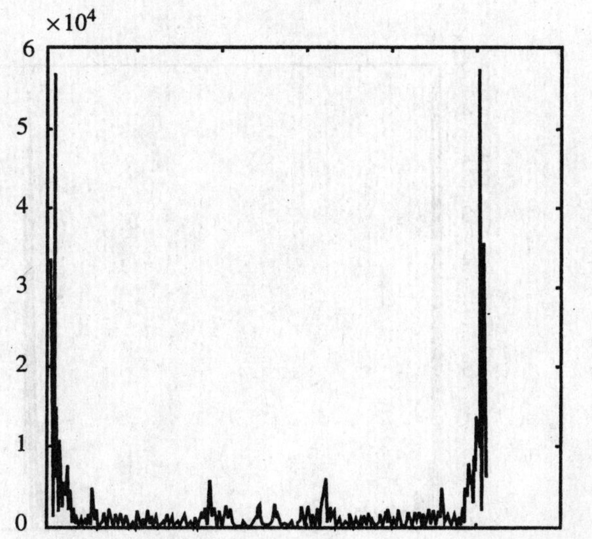

Figure 6.20 Power spectrum of RBF-filtered signal.

The ellipsoidal rule structure constrained our fuzzy systems. This need not be the case. Rules with highly irregular or impulsive shapes may better predict and filter impulsive noise. Or fuzzy systems might keep the ellipsoidal rule structure but estimate other statistics [13] of alpha-stable random variables. The search for such rules and estimators remains an open research area.

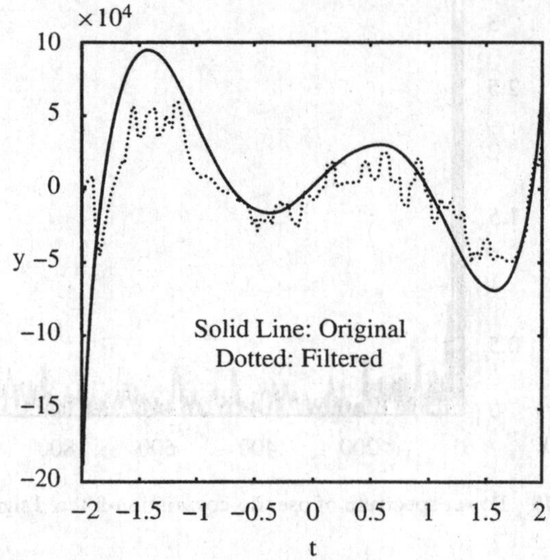

Figure 6.21 Smoothed pseudo-covariation filter of Cauchy noise.

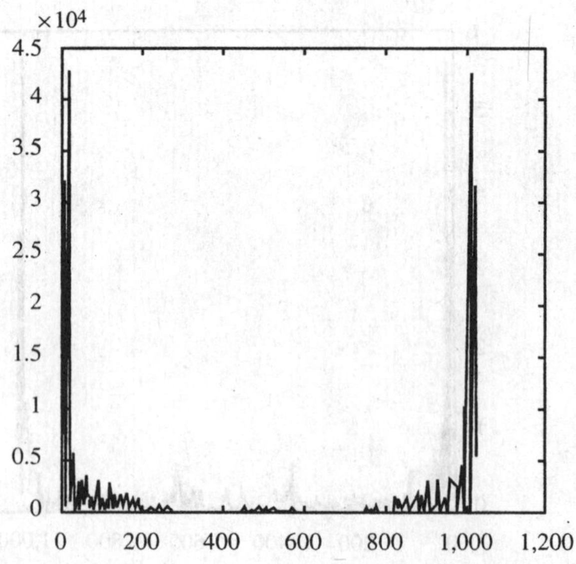

Figure 6.22 Power spectrum of covariation-filtered signal with smoothing.

REFERENCES

[1] Akgiray, V., and Lamoureux, C. G., "Estimation of Stable-law Parameters: A Comparative Study," *J. Business and Economic Statistics*, Vol. 7, 85–93, January 1989.

[2] Bergstrom, H., "On Some Expansions of Stable Distribution Functions," *Ark. Math.*, Vol. 2, 375–378, 1952.

[3] Chen, S., Cowan, C. F. N., and Grant, P. M., "Orthogonal Least Squares Learning Algorithm for Radial Basis Function Networks," *IEEE Transactions on Neural Networks*, Vol. 2, No. 2, 302–309,1991.

[4] Dickerson, J. A., and Kosko, B., "Fuzzy Function Learning with Covariance Ellipsoids," *IEEE International Conference on Neural Networks (IEEE ICNN–93)*, 1162–1167, March 1993.

[5] Dickerson, J. A., and Kosko, B., "Fuzzy Function Approximation with Supervised Ellipsoidal Learning," *World Congress on Neural Networks (WCNN '93)*, Vol. II, 9–13, July 1993.

[6] Hamming, R. W., *Digital Filters*, 2nd ed., Prentice Hall, 1983.

[7] Kong S. G., and Kosko, B., "Adaptive Fuzzy Systems for Backing up a Truck-and-trailer," *IEEE Transactions on Neural Networks*, Vol. 3, No. 2, 211–223, March 1991.

[8] Kosko, B., *Neural Networks and Fuzzy Systems*, Prentice Hall, 1991.

[9] Kosko, B., "Fuzzy Systems as Universal Approximators," *IEEE Transactions on Computers*, Vol. 43, No. 11, 1329–1333, November 1994; an early version appears in *Proceedings of the 1st IEEE International Conference on Fuzzy Systems (IEEE FUZZ–92)*, 1153–1162, March 1992.

[10] Moody, J., and Darken, C. J., "Fast Learning in Networks of Locally-tuned Processing Units," *Neural Computation*, Vol. 1, 281–294, 1989.

[11] Pacini, P. J., and Kosko, B., "Adaptive Fuzzy Frequency Hopper," *IEEE Transactions on Communications*, vol. 43, no. 6, 2111–2117, June 1995.

[12] Papoulis, A., *Probability, Random Variables, and Stochastic Processes*, 2nd ed., McGraw-Hill, 1984.

[13] Shao, M., and Nikias, C. L., "Signal Processing with Fractional Lower Order Moments: Stable Processes and Their Applications," *Proceedings of the IEEE*, Vol. 81, No. 7, 984–1010, July 1993.

APPENDIX 6.A: COMPETITIVE AVQ ALGORITHM FOR LOCAL MEANS AND COVARIATIONS

The following procedure describes the competitive adaptive vector quantization (AVQ) algorithm for local means and pseudo-covariations. Step 3 differs from the AVQ scheme in Chapter 3 since it updates a pseudo-covariation and not a covariance matrix.

1. Initialize cluster centers from sample data: $m_i(0) = x(i)$ for $i = 1, \ldots, p$.
2. Find the "winning" or closest synaptic vector $m_j(k)$ to sample vector $x(k)$:

$$\|m_j(k) - x(k)\| = \min_i \|m_i(k) - x(k)\|$$

where $\|\cdot\|$ denotes the Euclidean norm: $\|x\|^2 = x_1^2 + \cdots + x_n^2$.

3. Update the winner $m_j(k)$

$$m_j(k+1) = \begin{cases} m_j(k) + c_k[x(k) - m_j(k)] & \text{if the } j\text{th neuron wins} \\ m_j(k) & \text{if the } j\text{th neuron loses} \end{cases} \tag{6.23}$$

and its pseudo-covariation estimate $C_j(k)$

$$C_j(k+1) = \begin{cases} C_j(k) + d_k\left[(x(k) - m_j(k))^{\frac{p}{2}}((x(k) - m_j(k))^{\frac{p}{2}})^T - C_j(k)\right] & \text{if the } j\text{th neuron wins} \\ C_j(k) & \text{if the } j\text{th neuron loses} \end{cases} \tag{6.24}$$

where $(x(k))^{\frac{p}{2}} = (\text{sign}(x_1(k))|x_1(k)|^{\frac{p}{2}}, \ldots, \text{sign}(x_n(k))|x_n(k)|^{\frac{p}{2}})^T$. We used $p = 1$ in all simulations.

HOMEWORK PROBLEMS

6.1. The standard Cauchy density is $f(x) = \frac{1}{\pi(1+x^2)}$. Derive its characteristic function.

6.2. X and Y are independent Cauchy variables with the same probability density function $f(x) = \frac{1}{\pi(1+x^2)}$. Show that the sum $Z = \frac{X+Y}{2}$ is also Cauchy with the same probability density function.

6.3. Suppose random variable X has Cauchy density. Show that the variance $V(X)$ of X is infinite. Assume X has a zero mean.

6.4. Prove that the ratio $z = x/y$ of jointly Gaussian random variables x and y is Cauchy. Random variables x and y are jointly Gaussian in the sense that they have the joint probability density function

$$f(x, y) = \frac{1}{2\pi\sigma_1\sigma_2\sqrt{1-r^2}} \exp\left\{-\frac{1}{2(1-r^2)}\left(\frac{x^2}{\sigma_1^2} - 2r\frac{xy}{\sigma_1\sigma_2} + \frac{y^2}{\sigma_2^2}\right)\right\}$$

where r is the correlation coefficient of the random variables x and y defined as $r = \frac{\mu_{xy}}{\sigma_x\sigma_y}$. The covariance μ_{xy} is the mixed central moment defined as $\mu_{xy} = E\{(x - m_x)(y - m_y)\}$.

6.5. Define $S_n = \frac{\sum_{i=1}^{n} X_i}{n}$ for independent random variables X_i. Suppose each X_i has variance $V(X_i) = \sigma^2 < \infty$. Show that the variance obeys $V(S_n) = \frac{\sigma^2}{n}$. So $V(S_n)$ falls to zero as n grows.

6.6. Threshold Cauchy random variable X with positive number a to give Y:

$$Y = \begin{cases} 0 & \text{if } X \geq a \\ X & \text{if } |X| < a \\ 0 & \text{if } X \leq -a. \end{cases}$$

Show that Y has a finite variance.

6.7. For which numbers b is the matrix $\begin{bmatrix} 1 & b \\ b & 16 \end{bmatrix}$ positive definite?

6.8. Suppose $A = \begin{bmatrix} a & b \\ b & c \end{bmatrix}$ is positive definite. Prove that A^{-1} is positive definite.

6.9. $K = \begin{bmatrix} 5 & 4 \\ 4 & 5 \end{bmatrix}$ is a covariance matrix. Find the Mahalanobis distance from (1, 1) to the origin. Repeat for (−1, 1).

6.10. Plot the locus of $x^T K^{-1} x = 1$ for the matrix K in problem 6.9 to see why the Mahalanobis distances from (1, 1) to the origin and from (−1, 1) to the origin differ although the Euclidean distances are the same.

Signals leave changing footprints in the frequency domain as they change in time. Engineers have long observed that most time signals use only a small part of the frequency domain at any one time. This suggests a basic coding strategy to compress a signal. Use fewer bits or no bits at all to code those parts of the frequency spectrum that have little or no energy. Most compression schemes use some form of this energy-coding strategy.

Subband coding divides the frequency band into subbands. Then it allocates coding bits to those frequency bands with the most energy. Wavelet schemes further recursively split the frequency subbands. Other schemes may use some form of scalar or vector quantization to code the bands.

The first subband coders compressed 1-D signals like speech. They made direct use of 1-D quadrature mirror filters to suppress aliasing. Subband coding of 2-D signals like images requires the more cumbersome 2-D mirror filters in either joint or factored form. Subband image coding also requires much greater computation than does 1-D coding and often involves complex processing schemes that vary widely in their components.

Chapter 7 presents just one way an additive fuzzy system can assist the complex process of subband image coding. The fuzzy rules help prune a tree of tiles or frequency splits in the search for a high-energy subtree. Other fuzzy schemes could split or quantize frequency subbands in other ways or replace some of the many other processing boxes in the complex subband coding process. The *ad hoc* nature of such image compressors offers many opportunities for the use of fuzzy and other black-box techniques.

7

FUZZY SUBBAND IMAGE CODING

Hyun Mun Kim and Bart Kosko

PREVIEW

A simple fuzzy system can improve how well subband coding compresses images. A fuzzy system of if-then rules maps the coefficients of frequency subbands to virtual tiles of the time-frequency plane. The tilings are "virtual" in the sense that we extract some subbands from the set of fully decomposed subbands instead of really splitting the frequency band. This prunes the tiling tree and gives a high-energy subtree.

The new tiling subtree shapes the error spectrum and helps preserve edges in the image. The system acts much as do wavelet packets with an adaptive tree-structured bank of filters and exploits the directionality of the tile splitting. Vector quantization codes the tiles with subcodebooks for each tile. Each subcodebook has a low distortion level with less computational load for the code vector search.

The fuzzy virtual tiles do not compute a gradient for each image pixel and this speeds the computation. Simulations show that the fuzzy system improves on the signal-to-noise ratio of a standard subband coding system.

7.1 TILINGS OF THE TIME-FREQUENCY PLANE

Gabor [3] first proved the uncertainty principle between time and frequency that holds for all linear time-invariant systems [5, 18]: $\mathrm{Var}(s(t))\mathrm{Var}(S(f)) \geq \frac{\pi}{2}$. The variance of a signal $s(t)$ in the time domain varies inversely with the variance of its Fourier transform $S(f)$ in the frequency domain. No system can fully resolve an image in both time and frequency.

Different transforms tile or partition the time-frequency plane into different shapes. Gabor transforms or "short-time Fourier transforms" give minimal tiles or *logons*. They have the approximate shape of rectangles with constant time and frequency resolutions and arise from the lower-bound equality in the uncertainty principle. In math terms a Gaussian window modulates a sinusoid. Wavelets [15] give tiles of constant area but different shape in the time and frequency directions. Wavelet packets [6–7] include Gabor and wavelet transforms as special cases. They give binary frequency slicings that adapt to a signal and that do not change in time. More general wavelet packets [6] change in time and give adaptive or time-varying filter banks.

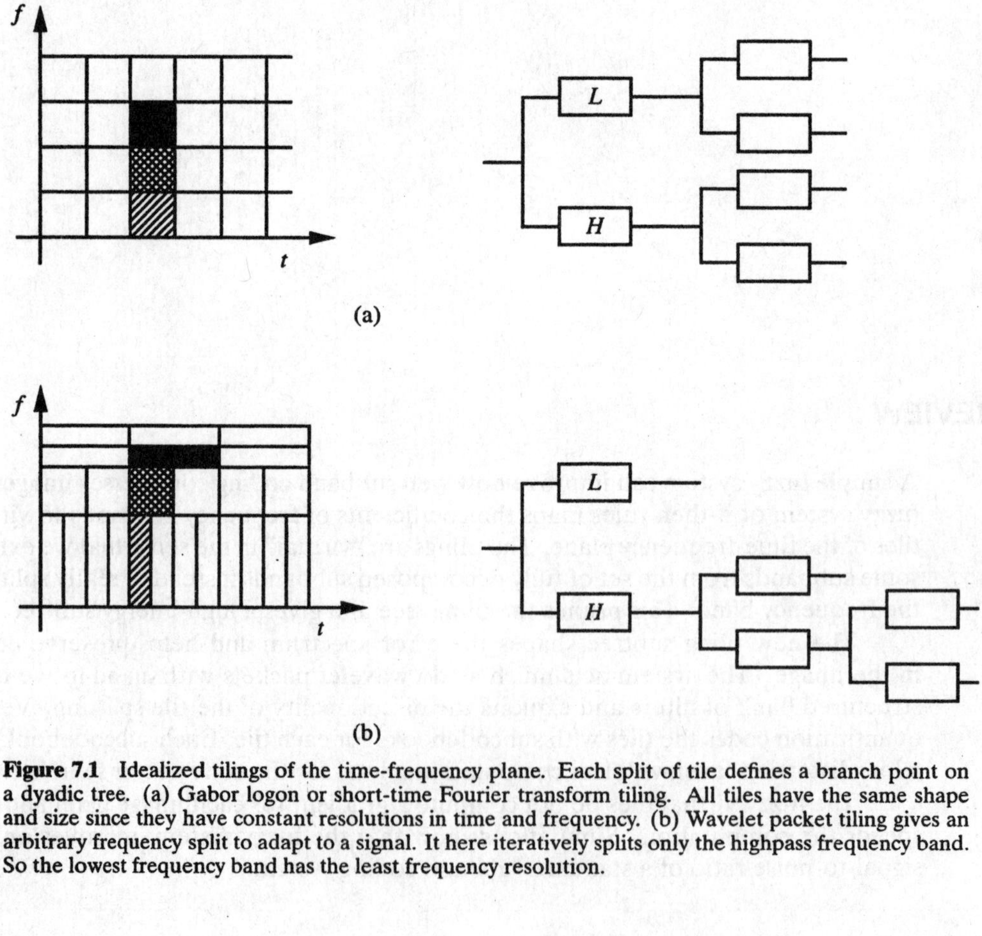

Figure 7.1 Idealized tilings of the time-frequency plane. Each split of tile defines a branch point on a dyadic tree. (a) Gabor logon or short-time Fourier transform tiling. All tiles have the same shape and size since they have constant resolutions in time and frequency. (b) Wavelet packet tiling gives an arbitrary frequency split to adapt to a signal. It here iteratively splits only the highpass frequency band. So the lowest frequency band has the least frequency resolution.

Subband coding trees define the tiling process. Subband coding [1] divides the frequency band of a signal into subbands and uses more bits to code subbands that contain more data. Many subbands contain little or no data at a given time. This makes subband coding an efficient way to compress signals and to shape their noise spectra. Each split of a subband or tile defines a node branching on a dyadic tree. Both speech and image signals admit such hierarchical source coding. Image subband coding ([21], [22], and [23]) often uses a 16-band filter system with quadrature mirror filters (QMFs) to suppress aliasing from the finite impulse response filters. The system carries out the reverse filtering process after it has encoded and sent and decoded the signal. The Woods [23] scheme applies 1-D QMFs recursively and allocates bits among subbands to minimize the mean-squared error of each subband's differential pulse coding modulation (DPCM). A later scheme [24] views signals as an n-vector and codes them with vector quantization (VQ). These schemes do not shape the error spectrum because the splittings remain fixed for all blocks and because VQ ignores each block's signal structure.

Vector quantization does promise high rates of image compression and plays a role in most subband coding schemes. VQ packs signal spheres better than does scalar quantization [4, 19] since it does not ignore the dependence among vector components and need not limit quantized regions to hyperrectangles. But VQ algorithms can be complex and can degrade the edges that make up much of the image information.

A fuzzy system can enhance the subband coding of images. Fuzzy systems might apply at many levels in the subband coding process where the process needs their "black box" function approximation power. In our case a system of fuzzy if-then rules maps the subband coefficients to four virtual tiles. Then VQ encodes each tile. The tilings are "virtual" in the sense that we extract some subbands from the set of fully decomposed subbands instead of really splitting the frequency band. This prunes the tiling tree and gives a new high-energy tiling subtree. It deletes the low-energy tiles or quantizes them to zero. Other schemes ([6] and [7]) choose a tiling tree and then quantize the entire tree. Our virtual tiling scheme changes with each image block to track the energy distribution of the tiling trees and so acts much as wavelet packets act that change in time. Simulations show that such a compressed system maintains edges well in images and has a favorable signal-to-noise ratio. More complex systems can use neural systems to tune the fuzzy rules and better track the changing signal statistics.

7.2 SUBBAND IMAGE CODING

This section reviews 2-D subband filtering. It also reviews the structure of the fully decomposed subband coding (SBC) system in [24].

7.2.1 Subband Filters in Two Dimensions

Figure 7.2 shows how the 16-band SBC system depends on 4-band splitting and recombination. The system splits the image signal and then demodulates it to baseband with (2,2) subsampling. The subsampler decimates by the factor 2 in each dimension. The new sequence $x_d[n] = x[nM]$ reduces the sampling rate. Then the discrete-time Fourier transform of $x_d[n]$ has the form [17]

$$X_d(e^{i\omega}) = \frac{1}{M}\sum_{k=0}^{M-1} X\left(e^{i(\omega/M - 2\pi k/M)}\right) \tag{7.1}$$

with $M = 2$. This process repeats to further split each subband into 4 more subbands to give a 16-band system.

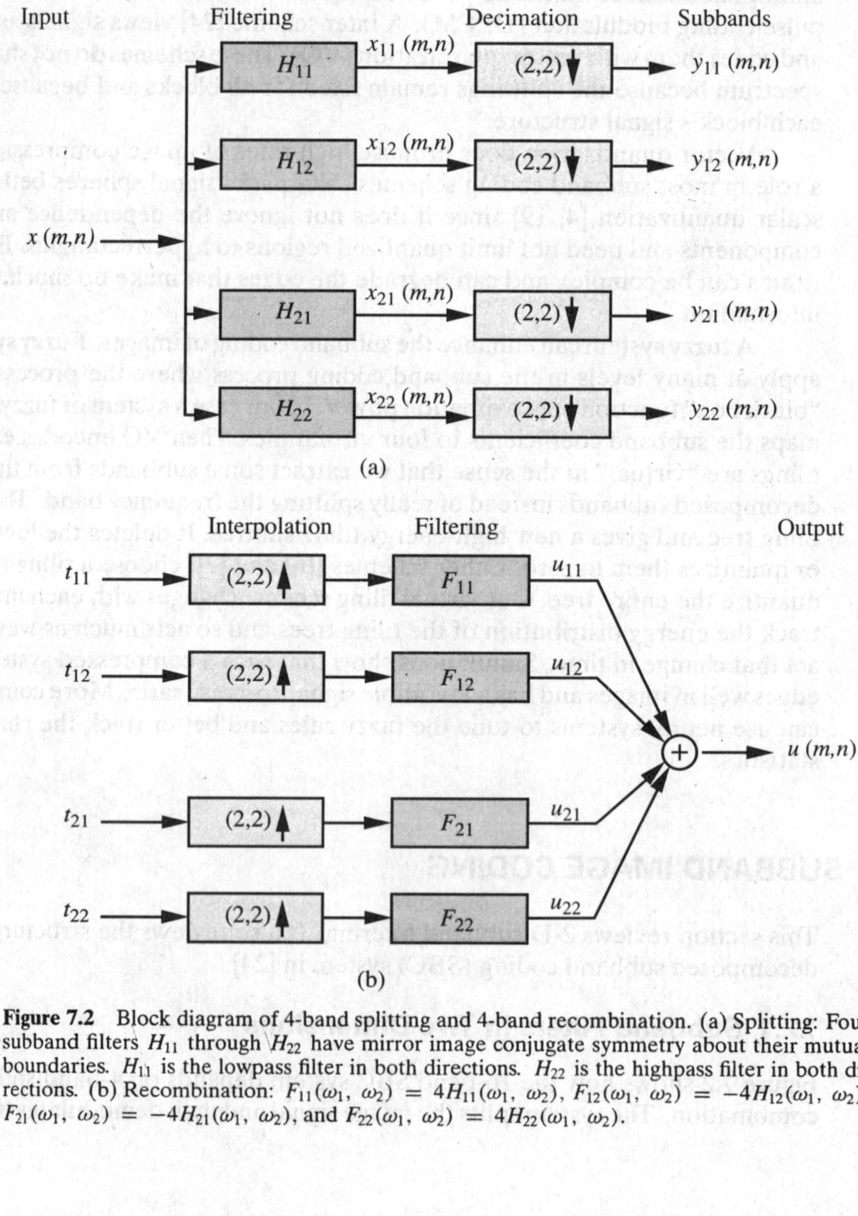

Figure 7.2 Block diagram of 4-band splitting and 4-band recombination. (a) Splitting: Four subband filters H_{11} through H_{22} have mirror image conjugate symmetry about their mutual boundaries. H_{11} is the lowpass filter in both directions. H_{22} is the highpass filter in both directions. (b) Recombination: $F_{11}(\omega_1, \omega_2) = 4H_{11}(\omega_1, \omega_2)$, $F_{12}(\omega_1, \omega_2) = -4H_{12}(\omega_1, \omega_2)$, $F_{21}(\omega_1, \omega_2) = -4H_{21}(\omega_1, \omega_2)$, and $F_{22}(\omega_1, \omega_2) = 4H_{22}(\omega_1, \omega_2)$.

The SBC system reverse filters the signal after it has encoded and sent and decoded it. The system upsamples each subband by (2,2). The new sequence

$$x_e[n] = \sum_{k=-\infty}^{\infty} x[k]\delta[n - kL]$$

expands the sampling rate. Then the discrete-time Fourier transform of $x_e[n]$ has the form [17]

$$X_e(e^{i\omega}) = X(e^{i\omega L}). \tag{7.2}$$

The SBC system reconstructs the signal by summing the sets of subbands as shown in Figure 7.2.

7.2.2 Quadrature Mirror Filters

Finite impulse response (FIR) filters cause aliasing error when they approximate the ideal subband filters [17]. Quadrature mirror filters (QMFs) can help cancel the aliasing.

QMFs extend to two dimensions as follows [23]. Figure 7.2 shows the four linear subband filters H_{11}, H_{12}, H_{21}, and H_{22} for the time invariant systems $x_{ij}(m, n)$: $x_{ij}(m, n) = x(m, n) \times h_{ij}(m, n)$ and thus $X_{ij}(\omega_1, \omega_2) = X(\omega_1, \omega_2)H_{ij}(\omega_1, \omega_2)$. We require that the four subband filters have mirror image conjugate symmetry about their mutual boundaries. This gives for real filter coefficients h_{ij} with Fourier transforms H_{ij}

$$H_{12}(\omega_1, \omega_2) = H_{11}(\omega_1, \omega_2 + \pi) \tag{7.3}$$

$$H_{21}(\omega_1, \omega_2) = H_{11}(\omega_1 + \pi, \omega_2) \tag{7.4}$$

$$H_{22}(\omega_1, \omega_2) = H_{11}(\omega_1 + \pi, \omega_2 + \pi) \tag{7.5}$$

Denote the outputs of the filters in Figure 7.2 as x_{11} through x_{22}. This gives the filter outputs y_{ij} after (2,2) downsampling and Fourier transforming as

$$Y_{ij}(\omega_1, \omega_2) = \frac{1}{4}\sum_{k=0}^{1}\sum_{l=0}^{1} H_{ij}\left(\frac{\omega_1 + 2k\pi}{2}, \frac{\omega_2 + 2l\pi}{2}\right) X\left(\frac{\omega_1 + 2k\pi}{2}, \frac{\omega_2 + 2l\pi}{2}\right)$$

$$0 \le i, j \le 1. \tag{7.6}$$

The output transforms U_{ij} of the filters in Figure 7.2 become

$$U_{ij}(\omega_1, \omega_2) = Y_{ij}(2\omega_1, 2\omega_2)F_{ij}(\omega_1, \omega_2) \qquad 0 \le i, j \le 1 \tag{7.7}$$

if we assume perfect transmission. The filters F_{ij} are not in general separable. The final output transform U becomes

$$U(\omega_1, \omega_2) = \frac{1}{4} \sum_{k,l} X(\omega_1 + k\pi, \omega_2 + l\pi) \cdot \left[\sum_{i,j} H_{ij}(\omega_1 + k\pi, \omega_2 + l\pi) F_{ij}(\omega_1, \omega_2) \right]. \tag{7.8}$$

The QMFs remove the aliased terms from the filters F_{ij} [23]:

$$F_{11}(\omega_1, \omega_2) = 4H_{11}(\omega_1, \omega_2) \tag{7.9}$$

$$F_{12}(\omega_1, \omega_2) = -4H_{12}(\omega_1, \omega_2) \tag{7.10}$$

$$F_{21}(\omega_1, \omega_2) = -4H_{21}(\omega_1, \omega_2) \tag{7.11}$$

$$F_{22}(\omega_1, \omega_2) = 4H_{22}(\omega_1, \omega_2) \tag{7.12}$$

$$H_{11}(\omega_1,\omega_2)H_{11}(\omega_1 + \pi,\omega_2 + \pi) = H_{11}(\omega_1,\omega_2 + \pi)H_{11}(\omega_1 + \pi,\omega_2). \tag{7.13}$$

1-D QMFs can remove the aliased terms and help avoid the hard task of designing 2-D QMFs. The product of two 1-D QMFs gives a 2-D QMF: $H_1 H_2 = H_{12}$. Figure 7.3 shows the separable 4-subband filter.

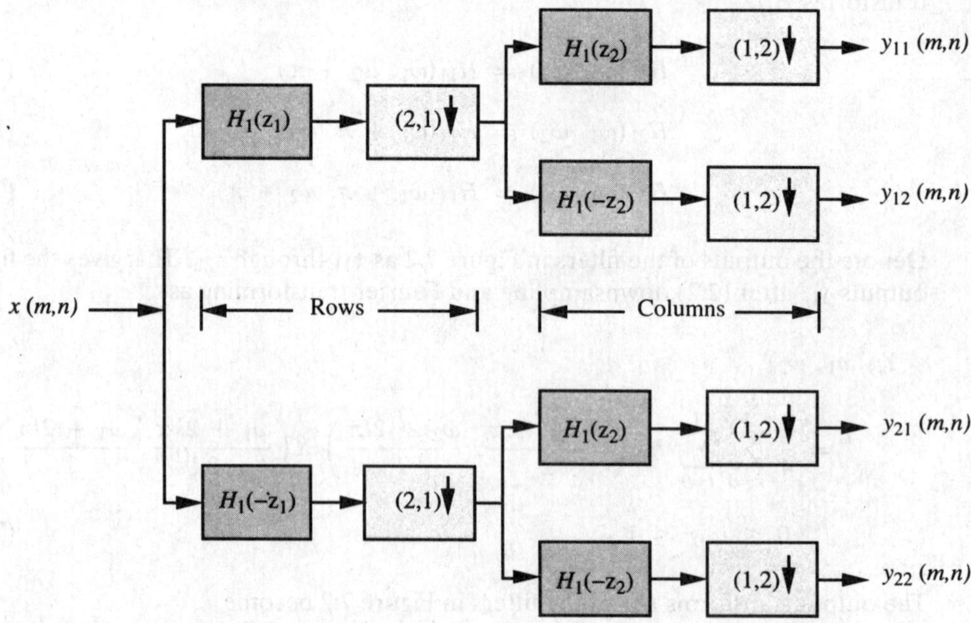

Figure 7.3 Separable 4-subband filter. 1-D QMF filters with the row-column approach [21] perform the 4-band decomposition.

We next design a low-pass filter that partially satisfies the constraint (7.14) of perfect signal reconstruction:

$$|H_{11}^2(\omega_1, \omega_2)| + |H_{11}^2(\omega_1, \omega_2 + \pi)| + |H_{11}^2(\omega_1 + \pi, \omega_2)| + |H_{11}^2(\omega_1 + \pi, \omega_2 + \pi)| = 1. \tag{7.14}$$

We used the 32-point QMF called 32D in [9]. Figure 7.4 shows the filter coefficients and frequency response. Figure 7.5 shows the frequency structure of the QMF filter that nearly satisfies (7.14).

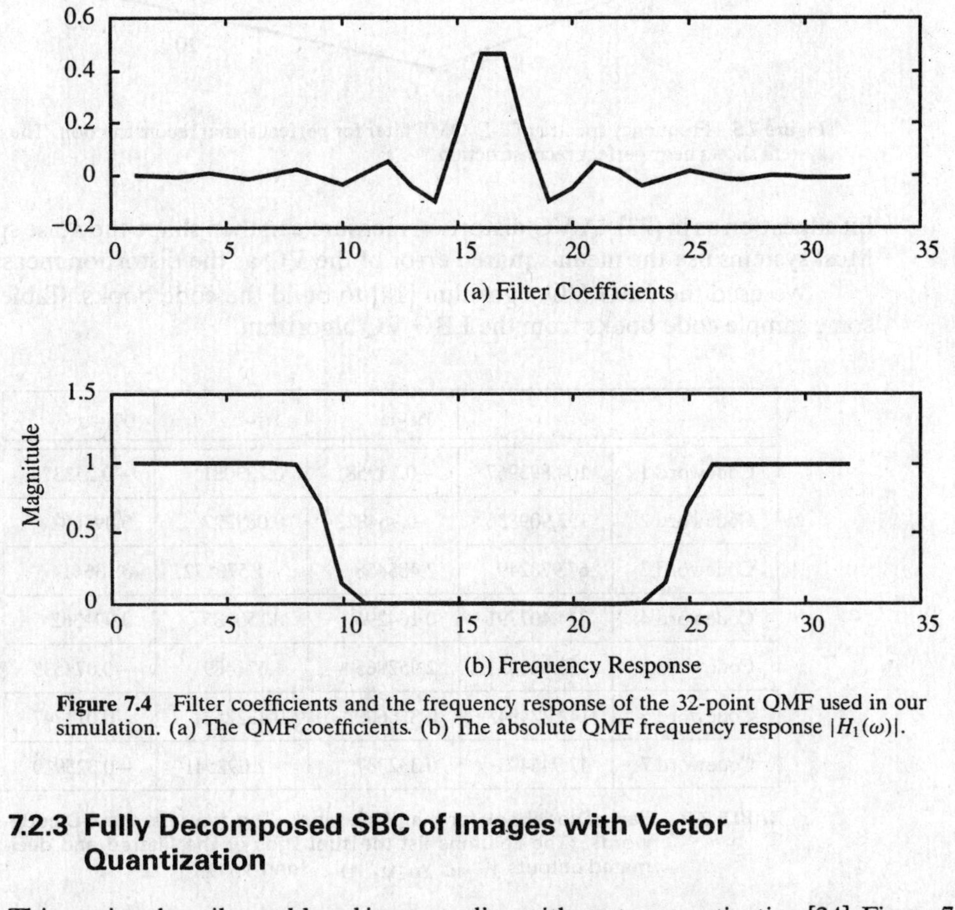

Figure 7.4 Filter coefficients and the frequency response of the 32-point QMF used in our simulation. (a) The QMF coefficients. (b) The absolute QMF frequency response $|H_1(\omega)|$.

7.2.3 Fully Decomposed SBC of Images with Vector Quantization

This section describes subband image coding with vector quantization [24]. Figure 7.7 shows the tree structure of a 16-band separable SBC. We call it a *fully* decomposed SBC (FSBC) system since it recursively applies the two-scale decomposition to all frequency channels. The output of the QMF bank consists of 16 signals sampled at the same rate. The FSBC treats scalar sample streams as vectors or points in a 16-dimensional space. A vector quantizer encodes each 16-D vector. FSBC uses no

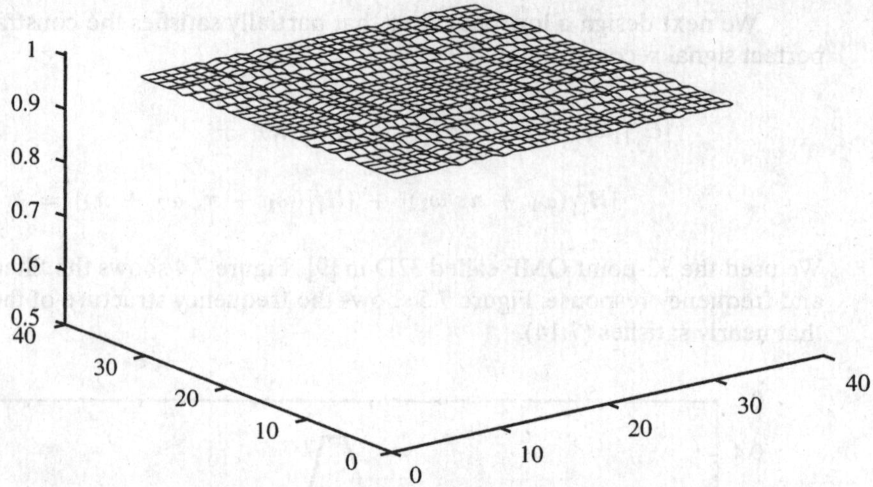

Figure 7.5 Frequency spectra of 2-D QMF filter for perfect signal reconstruction. The QMF system shows near perfect reconstruction.

bit allocation as in [23]. A VQ distortion measure can then shape the noise spectrum. Most systems use the mean-squared error of the VQ as the distortion measure.

We used the LBG VQ algorithm [14] to build the code books. Table 7.1 lists some sample code books from the LBG VQ algorithm.

	y_{11-11}	y_{11-12}	y_{11-21}	y_{11-22}
Codeword 1	104.893967	−0.139581	0.233081	−0.253378
Codeword 2	152.909836	−0.964823	0.089227	0.095682
Codeword 3	67.978249	2.488458	−1.570172	0.136414
Codeword 4	214.407791	0.164296	0.150285	0.091362
Codeword 5	123.820831	2.952965	1.671489	−0.074533
Codeword 6	142.224945	0.830600	0.432258	−0.011747
Codeword 7	47.946495	0.332387	−1.092541	−0.325910

TABLE 7.1: Part of codebook for the plane block. The rows list the 4-D codewords. The columns list the subbands or the filtered and decimated outputs y_{11-11}, y_{11-12}, y_{11-21}, and y_{11-22}.

LBG VQ Algorithm

1. Initialization: Given N = number of levels, distortion threshold $t \geq 0$, an initial codebook $\hat{A}_0$, and a training sequence $\{x_j\colon j = 0, \cdots, n-1\}$. Set $m = 0$ and $D_{-1} = \infty$.

2. Given $\hat{A}_m = \{y_i : i = 1, \cdots, N\}$. Find the minimum distortion partition $P(\hat{A}_m) = \{S_i : i = 1, \cdots, N\}$ of the training sequence: $x_j \in S_i$ if $d(x_j, y_i) \leq d(x_j, y_l)$ for all l. Compute the average distortion D_m:

$$D_m = D(\{\hat{A}_m, P(\hat{A}_m)\}) = n^{-1} \sum_{j=0}^{n-1} \min_{y \in \hat{A}_m} d(x_j, y).$$

3. Halt with $\hat{A}_m$ as the final reproduction alphabet if $(D_{m-1} - D_m)/D_m \leq t$. Else continue.

4. Find the optimal reproduction alphabet $\hat{x}(P(\hat{A}_m)) = \{\hat{x}(S_i) : i = 1, \cdots, N\}$ for $P(\hat{A}_m)$. Set $\hat{A}_{m+1} = \hat{x}(P(\hat{A}_m))$. Replace m by $m + 1$ and go to step 2.

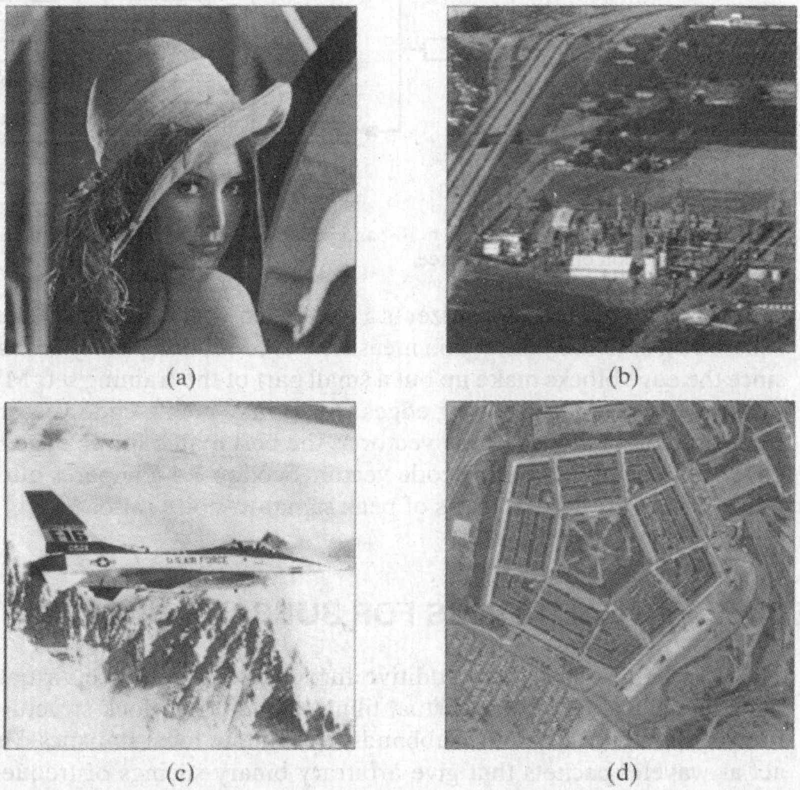

Figure 7.6 Four training images used for codebooks: (a) ***Lenna***. (b) ***chem***. (c) ***f16***. (d) ***pentl***. Figure 7.22 shows the 5th image ***Elaine***.

The splitting technique in [14] gave a code book of rate R from a code book of rate $R - 1$. We started from $R = 2$. We split each vector y_i into two close vectors $y_i + \epsilon$ and $y_i - \epsilon$ for fixed perturbation vector ϵ with $\epsilon = (0.02, 0.02, \ldots, 0.02)$. We used a training set of 5 images to build the codebooks. Figure 7.6 shows the training images *Lenna, chem, f16, and pentl*.

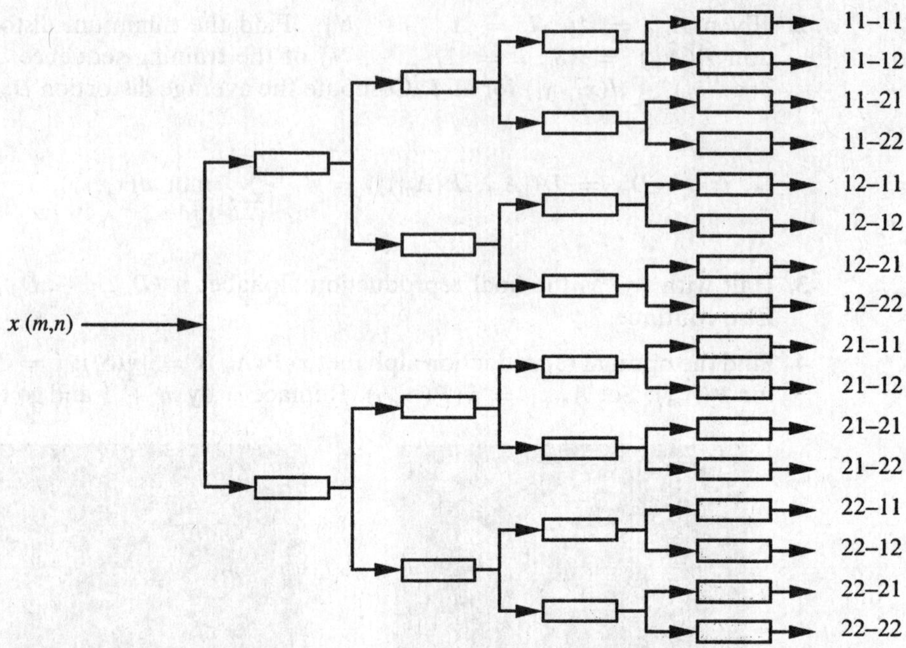

Figure 7.7 Tree structure for a 16-band separable subband coder. Repeating the 4-band decomposition once gives the tree.

The FSBC vector quantizer is a full search vector quantizer and uses the mean-squared error (MSE) distortion measure. Its codebook poorly codes the edge blocks since the edge blocks make up but a small part of the training set. MSE is also a poor distortion measure for coding edges. MSE may pick a code vector with a uniform shade instead of an edge code vector as the best match for an edge block even if the codebook contains the edge code vector. Section 7.4 compares our fuzzy SBC with the FSBC VQ system in terms of peak signal-to-noise ratio.

7.3 FUZZY VIRTUAL TILINGS FOR SUBBAND CODING

This section describes how additive fuzzy systems produce virtual tilings for subband image coding. Fuzzy virtual tilings exploit the block structure of edges. The fuzzy system virtually splits subbands to maintain local statistics. In that sense they act as wavelet packets that give arbitrary binary slicings of frequency bands. The main feature of wavelet packets is that their frequency slicings adapt to the sampled signal. But they do not change over time. A key problem is to find a wavelet packet decomposition that changes over time. Fuzzy virtual tilings approximate this behavior.

The frequency slicings should accept nonstationarity in the image signal. Fuzzy virtual tilings change over time depending on the image blocks. The system codes each image block with VQ based on each tiling. We built shorter subcodebooks for each tiling instead of designing a whole codebook. Each subcodebook has a low distortion level with less computational load for the code-vector search. This gives an efficient coding of the sampled image and helps match human visual sensitivity. And it helps maintain horizontal and vertical edges as it exploits the splitting direction of the row-column [23] approach in the subband decomposition.

7.3.1 Additive Fuzzy Systems

A scalar-valued fuzzy system $F: R^n \rightarrow R$ stores m rules of the word from "If $X = A_j$ then $Y = B_j$" or the patch form $A_j \times B_j \subset X \times Y = R^n \times R$. The if-part fuzzy sets $A_j \subset R^n$ and then-part fuzzy sets $B_j \subset R$ have arbitrary set functions $a_j: R^n \rightarrow [0, 1]$ and $b_j: R \rightarrow [0, 1]$. The system can use the joint set function a_j or some factored form such as $a_j(x) = a_j^1(x_1) \cdots a_j^n(x_n)$ or $a_j(x) = \min(a_j^1(x_1), \ldots, a_j^n(x_n))$ or any other conjunctive form for input vector $x = (x_1, \ldots, x_n) \in R^n$.

An additive fuzzy system [10–11] first sums the "fired" then part sets B_j'

$$B = \sum_{j=1}^{m} B_j' = \sum_{j=1}^{m} a_j(x) B_j \tag{7.15}$$

and then computes the output $F(x)$ as the centroid of B: $F(x) = \text{Centroid}(B)$. Figure 7.8 shows the parallel fire-and-sum structure of the SAM system. These systems can uniformly approximate any continuous (or bounded measurable) function f on a compact domain [11].

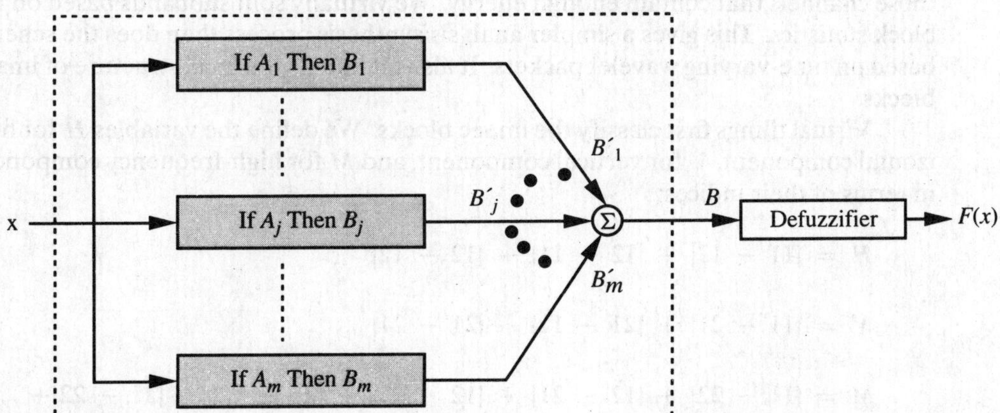

Figure 7.8 Architecture of an additive fuzzy system $F: R^n \rightarrow R^p$ with m rules. Each input $x_0 \in R^n$ enters the system F as a numerical vector. At the set level x_0 acts as a delta pulse $\delta(x - x_0)$ that combs the if-part fuzzy sets A_j and gives the m set values $a_j(x_0)$: $a_j(x_0) = \int_{R^n} \delta(x - x_0) a_j(x)dx$. The set values "fire" the then-part fuzzy sets B_j to give B_j'. A standard additive model (SAM) scales each B_j with $a_j(x)$. The system then sums the B_j' sets to give B. The system output $F(x_0)$ is the centroid of B.

The scaling choice $B_j' = a_j(x)B_j$ gives a *standard additive model* or SAM. Taking the centroid of B in (7.15) gives [10–12] the SAM ratio of Chapter 2:

$$F(x) = \frac{\sum_{j=1}^{m} a_j(x)V_jc_j}{\sum_{j=1}^{m} a_j(x)V_j} \tag{7.16}$$

where V_j is the volume or area of then-part set B_j and c_j is its centroid or center of mass.

The SAM theorem of (7.16) implies that the fuzzy structure of the then-part sets B_j does not matter. The ratio depends on just the volume V_j and location c_j of the then-part sets B_j. Our SAM has then-part sets of the same area and so the volume terms V_j cancel from (7.16). We need pick only the scalar centers c_j. More advanced systems tune V_j, c_j, or a_j with training data.

7.3.2 Fuzzy Virtual Tilings

Fuzzy virtual tilings can act as wavelet packet decompositions that change in time. The tilings are "virtual" since we extract some subbands from the FSBC instead of really splitting the frequency band. The additive fuzzy system classifies image blocks for virtual tilings. The system virtually splits subbands for tilings based on the classification. The filter banks do not change in time and so share the same tree structure with the FSBC system. But virtual tilings make the new filter banks act much as wavelet packets that do. This depends on how we deal with subband signals. FSBC treats whole subband signals as n-dimensional vectors. We select just those channels that contain enough energy. We virtually split subbands based on the block statistics. This gives a simpler analysis/synthesis process than does the scheme based on time-varying wavelet packets. It also maintains the local structure of image blocks.

Virtual tilings first classify the image blocks. We define the variables H for horizontal component, V for vertical component, and M for high-frequency component in terms of their indices:

$$H = |11 - 12| + |12 - 11| + |12 - 12|$$

$$V = |11 - 21| + |21 - 11| + |21 - 21|$$

$$M = |11 - 22| + |12 - 21| + |12 - 22| + |21 - 12| + |21 - 22| +$$

$$|22 - 11| + |22 - 12| + |22 - 21| + |22 - 22|$$

where $|\cdot|$ gives the absolute value of each frequency band as in Figure 7.7.

The next example shows how the system uses the above variables. Figure 7.10 shows four synthetic horizontal 4 × 4 edge blocks. The 16-band FSBC in Figure 7.7

computes the coefficients from 11 − 11 to 22 − 22 for each subband. We use a Haar basis [8] as the SBC filter set: $h_1 = [1/\sqrt{2} \quad 1/\sqrt{2}]$ and $h_2 = [1/\sqrt{2} \quad -1/\sqrt{2}]$. We arrange the coefficients taken from the FSBC in a square matrix with these indices:

$$R = \mathrm{SBC}(r) = \begin{bmatrix} 11-11 & 11-12 & 11-21 & 11-22 \\ 12-11 & 12-12 & 12-21 & 12-22 \\ 21-11 & 21-12 & 21-21 & 21-22 \\ 22-11 & 22-12 & 22-21 & 22-22 \end{bmatrix} \tag{7.17}$$

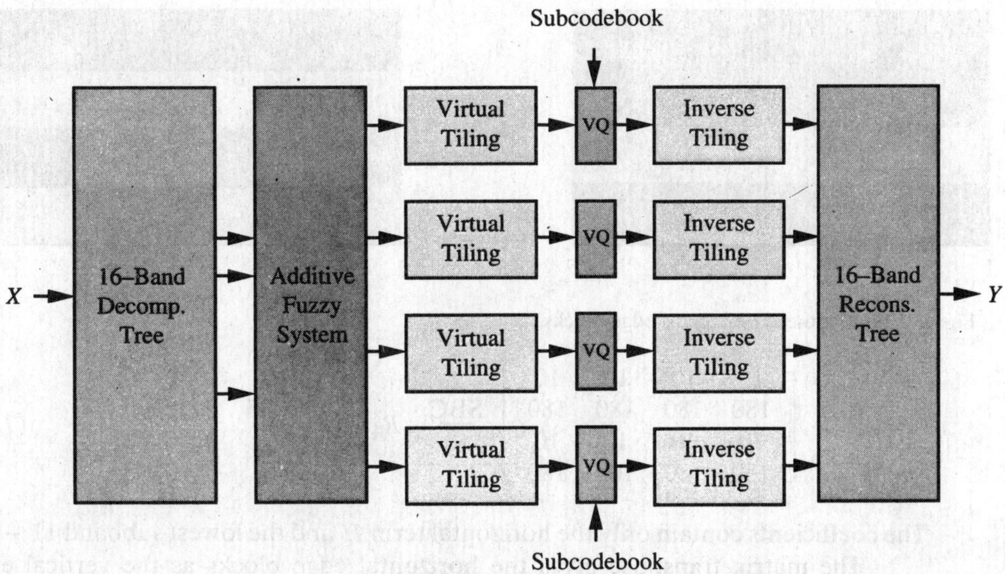

Figure 7.9 Block diagram of the subband coding system with fuzzy virtual tilings.

The matrices in (7.18) through (7.21) show the gray levels of the four blocks in Figure 7.10 and their SBC transform coefficients arranged as in (7.17).

$$r_1 = \begin{bmatrix} 10 & 10 & 10 & 10 \\ 180 & 180 & 180 & 180 \\ 180 & 180 & 180 & 180 \\ 180 & 180 & 180 & 180 \end{bmatrix} \overset{\mathrm{SBC}}{\Longrightarrow} R_1 = \begin{bmatrix} 137.5 & -42.5 & 0 & 0 \\ -42.5 & -42.5 & 0 & 0 \\ 0 & 0 & 0 & 0 \\ 0 & 0 & 0 & 0 \end{bmatrix} \tag{7.18}$$

$$r_2 = \begin{bmatrix} 10 & 10 & 10 & 10 \\ 10 & 10 & 10 & 10 \\ 180 & 180 & 180 & 180 \\ 180 & 180 & 180 & 180 \end{bmatrix} \overset{\mathrm{SBC}}{\Longrightarrow} R_2 = \begin{bmatrix} 95 & -85 & 0 & 0 \\ 0 & 0 & 0 & 0 \\ 0 & 0 & 0 & 0 \\ 0 & 0 & 0 & 0 \end{bmatrix} \tag{7.19}$$

$$r_3 = \begin{bmatrix} 10 & 10 & 10 & 10 \\ 10 & 10 & 10 & 10 \\ 180 & 180 & 180 & 180 \\ 10 & 10 & 10 & 10 \end{bmatrix} \overset{\mathrm{SBC}}{\Longrightarrow} R_3 = \begin{bmatrix} 52.5 & -42.5 & 0 & 0 \\ 42.5 & -42.5 & 0 & 0 \\ 0 & 0 & 0 & 0 \\ 0 & 0 & 0 & 0 \end{bmatrix} \tag{7.20}$$

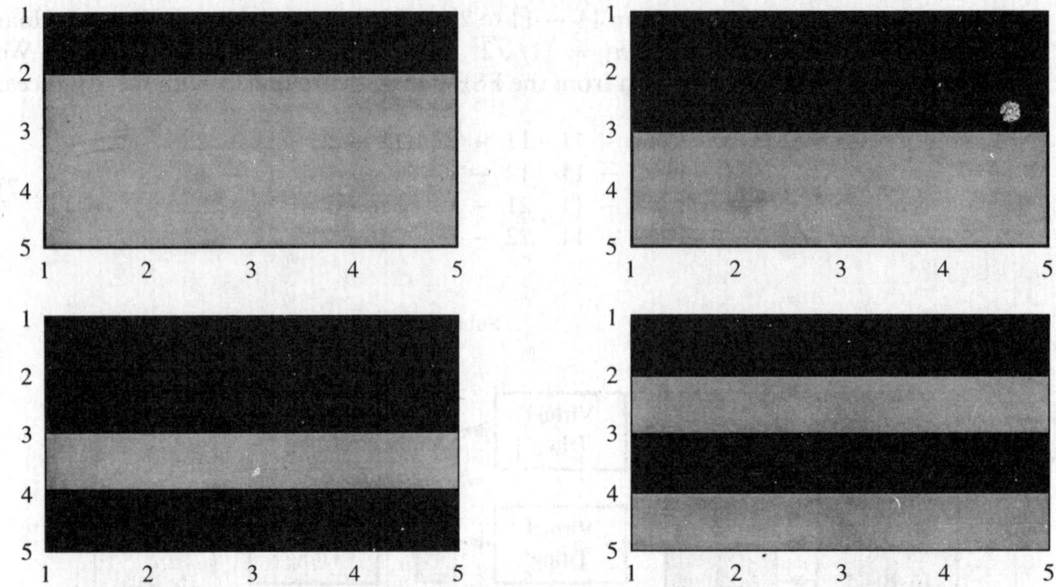

Figure 7.10 Synthetic horizontal edge blocks.

$$r_4 = \begin{bmatrix} 10 & 10 & 10 & 10 \\ 180 & 180 & 180 & 180 \\ 10 & 10 & 10 & 10 \\ 180 & 180 & 180 & 180 \end{bmatrix} \overset{\text{SBC}}{\Longrightarrow} R_4 = \begin{bmatrix} 95 & 0 & 0 & 0 \\ -85 & 0 & 0 & 0 \\ 0 & 0 & 0 & 0 \\ 0 & 0 & 0 & 0 \end{bmatrix} \tag{7.21}$$

The coefficients contain only the horizontal term H and the lowest subband $11 - 11$.

The matrix transpose gives the horizontal edge blocks as the vertical edge blocks. Using the FSBC gives

$$r_5 = r_1' = \begin{bmatrix} 10 & 10 & 10 & 10 \\ 180 & 180 & 180 & 180 \\ 180 & 180 & 180 & 180 \\ 180 & 180 & 180 & 180 \end{bmatrix}' \overset{\text{SBC}}{\Longrightarrow} R_5 = \begin{bmatrix} 137.5 & 0 & -42.5 & 0 \\ 0 & 0 & 0 & 0 \\ -42.5 & 0 & -42.5 & 0 \\ 0 & 0 & 0 & 0 \end{bmatrix} \tag{7.22}$$

$$r_6 = r_2' = \begin{bmatrix} 10 & 10 & 10 & 10 \\ 10 & 10 & 10 & 10 \\ 180 & 180 & 180 & 180 \\ 180 & 180 & 180 & 180 \end{bmatrix}' \overset{\text{SBC}}{\Longrightarrow} R_6 = \begin{bmatrix} 95 & 0 & -85 & 0 \\ 0 & 0 & 0 & 0 \\ 0 & 0 & 0 & 0 \\ 0 & 0 & 0 & 0 \end{bmatrix} \tag{7.23}$$

$$r_7 = r_3' = \begin{bmatrix} 10 & 10 & 10 & 10 \\ 10 & 10 & 10 & 10 \\ 180 & 180 & 180 & 180 \\ 10 & 10 & 10 & 10 \end{bmatrix}' \overset{\text{SBC}}{\Longrightarrow} R_7 = \begin{bmatrix} 52.5 & 0 & -42.5 & 0 \\ 0 & 0 & 0 & 0 \\ 42.5 & 0 & -42.5 & 0 \\ 0 & 0 & 0 & 0 \end{bmatrix} \tag{7.24}$$

$$r_8 = r_4' = \begin{bmatrix} 10 & 10 & 10 & 10 \\ 180 & 180 & 180 & 180 \\ 10 & 10 & 10 & 10 \\ 180 & 180 & 180 & 180 \end{bmatrix} \overset{\text{SBC}}{\Longrightarrow} R_8 = \begin{bmatrix} 95 & 0 & 0 & 0 \\ 0 & 0 & 0 & 0 \\ -85 & 0 & 0 & 0 \\ 0 & 0 & 0 & 0 \end{bmatrix}. \quad (7.25)$$

Again the coefficients contain only the vertical term V and the lowest subband $11 - 11$. The magnitudes stay the same but just switch positions based on the edge orientation. Next we add a Gaussian random number with mean 3 and variance 2 to r_1 and round to the nearest integer to give a realistic block. Then using the FSBC gives

$$r_9 = \begin{bmatrix} 13 & 12 & 13 & 10 \\ 182 & 180 & 184 & 183 \\ 184 & 183 & 183 & 181 \\ 183 & 182 & 181 & 183 \end{bmatrix} \overset{\text{SBC}}{\Longrightarrow} R_9 = \begin{bmatrix} 139.78 & -42.64 & 0.09 & -0.34 \\ -42.46 & -42.72 & 0.39 & 0.33 \\ 0.56 & 0.18 & 0.12 & -0.20 \\ 0.20 & -0.13 & -0.44 & 0.04 \end{bmatrix}. \quad (7.26)$$

The coefficients have slight energy except for the horizontal term H and the lowest subband $11 - 11$.

The fuzzy system of virtual tiles does not have the same effect as this simple example since it applies the FSBC to a 256×256 image and uses the 32-point QMF in [9]. Yet the example does give some guidelines for tilings.

We first classify the image blocks into mixed blocks and all others based on the high-frequency term M. We use virtual splitting for mixed blocks as in Figure 7.16 if $M > M_{th}$ for some threshold M_{th}. We used $M_{th} = 18$. The mixed block may have edge structure but few if any horizontal or vertical edges.

Next the fuzzy associative memory (FAM) bank of 4 rules in Figure 7.11 further classifies the other blocks. The 4 fuzzy rules speed the computation. More complex systems can use neural systems to tune the fuzzy sets and rules and thus better track the changing signal statistics if we can define an output error from fuzzy system. Our fuzzy system lacked such an error term. It only classified the incoming block vectors. Figure 7.12 shows the fuzzy-set values of fuzzy variable H. The horizontal term H is small or large to some degree. The vertical term V uses the same fuzzy sets for *small* and *large*. We based these sets on the statistics of the 5 training images. The sets can change in shape and number as a learning scheme samples more images. Simulations showed that fuzzy sets (triangles) tended to give better results than nonfuzzy sets (a rectangle partition). Simulations also showed that adding more rules did not significantly improve the classification even though they increased the burden of computation. We tried as many as 30 rules with varying if-part fuzzy sets.

Figures 7.13 through 7.16 show the virtual splittings for each category of the image-block classification. They apply the two-scale decomposition to channels that contain enough information based on the classification. This part of the system is symbolic or "virtual" since we extract just the named subbands from the FSBC. Figures 7.17 through 7.20 show the tilings that correspond to the above splittings. The vertical axis stands for the 2-D frequency that grows from $11 - 11$ to $22 - 22$ as in Figure 7.7. The horizontal axis stands for the 2-D sampling space that the

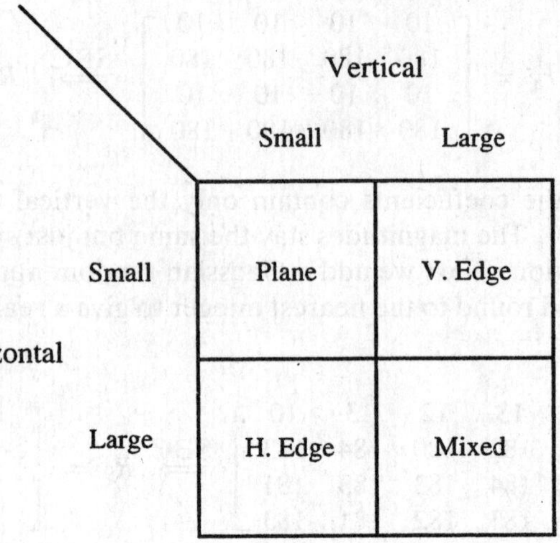

Figure 7.11 FAM bank of 4 rules to classify image blocks.

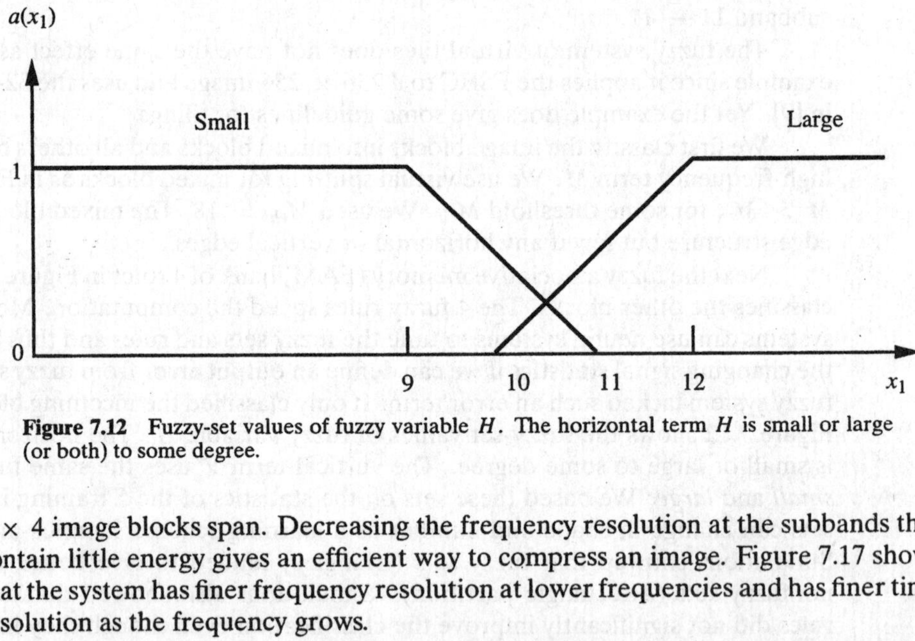

Figure 7.12 Fuzzy-set values of fuzzy variable H. The horizontal term H is small or large (or both) to some degree.

4 × 4 image blocks span. Decreasing the frequency resolution at the subbands that contain little energy gives an efficient way to compress an image. Figure 7.17 shows that the system has finer frequency resolution at lower frequencies and has finer time resolution as the frequency grows.

We build subcodebooks for each category based on the splittings rather than design a whole code book. Each subcodebook has a low distortion level with less computational load for the code-vector search. Such vector quantization can give high compression ratios but may degrade edges and have high complexity. Classified vector quantization (CVQ) [20] divides an image into smaller blocks to classify and design independent codebooks for each category. But the classification in CVQ needs a lot of preprocessing to compute the gradient for each pixel in each direction (horizontal and vertical). The classification scheme also works with many tables and

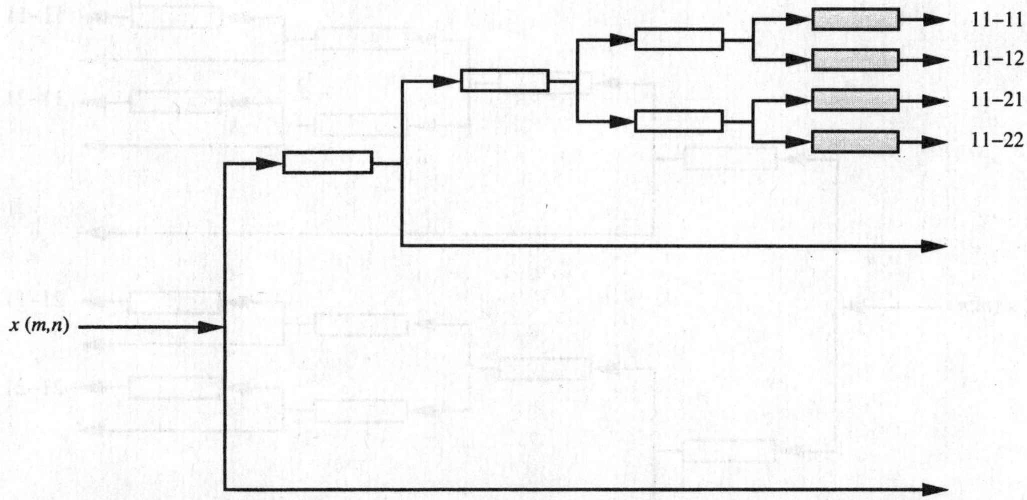

Figure 7.13 Virtual splitting for a plane block that splits only the lowpass band. This gives an octave-band subband structure.

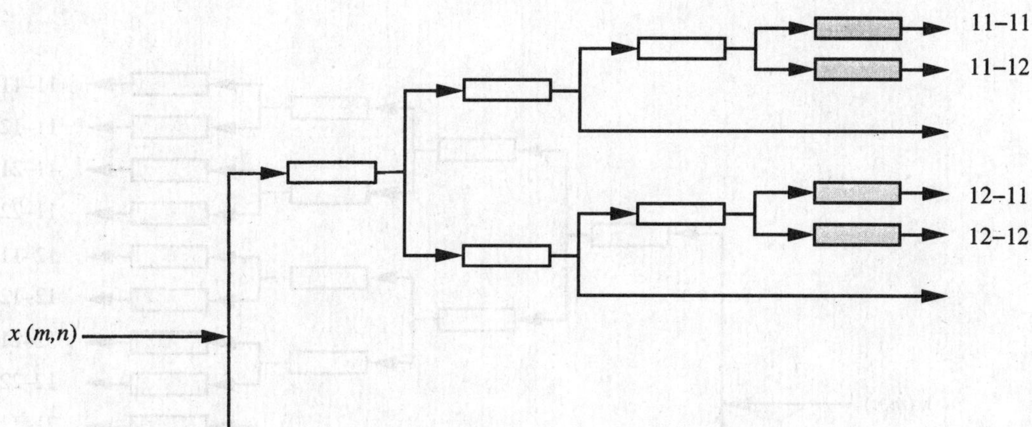

Figure 7.14 Virtual splitting for horizontal edge block.. Splits occur only on the columns of the input image.

counters that complicate the classification. Our system did not calculate the gradient for each pixel. It just used 4 if-then rules in a SAM system (7.16). This greatly improved the classification speed and yet maintained edge structure. More rules and splittings can track the changing signal statistics and reduce the mean-squared error of image reconstruction but at the cost of more complexity. Fuzzy systems can also use the neural system to tune their sets and rules for that purpose. Such adaptive fuzzy systems have more encoding complexity and require more bits per pixel.

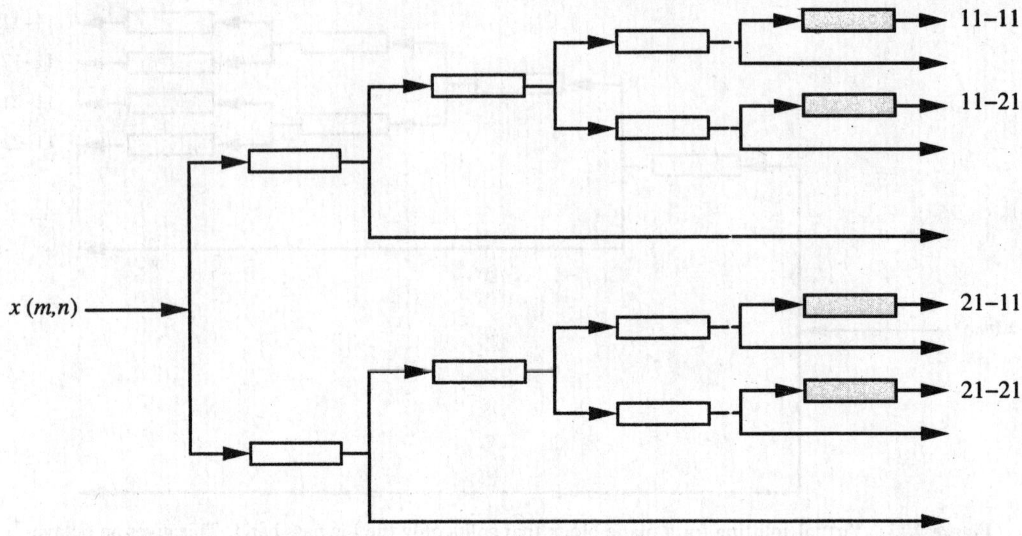

Figure 7.15 Virtual splitting for vertical edge block. Splits occur only on the rows of the input image.

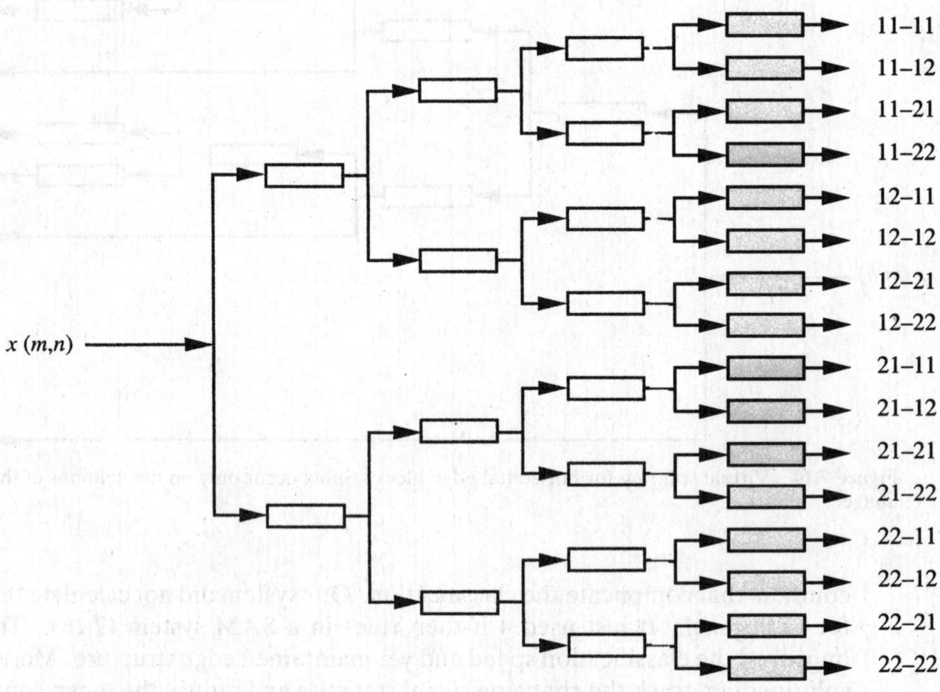

Figure 7.16 Virtual splitting for mixed block of the fully decomposed SBC.

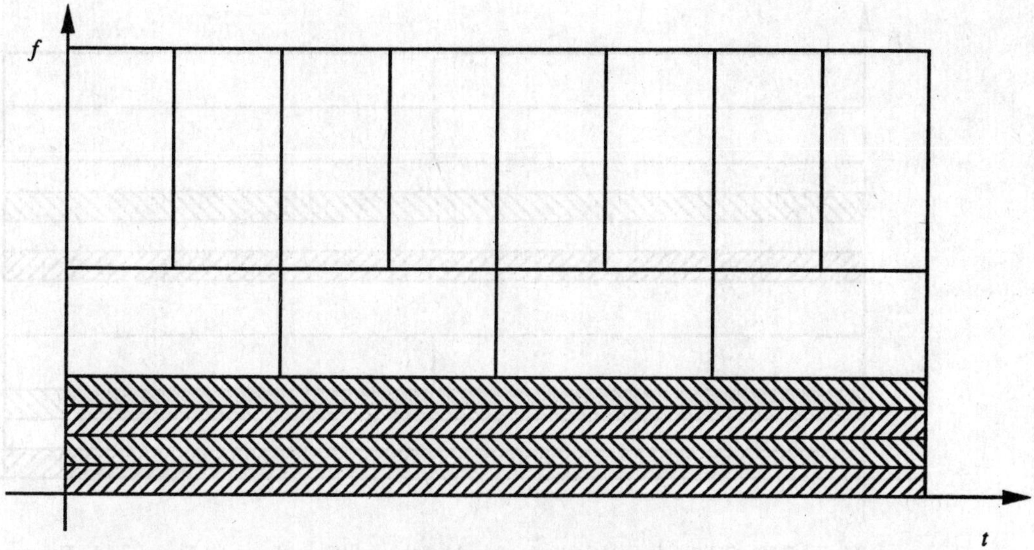

Figure 7.17. Tilings for plane block of the splittings shown in Figure 7.12. The lowpass band has high resolution in the frequency domain.

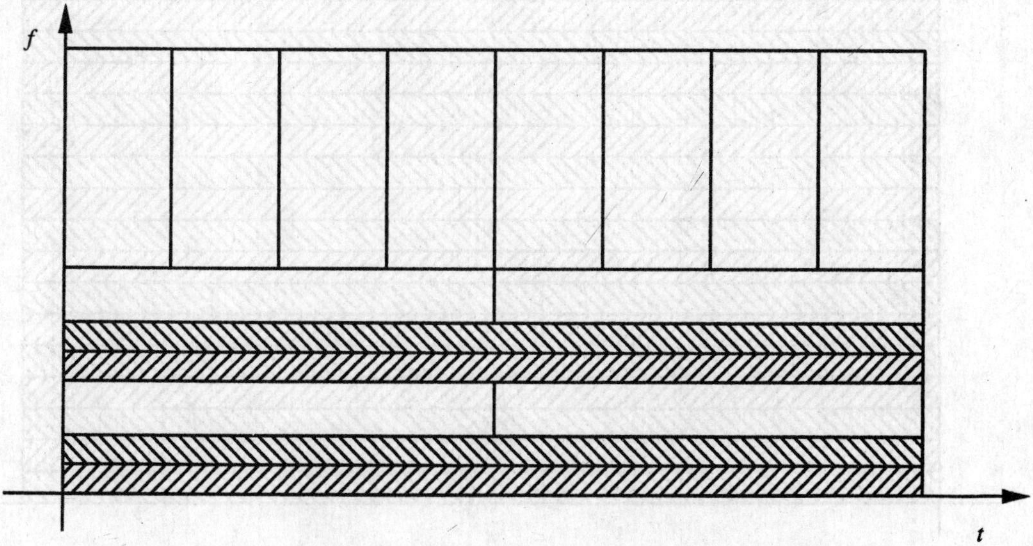

Figure 7.18 Tilings for horizontal edge block of the splittings shown in Figure 7.13. The components for horizontal edges have high resolution in the frequency domain.

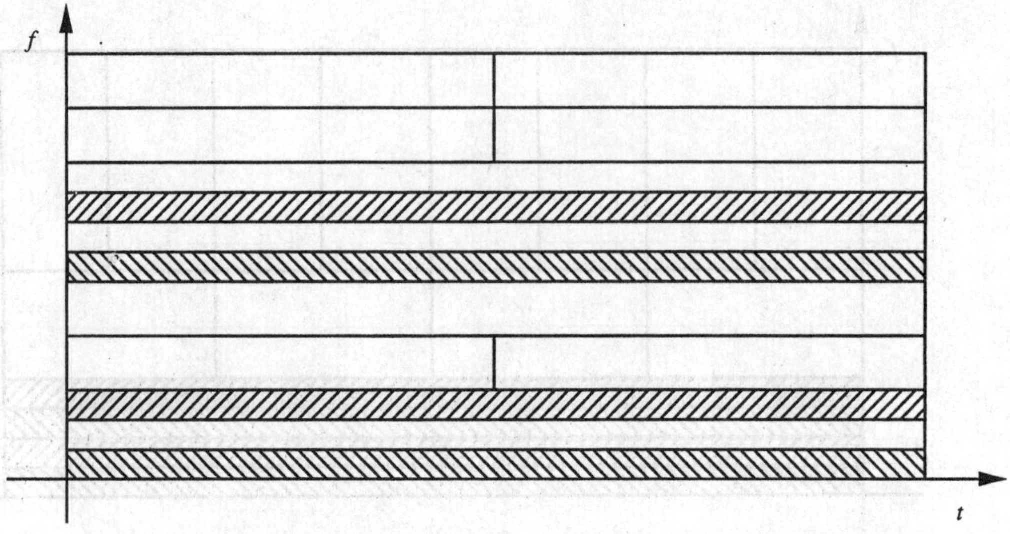

Figure 7.19 Tilings for vertical edge block of the splittings shown in Figure 7.14. The components for vertical edges have high resolution in the frequency domain.

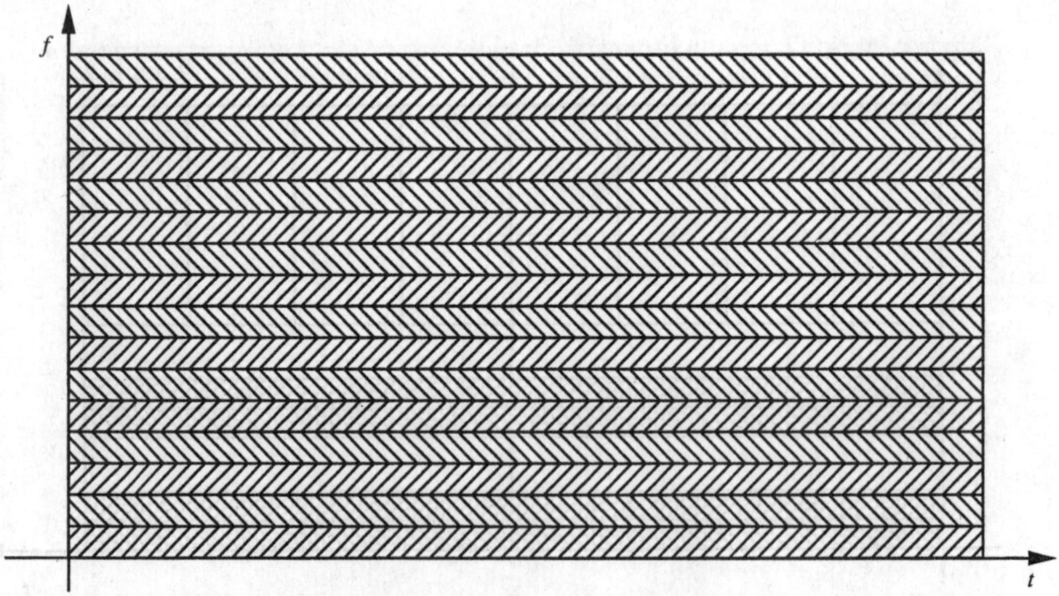

Figure 7.20 Tilings for mixed block of the splittings shown in Figure 7.15. All subbands have the same resolution in the frequency domain.

Seven steps give the subband coding scheme with fuzzy virtual tilings:

Subband coding with fuzzy virtual tilings

1. Take the 16-band decomposition of the input image as shown in Figure 7.7.
2. Compute the 3 variables H, V, and M as the inputs to the fuzzy system.
3. Use the fuzzy system of 4 rules in Figure 7.11 to give the 4 splits.
4. Choose code words from the subcodebooks based on the splittings.
5. Transmit the chosen codeword.
6. Inverse tiling: Zero-pad the missing coefficients of the decoded code-words.
7. Use the 16-band reconstruction to recover the image.

7.4 SIMULATION RESULTS

We compared the virtual tilings for subband coding with the VQ of FSBC in terms of peak signal-to-noise ratios (SNRs):

$$\text{SNR} = 10 \log_{10} \frac{(255)^2}{e_{ms}^2}. \tag{7.27}$$

Here e_{ms}^2 is the average mean-squared error:

$$e_{ms}^2 = \frac{1}{NM} \sum_{i=1}^{N} \sum_{j=1}^{M} (u_{i,j} - \hat{u}_{i,j})^2 \tag{7.28}$$

where $\{u_{i,j}\}$ and $\{\hat{u}_{i,j}\}$ stand for the $N \times M$ original and the reproduced image.

We built all codebooks based on a training set of 5 images for the coding simulations. All images had 256×256 pixels and had 8-bit gray levels. We built subcodebooks based on each splitting. The subcodebook for the plane block is 4-dimensional. The 4-D training vectors came from the first 4 subbands from $11 - 11$ to $11 - 22$. We used 64 code vectors for the plane block. The subcodebooks for the horizontal and vertical blocks were 6-dimensional. The training vectors came from the first 4 subbands from $11 - 11$ to $11 - 22$ with the remaining horizontal or vertical terms. We used 256 code vectors for both the horizontal and vertical blocks. The subcodebook for mixed block was 16-dimensional. The training vectors came from the whole 16 subbands. We used 1,024 code vectors for the mixed block and used 512 code vectors for the VQ of the FSBC. The training vectors for the VQ of the FSBC used the same 5 images we used for the SBC virtual tilings.

We used for the test the two images "Boat" and "Elaine" of 256×256 pixels and 8 bits per pixel. Boat was not in the training sequence. We used the Elaine part of the training sequence to build the codebooks. We encoded the class identity (from 1 to 4) separately from the code vector index. We encoded the code vector with fixed-length codewords and encoded the class label with Huffman coding [8]. Table 7.2 shows the results.

Figure 7.21 shows the simulation results for the Boat image. The classification shows that the numbers of plane blocks, horizontal blocks, vertical blocks, and mixed blocks were 2,068, 226, 787, and 1,015. So the bit rate per pixel was $(0.46 \times 7 + 0.17 \times 11 + 0.10 \times 11 + 0.25 \times 27)/16 = 0.589$ bits per pixel.

Image	FSBC with VQ	SBC with Fuzzy Tilings
Boat	27.4db, 14.2:1 (0.563 bpp)	28.7db, 13.8:1 (0.578 bpp)
Elaine	31.4db, 14.2:1 (0.563 bpp)	32.1db, 15.8:1 (0.506 bpp)

TABLE 7.2: Signal-to-noise ratio comparison for the FSBC with VQ and the SBC with fuzzy virtual tilings.

Figure 7.22 shows the simulation results for the Elaine image. The classification shows that the numbers of plane blocks, horizontal blocks, vertical blocks, and mixed blocks were 2,954, 469, 478, and 195. So the bit rate per pixel was 0.506 bits per pixel (bpp).

Subband coding with fuzzy virtual tilings gave somewhat better results than did FSBC with VQ both in SNR and in subjective quality. The SBC with fuzzy virtual tilings also greatly reduced the encoding complexity for the codevector search when compared with FSBC with VQ.

7.5 CONCLUSION

Fuzzy virtual tilings offer a new scheme for "intelligent" subband image coding. They do not change in time and yet act somewhat as wavelet packets that do. The fuzzy system has less encoding complexity and a better SNR than a fully decomposed standard subband coding system. The compression results still fall short of the results from more complex schemes such as JPEG coding.

Future fuzzy systems for subband coding may use neural systems to tune their sets and rules. These systems might apply to many steps in the subband coding process. The lack of error allows gradient descent to tune each parameter in the fuzzy SAM system [13] and reduce the mean-squared error of the image reconstruction or perhaps of the vector quantization. Unsupervised clustering ([2] and [16]) can give faster learning and does not need error data but may not perform as well as supervised gradient descent. The tradeoff between supervised and unsupervised learning can help adapt the subband filter bank in real time.

(a) (b) (c)

Figure 7.21 Subband coding of the "Boat" image. (a) Original image. (b) FSBC with VQ (0.563 bits per pixel and $SNR = 25.5db$). (c) SBC with fuzzy virtual tilings (0.589 bits per pixel and $SNR = 27.4db$).

REFERENCES

[1] Crochiere, R. E., Webber, S. A., and Flanagan, J. L., "Digital Coding of Speech in Subbands," *Bell Systems Technical Journal*, Vol. 55, 1069–1085, October 1976.

[2] Dickerson, J. A., and Kosko, B., "Fuzzy Function Learning with Covariance Ellipsoids," *IEEE International Conference on Neural Networks (IEEE ICNN–93)*, 1162–1167, March 1993.

[3] Gabor, D., "Theory of Communications," *Journal of the Institution of Electrical Engineers*, Vol. 93, 429–457, 1946.

[4] Gersho, A., and Gray, R. M., *Vector Quantization and Signal Compression*, Kluwer Academic Publishers, 1992.

[5] Hamming, R. W., *Digital Filters*, 2nd ed., Prentice Hall, 1983.

Figure 7.22 Subband coding of the "Elaine" image. (a) Original image. (b) FSBC with VQ (0.563 bits per pixel and $\boldsymbol{SNR} = 31.4db$). (c) SBC with fuzzy virtual tilings (0.506 bits per pixel and $\boldsymbol{SNR} = 31.4db$).

[6] Herley, C., Kovacevic, J., Ramchandran, K., and Vetterli, M., "Tilings of the Time-Frequency Plane: Construction of Arbitrary Orthogonal Bases and Fast Tiling Algorithms," *IEEE Transactions on Signal Processing*, Vol. 41, No. 12, 3341–3359, December 1993.

[7] Herley, C., and Vetterli, M., "Orthogonal TimeVarying Filter Banks and Wavelet Packets," *IEEE Transactions on Signal Processing*, Vol. 42, No. 10, 2650–2663, October 1994.

[8] Jain, A. K., *Fundamentals of Digital Image Coding*, Prentice Hall, 1989.

[9] Johnston, J. D., "A Filter Family Designed for Use in Quadrature Mirror Filter Banks," *Proceeding of International Conference on Acoustics, Speech and Signal Processing (ICASSP)*, 291–294, April 1980.

[10] Kosko, B., *Neural Networks and Fuzzy Systems*, Prentice Hall, 1991.

[11] Kosko, B., "Fuzzy Systems as Universal Approximators," *IEEE Transactions on Computers*, Vol. 43, No. 11, 1329–1333, November 1994; an early version appears in *Proceedings of the 1st IEEE International Conference on Fuzzy Systems (IEEE FUZZ–92)*, 1153–1162, March 1992.

[12] Kosko, B., "Optimal Fuzzy Rules Cover Extrema," *International Journal of Intelligent Systems*, Vol. 10, No. 2, 249–255, February 1995.

[13] Kosko, B., "Combining Fuzzy Systems," *Proceedings of the IEEE International Conference on Fuzzy Systems (IEEE FUZZ–95)*, Vol. IV, 1855–1863, March 1995.

[14] Linde, Y., Buzo, A., and Gray, R., "An Algorithm for Vector Quantizer Design," *IEEE Transactions on Communications*, Vol. COM–28, No. 1, 84–95, January 1980.

[15] Meyer, Y., *Wavelets Algorithms and Applications*, The Society for Industrial and Applied Mathematics (SIAM), 1993.

[16] Moody, J., and Darken, C. J., "Fast Learning in Networks of Locally-tuned Processing Units," *Neural Computation*, Vol. 1, 281–294, 1989.

[17] Oppenheim, A. V., and Schafer, R. W., *Discrete-Time Signal Processing*, Prentice Hall, 1989.

[18] Papoulis, A., *The Fourier Integral and Its Applications*, McGraw-Hill, 1962.

[19] Pratt, W. K., *Digital Image Processing*, 2nd ed., John Wiley & Sons, 1991.

[20] Ramamurthi, B., and Gersho A., "Classified Vector Quantization of Images," *IEEE Transactions on Communications*, Vol. COM–34, No. 11, 1105–1115, November 1986.

[21] Shapiro, J. M., "Embedded Image Coding Using Zerotrees of Wavelet Coefficients," *IEEE Transactions on Signal Processing*, Vol. 41, No. 12, 3445–3462, December 1993.

[22] Vetterli, M., "Multi-dimensional Sub-band Coding: Some Theory and Algorithms," *Signal Processing*, Vol. 6, 97–112, April 1984.

[23] Woods, J. W., and O'Neal, S. D., "Subband Coding of Images," *IEEE Transactions Acoustics, Speech and Signal Processing*, Vol. ASSP–34, No. 5, 1278–1288, October 1986.

[24] Westerink, P. H., Boekee, D. E., Biemond, J., and Woods, J. W., "Subband Coding of Images Using Vector Quantization," *IEEE Transactions on Communications*, Vol. COM–36, No. 6, 713–719, June 1988.

HOMEWORK PROBLEMS

7.1. Parseval's formula. Suppose the signal energy equals one:

$$\int_{-\infty}^{\infty} |f(t)|^2 dt = 1.$$

Show that

$$\frac{1}{2\pi} \int_{-\infty}^{\infty} |F(\omega)|^2 d\omega = 1.$$

7.2. Schwarz's inequality. Prove that

$$\left| \int_a^b g_1 \; g_2 \; dt \right|^2 \leq \int_a^b |g_1|^2 dt \int_a^b |g_2|^2 dt.$$

7.3. Define the variance of the time signal $f(t)$ as

$$V_t = \int_{-\infty}^{\infty} t^2 |f(t)|^2 dt$$

and the variance of its Fourier transform $F(\omega)$ as

$$V_\omega = \int_{-\infty}^{\infty} \omega^2 |F(\omega)|^2 d\omega.$$

Prove the "uncertainty principle"

$$V_t V_\omega \geq \frac{\pi}{2}.$$

Hint: Use Parseval's formula and Schwarz's inequality.

7.4. Show that equality in problem 7.3 holds only for Gaussian signals $f(t) = \sqrt{\frac{\alpha}{\pi}} e^{-\alpha t^2}$.

7.5. Suppose $\{y_n\}$ is a linear time invariant system:

$$y[n] = \sum_{k=-\infty}^{\infty} x[k]h[n-k] = x[n] * h[n].$$

Show that $Y(e^{i\omega}) = X(e^{i\omega})H(e^{i\omega})$. Here $X(e^{i\omega})$ is the Fourier transform of $x[n]$ and $i = \sqrt{-1}$ is the imaginary unit.

7.6. Reduce the sampling rate of $\{x[n]\}$ with the new sequence $x_d[n] = x[nM]$. Show that

$$X_d(e^{i\omega}) = \frac{1}{M} \sum_{k=0}^{M-1} X(e^{i(\omega/M - 2\pi k/M)}).$$

7.7. Expand the sampling rate of $\{x[n]\}$ with the new sequence $x_e[n] = \sum_{k=-\infty}^{\infty} x[k]\delta[n - kL]$.

Show that $X_e(e^{i\omega}) = X(e^{i\omega L})$.

7.8. Define a new sequence as

$$x_s[n] = \begin{cases} x[n], & n = Mk, \quad k = 0, \pm 1, \pm 2, \cdots \\ 0, & \text{otherwise.} \end{cases}$$

Express $X_s(e^{i\omega})$ in terms of $X(e^{i\omega})$.

7.9. Let $w[2n+1] = x[n]$ and $w[2n] = 0$, $n = 0, \pm 1, \pm 2, \ldots$. Derive the spectrum $W(e^{i\omega})$ of $w[n]$.

7.10. Prove that the conditions

$$F_{11}(\omega_1, \omega_2) = 4H_{11}(\omega_1, \omega_2)$$

$$F_{12}(\omega_1, \omega_2) = -4H_{12}(\omega_1, \omega_2)$$

$$F_{21}(\omega_1, \omega_2) = -4H_{21}(\omega_1, \omega_2)$$

$$F_{22}(\omega_1, \omega_2) = 4H_{22}(\omega_1, \omega_2)$$

$$H_{11}(\omega_1, \omega_2)H_{11}(\omega_1 + \pi, \omega_2 + \pi) = H_{11}(\omega_1, \omega_2 + \pi)H_{11}(\omega_1 + \pi, \omega_2)$$

remove the aliased terms for $(k, l) \neq (0, 0)$ in

$$\ddot{U}(\omega_1, \omega_2) = \frac{1}{4}\sum_{k,l} X(\omega_1 + k\pi, \omega_2 + l\pi)\left[\sum_{i,j} H_{ij}(\omega_1 + k\pi, \omega_2 + l\pi)F_{ij}(\omega_1, \omega_2)\right].$$

7.11. We have six codewords $s_1, \ldots, s_6$ with probabilities 0.40, 0.20, 0.15, 0.14, 0.06, 0.05. Encode them with Huffman coding with the following algorithm.

Huffman Coding Algorithm

1. Arrange the symbol probabilities p_i in a decreasing order. Consider them as leaf nodes of a tree.
2. Merge the two nodes with the smallest probability to form a new node whose probability is the sum of the two merged nodes.
3. Arbitrarily assign 1 and 0 to each pair of branches that merge into a node.
4. Read sequentially from the root node to the leaf node where the symbol lies.

Consider s_1, s_2, and s_3 with probabilities 0.7, 0.2, and 0.1. Merge the two nodes s_2 and s_3 to form a new node and assign 1 and 0 to each pair of branches. Repeat the merging and assign 1 and 0 to each new pair of branches. This gives $s_1 = \{1\}$, $s_2 = \{01\}$, and $s_3 = \{00\}$.

7.12. We have 8 codewords $s_1, \ldots, s_8$ with probabilities 0.26, 0.20, 0.15, 0.14, 0.0625, 0.0625, 0.0625, 0.0625. Encode them with Huffman coding.

PART V

FUZZY COMMUNICATION

Spread spectrum is a key technology in wireless communications. It spreads a signal over a wide range of frequencies before it sends the signal. A receiver despreads the signal with a synchronized key system. The sender can send more than one signal over the spread signal's bandwidth. This gives rise to many schemes for signal multiplexing. An eavesdropper receives the signal as random noise if he lacks the key system. He also needs to use a great deal of energy to jam the spread signal since it covers such a wide bandwidth.

Spread spectrum grew out of the war effort in the 1940s. The military classified most early research in the field as secret. Spread spectrum remains the basis of modern cockpit electronic warfare and has moved into wireless networks and cellular phone networks.

Frequency hopping is one way to spread and despread signals. A code sequence changes the transmission frequency many times per second. Actress Hedy Lamarr both starred in the 1949 film *Samson and Delilah* and co-filed the first patent on June 10th of 1941 for frequency hopping and applied it to anti-jamming secret communication.

Chapter 8 shows that an additive fuzzy system can learn to produce a "random" code sequence for a frequency hopper. The fuzzy system watches a random number generator to learn a set of "random rules." The fuzzy system gives frequency sequences that appear random in a statistical sense. The additive fuzzy system has a novel structure that depends on modular arithmetic. The system could apply to other spreading schemes and could learn sequences that reflect goals other than security.

8

ADAPTIVE FUZZY FREQUENCY HOPPING FOR SPREAD SPECTRUM

Peter J.Pacini and Bart Kosko

PREVIEW

An adaptive additive fuzzy system generates the frequency hopping sequence for a "wireless" spread spectrum communications system. The system learns rules from data and acts as a pseudorandom number generator. The (IMSL) uniform random number generator gives training samples. An adaptive scheme learns associations between previous samples and the current sample and encodes these as fuzzy rules.

The then-part fuzzy set for each rule acts as a conditional probability density function. The if-part of the rule states the conditions. At each step the system scans 30 prior outputs according to a fixed sampling pattern and gives a new sample distribution x_k. The vector x_k partially matches the if-part set of a fuzzy rule and partially fires that rule's output fuzzy set. With the estimated output fuzzy sets the fuzzy system computes the conditional density $p_{Y|X}(y|x_k)$ for input field X and output field Y. Defuzzification yields the next number in the frequency hopping sequence in the context of modular arithmetic.

The rules, sampling pattern, and initial conditions fix the output sequence. An eavesdropper who did not know all three could not predict the sequence. This fuzzy system can generate a sequence uniform over any number of frequencies. We tested the fuzzy system with 100 and 1,025 frequencies and compared it to a shift register with linear feedback. The fuzzy system had lower chi-squared values and thus gave a more uniform or more "random" spread than did the shift register. The fuzzy system was easier to change and harder to intercept.

8.1 FREQUENCY HOPPING SPREAD SPECTRUM

A frequency hopping system spreads a transmitted signal's energy across a bandwidth larger than the minimum required for the signal. It spreads the signal when it changes or "hops" the transmission frequency many times per second. The transmitter and receiver must stay synchronized. They must both hop to the same frequency at the same time. If an eavesdropper does not know the frequency sequence then the hopped signal looks like low-intensity noise spread over the entire bandwidth.

Frequency hopping systems use a pseudorandom number (PN) generator to produce a "random" sequence of frequencies. Figure 8.1 shows the signal flow. Both the transmitter and receiver must have a copy of any code used or the procedure to generate it.

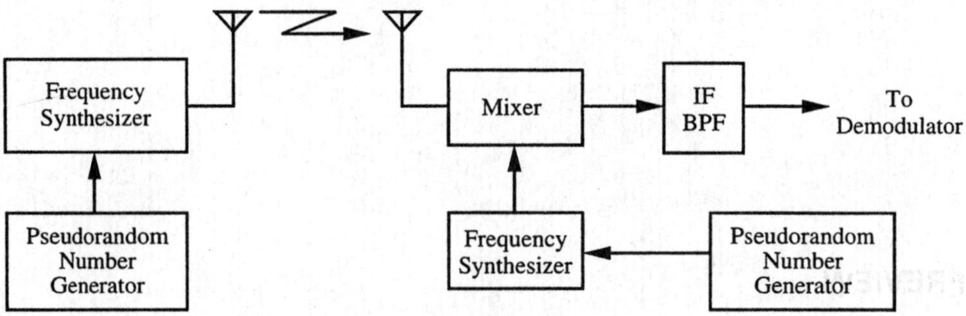

Figure 8.1 Spread spectrum communications system that uses frequency hopping.

The most common modulation scheme for frequency hopping is M-ary frequency shift keying or MFSK:

$$s_k \in \{0, m - 1\}, \qquad s(t) = \sin(f(y_k) + s_k \Delta f). \tag{8.1}$$

s_k is the data sent at time k. y_k is the output of the PN generator in the transmitter and receiver. $s(t)$ is the waveform sent. Equation (8.1) describes frequency shift keying or FSK when $m = 2$.

The transmitter's frequency synthesizer emits a frequency $f(y_k)$ that modulates the input. The receiver's frequency synthesizer emits the term $f(y_k) + f_{IF}$. This term translates the received signal to *intermediate frequency* f_{IF}. A bandpass filter (or several BPFs in the case of MFSK) and demodulator choose which value s_k the transmitter sent.

This chapter describes a *fuzzy rule-based* PN generator. Adaptive fuzzy rules map distributions of old output frequencies to new output frequencies in a standard additive model (SAM) fuzzy system with modular arithmetic. These rules fix the output sequence given some initial conditions and the sampling pattern that gives the previous outputs' distribution. Each sequence has an unknown length if it repeats at all. Such sequences are hard for an eavesdropper to predict if he does not know either the rules or the initial conditions or the sampling pattern.

A complete communication system includes more subsystems than the PN generator. How the system performs depends on source coding, encryption, channel coding, modulation schemes, and synchronization. These subsystems exceed the scope of this chapter. We limit our discussion to the design and performance of the PN generator.

8.2 FUZZY PSEUDORANDOM NUMBER GENERATOR

Our simulations assumed a frequency synthesizer that gives N distinct frequencies. We partitioned the bandwidth into n *frequency bins* that each contain N/n frequencies. Figure 8.2 shows the partitioned spectrum for $N = 100$ and $n = 5$. We numbered the frequencies $0, \ldots, N-1$. So the fuzzy system gave a sequence of integers in $\{0, \ldots, N-1\}$ such that each integer tended to occur equally often.

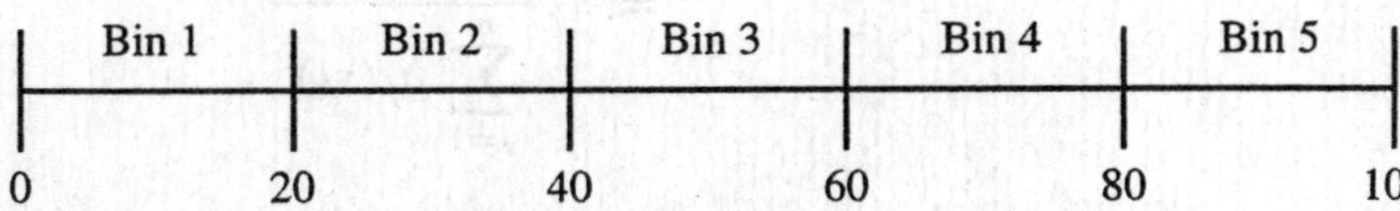

Figure 8.2 100 frequencies partitioned into 5 "frequency bins."

We chose $N = 100$ frequencies and $n = 5$ bins for initial simulations. After testing with 100 frequencies we reset N to 1,025 frequencies to compare with a shift register. Changing N does not affect the algorithm or the rules. So the system can use any number of frequencies. N should be a multiple of n (for any n) to give a balanced partition into bins. We chose the small number $n = 5$ to limit the number of rules. The adaptive scheme in Section 8.2.3 works for any n.

8.2.1 Inputs to the Fuzzy System

A fixed sampling pattern decides which 30 of the previous output frequencies will give the next output. At time k we might sample the output frequencies from times $k-1, k-6, k-8, k-14$, etc. Varying the time delays between samples makes the samples appear independent from one step to the next. The number of samples that fall in the 5 frequency bins are the 5 inputs to the fuzzy system. We store them as a length–5 vector x_k.

The decision to use 30 samples followed directly from two other design choices. We first chose 5 frequency bins to give about 200 rules as discussed in Section 8.2.3. Then we designed the input fuzzy sets shown in Figure 8.3. This process involved some trial and error. The Medium fuzzy set should have its maximum value at the expected number of hits per bin. Medium peaks at 6 and there are 5 bins. So the number of samples should be $6 \times 5 = 30$.

A simpler sampling pattern, such as the last 30 output frequencies, would not work. Fuzzy systems tend to map close inputs to close outputs. If the inputs at times k and $k+1$ correlate highly then so will the outputs at times k and $k+1$. We designed the sampling pattern to minimize the output sequence's autocorrelation and to minimize the maximum sample time delay (the time delay for the sample).

We use the standard additive model (SAM) system $F: R^n \longrightarrow R$ in Chapter 2:

$$\begin{aligned} F(x_k) &= Centroid(B(x_k)) \\ &= \frac{\sum_{j=1}^{m} a_j(x_k) V_j c_j}{\sum_{j=1}^{m} a_j(x_k) V_j} \\ &= \frac{\sum_{j=1}^{m} a_j(x_k) c_j}{\sum_{j=1}^{m} a_j(x_k)} \end{aligned}$$

for constant-volume then-part sets $B_j \subset R$ with set function $b_j: R \longrightarrow [0, 1]$. Min factors the if-part set functions $a_j: R^n \longrightarrow [0, 1]$ to give

$$a_j(x_k) = \min(a_j^1(x_k^1), \ldots, a_j^n(x_k^n)) \quad (8.8).$$

For each bin we define three fuzzy sets: Small, Medium, and Large. Figure 8.3 shows these fuzzy sets as set functions:

$$a_S : \{0, \ldots, 30\} \longrightarrow [0, 1], \quad (8.2)$$

$$a_M : \{0, \ldots, 30\} \longrightarrow [0, 1], \quad (8.3)$$

$$a_L : \{0, \ldots, 30\} \longrightarrow [0, 1] \quad (8.4)$$

$a_S(u)$ stands for the degree to which the number of hits u is Small. $a_M(u)$ and $a_L(u)$ stand for the degree to which u is Medium and Large. So the fuzzy sets in Figure 8.3 map the number of hits $x_k[j]$ in each bin into a length–3 *fit vector* [5] $(a_S(x_k[j]), a_M(x_k[j]), a_L(x_k[j]))$ of fit or *fuzzy unit* values. If 6 is the input value or expected value per bin then it maps to (0, 1, 0). 5 maps to (1/3, 2/3, 0) and

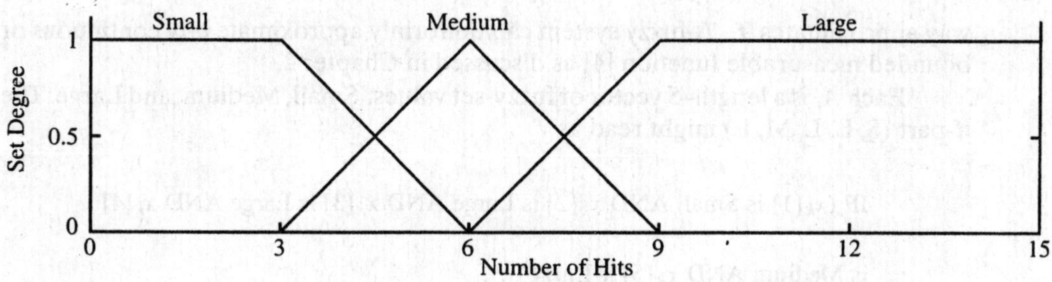

Figure 8.3 If-part fuzzy sets for each bin.

8 maps to (0, 1/3, 2/3). The fit vectors of the five integer inputs form a 5 × 3 matrix I_k:

$$I_k[j,\ 1] = a_S(x_k[j]) \tag{8.5}$$

$$I_k[j,\ 2] = a_M(x_k[j]) \tag{8.6}$$

$$I_k[j,\ 3] = a_L(x_k[j]) \tag{8.7}$$

The general case has n bins and p input fuzzy sets. Then I_k would have dimensions $n \times p$. I_k is the input to the fuzzy system in Figure 8.4.

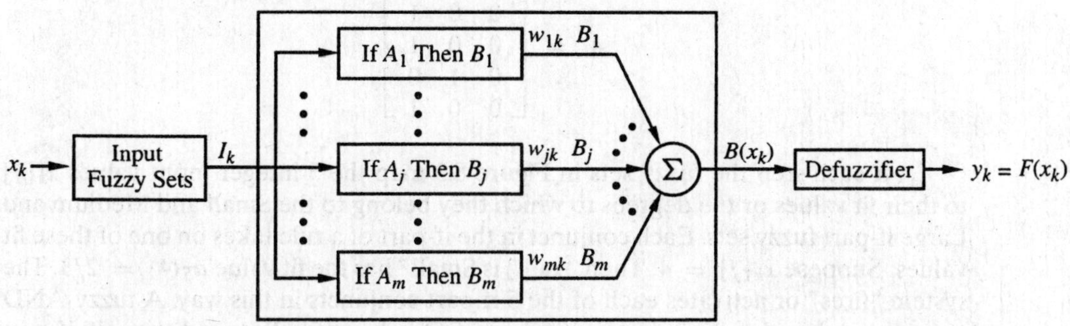

Figure 8.4 Parallel algorithmic structure of the standard additive fuzzy system of the frequency hopper. Input x_k fires all rules to some degree since x_k belongs to each A_i to some degree. The output fuzzy set $B(x_k)$ is the weighted sum of the then-part fuzzy sets B_i. The defuzzifier converts $B(x_k)$ to a number y_k or the centroid or center of mass of $B(x_k)$: $F(x_k) = \text{Centroid}(B(x_k))$.

8.2.2 Fuzzy Rules

Each fuzzy rule is a conditional:

$$\text{IF } x_k \text{ is } A_i \text{ THEN output is } B_i.$$

This rule equals the fuzzy patch or Cartesian product $A_i \times B_i$ in the state input–output space $X \times Y$. The fuzzy rules or patches cover the graph of some unknown function $f\colon X \longrightarrow Y$. The fuzzy system averages patches that overlap and in this way approximates f. A fuzzy system can uniformly approximate *any* continuous or bounded measurable function [4] as discussed in Chapter 2.

Each A_i is a length–5 vector of fuzzy-set values: Small, Medium, and Large. The if-part (S, L, L, M, L) might read as

IF ($x_k[1]$ is Small AND $x_k[2]$ is Large AND $x_k[3]$ is Large AND $x_k[4]$ is Medium AND $x_k[5]$ is Large) . . .

Or we can express A_i as a 5×3 matrix $\mathcal{A}_i$ similar to I_k. Each of the 5 fuzzy-set values in A_i maps to a unit bit vector:

$$\text{Small} \longrightarrow (1,\ 0,\ 0),$$

$$\text{Medium} \longrightarrow (0,\ 1,\ 0),$$

$$\text{Large} \longrightarrow (0,\ 0,\ 1).$$

So for the above if-part $A_i = (\text{S, L, L, M, L})$,

$$\mathcal{A}_i = \begin{bmatrix} 1 & 0 & 0 \\ 0 & 0 & 1 \\ 0 & 0 & 1 \\ 0 & 1 & 0 \\ 0 & 0 & 1 \end{bmatrix}$$

At each step the fuzzy sets in Figure 8.3 map the 5 integer input values $x_k[j]$ to their fit values or the degrees to which they belong to the Small and Medium and Large if-part fuzzy sets. Each conjunct in the if-part of a rule takes on one of these fit values. Suppose $x_k[j] = 4$. Then "$x_k[j]$ is Small" has the fit value $a_S(4) = 2/3$. The system "fires" or activates each of the 5 if-part conjuncts in this way. A fuzzy AND operation takes the minimum or product of fit values [5]. We used the min if-part combiner but could just as well have used the product if-part combiner as we do in the next chapter. So the complete if-part "x_k is A_i" takes on the fit value

$$a_{A_i}(x_k) = \min_j (a_{A_i[j]}(x_k[j])) \tag{8.8}$$

$$= \min_j (I_k \mathcal{A}_i^T [j,\ j]). \tag{8.9}$$

The ith rule *fires* or *activates* to degree w_{ik}:

$$w_{ik} = a_{A_i}(x_k). \tag{8.10}$$

Terms (8.8) and (8.9) give different views of a fuzzy rule. (8.8) includes the if-part fuzzy sets as part of each rule. So each rule looks at how much a numerical input x_k belongs to fuzzy set A_i in the if-part of the rule. In (8.9) each conjunct's fit value correlates the two fuzzy sets and $\mathcal{A}_i$ and the jth rows of I_k. The higher the correlation between I_k and $\mathcal{A}_i$ the higher the firing level w_{ik}. Figure 8.4 shows that the firing level w_{ik} scales the then-part B_i of each rule. So (8.9) implies that each rule is a simple type of *fuzzy associative memory* [5] or FAM. The FAM rule (A_i, B_i) or patch $A_i \times B_i$ takes as input the fuzzy set I_k for if-part $\mathcal{A}_i$ (the matrix equivalent of A_i) and gives as output the scaled fuzzy set $B_i' = w_{ik}B_i$. If $I_k = \mathcal{A}_i$ then $B_i' = B_i$.

8.2.3 Adaptive Rule Generation

The then-part of each rule is a length–5 fit vector. The fuzzy system learns these terms as the system trains with 10,000 samples from the IMSL (International Mathematical and Statistical Library) uniform random number generator. We scale and truncate these samples to give integers between 1 and 5. These integers are the bin numbers. The fuzzy system's sampling pattern gives the distribution x_k of previous samples. Then we find the if-part term A_i closest to x_k (the one with maximum w_{ik}) and associate the next sample with A_i.

Example: Observe $x_k = (5, 6, 3, 9, 7)$ and the next output is 4. The closest if-part term is (M, M, S, L, M). So associate output bin 4 with (M, M, S, L, M).

Processing all 10,000 samples in this way gives a histogram of output bin values for each if-part term. Normalizing these histograms gives then-part fit vectors B_i with Hamming norm 1. So the value of the jth component $B_i[j]$ gives the fraction of training samples that fall in bin j when the input distribution matches A_i. Each B_i should be nearly uniform since the samples are "random" and uniformly distributed in $\{1, \ldots, 5\}$. The B_i closest to (.2, .2, .2, .2, .2) correspond to the A_i that occur most often.

We want a uniform output distribution. Symmetry helps reduce the number of training samples. Each step gives not 1 but 5 input–output pairs if we rotate the inputs and outputs. Consider the example above. From the observation we can derive the 5 associations:

$$
\begin{aligned}
(\text{M, M, S, L, M}) &\longleftrightarrow \text{bin } 4\\
(\text{M, M, M, S, L}) &\longleftrightarrow \text{bin } 5\\
(\text{L, M, M, M, S}) &\longleftrightarrow \text{bin } 1\\
(\text{S, L, M, M, M}) &\longleftrightarrow \text{bin } 2\\
(\text{M, S, L, M, M}) &\longleftrightarrow \text{bin } 3.
\end{aligned}
$$

Thus 10,000 training samples with rotations give about the same results as 50,000 samples with no rotations.

In general the spectrum partition in Figure 8.2 has n bins with p fuzzy-set values each. So there are p^n possible if-part terms A_i. Each one may give rise to a fuzzy rule. We used $n = 5$ and $p = 3$ for 243 rules maximum. 57 of the if-part terms cannot have firing levels above 1/3. This adaptive scheme does not associate these if-parts with any output bins. So each fuzzy system can have at most 186 fuzzy rules.

Some rules give better results than do others. A simple screening process checks the covariance of the rules' then-part fuzzy sets. Rules tend to perform well if their covariance looks like that of white noise. The rules that produced the simulation results in Sections 8.4 and 8.5.1 have the following "white-like"covariance matrix:

$$0.0172 \times \begin{bmatrix} 1.000 & -.247 & -.253 & -.253 & -.247 \\ -.247 & 1.000 & -.247 & -.253 & -.253 \\ -.253 & -.247 & 1.000 & -.247 & -.253 \\ -.253 & -.253 & -.247 & 1.000 & -.247 \\ -.247 & -.253 & -.253 & -.247 & 1.000 \end{bmatrix}$$

8.2.4 Combined Output Fuzzy Set

The *combined output fuzzy set* $B(x_k)$ equals the sum of the weighted rule then-parts B_i where the weights are the input-based firing levels w_{ik}:

$$B(x_k) = \sum_i w_{ik} B_i. \tag{8.11}$$

This fuzzy inference method is *correlation-product inference* [5] or the "standard" part of a SAM fuzzy system. This is a generalized "set" since its fit values can exceed 1.

Only a few rules (32 maximum) can fire at any time. Figure 8.4 shows that the system fires all rules in parallel. This gives a fast map from an integer input vector x_k to a combined output fuzzy set $B(x_k)$.

8.2.5 Defuzzification

The combined output fuzzy set $B(x_k)$ defuzzifies or maps to a number or output frequency y_k where $y_k = F(x_k)$. In additive fuzzy systems $B(x_k)$ tends to be unimodal and easy to defuzzify. The centroid of the output fuzzy set gives a value near the mode. But we designed the fuzzy frequency hopper so that $B(x_k)$ is nearly uniform for each iteration k. The centroid might give a value near the middle of the spectrum each time. So instead we use a new type of centroid defuzzification.

Figures 8.5 through 8.7 show the extra steps that precede the centroid computation. First find the indices of the largest and smallest terms. Call them $j_{\max}$ and $j_{\min}$. Rotate $B(x_k)$ so that $j_{\max}$ is the center bin. Then the centroid of this fuzzy set will be near the center of bin $j_{\max}$. Next subtract $B_k[j_{\min}]$ from each term. This step subtracts the uniform part and isolates the "noise" part of the output. Call this fuzzy set $B_{\text{noise},k}$.

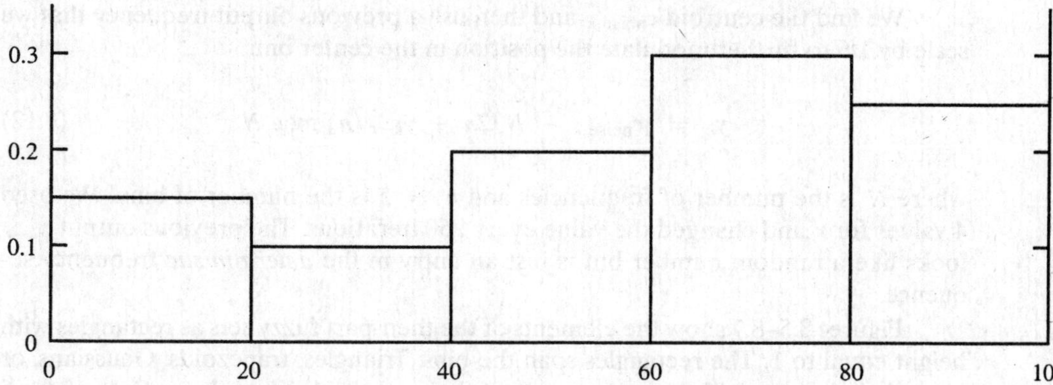

Figure 8.5 Combined output fuzzy set $B(x_k) = (.15, .10, .20, .30, .25)$.

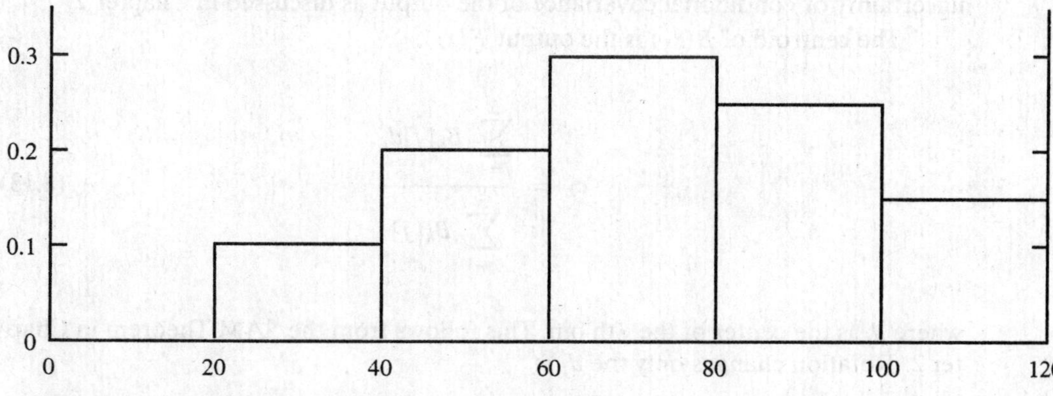

Figure 8.6 Rotated output fuzzy set $B(x_k)$ for Figure 8.5.

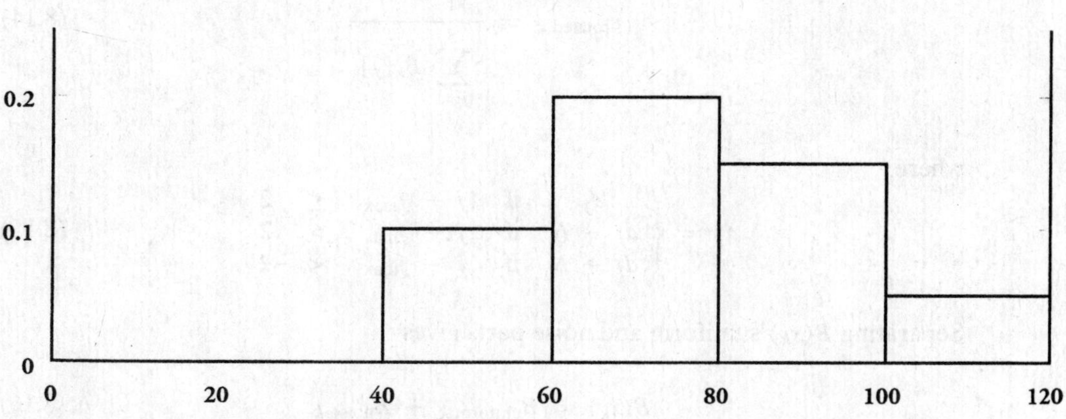

Figure 8.7 "Noise" component of output fuzzy set $B(x_k)$ in Figure 8.5.

We find the centroid $c_{\text{noise},k}$ and then use a previous output frequency that we scale by 1/5 to further modulate the position in the center bin:

$$y_k = \lfloor c_{\text{noise},k} - N/2n + y_{k-r}/n \rfloor \bmod N \tag{8.12}$$

where N is the number of frequencies and $n = 5$ is the number of bins. We used 4 values for r and changed the value every 250 iterations. The previous output y_{k-r} looks like a random number but is just an entry in the *deterministic* frequency sequence.

Figures 8.5–8.7 show the elements of the then-part fuzzy sets as rectangles with height equal to 1. The rectangles span the bins. Triangles, trapezoids, Gaussians, or any other shapes yield the same results if their centroids lie at the centers of the 5 bins and if they have the same area or volume V_j in the SAM ratio. The shape of the then-part fuzzy-set elements does not affect the defuzzified result for correlation-product inference and centroid defuzzification [7]. Here we ignore the second-order uncertainty or conditional covariance of the output as discussed in Chapter 2.

The centroid of $B(x_k)$ is the output $F(x_k)$:

$$c_k = \frac{\sum_{j=1}^{5} B_k[j] d_j}{\sum_{j=1}^{5} B_k[j]} \tag{8.13}$$

where d_j is the center of the jth bin. This follows from the SAM Theorem in Chapter 2. Rotation changes only the d_j's:

$$c_{\text{rotated},k} = \frac{\sum_{j=1}^{5} B_k[j] d'_j}{\sum_{j=1}^{5} B_k[j]} \tag{8.14}$$

where

$$d'_j = \begin{cases} d_j & \text{if } |j - j_{\max}| \le 2 \\ d_j - N & \text{if } j - j_{\max} > 2 \\ d_j + N & \text{if } j - j_{\max} < -2 \end{cases} \tag{8.15}$$

Separating $B(x_k)$'s uniform and noise parts gives

$$B(x_k) = B_{\text{uniform},k} + B_{\text{noise},k} \tag{8.16}$$

or

$$B_k[j] = B_k[j_{\min}] + B_{\text{noise},k}[j]. \tag{8.17}$$

The uniform part has centroid

$$c_{\text{uniform},k} = \frac{\sum_{j=1}^{5} B_k[j_{\min}] d'_j}{\sum_{j=1}^{5} B_k[j_{\min}]} \tag{8.18}$$

$$= \frac{B_k[j_{\min}] \sum_{j=1}^{5} d'_j}{5 B_k[j_{\min}]} \tag{8.19}$$

$$= \frac{1}{5} \sum_{j=1}^{5} d'_j. \tag{8.20}$$

where $c_{\text{uniform},k}$ does not depend on $B(x_k)$. So we drop the subscript k and call this centroid c_{uniform}. If the bins have equal width then c_{uniform} equals the center of the rotated spectrum.

The noise component has centroid

$$c_{\text{noise},k} = \frac{\sum_{j=1}^{5} B_{\text{noise},k}[j] d'_j}{\sum_{j=1}^{5} B_{\text{noise},k}[j]} \tag{8.21}$$

$$= \frac{c_{\text{rotated},k} - \left(\dfrac{5 B_k[j_{\min}]}{\sum_{j=1}^{5} B_k[j]} \right) c_{\text{uniform}}}{1 - \dfrac{5 B_k[j_{\min}]}{\sum_{j=1}^{5} B_k[j]}} \tag{8.22}$$

from (8.14) and (8.17) and (8.21).

Rewriting (8.22) gives

$$c_{\text{rotated},k} = \left(\frac{5B_k[j_{\min}]}{\sum_{j=1}^{5} B_k[j]} \right) c_{\text{uniform}} + \left(1 - \frac{5B_k[j_{\min}]}{\sum_{j=1}^{5} B_k[j]} \right) c_{\text{noise},k}. \tag{8.23}$$

The first term in (8.23) draws $c_{\text{rotated},k}$ toward the center of the rotated spectrum. So $c_{\text{noise},k}$ varies more from the center than $c_{\text{rotated},k}$ does. The difference between them increases as $B(x_k)$ becomes more uniform and as the first term dominates. So using $c_{\text{noise},k}$ instead of $c_{\text{rotated},k}$ in (8.12) gives better results.

8.3 FUNCTIONAL DESCRIPTION OF THE FUZZY SYSTEM

The fuzzy system shown in Figure 8.4 is a function $f\colon X \longrightarrow Y$. Here X consists of length–5 vectors of integers and $Y = [0, N)$ where N is the number of output frequencies. Truncation of the defuzzified output gives integers $y_k \in \{0, \ldots, N-1\}$. Fuzzy sets map inputs to matrices of fuzzy values in [0, 1] or $g : X \longrightarrow Z$. The fuzzy rules $h_i : Z \longrightarrow \phi$ map these fuzzy matrices to then-part fuzzy sets. Defuzzification maps those scaled then-part fuzzy sets to output values or $e\colon \phi \longrightarrow Y$ or simply $y = F(x_k)$.

8.3.1 Output Fuzzy Sets as Conditional Densities

We can view output fuzzy sets as conditional probability density functions

$$B_i(y) = p_{Y|Z}(y|A_i) \tag{8.24}$$

where A_i and B_i are the ith rule's if-part and then-part (both may be *random sets*). Then the degree $a_{A_i}(x_k)$ to which x_k belongs to A_i is the conditional probability $p(x_k \in A_i|X = x)$ that the input is in A_i given the value x_k. So A_i defines the locus of these two-point conditional densities. The antecedent should be $\mathcal{A}_i$ or the matrix equivalent of A_i. The mapping from A_i to $\mathcal{A}_i$ is a bijection and so we just use the term A_i.

A probabilistic view of the firing levels w_{ik} uses the next definition.

$$\text{Definition:} \quad \Pr(A_i|X = x) = \frac{a_{A_i}(x)}{\sum_{A_j \in Z} a_{A_j}(x)} \tag{8.25}$$

where $a_{A_i}(x)$ stands for the degree to which x belongs to A_i and where A_i is a fuzzy set or vector of fuzzy sets.

Equations (8.10) and (8.25) give

$$w_{ik} = \left(\sum_{j=1}^{M} w_{jk}\right) \Pr(A_i|X = x_k). \tag{8.26}$$

Note that

$$\Pr(A_i) = \int \Pr(A_i|X = x)\ p_X(x)dx \qquad \text{from Bayes's rule} \tag{8.27}$$

$$= \int \left[\frac{a_{A_i}(x)}{\sum_{A_j \in Z} a_{A_j}(x)}\right] p_X(x)\ dx \tag{8.28}$$

$$= E_X\left[\frac{a_{A_i}(x)}{\sum_{A_j \in Z} a_{A_j}(x)}\right] \tag{8.29}$$

from the definition of the expected value operator as Zadeh [9] suggested in his model of the probability of a fuzzy event.

The then-part fuzzy sets B_i do not depend on x_k. So we can rewrite (8.24) as

$$B_i(y) = p_{Y|ZX}(y|A_i,\ x_k). \tag{8.30}$$

Putting (8.26) and (8.30) into (8.11) gives

$$B_k(x_k,\ y) = \sum_i \left(\sum_j w_{jk}\right) \Pr(A_i|X = x_k)\ p_{Y|ZX}(y|A_i,\ x_k) \tag{8.31}$$

$$= \left(\sum_j w_{jk}\right) \sum_i p_{YZ|X}(y,\ A_i|x_k) \tag{8.32}$$

$$= \left(\sum_j w_{jk}\right) p_{Y|X}(y|x_k). \tag{8.33}$$

So the fuzzy system uses the conditional densities $p_{Y|Z}(y|A_i)$ to compute $p_{Y|X}(y|x_k)$. Here $B_k(x_k,\ y)$ estimates the uniform distribution. In the ideal case each $p_{Y|Z}(y|A_i)$ would be uniform and so would be $p_{Y|X}(y|x_k)$. This result shows that $p_{Y|X}(y|x_k)$ behaves as a uniform distribution. Simulations show that the covariance of $p_{Y|X}(y|x_k)$ is "white" in time. The variance of the additive white noise depends on the covariance of the then-part fuzzy sets.

8.3.2 Centroid Defuzzification

The centroid map acts as a "defuzzifier" of output set $B(x_k)$ to give the SAM output $F(x_k)$. Equation (8.33) shows why the centroid tends to perform well. The centroid of $B(x_k, y)$ is $F(x_k)$ or

$$c_k(x_k) = \frac{\int y \; B_k(x_k, y) \; dy}{\int B_k(x_k, y) \; dy} \tag{8.34}$$

$$= E[Y|X = x_k] \tag{8.35}$$

as derived in Appendix 8.A and in general in Chapter 2. The mean-squared optimality properties of the conditional mean [6] suggest the centroid as the preferred way to defuzzify an output set.

8.4 SIMULATION RESULTS

Figure 8.8 shows the results for 100 frequencies averaged over 4 runs. The standard chi-squared test [1] compares the fuzzy system's output distribution to the desired uniform distribution. We compared this with the IMSL random number generator that gave the sample data to train the fuzzy system. A solid line marks the mean chi-squared values for the fuzzy system. A dotted line shows those for the IMSL random number generator. The lower the chi-squared value the more closely the output distribution matches the uniform distribution.

Figure 8.9 shows the results for the fuzzy system's four runs. Three of the four runs used random initial conditions. The fourth run (shown in the solid line) used all zero initial conditions. This extreme case shows the fuzzy system's robustness to initial conditions.

We also looked at the autocorrelation of sequences to see how predictable they were. In the best case the bin number of the output at time k would not depend on the bin number at time $k - \ell$ for any ℓ. But the time delays in the sampling pattern gave slightly higher correlations than did other values of ℓ. So a clever eavesdropper who could detect the frequency sequence for 25,000–50,000 iterations could learn the sampling pattern. But he could not learn the fuzzy rules from just the frequency sequence. He might learn rules that worked as well. But he would not learn the same rules. He also could not find the time offsets r in (8.12) from the frequency sequence.

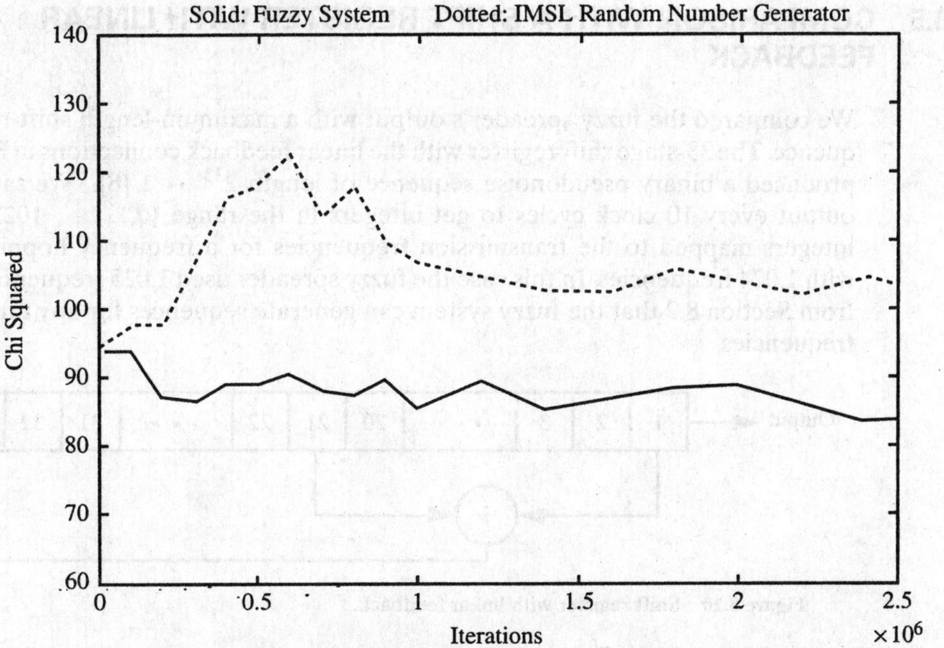

Figure 8.8 Mean chi-squared values for the fuzzy system and the IMSL random number generator for 100 frequencies. The lower the chi-squared value the more closely the output distribution matches the uniform distribution.

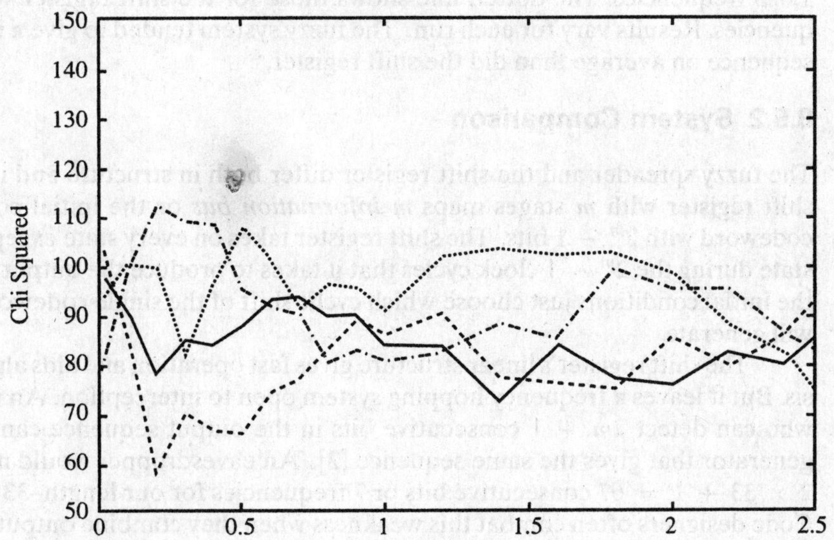

Figure 8.9 Chi-squared values for the fuzzy system for 100 frequencies. The solid-line run used all zero initial conditions. The other three used random initial conditions.

8.5 COMPARISON WITH A SHIFT REGISTER WITH LINEAR FEEDBACK

We compared the fuzzy spreader's output with a maximum-length shift-register sequence. The 33-stage shift register with the linear feedback connections in Figure 8.10 produced a binary pseudonoise sequence of length $2^{33} - 1$ [8]. We sampled the output every 10 clock cycles to get integers in the range $\{0, \ldots, 1023\}$. These integers mapped to the transmission frequencies for a frequency hopping system with 1,024 frequencies. In this case the fuzzy spreader used 1,025 frequencies. Recall from Section 8.2 that the fuzzy system can generate sequences for any multiple of 5 frequencies.

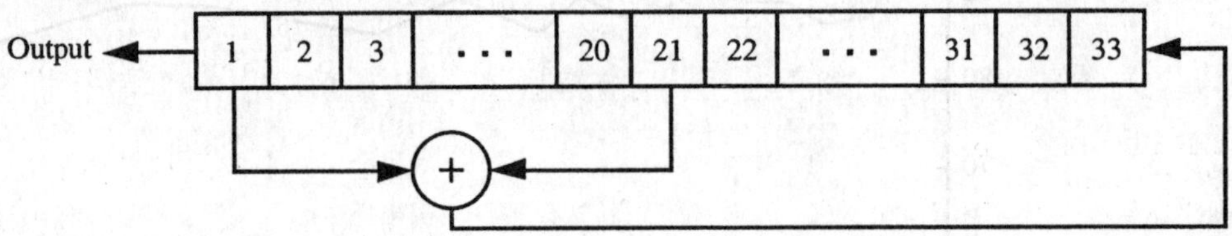

Figure 8.10 Shift register with linear feedback.

8.5.1 Simulation Results

We performed a chi-squared test of both systems. Again lower chi-squared values indicated more uniform sequences. Figure 8.11 shows the results averaged over 5 runs. The solid line marks the mean chi-squared values for the fuzzy system with 1,025 frequencies. The dotted line shows those for the shift register with 1,024 frequencies. Results vary for each run. The fuzzy system tended to give a more uniform sequence on average than did the shift register.

8.5.2 System Comparison

The fuzzy spreader and the shift register differ both in structure and in function. A shift register with m stages maps m *information bits* or the initial conditions to a codeword with $2^m - 1$ bits. The shift register takes on every state except the all-zero state during the $2^m - 1$ clock cycles that it takes to produce the output sequence. So the initial conditions just choose which cyclic shift of the single codeword the system will generate.

The shift register's linear structure gives fast operation and aids algebraic analysis. But it leaves a frequency hopping system open to interception. An eavesdropper who can detect $2m + 1$ consecutive bits in the output sequence can build a code generator that gives the same sequence [2]. An eavesdropper would need to detect $2 \times 33 + 1 = 67$ consecutive bits or 7 frequencies for our length–33 shift register. Code designers often combat this weakness when they combine output sequences in nonlinear ways from many stages of the shift register or from multiple shift registers [3].

The above fuzzy system is nonlinear. At each step it estimates the uniform distribution conditioned on prior sampled outputs. An eavesdropper cannot predict

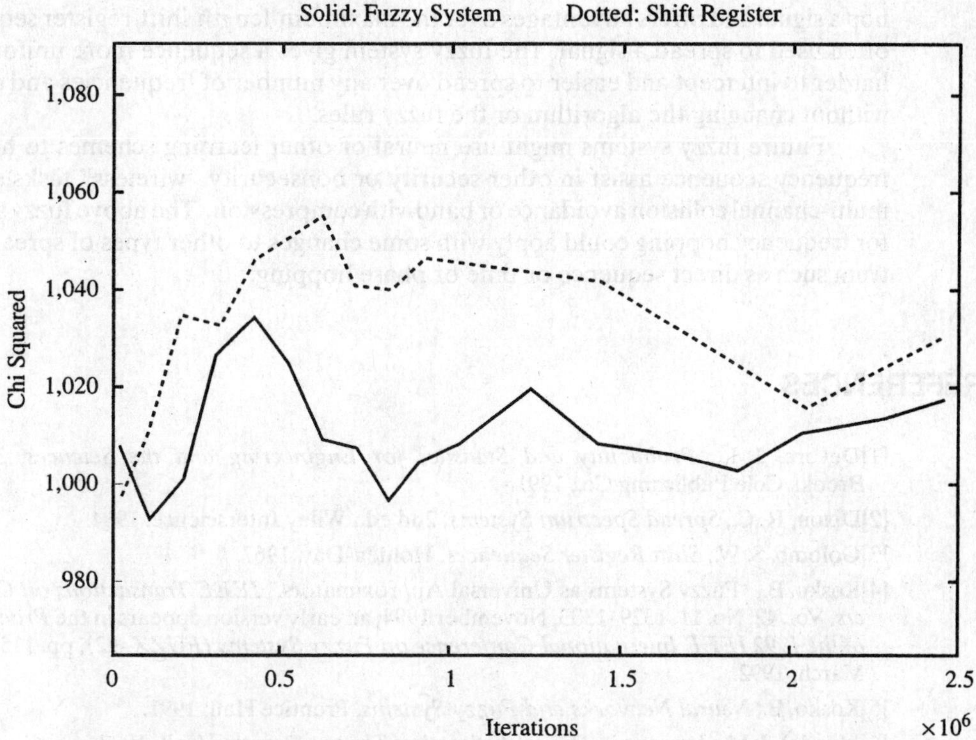

Figure 8.11 Mean chi-squared values for the fuzzy system for 1,025 frequencies and the shift register for 1,024 frequencies.

the output sequence if he does not know the fuzzy rules that drive the system. And the lack of algebraic structure and adaptive rule generation give a sequence of unknown length. The shift-register sequence repeats every $2^m - 1$ bits.

The shift register also limits the number of frequencies to a power of 2. The fuzzy system's rule structure works the same for any number of frequencies. We chose a multiple of 5 to allow an equal partition into 5 bins. We can relax this constraint as the frequencies grow in number.

8.6 CONCLUSION

The fuzzy spreader is deterministic and suggests that we do not need "randomness" to securely spread and despread a signal. The adaptive scheme captures the autocorrelation structure of a uniform random number generator in a set of fuzzy rules. The additive fuzzy system computes the density conditioned on the current input in the context of modular arithmetic. SAM centroid defuzzification computes the conditional mean. It gives in our case the next entry in the output sequence. The distribution of this sequence is at least as uniform as the distribution of the system that produced the training samples. The fuzzy system's output sequence can frequency hop a signal and offers advantages over the maximum-length shift-register sequences often used to spread a signal. The fuzzy system gives a sequence more uniform and harder to intercept and easier to spread over any number of frequencies and does so without changing the algorithm or the fuzzy rules.

Future fuzzy systems might use neural or other learning schemes to help the frequency sequence assist in other security or nonsecurity "wireless" tasks such as multi-channel collision avoidance or bandwith compression. The above fuzzy scheme for frequency hopping could apply with some changes to other types of spread spectrum such as direct sequence or time or phase hopping.

REFERENCES

[1]Devore, J. L., *Probability and Statistics for Engineering and the Sciences*, 3rd ed., Brooks/Cole Publishing Co., 1991.

[2]Dixon, R. C., *Spread Spectrum Systems*, 2nd ed., Wiley Interscience, 1984.

[3]Golomb, S. W., *Shift Register Sequences*, Holden-Day, 1967.

[4]Kosko, B., "Fuzzy Systems as Universal Approximators," *IEEE Transactions on Computers*, Vol. 42, No. 11, 1329–1333, November 1994; an early version appears in the *Proceedings of the 1992 IEEE International Conference on Fuzzy Systems (FUZZ-92)*, pp. 1153–1162, March, 1992.

[5]Kosko, B., *Neural Networks and Fuzzy Systems*, Prentice Hall, 1991.

[6]Mendel, J. M., *Lessons in Digital Estimation Theory*, Prentice Hall, 1987.

[7]Pacini, P. J., and Kosko, B., "Adaptive Fuzzy System for Target Tracking," *Intelligent Systems Engineering*, Vol. 1, No. 1, pp. 3–21, Autumn 1992.

[8]Proakis, J. G., *Digital Communications*, McGraw-Hill, 1983.

[9]Zadeh, L. A., "Probability Measures of Fuzzy Events," *Journal of Mathematical Analysis and Applications*, pp. 421–427, 1968.

APPENDIX 8.A: DERIVATION OF THE CENTROID AS THE CONDITIONAL MEAN FOR THE FREQUENCY-HOPPING SAM

The centroid of $B(x_k)$ is

$$c_k(x_k) = \frac{\int y\, B_k(x_k,\, y)\, dy}{\int B_k(x_k,\, y)\, dy} \tag{8.36}$$

$$\int y\, B_k(x_k,\, y)\, dy = \int y \left(\sum_j w_{jk} \right) p(y|x_k)\, dy \tag{8.37}$$

$$= \left(\sum_j w_{jk} \right) \int y \; p(y|x_k) \; dy \tag{8.38}$$

$$= \left(\sum_j w_{jk} \right) E[Y|X = x_k] \tag{8.39}$$

$$\int B_k(x_k, y) \; dy = \int \left(\sum_j w_{jk} \right) p(y|x_k) \; dy \tag{8.40}$$

$$= \left(\sum_j w_{jk} \right) \int p(y|x_k) \; dy \tag{8.41}$$

$$= \sum_j w_{jk} \tag{8.42}$$

Putting (8.39) and (8.42) into (8.36) gives

$$c_k(x_k) = E[Y|X = x_k] \tag{8.43}$$

in agreement with [4]. Chapter 2 further shows that (8.43) decomposes to a convex sum of local conditional means or then-part centroids c_j:

$$c_k(x_k) = F(x_k) = \sum_{j=1}^{m} p_j(x_k) c_j \tag{8.44}$$

for the convex coeffients

$$p_j(x_k) = \frac{a_j(x_k)}{\sum_{j=1}^{m} a_j(x_k)}. \tag{8.45}$$

HOMEWORK PROBLEMS

8.1. Mod-N arithmetic divides a number by N and keeps the remainder if N is a prime number. Use mod–2 arithmetic to compute the vector sums $A + B$, $B + C$, and $A + B + C$ for the bit vectors

$$A = (1\,0\,1\,1\,1\,0)$$

$$B = (0\,0\,1\,1\,0\,0)$$

$$C = (1\,1\,1\,0\,0\,0).$$

Compute the sums with mod–3 arithmetic.

8.2. A bivalent set A has binary set function I_A: $X \longrightarrow \{0, 1\}$. Let $p(y)$ be a continuous probability density function and define the expectation $E[Y]$ of random variable Y as

$$E[Y] = \int_{-\infty}^{\infty} y\, p(y)\, dy.$$

Show that the probability $P(A)$ of binary event A is just its expected set function:

$$P(A) = E[I_A].$$

Show how to extend this result to a fuzzy or multivalued set function a: $X \longrightarrow [0, 1]$ to find the probability $P(A)$ of the fuzzy event A. Then $P(A)$ is a compound measure of uncertainty. Set A measures the degree to which some event occurs. Measure value $P(A)$ measures whether the vague event A occurs.

8.3. Let D stand for some interval of the real line. Y is a random variable that takes values in D. Let constant d approximate Y on D. This gives the mean-squared error

$$\int_D (y - d)^2 p(y)\, dy.$$

Show that the value of d that minimizes the error on D is the centroid c on D. Then show directly that c equals the realization $E[Y|D]$ of the conditional expectation $E[Y|\bullet]$:

$$c = E[Y|D].$$

Data fusion combines measurements from more than one source. Fuzzy systems can do this in at least two ways. One first way is to use product rules that map all the sensors to a common output space. But adding sensors adds to the input dimension and so exponential rule explosion will soon take its toll. Experts may also find it hard to give rules that jointly describe disparate sensors.

The second way is to divide and conquer. Each sensor can have its own fuzzy system that models how the sensor maps data to features or to measurements or even to output decisions. Then a global additive system can combine these fuzzy systems as discussed in Section 2.8 of Chapter 2. The global system combines the rule firings of the local fuzzy systems and so combines before it defuzzifies.

The additive fuzzy system of Chapter 8 combines and weights two additive fuzzy systems to detect signals in noise. It uses the mode to defuzzify the combined output because of the discrete nature of the signals. The bit-vector signals code the letters and spaces in a pangram text. One fuzzy system maps the current signal at time t to then-part fuzzy sets. The other fuzzy system describes the left-right context of the current signal. It maps the prior signal at time $t - 1$ and the future signal at time $t + 1$ to other then-part fuzzy sets. These then-part fuzzy sets act as conditional probability densities.

The combined fuzzy system weights the two fuzzy systems based on the fuzzy cube structure discussed in Chapters 12–15. Each of the 32 5-bit signals lies at one of the 32 binary corners of a fuzzy 5-cube. Alpha-stable impulsive noise knocks a signal point off its corner and moves it somewhere inside the cube or outside the cube. The perturbed signal falls in clarity or grows in fuzziness as it moves away from the binary corner. Fuzziness holds inside the cube and grows as a cube point moves from one of the binary corners to the maximally fuzzy cube midpoint.

Clarity is the additive inverse of fuzziness. It holds both inside the cube and outside the cube in a type of inverted cube. The combined fuzzy system uses these clarity values to weight the two systems' rule firings. The result blends the geometry of fuzzy cubes with the function-approximation power of feedforward additive fuzzy systems.

Data fusion combines measurements from more than one source. Fuzzy systems can do this in at least two ways. The first way is to use enough rules that map all the sensors to a common output space. But adding sensors adds to the input dimension and so exponential rule explosion will soon take its toll. Experts may also find it hard to give rules that jointly describe disparate sensors.

The second way is to divide and conquer. Each sensor can have its own fuzzy system that models how the sensor maps data to features or to measurements or even to output decisions. Then a global additive system can combine these fuzzy systems as discussed in Section 2.8 of Chapter 2. The global system combines the rule firings of the lone fuzzy systems and so combines before it defuzzifies.

The additive fuzzy system of Chapter 9 combines and weights two additive fuzzy systems to detect signals in noise. It uses the mode to defuzzify the combined output because of the discrete nature of the signals. The bit-vector signals code the letters and spaces in a page of text. One fuzzy system maps the current signal at time t to then-part fuzzy sets. The other fuzzy system describes the left-right context of the current signal. It maps the prior signal at time $t - 1$ and the future signal at time $t + 1$ to other then-part fuzzy sets. These then-part fuzzy sets act as conditional probability densities.

The combined fuzzy system weights the two fuzzy systems based on the fuzzy cube structure discussed in Chapters 12–15. Each of the 36 6-bit signals lies at one of the 64 binary corners of a fuzzy 6-cube. Alpha-stable impulsive noise knocks a signal point off the corner and moves it somewhere inside the cube or outside the cube. The perturbed signal falls in clarity or grows in fuzziness as it moves away from the binary corner. Fuzziness holds inside the cube and grows as a set point moves from one of the binary corners to the maximally fuzzy cube midpoint.

Clarity is the additive inverse of fuzziness. It holds both inside the cube and outside the cube in a type of inverted cube. The combined fuzzy system uses these clarity values to weight the two systems' rule firings. The result blends the geometry of fuzzy cubes with the function-approximation power of feedforward additive fuzzy systems.

9

FUZZY SIGNAL DETECTION IN IMPULSIVE NOISE

Peter J. Pacini and Bart Kosko

PREVIEW

An additive fuzzy system can detect noisy signals in an uncoded digital communication system. Additive alpha-stable noise perturbs bipolar signals. Demodulated data vectors are fuzzy sets or points in or near a hypercube with edges of length 2. Sample data give the probabilities of character strings. The learned probability density functions define the then-part fuzzy sets of the if-then fuzzy rules.

At each iteration the fuzzy system computes two conditional densities based on the context and the fuzziness of the received vector. The detector output is the symbol that maximizes a weighted sum of these densities. The weight assigned to each conditional density at time t depends on the relative fuzziness of the received data vector x_t and its context vectors x_{t-1} and x_{t+1}. The fuzziness depends on where the signal falls in a fuzzy cube of high dimension.

Graphs show the probability of message symbol error for the fuzzy detector, the maximum a posteriori (MAP) detector, and the maximum likelihood (ML) detector. In these tests the fuzzy detector had less error probability than did the MAP or ML detectors that ignored the data in the context vectors x_{t-1} and x_{t+1}.

9.1 FUZZY SIGNAL DETECTION IN NOISE

Communication theory gives the standard optimum detector for an uncoded system with known source and channel models [8]. Figure 9.1 shows a block diagram of an uncoded digital communication system with additive channel noise. For a discrete memoryless source the optimum Bayesian detector finds the message symbol y_i with the maximum posterior probability $p(y_i|x_t)$. The term x_t is the noisy received vector at time t. The detector output at time t is $y_t = y_i$ such that

$$p(y_i|x_t) = \max_j p(y_j|x_t). \tag{9.1}$$

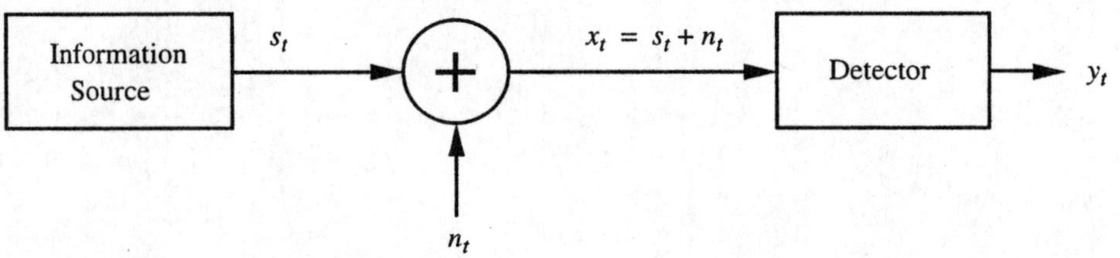

Figure 9.1 Uncoded digital communication system with additive channel noise.

In this chapter we describe a *fuzzy detector* for an uncoded system. The fuzzy detector computes *two* conditional densities $p(y|x_t)$ and $p(y|x_{t-1}, x_{t+1})$ that act as output fuzzy sets $B(x_t)$ and $B(x_{t-1}, x_{t+1})$ in the sense of the additive combiner

$$B(x) = \sum_{j=1}^{m} w_j B_j(x)$$

of two fuzzy systems as discussed in Chapter 2. Bayesian detection uses only the first density. The fuzzy detector weights the two densities by the clarity of their if-parts. So this method combines fuzzy and probabilistic concepts. The detector output is the message symbol that maximizes the weighted sum of the conditional densities.

The fuzzy detector has a lower probability of message symbol error than does the optimum Bayesian detector since it uses the contextual information in $p(y|x_{t-1}, x_{t+1})$. It reduces the error rate without source coding or channel coding. A fuzzy detector device could use simple hardware such as adders and multipliers and comparators. The fuzzy rules fire in parallel for fast execution and apply all the knowledge in the rules to each decision.

9.2 SOURCE MODEL

We assume a discrete signal source. All signal vectors have equal length. The number of distinct source symbols fixes the number of fuzzy rules and the minimum length for the uncoded message vectors. Table 9.1 shows the 36-symbol alphabet we used and the binary message vectors. Each source symbol maps to a 6-bit vector s. We use bipolar or antipodal signaling: Each bit $s[k]$ lies in $\{-1, +1\}$. Uppercase and lowercase letters map to the same messages. Source data are prose texts with no special characters and a few punctuation symbols. To symbols that occur in similar contexts we assign dissimilar message vectors. The binary vectors for any two vowels have Hamming distance 3.

Message Symbol	Message Vector	Message Symbol	Message Vector
SPACE	000000	r	001001
a	000111	s	010010
b	010101	t	100100
c	101101	u	010001
d	011110	v	110011
e	011100	w	111010
f	110100	x	010000
g	001110	y	001010
h	101000	z	111101
i	110010	'	101111
j	100101	(	000010
k	101100	)	000011
l	110001	,	011000
m	111000	-	111111
n	000001	.	000110
o	101001	:	110110
p	010111	;	100001
q	011011	?	111001

TABLE 9.1: Alphabet with 36 symbols.

We denote the alphabet of Table 9.1 as the space Y and the ith symbol in the alphabet as y_i. So y_1 = 'SPACE', y_2 = 'a', y_3 = 'b', etc.

9.3 SCALAR CHANNEL MODEL

The additive Gaussian noise model dominates the literature of communications and signal processing. It models some noise sources well and has simple math. But many communication links have additive "impulsive" noise or large interference levels as discussed in Chapter 6. Sources of impulsive noise include atmospheric noise in radio links, underwater acoustic noise, lightning, switching transients, and accidental hits in telephone lines [10]. Real data may suggest a noise process with a smooth symmetric bell-shaped density but with thicker tails than the Gaussian curve. This suggests that we should model impulsive noise as an *alpha-stable* process [9].

We assume that the communication channel adds uncorrelated symmetric alpha-stable (SαS) noise to each transmitted bit. The SαS densities have the form [9]

$$f_\alpha(\gamma,\ \delta;\ x) = \frac{1}{2\pi}\int_{-\infty}^{\infty} \exp(i\delta\omega - \gamma|\omega|^\alpha)e^{-i\omega x}\, d\omega \tag{9.2}$$

where α is the *characteristic exponent* ($0 < \alpha \leq 2$), δ is the *location parameter* ($-\infty < \delta < \infty$) or median, and γ is the *dispersion* ($\gamma > 0$). Chapter 6 gives a more general form for an alpha-stable density and its characteristic function. The dispersion measures the spread of the SαS density around its median. It is similar to the variance but more general in that the SαS densities for $\alpha < 2$ have infinite variance. This improves the accuracy of the noise model but reduces its tractability.

Closed form expressions for (9.2) exist for two values of α. These are the Cauchy ($\alpha = 1$) and Gaussian ($\alpha = 2$) densities:

$$f_1(\gamma,\ \delta;\ x) = \frac{1}{\pi\gamma}\frac{1}{1 + \left(\frac{x-\delta}{\gamma}\right)^2} \tag{9.3}$$

$$f_2(\gamma,\ \delta;\ x) = \frac{1}{2\sqrt{\pi\gamma}}\exp\left(-\frac{(x-\delta)^2}{4\gamma}\right). \tag{9.4}$$

The standard SαS density has $\gamma = 1$ and $\delta = 0$:

$$f_\alpha(\gamma = 1,\ \delta = 0;\ x) = \int_{-\infty}^{\infty} e^{-|\omega|^\alpha - i\omega x}\, d\omega. \tag{9.5}$$

Change of variables gives (9.2) in terms of the standard density:

$$f_\alpha(\gamma,\ \delta;\ x) = \gamma^{-\frac{1}{\alpha}} f_\alpha(1,\ 0;\ (x-\delta)\gamma^{-\frac{1}{\alpha}}). \tag{9.6}$$

Equation (9.4) shows that in the Gaussian case dispersion and variance differ by a factor of 2:

$$\gamma = \frac{\sigma^2}{2}. \tag{9.7}$$

So the standard ($\gamma = \frac{1}{2}$) Gaussian distribution has a variance of 1. The variance

$$V\left(\frac{1}{n}\sum_{i=1}^{n} x_i\right) = \frac{\sigma^2}{n}$$

falls to zero as the number of independent Gaussian variables x_i grows. But

$$V\left(\frac{1}{n}\sum_{i=1}^{n} x_i\right) = \infty$$

for all n for independent Cauchy or other SαS variables with $\alpha < 2$.

We do not use a lengthy series expansion for $f_{1.5}(x)$. Instead we approximate it as

$$\hat{f}_{1.5}(1,\ 0;\ x) = \frac{1}{2\sqrt{\pi} + \pi}\left(e^{-\frac{x^2}{4}} + \frac{1}{1+x^2}\right) \tag{9.8}$$

$$= 0.53\ f_2(1,\ 0;\ x) + 0.47\ f_1(1,\ 0;\ x). \tag{9.9}$$

Figure 9.2 plots $f_1(1,\ 0;\ x)$, $\hat{f}_{1.5}(1,\ 0;\ x)$, and $f_2(1,\ 0;\ x)$.

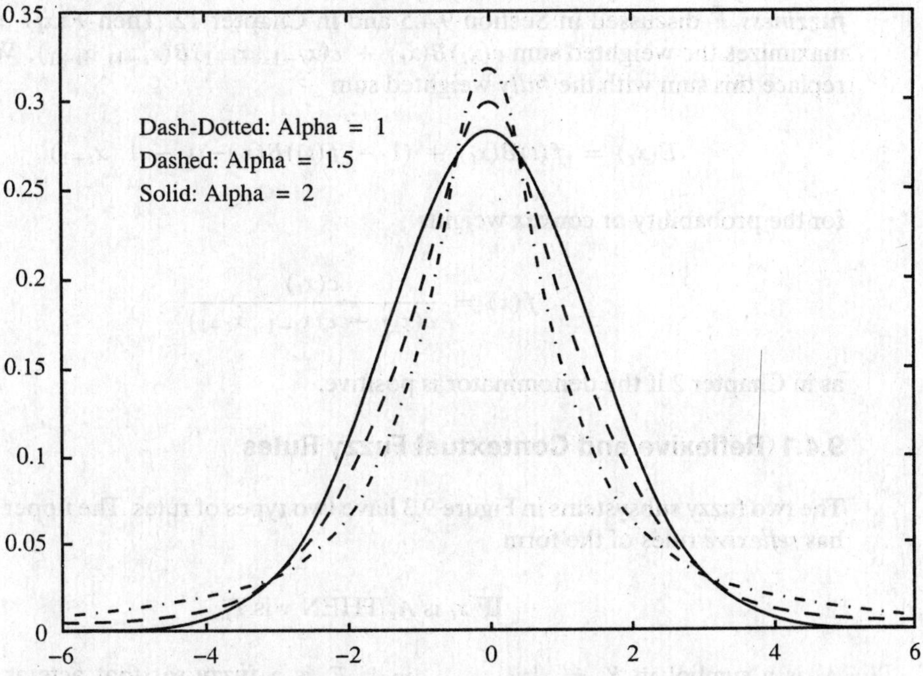

Figure 9.2 Standard SαS densities for $\alpha = 1$ (Cauchy), $\alpha = 1.5$, and $\alpha = 2$ (Gaussian). The dashed line ($\alpha = 1.5$) is an approximation. All standard SαS curves with $\alpha < 2$ have thicker tails than the Gaussian curve has and so impulsive events are more frequent.

9.4 FUZZY DETECTOR

A fuzzy system is a set of fuzzy if-then rules that maps inputs to outputs. In an additive fuzzy system [3] each input fires the antecedent or *if-part* of all rules to some degree and then weights and adds the consequent or *then-part* fuzzy sets. A *defuzzifier* converts the weighted sum into a number by picking its centroid or its mode. *Centroidal* additive fuzzy systems can uniformly approximate any continuous or bounded measurable function on a compact domain [4]. We use a *noncentroidal* additive system as discussed in Chapter 2.

Figure 9.3 shows a block diagram of the fuzzy detector. It uses two additive fuzzy subsystems in parallel. It weights and sums their output sets $B(x_t)$ and $B(x_{t-1}, x_{t+1})$ and then picks the mode y_t as the fuzzy system output $F(x_t)$. The upper subsystem takes as input the length–6 vector x_t formed from 6 consecutive samples:

$$x_t = (x_t^1, \ldots, x_t^6). \tag{9.10}$$

It gives as output $B(x_t)$ or the conditional density $p(y|x_t)$ [6] as discussed in Chapter 2. The lower subsystem takes as input the concatenated vector (x_{t-1}, x_{t+1}) and gives as output $B(x_{t-1}, x_{t+1})$ or $p(y|x_{t-1}, x_{t+1})$. The function c: $X \longrightarrow [0, 1]$ measures the *clarity* of a fuzzy set. The clarity is the additive inverse ($c = 1 - \bar{F}$) of the *fuzziness* $\bar{F}$ discussed in Section 9.4.5 and in Chapter 12. Then $F(x_t) = y_t$ iff y_t maximizes the weighted sum $c(x_t)B(x_t) + c(x_{t-1}, x_{t+1})B(x_{t-1}, x_{t+1})$. We can also replace this sum with the *fully* weighted sum

$$\mathcal{B}(x_t) = f(t)B(x_t) + (1 - f(t))B(x - t - 1, x_{t+1})$$

for the probability or convex weights

$$f(x) = \frac{c(x_t)}{c(x_t) + c(x_{t-1}, x_{t+1})}$$

as in Chapter 2 if the denominator is positive.

9.4.1 Reflexive and Contextual Fuzzy Rules

The two fuzzy subsystems in Figure 9.3 have two types of rules. The upper subsystem has *reflexive* rules of the form

$$\text{IF } x_t \text{ is } A_i \text{ THEN } y \text{ is } F_i.$$

A_i is a symbol in $Y = \{y_1, \ldots, y_{36}\}$. F_i is a fuzzy set that acts as a discrete probability density function [4], [6] with domain Y. These rules are "reflexive" because each F_i has 35 zero values and a single unit spike at the index corresponding to A_i. So each of the 36 reflexive rule then-parts F_i is a degenerate conditional density $p(y|A_i)$.

The lower subsystem has *contextual* rules of the form

$$\text{IF } x_{t-1} \text{ is } B_j \text{ AND } x_{t+1} \text{ is } C_j \text{ THEN } y \text{ is } G_j$$

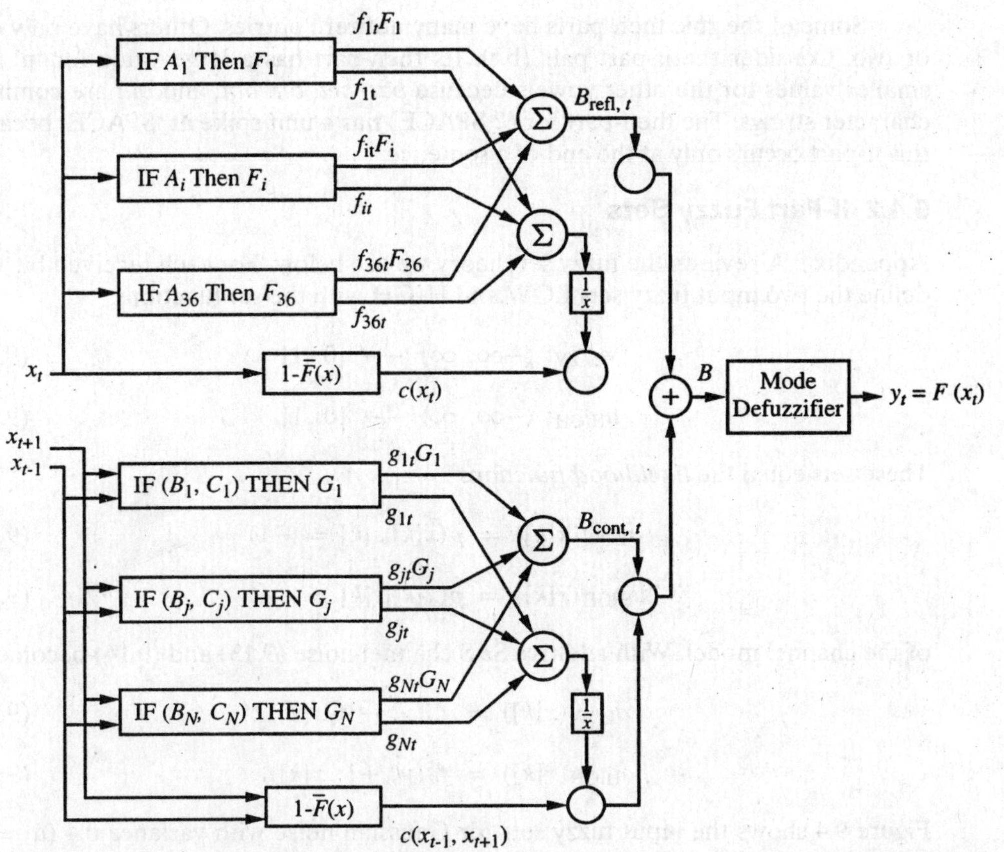

Figure 9.3 Block diagram of the fuzzy detector. The upper additive fuzzy subsystem contains the reflexive rules. The lower subsystem contains the contextual rules. Here $F(x_t) = y_t$ iff y_t maximizes the weighted sum $c(x_t)B(x_t) + c(x_{t-1}, x_{t+1})B(x_{t-1}, x_{t+1})$. The clarity $c(x_t)$ of input x_t is the additive inverse of its fuzziness $\bar{F}(x_t)$: $c(x_t) = 1 - \bar{F}(x_t)$. This gives a combined fuzzy system that combines before it defuzzifies. It combines weighted throughputs rather than weighted outputs.

or

$$\text{IF } (x_{t-1},\ x_{t+1}) \text{ is } (B_j,\ C_j) \text{ THEN } y \text{ is } G_j.$$

$(B_j,\ C_j)$ is a symbol pair in Y^2. G_j is a discrete probability density function with domain Y. So each contextual rule then-part G_j is a conditional density $p(y|(B_j,\ C_j))$. There could be up to $36^2 = 1{,}296$ contextual rules. Many pairs $(B_j,\ C_j)$ do not occur in practice. For example: (q,z) and (q,j) need not appear as rule if-parts.

Contextual rule then-parts depend on the type of source data. Language, writing style, and subject matter all affect the conditional densities $p(y|(B_j,\ C_j))$. These three elements define a local map Ψ: $Y^2 \longrightarrow \Phi$ where $F_i \in \Phi$ and $G_j \in \Phi$. Sample data estimate Ψ. Associations learned from those samples give the contextual rule then-parts G_j. So the lower fuzzy subsystem approximates the function Ψ.

Some of the rule then-parts have many nonzero entries. Others have only one or two. Consider the if-part pair (b,t). Its then-part has a large value for 'u' and smaller values for the other vowels because *bat, bet, bit, bot,* and *but* are common character strings. The then-part for (?,SPACE) has a unit spike at 'SPACE' because this if-part occurs only at the end of a sentence.

9.4.2 If-Part Fuzzy Sets

Appendix 9.A reviews the fuzzy set theory we use below. For each received bit $x[k]$ define the two input fuzzy sets LOW and HIGH with the set functions

$$a_{\text{LOW}}: (-\infty, \infty) \longrightarrow [0, 1] \tag{9.11}$$

$$a_{\text{HIGH}}: (-\infty, \infty) \longrightarrow [0, 1]. \tag{9.12}$$

These sets equal the *likelihood functions*

$$a_{\text{LOW}}(x[k]) = p(x[k]|s[k] = -1) \tag{9.13}$$

$$a_{\text{HIGH}}(x[k]) = p(x[k]|s[k] = +1) \tag{9.14}$$

of the channel model. With additive SαS channel noise (9.13) and (9.14) become

$$a_{\text{LOW}}(x[k]) = f_\alpha(\gamma, -1; x[k]) \tag{9.15}$$

$$a_{\text{HIGH}}(x[k]) = f_\alpha(\gamma, +1; x[k]). \tag{9.16}$$

Figure 9.4 shows the input fuzzy sets for Gaussian noise with variance 0.4 ($\alpha = 2$ and $\gamma = 0.2$). This is the standard AWGN channel model.

The if-part fuzzy sets map received bit $x[k]$ to two *fit values* $a_{\text{LOW}}(x[k])$ and $a_{\text{HIGH}}(x[k])$. Input $x[k]$ belongs to the fuzzy set LOW to degree $a_{\text{LOW}}(x[k])$ and to the fuzzy set HIGH to degree $a_{\text{HIGH}}(x[k])$.

The joint symbol fit value $a_{y_i}(x)$ equals the product of the component fit values:

$$a_{y_i}(x) = \prod_{k=1}^{6} a_{y_i[k]}(x[k]), \qquad y_i[k] \in \{\text{LOW, HIGH}\}. \tag{9.17}$$

For the reflexive rules (9.17) becomes

$$a_i(x_t) = \prod_{k=1}^{6} a_i[k](x_t[k]), \qquad A_i[k] \in \{\text{LOW, HIGH}\}. \tag{9.18}$$

Contextual rule if-parts have the two symbols B_j and C_j. We treat (B_j, C_j) and (x_{t-1}, x_{t+1}) as concatenated length–12 vectors. So the contextual rules have fit values

$$a_{(B_j,C_j)}(x_{t-1}, x_{t+1}) = \left[\prod_{k=1}^{6} a_{B_j[k]}(x_{t-1}[k])\right]\left[\prod_{k=1}^{6} a_{C_j[k]}(x_{t+1}[k])\right] \tag{9.19}$$

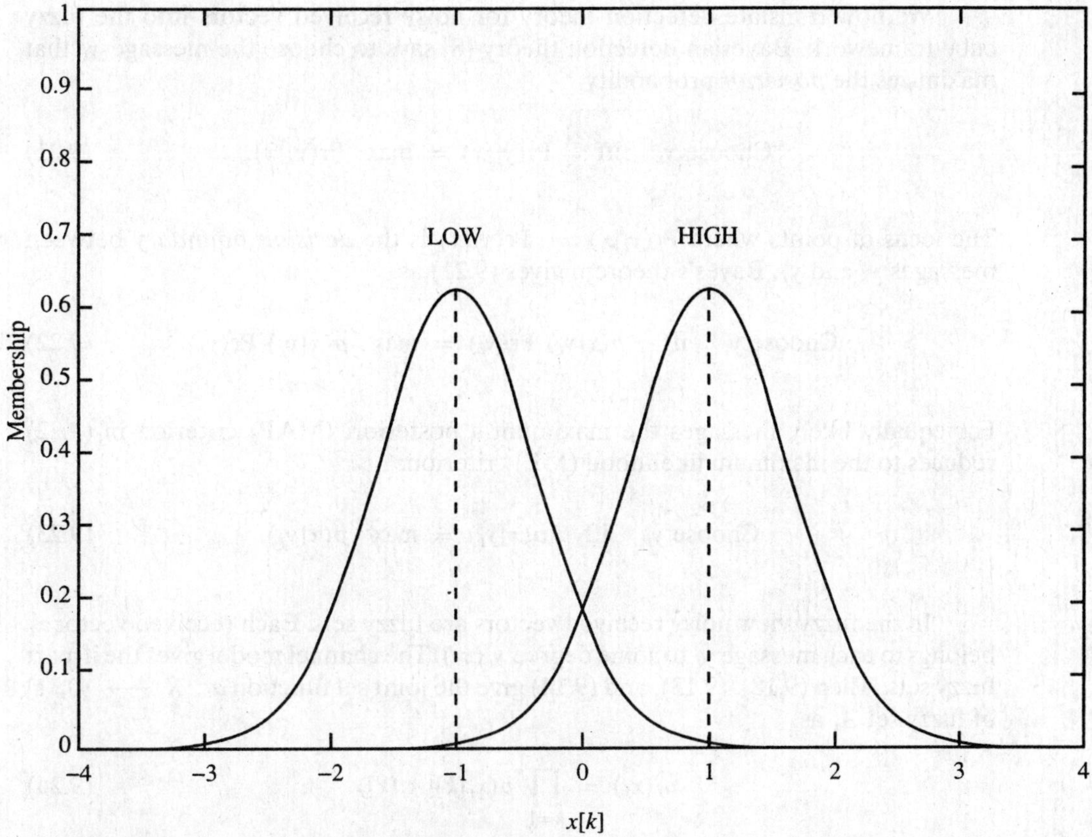

Figure 9.4 If-part fuzzy sets for Gaussian noise with variance 0.4. The if-part sets define the channel likelihood functions: $a_{\text{LOW}}(x[k]) = p(x[k]|s[k] = -1)$ and $a_{\text{HIGH}}(x[k]) = p(x[k]|s[k] = +1)$.

$$= a_{B_j}(x_{t-1})\ a_{C_j}(x_{t+1}). \tag{9.20}$$

The symbol fit values in (9.18) determine the *firing levels* f_{it} for the reflexive rules. The symbol fit values in (9.20) determine the firing levels g_{jt} for the contextual rules. Section 9.4.4 explains the firing or inferencing process.

9.4.3 Detection in the Fuzzy Cube

Appendix 9.A describes fuzzy sets as points in the unit n-cube $[0,\ 1]^n$. The vertices are bivalent or nonfuzzy sets. All other points are multivalent or fuzzy sets. All theorems derived from this sets-as-points definition hold for the *fuzzy Hamming n-cube* $[-1,\ 1]^n$. Fuzzy sets still fill in the lattice $\{-1,\ 1\}^n$. Here vertices are nonfuzzy *bipolar* vectors. The maximally fuzzy set is the origin. In the reflexive subsystem in Figure 9.3 the message vectors A_i are vertices of the fuzzy Hamming 6-cube. In the contextual subsystem the concatenated message vectors $(B_j,\ C_j)$ are vertices of the fuzzy Hamming 12-cube.

We now translate detection theory for noisy received vectors into the fuzzy cube framework. Bayesian detection theory [8] says to choose the message y_i that maximizes the *posterior* probability:

$$\text{Choose } y_i \quad \text{iff} \quad \Pr(y_i|x) = \max_j \Pr(y_j|x). \tag{9.21}$$

The locus of points where $\Pr(y_i|x) = \Pr(y_j|x)$ is the *decision boundary* between messages y_i and y_j. Bayes's theorem gives (9.21) as

$$\text{Choose } y_i \quad \text{iff} \quad p(x|y_i)\Pr(y_i) = \max_j p(x|y_j)\Pr(y_j). \tag{9.22}$$

For equally likely messages the maximum a posteriori (MAP) criterion in (9.22) reduces to the maximum likelihood (ML) criterion:

$$\text{Choose } y_i \quad \text{iff} \quad p(x|y_i) = \max_j p(x|y_j). \tag{9.23}$$

In the fuzzy view noisy received vectors are fuzzy sets. Each received vector x_t belongs to each message y_i to some degree $y_i(x_t)$. The channel model gives the if-part fuzzy sets. Then (9.18), (9.13), and (9.14) give the joint set function $a_i: X \longrightarrow [0, 1]$ of fuzzy set A_i as

$$a_i(x_t) = \prod_{k=1}^{6} p(x_t[k]|A_i[k]). \tag{9.24}$$

The noise is uncorrelated and independent of the signal. The bits of the noiseless message vector s_t are largely uncorrelated. This gives the approximations

$$a_i(x_t) \approx p(x_t|A_i) \tag{9.25}$$

and

$$a_{(B_j,C_j)}(x_{t-1}, x_{t+1}) \approx p(x_{t-1}|B_j)\, p(x_{t+1}|C_j) \tag{9.26}$$

$$\approx p(x_{t-1}, x_{t+1}|(B_j, C_j)). \tag{9.27}$$

Putting (9.25) into (9.22) and (9.23) gives the fuzzy versions of the MAP and ML decision rules:

$$\text{Fuzzy MAP: Choose } y_i \quad \text{iff} \quad y_i(x)\Pr(y_i) = \max_j y_j(x)\Pr(y_j) \tag{9.28}$$

$$\text{Fuzzy ML: Choose } y_i \quad \text{iff} \quad y_i(x) = \max_j y_j(x). \tag{9.29}$$

9.4.4 Computing the Conditional Densities

An input x *fires* or *activates* the fuzzy rule "If x is A_i then y is B_i" to degree $a_i(x)$. The if-part sets can act as joint probability densities $a_i(x) = p(x, A_i)$ [7] that help compute the posterior density $p(y|x)$. Here the if-part sets are conditional densities. So we scale $a_i(x)$ by the prior probability $\Pr(A_i)$. The algorithm that learns the contextual rule then-parts G_j also finds $\Pr(y_i)$ and $\Pr(y_i, y_j)$ for all y_i and y_j in Y.

At time t the ith reflexive rule fires to degree

$$f_{it} = a_i(x_t) \; \Pr(A_i) \tag{9.30}$$

$$= p(x_t|A_i) \; \Pr(A_i) \quad \text{from (9.25)} \tag{9.31}$$

$$= p(x_t, A_i). \tag{9.32}$$

The jth contextual rule fires to degree

$$g_{jt} = a_{(B_j, C_j)}(x_{t-1}, x_{t+1}) \; \Pr(B_j, C_j) \tag{9.33}$$

$$= p(x_{t-1}, x_{t+1}|(B_j, C_j)) \; \Pr(B_j, C_j) \quad \text{from (9.27)} \tag{9.34}$$

$$= p(x_{t-1}, x_{t+1}, B_j, C_j). \tag{9.35}$$

The firing levels f_{it} and g_{jt} scale the then-part fuzzy sets F_i and G_j.

The theorem of total probability gives the marginal densities as

$$p(x_t) = \sum_{A_i \in Y} p(x_t, A_i) = \sum_i f_{it} \tag{9.36}$$

$$p(x_{t-1}, x_{t+1}) = \sum_{(B_j, C_j) \in Y^2} p(x_{t-1}, x_{t+1}, B_j, C_j) = \sum_j g_{jt}. \tag{9.37}$$

We can also write the posterior probabilities $\Pr(A_i|x_t)$ and $\Pr((B_j, C_j)|x_{t-1}, x_{t+1})$ in terms of the firing levels:

$$\Pr(A_i|x_t) = \frac{p(x_t, A_i)}{p(x_t)} \tag{9.38}$$

$$= \frac{f_{it}}{\sum_n f_{nt}} \tag{9.39}$$

$$\Pr((B_j, C_j)|x_{t-1}, x_{t+1}) = \frac{p(x_{t-1}, x_{t+1}, B_j, C_j)}{p(x_{t-1}, x_{t+1})} \tag{9.40}$$

$$= \frac{g_{jt}}{\sum_n g_{nt}}. \tag{9.41}$$

The then-part fuzzy sets do not depend on the inputs. So we can write

$$F_i(y) = p(y|A_i,\ x_t) \tag{9.42}$$

$$G_j(y) = p(y|(B_j,\ C_j),\ x_{t-1},\ x_{t+1}). \tag{9.43}$$

The nodes labeled $B_{\text{refl},t}$ and $B_{\text{cont},t}$ in Figure 9.3 stand for weighted sums of these rule then-parts:

$$B_{\text{refl},t} = B(x_t) = \sum_i f_{it} F_i \tag{9.44}$$

$$= \sum_i p(x_t,\ A_i)\ p(y|A_i,\ x_t) \quad \text{from (9.32) and (9.42)} \tag{9.45}$$

$$= \sum_i p(y,\ A_i,\ x_t) \tag{9.46}$$

$$= p(y,\ x_t) \tag{9.47}$$

$$= p(x_t)\ p(y|x_t) \tag{9.48}$$

$$= \left(\sum_i f_{it}\right) p(y|x_t) \tag{9.49}$$

from (9.36) and

$$B_{\text{cont},t} = B(x_{t-1},\ x_{t+1}) = \sum_j g_{jt} G_j \tag{9.50}$$

$$= \sum_j p(x_{t-1},\ x_{t+1},\ B_j,\ C_j)\ p(y|(B_j,\ C_j),\ x_{t-1},\ x_{t+1}) \tag{9.51}$$

from (9.35) and (9.43) and

$$= \sum_j p(y,\ (B_j,\ C_j),\ x_{t-1},\ x_{t+1}) \tag{9.52}$$

$$= p(y,\ x_{t-1},\ x_{t+1}) \tag{9.53}$$

$$= p(x_{t-1},\ x_{t+1})\ p(y|x_{t-1},\ x_{t+1}) \tag{9.54}$$

$$= \left(\sum_j g_{jt}\right) p(y|x_{t-1},\ x_{t+1}) \tag{9.55}$$

from (9.37).

We normalize the weighted sums in (9.49) and (9.55) by $\sum_i f_{it}$ and $\sum_j g_{jt}$ to give the desired conditional densities $p(y|x_t)$ and $p(y|x_{t-1}, x_{t+1})$.

9.4.5 Clarity of a Fuzzy Set

The weights that scale the conditional densities $p(y|x_t)$ and $p(y|x_{t-1}, x_{t+1})$ should reflect the relative fuzziness of x_t and the concatenated vector (x_{t-1}, x_{t+1}). If x_t is much closer to a message vector than are x_{t-1} and x_{t+1} then $p(y|x_t)$ should exceed $p(y|x_{t-1}, x_{t+1})$ and vice versa.

Appendix 9.B gives the fuzziness measure as the ratio of the counted fuzzy overlap $A \cap A^c$ and underlap $A \cup A^c$ in a fuzzy cube. We modify this to the ratio of fuzzy Hamming distances

$$\bar{F}(x) = \frac{\min_i \ \ell^1(x, y_i)}{\max_i \ \ell^1(x, y_i)}. \tag{9.56}$$

$\bar{F}$ in (9.56) differs from the fuzziness operator F in (9.76) in that the nearest and farthest cube vertices may not be message vectors.

We define the *clarity* $c(x)$ of fuzzy set x as the negation or opposite of its measure of fuzziness $\bar{F}(x)$:

$$c(x) = 1 - \bar{F}(x). \tag{9.57}$$

Message vector y_i and nonfuzzy set $x = y_j$ for some j give

$$\min_i \ \ell^1(x, y_i) = 0, \tag{9.58}$$

$$c(x) = 1. \tag{9.59}$$

When x is maximally fuzzy it lies at the center of the fuzzy cube and is equidistant from all message vectors. Then

$$\min_i \ \ell^1(x, y_i) = \max_i \ \ell^1(x, y_i), \tag{9.60}$$

$$c(x) = 0. \tag{9.61}$$

The clarity $c(x)$ falls to zero as x moves far outside the cube and thus as all the distances $\ell^1(x, y_i)$ grow large.

Figure 9.5 shows lines of equal clarity for a 2-dimensional case with message vectors at all four vertices. Fuzzy sets near any of the corners have high clarity or low fuzziness. Fuzzy sets near the midpoint or far outside the cube have low clarity or high fuzziness. Figure 9.6 plots $c(x)$ versus the ℓ^1 or fuzzy Hamming distance from the midpoint for fuzzy sets in the cube.

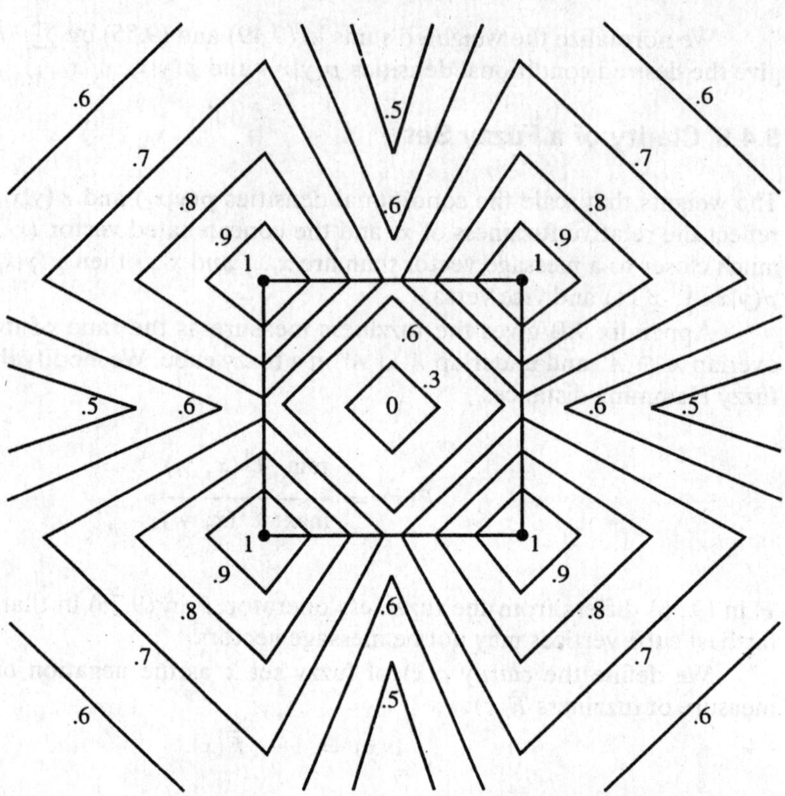

Figure 9.5 Lines of equal clarity for a 2-dimensional case with message vectors at all four vertices. Fuzzy sets near any of the corners have high clarity. Fuzzy sets near the midpoint or far outside the cube have low clarity.

The curve in Figure 9.6 may not be optimal. It depends on how we define clarity in (9.57). Functions such as $(1 - \bar{F}(x))^2$ and $(1 - \bar{F}(x))^3$ also satisfy the conditions (9.59) and (9.61) at the vertices and midpoint of the cube but give different curves. Trial and error showed that the shape of the curve had little effect on the error rate. The function $c(x) = 1 - \bar{F}(x)$ is the simplest choice and gives better results than the other functions we tried.

We define the joint clarity for two fuzzy sets x_1 and x_2 as the minimum of their clarities:

$$c(x_1, x_2) = \min(c(x_1), c(x_2)). \tag{9.62}$$

(A maximum or OR-operator would be less stringent. But the context of x_t depends on x_{t-1} *and* x_{t+1}.) A more stringent clarity measure might replace the min in (9.62) with product or some other t-norm operator as discussed in the homework problems of Chapter 1. This gives for the contextual subsystem

$$c(x_{t-1}, x_{t+1}) = \min(c(x_{t-1}), c(x_{t+1})). \tag{9.63}$$

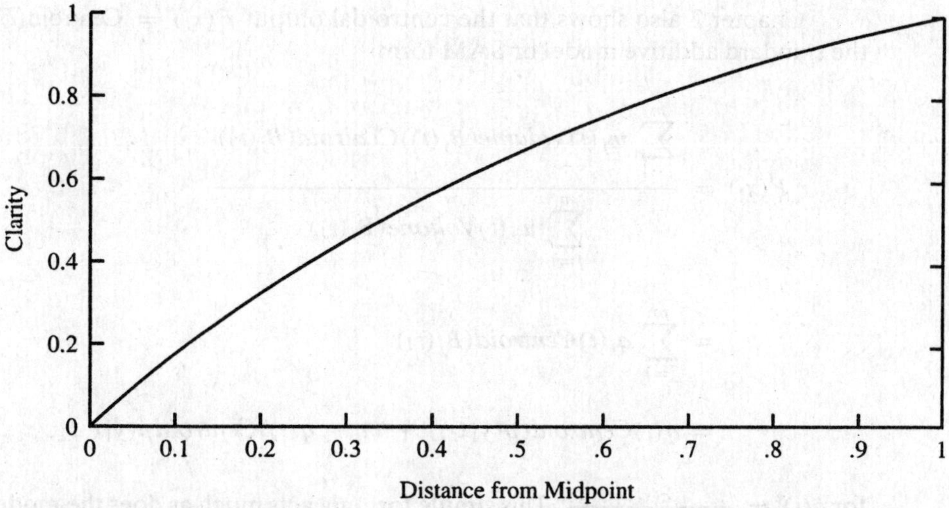

Figure 9.6 Clarity versus the ℓ^1 distance from the center of the cube for fuzzy sets in the cube.

9.4.6 Combined Additive Output Fuzzy Set: Combining Throughputs

The combined output fuzzy set B_t serves as input to the defuzzifier. Figure 9.3 shows that

$$\mathcal{B}(x_{t-1},\ x_t,\ x_{t+1}) = \mathcal{B}(x_t)$$

$$= \frac{c(x_t)}{\sum_i f_{it}} B_{\text{refl},t} + \frac{c(x_{t-1},\ x_{t+1})}{\sum_j g_{jt}} B_{\text{cont},t} \tag{9.64}$$

$$= c(x_t)\ p(y|x_t) + c(x_{t-1},\ x_{t+1})\ p(y|x_{t-1},\ x_{t+1}) \tag{9.65}$$

from (9.49) and (9.55). So $B(x_t)$ is a weighted sum of the conditional densities and gives more weight to the less fuzzy input.

The combined set $\mathcal{B}(x_t) = c(x_t)B(x_t) + c(x_{t-1},\ x_{t+1})B(x_{t-1},\ x_{t+1})$ is a special case of the additive combination scheme

$$B(x) = \sum_{j=1}^{m} w_j B_j(x)$$

for combining m fuzzy systems $F_1, \ldots, F_m$ as discussed in Chapter 2. We did not use the convex weights $w(t)$ and $1 - w(t)$ with $w(t) = \frac{c(x_t)}{c(x_t)+c(x_{t-1},x_{t+1})}$ since $c(x_t) = 0 = c(x_{t-1},\ x_{t+1})$ can hold in the maximally fuzzy case when x_t and either x_{t-1} or x_{t+1} lie at the midpoint of the fuzzy cube.

Chapter 2 also shows that the centroidal output $F(x_t) = \text{Centroid}(\mathcal{B}(x_t))$ has the standard additive model or SAM form

$$F(x_t) = \frac{\sum_{j=1}^{m} w_j(t) Volume(B_j(t)) Centroid(B_j(t))}{\sum_{j=1}^{m} w_j(t) Volume(B_j(t))} \tag{9.66}$$

$$= \sum_{j=1}^{m} q_j(t) Centroid(B_j(t)) \tag{9.67}$$

$$= q(t) Centroid(p(y|x_t)) + (1 - q(t)) Centroid\, p(y|x_{t-1}, x_{t+1}) \tag{9.68}$$

for $q(t) = \frac{c(x_t)}{c(x_t)+c(x_{t-1}, x_{t+1})}$. This simple formula acts much as does the mode of $\mathcal{B}(x_t)$ but is not defined when $c(x_t) = 0 = c(x_{t-1}, x_{t+1})$. The centroid of $\mathcal{B}(x_t)$ can also give nonsense outputs on the discrete symbol space Y. For this reason we let the mode of $\mathcal{B}(x_t)$ define the overall additive fuzzy output $F(x_t)$.

9.4.7 Mode Defuzzification and Detection

Mode defuzzification picks the message symbol $y_i \in Y$ with the largest fit value in $\mathcal{B}(x_t)$. Note from (9.49) that mode defuzzification of $B_{\text{refl},t}$ gives the MAP decision

$$y_i \text{ such that } \Pr(y_i|x_t) = \max_j \Pr(y_j|x_t). \tag{9.69}$$

So the reflexive subsystem and mode defuzzifier give a memoryless MAP detector.

From the reflexive subsystem we can compute

$$\sum_i a_i(x_t)\, F_i = \begin{bmatrix} a_1(x_t) \\ \vdots \\ a_{36}(x_t) \end{bmatrix} \tag{9.70}$$

$$\tag{9.71}$$

$$= \begin{bmatrix} p(x_t|A_1) \\ \vdots \\ p(x_t|A_{36}) \end{bmatrix} \tag{9.72}$$

from (9.25). Mode defuzzification then gives the maximum likelihood (ML) decision

$$y_i \text{ such that } p(x_t|y_i) = \max_j p(x_t|y_j). \tag{9.73}$$

9.5 SIMULATION RESULTS

Sample data gave the contextual rule then-parts G_j and the prior message probabilities $\Pr(y_i)$ and $\Pr(y_i, y_j)$. We sampled Chapter 1 of James Gleick's *Chaos* [1] after we edited it to fit the small alphabet in Table 9.1. So the number "10" that has two characters not in the alphabet became the word "ten." The edited text had 39,655 characters. This sample gave 768 contextual rules ($N = 768$ in Figure 9.3) and the prior message probabilities in Table 9.2. We selected at random 7 paragraphs from the sample text to test the fuzzy detector. The test file had 5,026 characters.

Message Symbol	Prior Probability	Message Symbol	Prior Probability
SPACE	.155995	r	.047812
a	.065288	s	.051948
b	.011121	t	.081856
c	.027437	u	.025091
d	.030816	v	.007792
e	.104627	w	.017148
f	.017930	x	.001816
g	.013643	y	.014021
h	.041609	z	.002118
i	.056033	'	.004312
j	.000656	(	.000000
k	.004665	)	.000000
l	.035456	,	.009305
m	.021536	-	.003883
n	.055731	.	.009356
o	.061430	:	.000454
p	.017526	;	.000177
q	.001337	?	.000076

TABLE 9.2: Prior probabilities for the 36 message symbols.

We varied the noise parameters α and γ to test the fuzzy detector. The IMSL (International Mathematical and Statistical Library) random number generator gave the alpha-stable noise samples. Figure 9.8 shows the probability of message symbol error versus the bit signal-to-noise ratio for the Gaussian case ($\alpha = 2$). Each message bit is either +1 or −1 since we used bipolar signaling. So the bit signal-to-noise ratio is

$$\text{SNR} = \frac{1}{\sigma^2} = \frac{1}{2\gamma} \tag{9.74}$$

from (9.7).

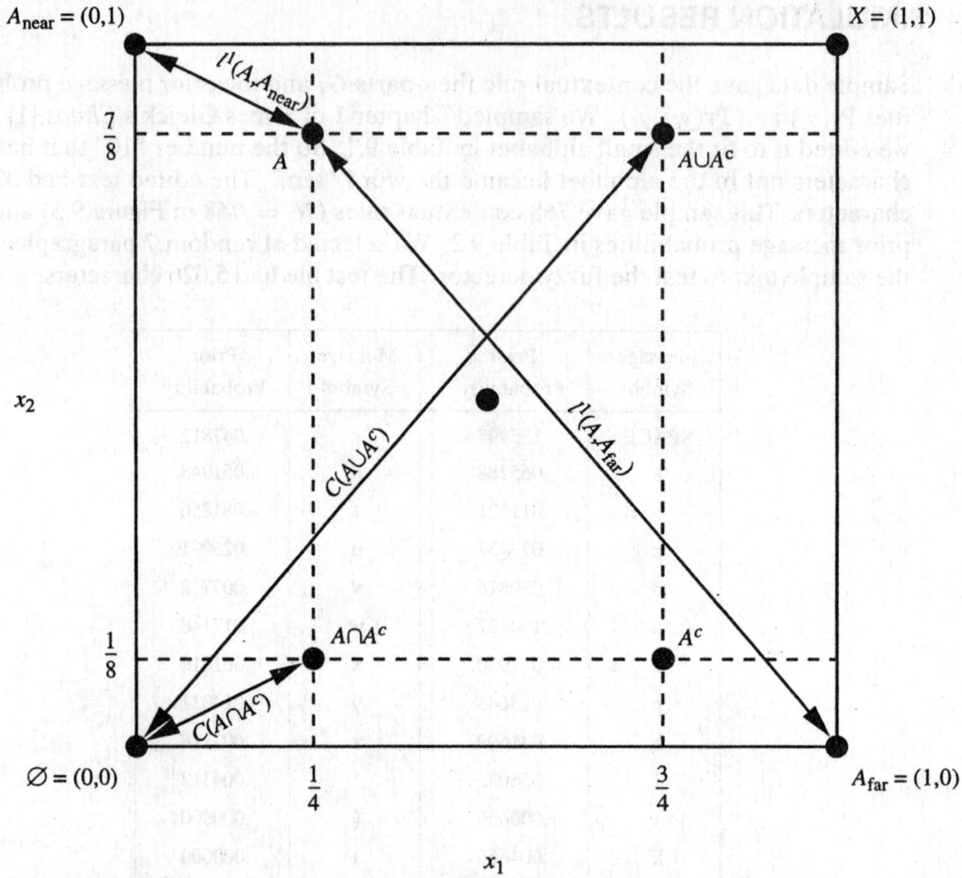

Figure 9.7 Fuzzy sets as points in the fuzzy 2-cube. The fuzziness $\bar{F}(A)$ of A grows as A moves toward the midpoint. Here $\bar{F}(A) = \frac{C(A \cap A^c)}{C(A \cup A^c)} = \frac{\frac{1}{4}+\frac{1}{8}}{\frac{3}{4}+\frac{7}{8}} = \frac{3}{13}$. The 2^n dual fuzzy sets A, A^c, $A \cap A^c$, and $A \cup A^c$ lie at the vertices of a hyperrectangle centered at the cube midpoint. The cube itself is the largest such hyperrectangle. The 2^n-fold symmetry of the fuzzy cube I^n shows that the fuzziness ratio $\frac{\ell^1(A, A_{near})}{\ell^1(A, A_{far})}$ equals the counting ratio $\frac{\ell^1(\varnothing, A \cap A^c)}{\ell^1(\varnothing, A \cup A^c)} = \frac{C(A \cap A^c)}{C(A \cup A^c)}$.

The noise process has infinite variance and hence infinite power when $\alpha < 2$. So we cannot compute a signal-to-noise ratio when $\alpha < 2$. But we can plot dispersion γ on the x-axis. For comparison we include the Gaussian case. Figures 9.9–9.11 show the probability of message symbol error versus the dispersion for $\alpha = 2$, $\alpha = 1.5$, and $\alpha = 1$. Each data point stands for the average of five runs with the same parameters. The dash-dotted and dashed lines show the maximum likelihood (ML) and maximum a posteriori (MAP) decision rules in (9.29) and (9.28). The solid lines show the fuzzy detector. The ML detector had the highest probability of error because it ignored the prior message probabilities. The ML and MAP detectors are the same for equally likely messages. Table 9.2 shows that the prior probabilities differed greatly in this study. The MAP detector used the known prior probabilities to shift the decision boundaries away from the most likely message vectors. This

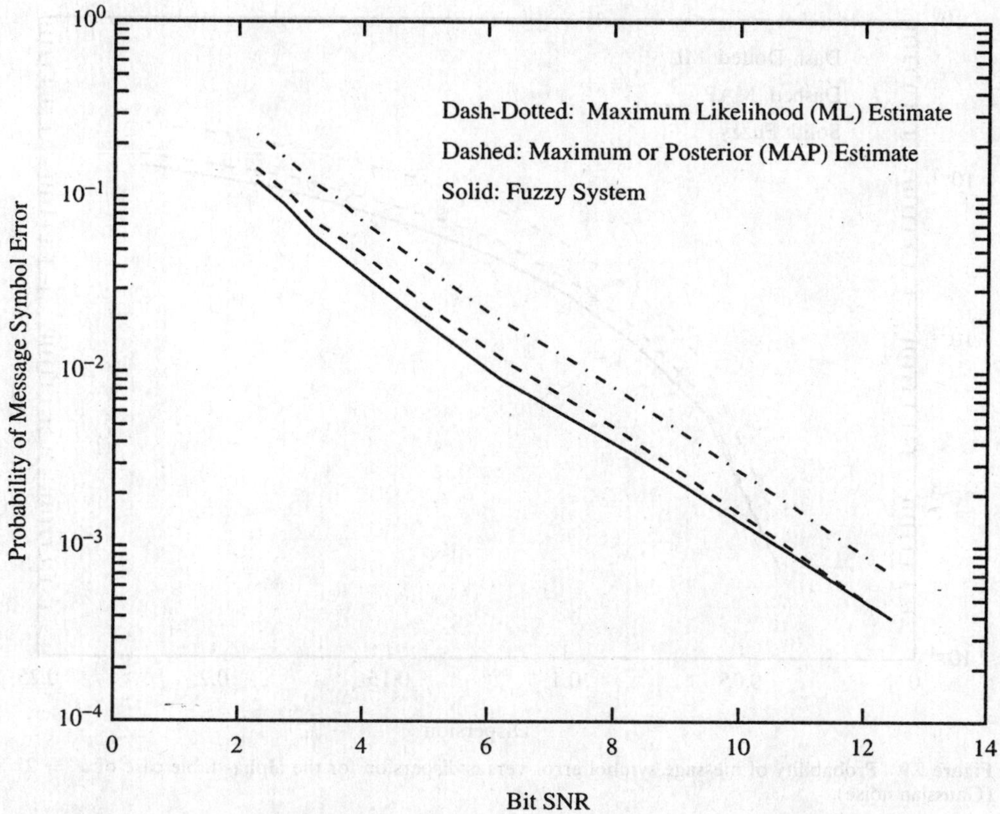

Figure 9.8 Probability of message symbol error versus bit signal-to-noise ratio for the Gaussian case ($\alpha = 2$) of alpha-stable noise. The fuzzy system used context data that the nonfuzzy systems did not.

shrinks the decision regions of the least likely messages and so lowers the error rate. The fuzzy detector used contextual information to reduce the probability of error still further. We did not compare the contextual fuzzy detector with contextual versions of the ML and MAP detectors.

9.6 CONCLUSION

The fuzzy detector combines Bayesian detection theory with contextual fuzzy rules learned from sample data. The sets-as-points view of fuzzy sets models received data vectors as fuzzy sets in or near the Hamming 6-cube. The noiseless message vectors are nonfuzzy sets at the vertices of the cube. The detector's reflexive additive fuzzy subsystem acts as a MAP detector. It estimates the probabilities of the message symbols in the alphabet conditioned on the current received vector. The fuzzy rules in the contextual additive subsystem associate pairs of symbols with probability density functions conditioned on those symbol pairs. Sample data estimate the rule then-parts. The output from the contextual subsystem is a density conditioned on the

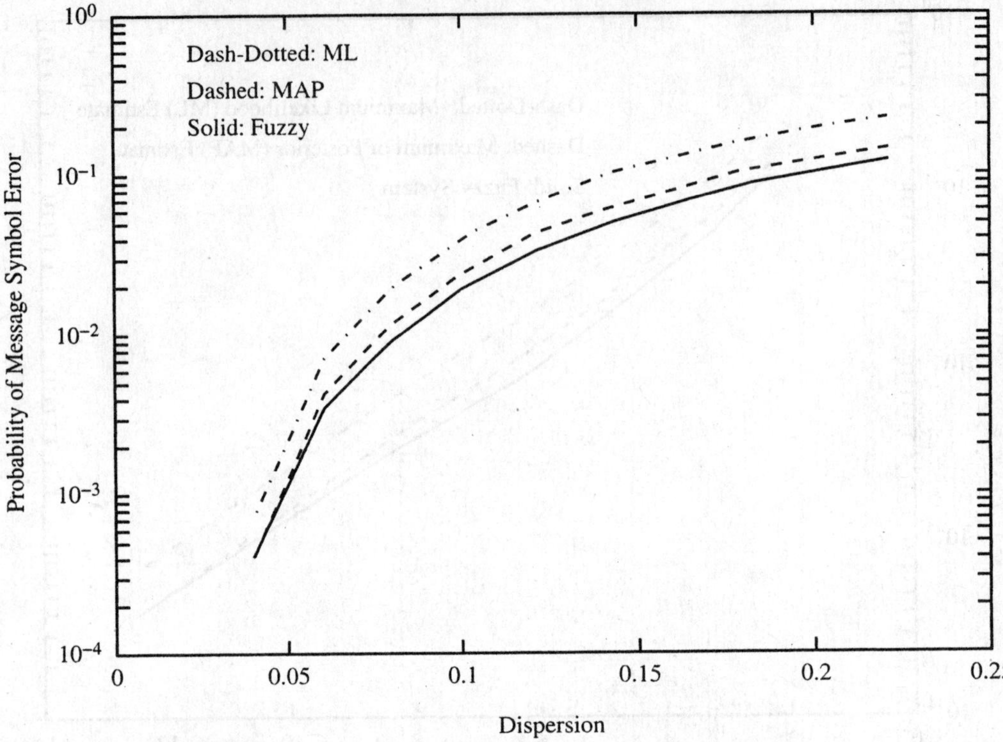

Figure 9.9 Probability of message symbol error versus dispersion for the alpha-stable case of $\alpha = 2$ (Gaussian noise).

context of the current received vector. The combined output fuzzy set is a clarity-weighted sum of the two densities.

The fuzzy detector works well in the presence of additive symmetric alpha-stable noise for many values of the characteristic exponent and dispersion. The combined two estimated densities give a lower message symbol error rate than does the MAP detector value as we would expect from using the extra contextual data. The results suggest that the cube-based clarity measure gives an appropriate weighting function.

The fuzzy detector extends to sample windows larger than three symbols. The current design has two subsystems to estimate $p(y|x_t)$ and $p(y|x_{t-1}, x_{t+1})$. A third subsystem could estimate $p(y|x_{t-2}, x_{t-1}, x_{t+1}, x_{t+2})$. The combined output fuzzy set would be a weighted sum of the *three* densities scaled by the clarity of their if-parts or

$$B(x_t) = \sum_{j=1}^{3} c_j(t) B_j(t)$$

as discussed above. The inclusion of more contextual information may further reduce the error rate. But a fourth or fifth subsystem may not reduce the error rate enough to justify the added cost and complexity.

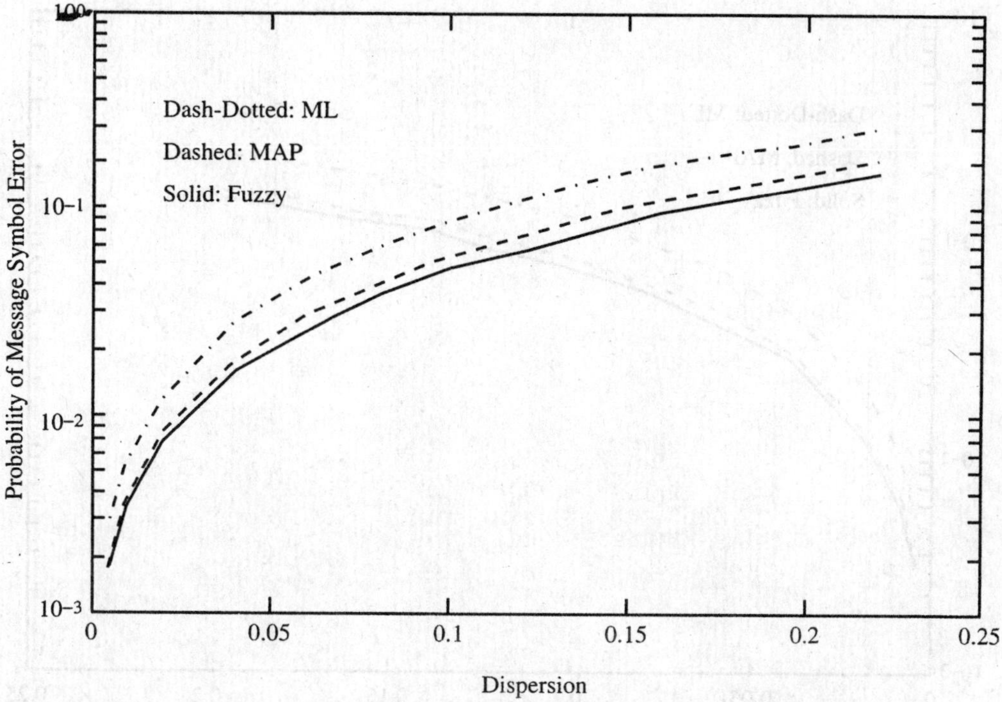

Figure 9.10 Probability of message symbol error versus dispersion for the alpha-stable case of $\alpha = 1.5$.

REFERENCES

[1] Gleick, J., *Chaos*, Penguin Books, 1987.

[2] Kosko, B., "Fuzzy Entropy and Conditioning," *Information Sciences*, vol. 40, 165–174, 1986.

[3] Kosko, B., *Neural Networks and Fuzzy Systems*, Prentice Hall, 1991.

[4] Kosko, B., "Fuzzy Systems as Universal Approximators," *IEEE Transactions on Computers*, vol. 42, no. 11, 1329–1333, November 1994; an early version appears in the *Proceedings of the 1992 IEEE International Conference on Fuzzy Systems (FUZZ–92)*, pp. 1153–1162, March 1992.

[5] Kosko, B., "Addition as Fuzzy Mutual Entropy," *Information Sciences*, vol. 73, no. 3, 273–284, October 1993.

[6] Pacini, P. J., and Kosko, B., "Adaptive Fuzzy Frequency Hopper," *IEEE Transactions on Communications*, vol. 43, no. 6, 2111–2117, June 1995.

[7] Pacini, P. J., "Fuzzy Communications," Ph.D. dissertation, Department of Electrical Engineering–Systems, University of Southern California, December 1993.

[8] Proakis, J. G., *Digital Communications*, McGraw-Hill, 1983.

[9] Shao, M., and Nikias, C. L., "Signal Processing with Fractional Lower-Order Moments: Stable Processes and Their Applications," *Proceedings of the IEEE*, vol. 81, no. 7, 984–1010, July 1993.

[10] Wegman, E. J., Schwartz, S. G., and Thomas, J. B., *Topics in Non-Gaussian Signal Processing*, Academic Press, 1989.

[11] Zadeh, L. A., "Fuzzy Sets," *Information and Control*, vol. 8, 338–353, 1965.

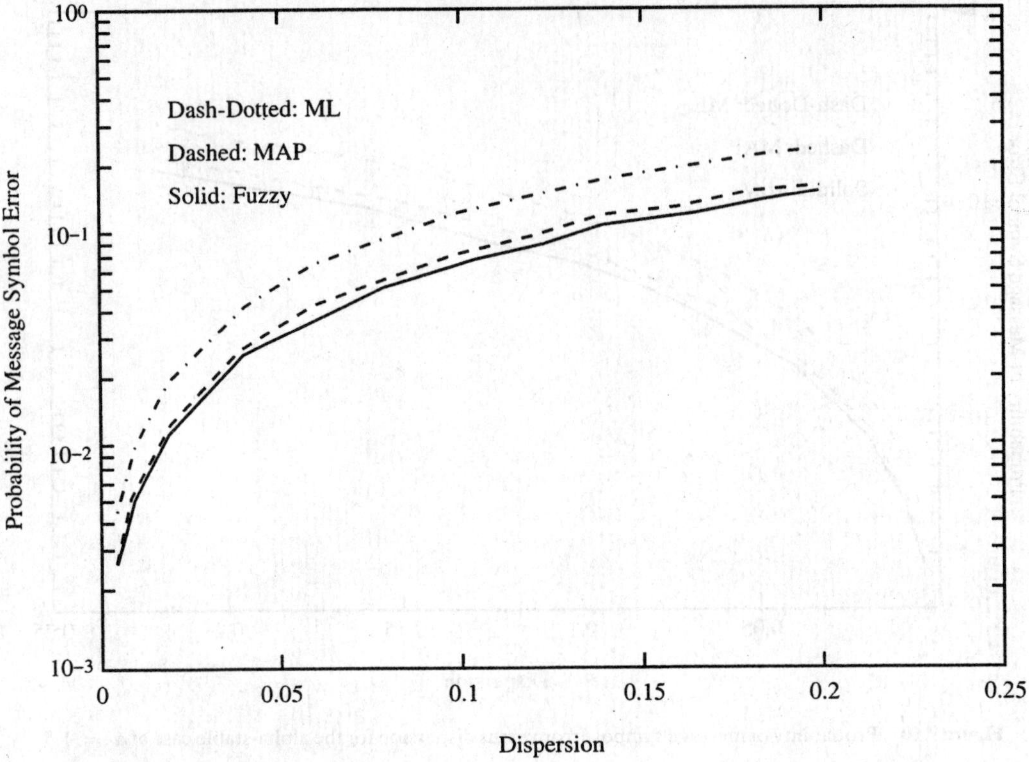

Figure 9.11 Probability of message symbol error versus dispersion for the alpha-stable case of $\alpha = 1$ (Cauchy noise).

APPENDIX 9.A: FUNCTIONAL AND GEOMETRICAL VIEW OF FUZZY SETS

Fuzzy sets are multivalued sets. They have multivalued indicator or *set functions* $a: X \longrightarrow [0, 1]$ that map a domain X into the unit interval [0, 1]. Element $x \in X$ belongs to fuzzy subset A to degree $a(x)$. This is the *sets-as-functions* definition of a fuzzy set [11]. The set values can be continuous or discrete.

Let X be a finite (or countable) domain: $X = \{x_1, x_2, \ldots, x_n\}$. Its power set 2^X contains the 2^n nonfuzzy subsets. These sets correspond to the 2^n length-n bit vectors or the vertices of the unit n-cube $I^n = [0, 1]^n$. Let $A \subset X$ be a fuzzy set with membership function a. Then element x_1 belongs to A to degree $a(x_1)$, x_2 belongs to degree $a(x_n)$, and so on for all n elements. So A is a length-n vector of these membership values:

$$A = (a(x_1), a(x_2), \ldots, a(x_n)). \tag{9.75}$$

A is a *fit vector* in (9.75). Each element $a(x_i)$ or just a_i is a *fit* or *fuzzy unit* [2] that states how much x_i belongs to or fits in A. So the set A is a point *in* the unit cube I^n. This is the *sets-as-points* definition of fuzzy sets [2,3] and holds for countable sets X. In this geometric view the *fuzzy power set* is the entire cube: $\mathcal{F}(2^X) = I^n$. Chapter 12 discusses the structure of such fuzzy cubes. Here we review just those properties that lead to the clarity measure.

Set A is fuzzy if and only if $A \cap A^c \neq \emptyset$ and $A \cup A^c \neq X$. Here $a^c = 1 - a$, $a \cap b = \min(a, b)$, and $a \cup b = \max(a, b)$. So in the fuzzy cube I^n the set A is fuzzy if and only if it does not lie at one of the 2^n vertices.

Figure 9.7 shows the geometry of fuzzy sets in the fuzzy 2-cube. The n-cube consists of 2^n symmetric subcubes. Each fuzzy set A has a dual set in the other $2^n - 1$ subcubes. The 2^n dual sets have fit values $a(x_i)$ or $1 - a(x_i)$ and so are all the same distance from the cube midpoint and from the nearest vertex. The dual sets are the 2^n vertices of a hyperrectangle centered at the cube midpoint. The 2^{n-1} long diagonals of this hyperrectangle pass through the cube midpoint. A long diagonal connects each dual fuzzy set to its unique complement just as in the special case where the 2^{n-1} long diagonals of the whole cube connect the 2^n nonfuzzy sets with their complements at the cube corners.

APPENDIX 9.B: FUZZINESS AND CLARITY MEASURES

A *fuzziness measure* $\bar{F}(A)$: $\mathcal{F}(2^X) \longrightarrow [0, 1]$ maps fuzzy sets to values between 0 and 1. It measures how fuzzy or vague the fuzzy set A is or how clear it is not. $\bar{F}(A) = 0$ at the vertices where $A \cap A^c = \emptyset$ and $A \cup A^c = X$ hold. $\bar{F}(A) > 0$ when these binary laws do not hold. $\bar{F}(A)$ grows as A moves toward the cube midpoint. $\bar{F}(A) = 1$ iff A is the cube midpoint iff $A = A \cap A^c = A \cup A^c = A^c$. The midpoint alone is equidistant from all 2^n vertices. A *clarity* measure is the negation or additive inverse of a fuzziness measure: $c(A) = 1 - F(A)$.

A fuzziness measure relates the "overlap" $A \cap A^c$ and the "underlap" $A \cup A^c$. The simplest measure is the ratio [2,5] discussed in Chapter 12:

$$\bar{F}(A) = \frac{\ell^1(A, A_{\text{near}})}{\ell^1(A, A_{\text{far}})}. \tag{9.76}$$

A_{near} is the vertex closest to A. A_{far} is the vertex farthest from A. And

$$\ell^1(A, B) = \sum_{i=1}^{n} |a_i - B_i|$$

is the fuzzy Hamming distance between A and B in the n-cube. A_{near} and A_{far} lie at opposite ends of a long diagonal. At each vertex $\ell^1(A, A_{\text{near}}) = 0$ and $\ell^1(A, A_{\text{far}}) = n$. So $\bar{F}(A) = 0$. At the midpoint $\ell^1(A, A_{\text{near}}) = \ell^1(A, A_{\text{far}})$. So $\bar{F}(A) = 1$.

Figure 9.7 shows that $\bar{F}(A)$ in (9.76) equals [2] the ratio of counted overlap to underlap:

$$\bar{F}(A) = \frac{C(A \cap A^c)}{C(A \cup A^c)}. \tag{9.77}$$

This shows [2] that $\bar{F}(A)$ measures the degree that the subset $A \cap A^c$ contains the superset $A \cup A^c$. Note that

$$C(A) = \sum_{i=1}^{n} a_i$$

$$= \sum_{i=1}^{n} |0 - a_i|$$

$$= \ell^1(\emptyset, A). \tag{9.78}$$

So the count or size of fuzzy set A is just its ℓ^1 norm or fuzzy Hamming distance from the cube origin. So the fuzziness measure divides the ℓ^1 length $\ell^1(\emptyset, A \cap A^c)$ by the longer ℓ^1 length $\ell^1(\emptyset, A \cup A^c)$. The clarity measure $c(A)$ also relates the overlap $A \cap A^c$ and the underlap $A \cup A^c$:

$$c(A) = 1 - \bar{F}(A) \tag{9.79}$$

$$= \frac{C(A \cup A^c) - C(A \cap A^c)}{C(A \cup A^c)}. \tag{9.80}$$

HOMEWORK PROBLEMS

9.1. Derive the Cauchy density

$$f_1(\gamma, \delta; x) = \frac{1}{\pi\gamma\left(1 + \left(\frac{x-\delta}{\gamma}\right)^2\right)}$$

and the Gaussian density

$$f_2(\gamma, \delta; x) = \frac{1}{2\sqrt{\pi\gamma}} \exp\left(-\frac{(x-\delta)^2}{4\gamma}\right)$$

from the general symmetric alpha-stable density

$$f_\alpha(\gamma, \delta; x) = \frac{1}{2\pi} \int_{-\infty}^{\infty} \exp(i\delta\omega - \gamma|\omega|^\alpha) e^{-i\omega x}\, d\omega.$$

9.2. Prove that the Cauchy density in problem 9.1 has infinite variance.

9.3. Derive

$$f_\alpha(\gamma, \delta; x) = \gamma^{-\frac{1}{\alpha}} f_\alpha(1, 0; (x - \delta)\gamma^{-\frac{1}{\alpha}})$$

from the general alpha-stable density in problem 9.1.

9.4. The volume $v(R_A)$ is a clarity measure when applied to the hyperrectangle R_A whose 2^n vertices are the 2^n sets $(a_1 \text{ or } 1 - a_1, \ldots, a_n \text{ or } 1 - a_n)$. Assume $a_i \leq 1 - a_i$ for all i. Show that $v(R_A)$ gives a lower bound on the clarity measure $c(A)$ in (9.79):

$$v(R_A) \leq c(A)$$

$$\prod_{i=1}^{n} (1 - 2a_i) \leq 1 - \bar{F}(A).$$

PART VI

FUZZY HARDWARE

Engineers have designed chips to house fuzzy systems since the late 1980s. The chips processed the math of a center-of-gravity fuzzy system or simple standard additive model. These "fuzzy" chips were not fuzzy in a design sense. Engineers simply recast standard VLSI digital designs to run fuzzy algorithms. The chips could in theory cycle through a few rules millions of times per second or cycle through more rules thousands of times per second. This led for a time to a contest of who could produce the fastest chip with the most FLIPS or fuzzy logical inferences per second.

Most of these chips proved far more powerful than control applications required. The chips soon met the fate of many new technical tools that search for problems to solve. They fell into disuse and then dropped out of the market. Engineers could build real-time systems without them. The fuzzy algorithms were simple enough that engineers could just reprogram the chip in a camcorder or braking system or write their own short software program to run the fuzzy system.

Fuzzy rule explosion and learning present new challenges to the digital or analog chip designer. Video and communication systems may use many input variables and process them at much higher speeds than most control systems require. These systems may need custom hardware to run in real time. Learning compounds these processing needs as it tunes some or all of the parameters in the standard additive model.

Chapter 10 presents digital designs for many of the SAM algorithms in Chapter 2. It presents the first VLSI designs for adaptive fuzzy systems that tune both their if-part and then-part fuzzy sets as well as their rule weights. It extends these designs to set SAMs that take whole fuzzy sets as inputs and then match the input sets to the stored if-part sets. The design scheme approximates the input sets as triangles to reduce the match computation from a correlation integral to a simple count of triangle areas.

10

ADAPTIVE VLSI ADDITIVE FUZZY SYSTEMS

Rodney Corder

PREVIEW: SAMS IN SILICON

This chapter shows how to convert a standard additive model (SAM) fuzzy system to a very large scale integrated (VLSI) chip. A SAM converts each input to an output with a fixed number of products and sums and with one division. Digital VLSI can produce precise and flexible SAM chips. Digital VLSI can also tune a SAM system through supervised learning. The gradient descent algorithms reduce to products and sums with at most one division per training cycle.

This chapter shows how adaptive VLSI can tune Gaussian, Cauchy, and exponential if-part set functions. These schemes can also tune triangular and trapezoidal set functions. Learning compounds the chip's rule complexity and can greatly reduce its speed. Set SAMs take entire fuzzy sets as inputs and lead to more complex digital designs.

10.1 VLSI FOR FUZZY SYSTEMS

10.1.1 Function Approximation with Electrons

The math of fuzzy systems is not fuzzy. Math used for fuzzy systems is bivalent as is all math. A VLSI chip can map the math of a fuzzy system to the structured flow and storage of electrons. This mapping can occur in either the continuous or discrete domains. Each domain has its advantages and disadvantages. The continuous domain leaves the inputs continuous and operates on them with the physics of semiconductor devices. The discrete domain maps the inputs into digital symbols and operates on them with Boolean functions. The discrete domain can also implement programmable systems as well as adaptive systems.

We begin with a quick review of the additive fuzzy systems in Chapter 2. An additive fuzzy system sums the weighted then-parts B'_j of its rules and then converts this summed set B to an output $F(x)$ by taking the centroid of B: $F(x) = \text{Centroid}(B)$. The m if-then rules have the word form "*If* $X = A_j$ *then* $Y = B_j$" or patch form $A_j \times B_j \subset R^n \times R^p$. The rule firing scales the then-part for each of the rules:

$$B'_j = a(x) B_j \tag{10.1}$$

The SAM (standard additive model) architecture also allows independent rule sets to function cooperatively. Each scaled then-part set B'_j can have its own weight w_j that reflects its importance. The weighting may even reflect the credibility of the knowledge source in relation to other knowledge sources. Then the system sums the "fired" and weighted then-parts B'_j to produce the output set $B(x)$:

$$B = \sum_{j=1}^{m} w_j B'_j = \sum_{j=1}^{m} w_j \; a_j(x) \; B_j \tag{10.2}$$

Taking the centroid of the output set B provides the output value $F(x) \in R^p$ when the vector input $x \in R^n$ fires the rules:

$$F(x) = \text{Centroid}(B(x)) = \text{Centroid}\left(\sum_{j=1}^{m} w_j \; a_j(x) \; B_j\right) \tag{10.3}$$

The rule weights w_j cancel from the SAM Theorem if they all equal the same value $w_1 = \ldots = w_m > 0$ since

$$F(x) = \frac{\sum_{j=1}^{m} w_j \; a_j(x) \; V_j \; c_j}{\sum_{j=1}^{m} w_j \; a_j(x) \; V_j} = \sum_{j=1}^{m} p_j(x) \; c_j \tag{10.4}$$

This basic SAM structure requires only 3 simple math operators: add, multiply, and divide. The add and multiply operators are common in both analog and digital systems. The SAM structure uses the divide operator only once to compute the output of the system. Iterative division approaches can give sufficient system throughput for both analog and digital systems.

Many developers and vendors have designed VLSI devices for fuzzy applications. Table 10.1 shows some of these devices as well as their functionality and manufacturers.

Device Name (Reference number in Brackets)	Vendor/Developer	Functionality
FC110 [5]	Togai InfraLogic Inc.	Optimized instruction set
VY86C500 (Fuzzy Computation Array) [9]	VLSI Technology Inc.	Coprocessor
81C99 [6]	Siemens	Coprocessor
MSM91U112 [6]	OKI Semiconductor	In-line accelerator
NLX220 [2]	American Neuralogix Inc.	Dedicated controller
FP–9000/FP–9001 [7]	Omron	Analog controller

TABLE 10.1: Survey of fuzzy chips

Table 10.1 shows the four basic approaches to using VLSI for fuzzy systems. These approaches include central processing units (CPUs) with optimized instruction sets, co-processors for general purpose CPUs, in-line accelerators, and dedicated controllers. The optimized instruction set speeds up fuzzy systems within the general purpose computer framework. Adding special purpose instructions reduces the repetitive operations of the fuzzy inference. Coprocessors act as peripheral accelerators to general purpose computers. They act as slave processors on the information from the main system CPU. In-line accelerators reduce the interaction between the system and the general purpose CPU. They attach to the system inputs and outputs to replace the latency of the general purpose CPU shuttling data back and forth. Dedicated controllers do away with the general purpose CPU. They connect to the controlled system and perform only a fixed function. The more flexible designs can implement a wider variety of approaches [4]. Often the limitations stem not from the chips but from the systems themselves.

There is an old debate on the benefits of digital versus analog processing. Digital or discrete processing tends to be more immune to processing noise. Changes in semiconductor processing often do not change the circuit. But in the digital approach the sampling of the signal creates quantization noise. We can change the algorithm as well as the data but operations on the inputs tend to occur serially. Analog or continuous processing tends to be much faster and its operations tend to occur in parallel. But the analog approach is more prone to noise at each step of the processing. It is also much more sensitive to changes in semiconductor processing than is the digital approach. The algorithm lies in the connection of the processing elements and is not flexible. Which method to use depends on more than just a surface view of the two methods. A designer must know how each method will affect the system. Figure 10.1 shows the basic tradeoffs for analog and digital systems in terms of chip speed and flexibility.

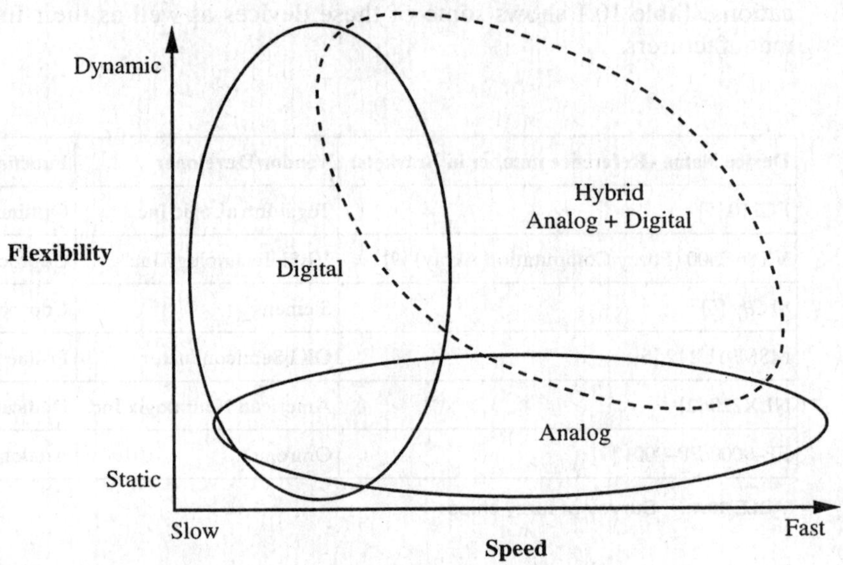

Figure 10.1 Tradeoffs in the design of analog and digital chips.

10.1.2 Rule Explosion

Rule explosion in a fuzzy system can quickly outstrip a chip's ability to store and process the rules. This may not be a significant problem for systems with only a few input variables. As the dimensionality of the problem grows so grows the number of rules. For the simple approximand $f: R^2 \rightarrow R$ a fuzzy approximator needs on the order of k^2 rule patches $A_j \times B_j$ to cover the 3-D surface of the approximand f. For $f: R^n \rightarrow R^p$ a fuzzy appromator needs on the order of k^{n+p-1} rules to cover the graph of f. This rapidly leads to problems implementing real-world higher-order problems. Lone optimal rules cover the extrema of f but the user may not know how to find them.

Learning systems can help with this "curse of dimensionality." Adaptive SAM systems can learn or search for the extrema of the system and allocate rule patches to and between these "bumps". The SAM has many parameters that data can tune to optimize the set of rules. Functions, vectors, and scalars can change the fuzzy output $F(x)$. Learning can tune the shape, center, and orientation of the if-part set functions a_j: $R^n \rightarrow [0, 1]$ or the then-part set functions b_j: $R^p \rightarrow [0, 1]$. Learning can also tune the then-part set volumes V_j or centroids c_j or the rule weights w_j. These functions and weights can change the shapes and placement of the rule patches to help minimize the function error.

10.1.3 Evaluating Implementations: The Time/Area Product

We need a tool to compare one algorithm with another early in the design cycle. The time/area product measures the resources required for an algorithm. We normalize the operations in the algorithm to the time and silicon area required for a 1-bit adder. This shows how much time and area the chip will need to implement the design and gives a way to compare architectures. The time/area product does not depend on the exact VLSI technology or the feature size.

Consider an n-bit system. The time/area product is n for an addition, subtraction, or comparison. The time/area product is n^2 for a multiplication or division. Then we can decompose the algorithm into the sums and products to estimate the physical complexity. The time/area product can be a poor measure for noncomputational structures such as registers and memories. A rough estimate for a dedicated register is $n/4$. The estimate for random access memory (RAM) is the number of words times $n/8$. There is no tradeoff between time and area for memory. Area tends to measure it. Often we can assume a basic register set that is uniform across most designs and can ignore the calculation when we look for tradeoffs.

We can optimize digital systems in the time domain or the area domain. We can design them to run as fast as possible or to be as small as possible or somewhere in between. We make these tradeoffs when we consider the performance of the design or its cost. The time/area product measures these parameters. To optimize for time we can arrange each single adder equivalent on the silicon to operate at the same time. That gives a small amount of time to complete the task with a large area. If there is only one adder-equivalent cell on the chip then the chip must compute each function serially. This leads to a long time to complete the task with a very small area. The time/area product is not an exact measure but it does give a quick way to estimate the performance and area tradeoffs between algorithms.

10.2 THE SAM ARCHITECTURE AND VLSI

The key to implementing the fuzzy SAM is to organize the flow of information. Good flow and data structures yield realizable silicon. The flow of the SAM consists of "firing" the rules (applying the input x to the if-part sets A_j), weighting the then-part fuzzy sets, combining the fuzzy output sets, and then defuzzifying. The SAM Theorem compresses this flow into the simple ratio in (10.4).

We can apply inputs x to the fuzzy system at some fixed time or on demand. The fixed approach samples the input n times each second. It can use a single input, all inputs, or some mix of the inputs. Evaluation of the fuzzy system occurs at the sampling of one or more inputs. The system need not sample all of the inputs at the same time. This approach works well for systems where the inputs change continuously. The demand approach evaluates a portion of the fuzzy system each time one or more inputs change. How the input change affects an output change fixes how much of the system we must recompute. If the input change does not cause the output to change then evaluation stops quickly.

The nonzero domain or support of each if-part set function a_j: $R^n \rightarrow [0, 1]$ helps choose how much calculation the system needs. The support is the width of the nonzero portion of the set A_j. If the input x belongs to a_j to a nonzero degree ($a_j(x) > 0$) then the system must compute the effects of the input on the output. Less overlap between the sets means the system needs less effort to compute the output $F(x)$ of the SAM.

The output of the SAM need change only in relation to a change in one or more of the inputs that change $a_j(x)$. The demand approach works well for systems that have slowly changing inputs or for systems that do not have a lot of free computing resources. It propagates only the changes inside the SAM and so conserves computing resources at the cost of a more complex data structure. We work here and often in practice with only scalar or factored set functions a_j^i: $R \rightarrow [0, 1]$.

10.2.1 Scalar If-Part Set Functions

The input $x \in X$ fires the rule "If X is A_j then Y is B_j" if $a_j(x) > 0$. The rule does not affect the output if $a_j(x) = 0$. The joint set function a_j: $X \rightarrow [0, 1]$ can in theory have any shape. Some common set functions are Gaussian functions, Cauchy functions, and exponential functions. We seek a scalar set function that is easy to implement and that maintains the desired system structure.

10.2.1.1 Gaussian Set Functions

The scalar Gaussian set function has the exponential form

$$a_j(x) = \exp\left(-\frac{1}{2}\left(\frac{x - m_j}{\sigma_j}\right)^2\right) \tag{10.5}$$

We suppress the superscript in a_j^i: $R \rightarrow [0, 1]$ for simplicity.

The Gaussian set function depends on just the mean m_j and the standard deviation σ_j. In theory the Gaussian has infinite support. In a digital system the support can extend only to the smallest quantizable value. In practice the support of the Gaussian set function extends about 3σ on either side of the mean. Figure 10.2 shows a Gaussian scalar set function over the domain [0,63] with a mean of 32 and a standard deviation of 5.

It is not practical to compute the exponential at runtime in a high-speed system. This need not limit how we apply the exponential because the system can use look-up tables. This greatly reduces the overhead of the evaluation. The table can be fairly compact since the resolution of the look-up table need not exceed the resolution of a_j. Eight bits of resolution should suffice for most applications. This leads to a fairly small look-up table of about 256 bytes.

The contents of the look-up table need not be just of the form

$$a_j(x) = e^y \tag{10.6}$$

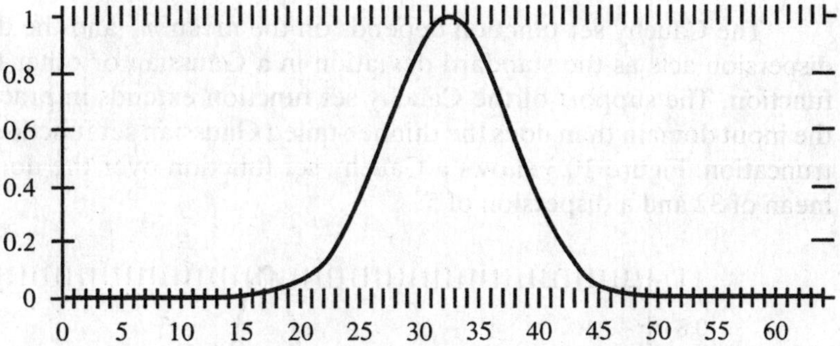

Figure 10.2 Gaussian scalar set function. The bell curve has a mean of 32 and a standard deviation of 5.

where

$$y = -\frac{1}{2}\left(\frac{x - m_j}{\sigma_j}\right)^2.$$

It is better to use a look-up table whose contents combine all of the elements that do not depend on the variables. This gives

$$a_j(x) = \exp\left(-\frac{1}{2}(y)^2\right) \tag{10.7}$$

where

$$y = \frac{x - m_j}{\sigma_j}.$$

Then we can compute y with only one subtraction and one division with one look-up to produce the fit values $a_j(x)$. This is efficient for high-speed evaluation and gives a time/area product of 328 for an 8-bit system. The computational structure accounts for only 72 units of this. The remaining 256 units are due to the look-up table in RAM. The time/area product for the Gaussian set function is 72 if the system already has access to shared RAM.

10.2.1.2 Cauchy Set Functions

The Cauchy set function bell curve looks much like the Gaussian set function but has no finite variance. The Cauchy bell curve has thicker tails than the Gaussian bell curve as discussed in Chapters 6 and 9. These two bell curves belong to the alpha-stable family of bell-curve probability densities. The thicker tails give rise to more frequent and more extreme outcomes or "outliers." Cauchy set functions also tend to converge faster than do Gaussian set functions in adaptive fuzzy systems. The scalar Cauchy set function has the form

$$a_j(x) = \frac{d_j^2}{(x - m_j)^2 + d_j^2} \tag{10.8}$$

The Cauchy set function depends on the mean m_j and the dispersion d_j. The dispersion acts as the standard deviation in a Gaussian or other finite-variance set function. The support of the Cauchy set function extends in practice over more of the input domain than does the thinner-tailed Gaussian set function and may require truncation. Figure 10.3 shows a Cauchy set function over the domain [0,63] with a mean of 32 and a dispersion of 5.

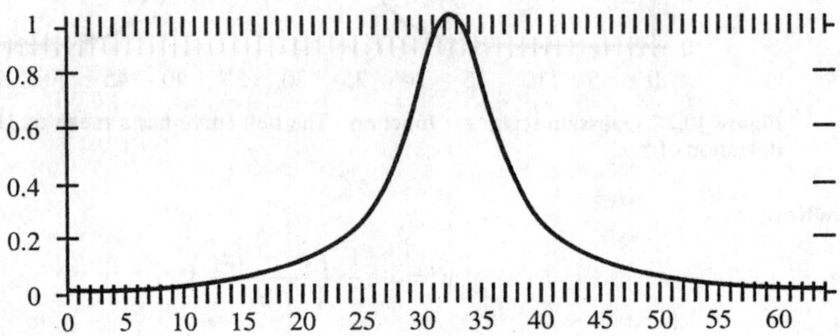

Figure 10.3 Cauchy scalar set function. The bell curve has a mean of 32 and a dispersion of 5.

The Cauchy set function needs 2 additions, 2 multiplications, and 1 division to compute. It needs only 2 multiplications because the chip can store and reuse d_j^2 once it computes the term. This removes 1 multiply at the cost of a register. The algorithm is compute intensive but does not require the addition of a look-up table as does the Gaussian set function. The time/area product for an 8-bit Cauchy function is 208.

10.2.1.3 Exponential Set Functions

The exponential set function decreases monotonically. It arises in physics as a key term in the Gibbs probability density and occurs in probability models that range from queuing theory to game theory. Exponential set functions have one of the simplest learning laws and often give good function approximations. The exponential set function has the form

$$a_j(x) = \begin{cases} \exp(u_j(v_j x_i)) & \text{if } x \geq v_j \\ 0 & \text{if } x < v_j. \end{cases} \tag{10.9}$$

The exponential set function depends on the initial point v_j and the decay factor u_j. The terms v_j and u_j may not be of the same orders of magnitude. The term v_j is an integer when u_j is a fraction and vice versa. We can replace u_j with $\frac{1}{z_j}$. Then v_j and z_j have the same order of magnitude. Figure 10.4 shows an exponential set function over the domain [0,63] with an initial point of 32 and a decay factor of 0.2 (or $\frac{1}{5}$).

The exponential set function can exploit a look-up table as can the Gaussian set function. The exponential set function needs only one addition, one multiply, and one look-up to compute. This is the same cost as the Gaussian function. They differ only in the contents of the look-up table for $a_j(x) = \exp(y)$.

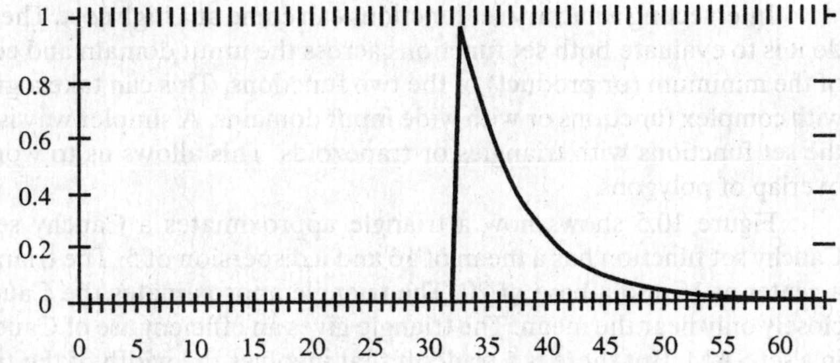

Figure 10.4 Exponential scalar set function. The curve has an initial point of 32 and a decay factor of 0.2.

10.3 SET SAMS AND SECOND-ORDER STATISTICS

10.3.1 Set SAMs

The set SAM of Chapter 2 takes the fuzzy set A as input instead of the scalar or vector x. The set input acts much like a random vector with a probability density centered at x.

The fit value $a_j(x)$ becomes $a_j(\delta(x - x_0))$ in a set SAM. The delta function acts as a set centered at x_0 that has an area of 1. Then the sifting theorem on delta functions gives

$$a_j(x_0) = \int_{R^n} a_j(x)\ \delta(x - x_0)\ dx. \tag{10.10}$$

Replacing vector x with fuzzy set A gives the generalized set value $a_j(A)$ as a standard correlation:

$$a_j(A) = \int_{R^n} a_j(x)\ a(x)\ dx. \tag{10.11}$$

This measures the area of the intersection $A_j \cap A$. This leads in turn to the set SAM theorem:

$$F(A) = \frac{\sum_{j=1}^{m} w_j\ a_j(A)\ V_j\ c_j}{\sum_{j=1}^{m} w_j\ a_j(A)\ V_j} = \sum_{j=1}^{m} p_j(A)\ c_j. \tag{10.12}$$

The range of $a_j(x)$ is [0,1] for the point SAM. The set SAM gives the range of $a_j(A)$ as [0, Volume(A)] since $a(x) \leq 1$ for all x in (10.11). The range does not change other SAM calculations because $p_j(A)$ still lies in [0,1].

Intersecting arbitrary set functions can come at a high cost. The only real way to do it is to evaluate both set functions across the input domain and compute the area of the minimum (or product) of the two functions. This can take a great deal of time with complex functions or with wide input domains. A simpler way is to approximate the set functions with triangles or trapezoids. This allows us to work only with the overlap of polygons.

Figure 10.5 shows how a triangle approximates a Cauchy set function. The Cauchy set function has a mean of 16 and a dispersion of 5. The triangle function has a center at 16 and a base of 20. The triangle approximates the Cauchy set function closely only near the mean. The triangle gives an efficient use of Cauchy set functions in a set SAM. But there is a tradeoff that involves the width of the triangle base. We can choose to minimize the error around the mean or to minimize the maximum error.

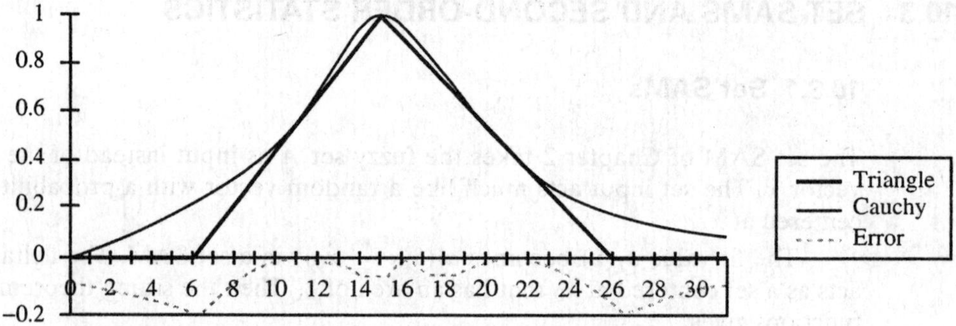

Figure 10.5 Isosceles triangle approximates a Cauchy bell curve.

A trapezoid gives a better approximation of a Gaussian set function than does a triangle. Figure 10.6 shows how a trapezoid approximates a Gaussian set function. The Gaussian set function has a mean of 16 and a standard deviation of 5. The trapezoidal set function has a center at 16, a plateau of 2, and a base of 22. The trapezoid approximates the Gaussian set function closely out to about 2 standard deviations from the mean.

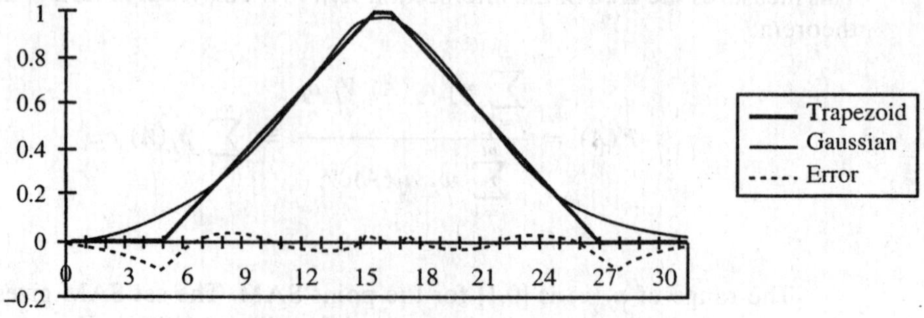

Figure 10.6 Trapezoid approximates a Gaussian bell curve.

A right triangle can approximate an exponential set function as shown in Figure 10.7. The exponential set function has an initial point at 8 and a decay factor of $\frac{1}{8}$. The right triangle has a corner at 8 and a base of 12. The approximation is poor through the entire domain of the exponential set function. A higher-order polynomial would give a better approximation but at a much higher cost of computation.

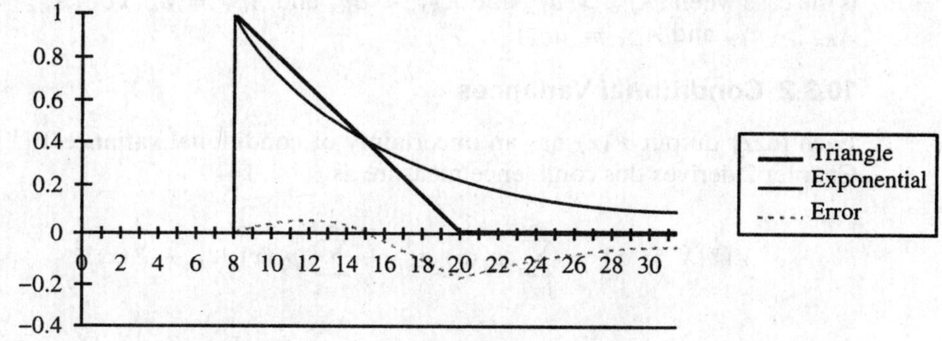

Figure 10.7 Right triangle approximates an exponential set function.

Approximating set function a_j with simple polygons simplifies the area calculation. We need find only the intersection of two triangles to compute $a_j(A)$. Figure 10.8 shows the 4 ways that 2 triangles can intersect. For ease of calculation we use 3 ordered points to represent the set. These points are the left base, the center, and the right base. We represent these points for the input set A as $[A_{Lx}, 0]$, $[A_{Cx}, 1]$, and $[A_{Rx}, 0]$ and for the stored set or set function a_j as $[a_{Lx}, 0]$, $[a_{Cx}, 1]$, and $[a_{Rx}, 0]$. Keeping these points in ascending order helps minimize the number of comparisons we need to perform the intersections.

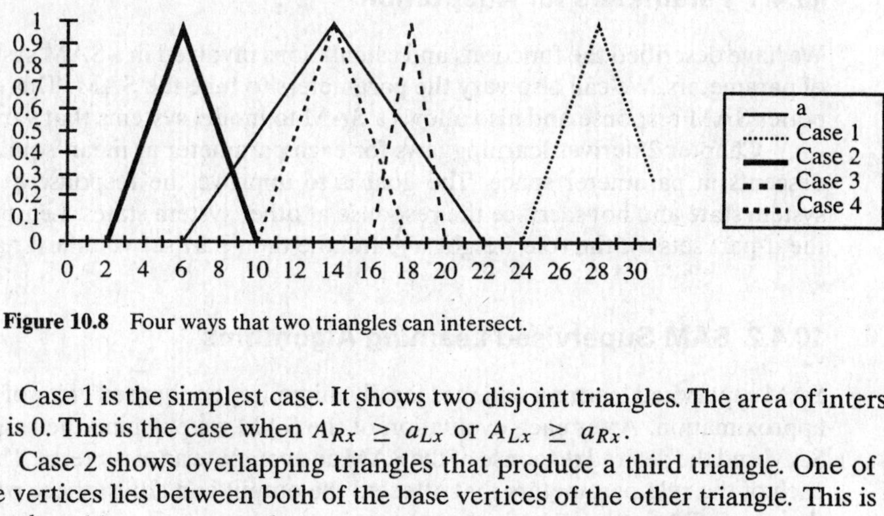

Figure 10.8 Four ways that two triangles can intersect.

Case 1 is the simplest case. It shows two disjoint triangles. The area of intersection is 0. This is the case when $A_{Rx} \geq a_{Lx}$ or $A_{Lx} \geq a_{Rx}$.

Case 2 shows overlapping triangles that produce a third triangle. One of the base vertices lies between both of the base vertices of the other triangle. This is the case when $(A_{Rx} > a_{Lx}$ and $A_{Lx} < a_{Lx})$ or $(A_{Lx} < a_{Rx}$ and $A_{Rx} > a_{Rx})$.

Case 3 shows overlapping triangles that produce a polygon. Both of the base vertices lie between the base vertices of the other triangle. This is the case when ($A_{Lx} > a_{Lx}$ and $A_{Rx} < a_{Rx}$ and $A_{Cx} \neq a_{Cx}$) or ($A_{Lx} < a_{Lx}$ and $A_{Rx} > a_{Rx}$ and $A_{Cx} \neq a_{Cx}$).

Case 4 is the special condition of case 3 where the apexes of the two triangles lie at the same point. This overlap of this case is just the area of the smaller triangle. This is the case when ($A_{Lx} > a_{Lx}$ and $A_{Rx} < a_{Rx}$ and $A_{Cx} = a_{Cx}$) or ($A_{Lx} < a_{Lx}$ and $A_{Rx} > a_{Rx}$ and $A_{Cx} = a_{Cx}$).

10.3.2 Conditional Variances

Each fuzzy output $F(x)$ has an uncertainty or conditional variance $V[Y|X = x]$. Chapter 2 derives this confidence measure as

$$V[Y|X = x] = \sum_{j=1}^{m} p_j(x)\, \sigma_{B_j}^2 + \sum_{j=1}^{m} p_j(x) \left(c_j - F(x)\right)^2 \tag{10.13}$$

$$\sigma_{B_j}^2 = \int_{-\infty}^{\infty} (y - c_j)^2 p_{B_j}(y) dy. \tag{10.14}$$

This measure tells in the COG case how closely the output $F(x)$ lies to the then-part set centroids of the m rules. The closer $F(x)$ lies to c_j then the smaller $V[Y|X = x]$ is. Learning laws can take this into account to adjust the convergence rate based on how well the adaptive SAM (ASAM) approximates the function.

10.4 ADAPTIVE DIGITAL SAM

10.4.1 Parameters for Adaptation

We have described the functions and calculations involved in a SAM system in terms of parameters. We can also vary the parameters to tune the SAM. This can lead to a better SAM response and also allows a SAM to model systems that vary slowly.

Chapter 2 derives learning laws for each parameter as mean-squared gradient descents in parameter space. The goal is to improve the response at the current system state and not sacrifice the response at other system states. Learning can tune the if-part sets a_j, the rule weight w_j, and the then-part set volume V_j and centroid c_j.

10.4.2 SAM Supervised Learning Algorithms

SAM supervised learning schemes locally minimize the squared error of the function approximation. After each evaluation of the SAM we examine the response of the SAM and the desired response of the SAM and compute an error term. Then we check each of the rule parameters that affected the output and adjust those parameters so that the SAM output $F(x)$ better approximates the function.

Chapter 2 gives the basic form of the gradient descent algorithm for an arbitrary SAM if-part set parameter q as

$$q(t+1) = q(t) + \mu_t \ \varepsilon_t \ \frac{p_j(x_t)}{a_j(x_t)} \ (c_j - F(x)) \frac{\partial a_j}{\partial q}. \tag{10.15}$$

Here μ_t controls the learning or convergence rate. The error $\varepsilon_t = (d_t - F(x_t))$ is the difference between the desired response and the actual response at time t. The convex coefficient

$$p_j(x_t) = \frac{w_j \ a_j \ V_j}{\sum_{i=1}^{m} w_i \ a_i \ V_i}$$

measures the degree to which the jth rule fires. The difference $(c_j - F(x))$ measures how well the local rule output or then-part centroid c_j matches the global SAM output $F(x)$. The term $\frac{\partial F}{\partial q}$ is the derivative of the SAM system with respect to parameter q.

10.4.2.1 Gaussian Set Learning

The Gaussian set function has as its parameters the mean m_j and the standard deviation σ_j. Chapter 2 derives the learning laws as

$$m_j(t+1) = m_j(t) + \mu_t \ \varepsilon_t \ p_j(x_t) \ (c_j - F(x)) \left(\frac{x - m_j}{(\sigma_j)^2} \right) \tag{10.16}$$

$$\sigma_j(t+1) = \sigma_j(t) + \mu_t \ \varepsilon_t \ p_j(x_t) \ (c_j - F(x)) \left(\frac{(x - m_j)^2}{(\sigma_j)^3} \right) \tag{10.17}$$

10.4.2.2 Cauchy Set Learning

The Cauchy set function has as its parameters the mean m_j and the dispersion d_j. Their learning laws have the form

$$m_j(t+1) = m_j(t) + \mu_t \ \varepsilon_t \ \frac{p_j(x_t)}{a_j(x_t)} (c_j - F(x)) \left(\frac{2d_j^2(x_t - m_j)}{[(x_t - m_j)^2 + d_j^2]^2} \right) \tag{10.18}$$

$$d_j(t+1) = d_j(t) + \mu_t \ \varepsilon_t \ \frac{p_j(x_t)}{a_j(x_t)} (c_j - F(x)) \left(\frac{2d_j(x_t - m_j)^2}{[(x_t - m_j)^2 + d_j^2]^2} \right). \tag{10.19}$$

10.4.2.3 Exponential Set Learning

The exponential set function has as its parameters the initial point v_j and the decay rate u_j. Their learning laws have the form

$$v_j(t+1) = v_j(t) + \mu_t\ \varepsilon_t\ p_j(x_t)\ (c_j - F(x_t))(u_j) \tag{10.20}$$

$$u_j(t+1) = u_j(t) + \mu_t\ \varepsilon_t\ p_j(x_t)\ (c_j - F(x_t))(v_j - x_t). \tag{10.21}$$

10.4.2.4 Then-part Set Learning

Learning can also tune the rule weight w_j and the volume V_j and centroid c_j of the then-part set B_j. Chapter 2 derives their learning laws as

$$w_j(t+1) = w_j(t) - \mu_t\ \varepsilon_t\ \frac{p_j(x_t)}{w_j(t)}[(c_j - F(x_t)] \tag{10.22}$$

$$V_j(t+1) = V_j(t) - \mu_t\ \varepsilon_t\ \frac{p_j(x_t)}{V_j(t)}[(c_j - F(x_t)] \tag{10.23}$$

$$c_j(t+1) = c_j(t) - \mu_t\ \varepsilon_t\ p_j(x_t). \tag{10.24}$$

10.4.3 Learning Rates

The gradient descent algorithm tries to minimize the error by moving the parameters closer and closer to the unknown optimal values. This constant tuning may work well when the system parameters lie far from the optimal parameters. Constant tuning may lead the learned parameters to overshoot the optimal parameters as the learned parameters move close to the optimal parameters. One way to deal with this is to decrease the learning rate μ_t over time. This slows the learning process and allows the system to stabilize. The term μ_t can rely solely on time. Other ways can decide when to change the learning rate. We can even use a second SAM system that watches the first SAM system and adjusts its μ_t over time.

The learning laws may be costly to compute. The number of computations it takes to adjust all of the parameters for each of the active rules may well exceed the computation it takes to evaluate the SAM output $F(x)$ in the first place. This can cause the system speed to fall by nearly a full order of magnitude. Suboptimal learning laws can help reduce the computation. One simple way is to find just the sign of the change and move the parameter in that direction by some small or scaled constant. Finding the direction of the change is much simpler than evaluating the complete expressions. A suboptimal but faster approach to such supervised learning can have the form

$$q(t+1) = q(t) + k\ \mathrm{sgn}(\varepsilon_t)\ p_j(x)\ \mathrm{sgn}(c_j - F(x))\ \mathrm{sgn}\left(\frac{\partial a_j}{\partial q}\right) \tag{10.25}$$

where

$$\text{sgn}(x) = \begin{cases} 1 & \text{if } x > 0 \\ 0 & \text{if } x = 0 \\ -1 & \text{if } x < 0. \end{cases}$$

This equation includes the learning factor k as well as the signs of all of the learning terms. The factor k need not be a constant. Then the change in the parameter relates to the magnitude of the parameter. The next equation shows how this can work:

$$q(t+1) = q(t)\left(1 + k \ \text{sgn}(\varepsilon_t) \ p_j(x) \ \text{sgn}(c_j - F(x)) \ \text{sgn}\left(\frac{\partial a_j}{\partial q}\right)\right). \quad (10.26)$$

These suboptimal approaches may have problems with their convergence rates and the stability of the parameters after convergence. But they require only a fraction of the computation that the optimal approaches do. The advantage is the ability to do more updates with fewer resources.

10.5 ADAPTIVE SAM VLSI

We now present the major steps of designing the ASAM architecture but stop short of the silicon implementation. These steps define the functional requirements of the system, the algorithms used in the system, and the approach taken for the silicon. If we built a real system then we would also include how to define the interface requirements and how to define the implementation specification for the system and the silicon.

10.5.1 Functional Requirements

The example is a SAM system that gives a flexible base for basic ASAM research. We require that it implement a high-speed 1-input/1-output ASAM. We also want the architecture flexible enough so that we can experiment with multiple-input/multiple-output ASAM systems with a large number of rules. The ASAM should also support a wide range of algorithms without physical reconfiguration. We must be able to program the ASAM system.

10.5.2 Algorithms

We design the first ASAM architecture for Gaussian set functions. This gives a good base that includes the requirements for not only the Cauchy and exponential set functions but for the triangular and trapezoidal approximations as well. We also need to tune the learning parameters in the system. Learning can tune 5 parameters: the 2 parameters for the if-part Gaussian set function m_j and σ_j, the rule weight w_j, and the 2 parameters for the then-part set V_j and c_j. This gives $5n$ parameters for a Gaussian or Cauchy or exponential ASAM system with n rules.

10.5.2.1 Numerical Representation

The chip's numbering system affects the system accuracy. We want the SAM to take in 8-bit numbers and return 8-bit numbers and not sacrifice accuracy. A simple way to do this is to use a fractional representation for all parameters. Suppose the input x, the parameters, and the output $F(x)$ take values in [0,1]. ASAM values may go outside the 8-bit range and may even extend beyond the range [0,1]. We can accept this if we can maintain the accuracy of the output at about 8-bits.

Repeated multiplications and divisions may explode the dynamic range. Multiplying two 8-bit numbers gives a 16-bit result. Do this 5 times in a row and the system must use a 40-bit dynamic range. This quickly goes out of bounds not only in terms of storage of these long numbers but also in terms of performing the multiplication. Multiplying a 32-bit number by an 8-bit number to get a 40-bit may prove taxing in terms of the time/area product. A block floating point approach is a more attractive solution. We keep an 8-bit mantissa and an 8-bit exponent. Then the time/area product is 72 instead of 256 for the brute-force multiplication. The normalization function of at most 8 cycles brings the total up to about 88. This is still about one third of the cost of the brute-force multiplication.

10.5.2.2 Evaluation Phase

Figure 10.9 shows the data flow graph for the Gaussian set function.

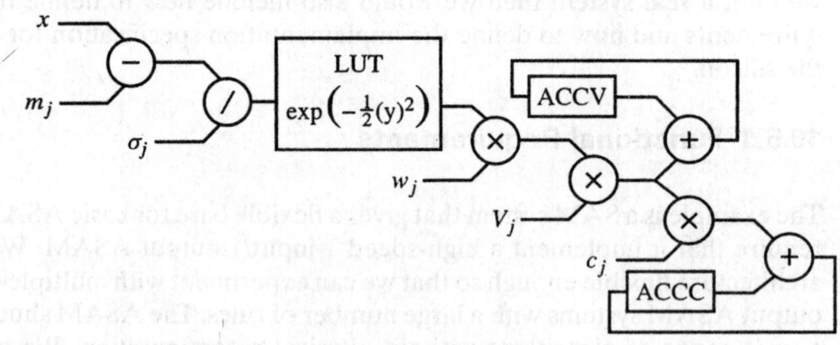

Figure 10.9 Data flow for a scalar Gaussian set function.

The data flow graph shows each of the operations required to perform the function. For the evaluation phase most operations are linearly dependent and have few places for parallelism. This type of algorithm is ideal for a pipelined approach. A functional unit can perform its operation and pass the results on to the next functional unit. This way the system can bring to bear more functional units on the single task.

We represent the inputs and parameters of the system as 8-bit positive fixed-point fractions. The result of the first division $\frac{x-m_j}{\sigma_j}$ ends with a 17-bit result in the range [−255, +255] with 1 bit for sign, 8 bits for the integer part, and 8 bits for the fractional part. The Gaussian exponential function is symmetric. So we can exchange the $(x - m_j)$ calculation with $\text{abs}(x - m_j)$. This adds a comparison and perhaps a second subtraction to the algorithm. But it keeps the result of this calculation to an 8-bit positive fraction. After the division we now have a 16-bit result in the domain [0,255]. The exponential function $\exp\left(-\frac{1}{2}(y)^2\right)$ tends to 0 for any $y > 4$. We can use an 8-bit number with 2 bits of integer and 6 bits of fraction as the index to the look-up table if we add a step to restrict the range of the division to [0,4]. This requires a look-up table of 256 bytes where an 8-bit value codes each entry in the table.

The values stored in ACCV and ACCC are the sums

$$\text{ACCV} = \sum_{j=1}^{m} w_j \; a_j(x) \; V_j \tag{10.27}$$

$$\text{ACCC} = \sum_{j=1}^{m} w_j \; a_j(x) \; V_j \; c_j. \tag{10.28}$$

These sums mean we need more dynamic range than just 16 bits. A series of m additions can increase the dynamic range by $\log_2(m)$ bits. We must adjust the length of ACCV and ACCC to fit m. The SAM output $F(x)$ has the form

$$F(x) = \frac{\text{ACCC}}{\text{ACCV}}. \tag{10.29}$$

10.5.2.3 Update Phase

It is tempting to factor the learning laws and attempt to minimize the total number of calculations required. This will work for the above learning laws. If we design an architecture based on this minimization then we may run into problems when we try other learning laws in the future. It helps to understand the basics of the learning laws and work on an architecture that can implement both the feature of the above learning laws as well as those of other learning laws.

The learning law takes the environment of a variable and adjusts the variable in the direction that produces a "better" result. Each variable has its own computation. They all have a common set of parameters for the equation. This is what we design the architecture around. Each of the equations of the learning laws uses global variables and local variables. The global variables are those common to all of the updates. The local variables are those specific to the jth rule. This type of problem lends itself to a parallel processing scheme where each of the functional units focuses on updating one variable using broadcast variables.

The long strings of multiplications and divisions in the learning laws cause problems with fixed-point representations. The dynamic range grows each time we multiply or divide two numbers. The number of bits for the product is the sum of the number of bits that represent the multiplier and the multiplicand. We need a floating-point format to keep any kind of accuracy with such a large dynamic range. Then the only requirement is to pick the size of the mantissa and the exponent. The accuracy of the mantissa does not need to be more than the accuracy of the variable. That is 8 bits in the SAM system. The exponent must handle the dynamic range of the calculation. The above learning laws require about 48 bits of dynamic range because of the repeated multiplications and divisions. An 8-bit exponent will give 256 bits of dynamic range and is more than adequate.

10.5.3 Silicon Architecture

We need an architecture that can function in both a pipelined and in a parallel configuration. Each of the functional units needs to communicate with both its neighbor in a pipelined manner and globally in the parallel manner. An architecture along the lines of Figure 10.10 achieves these goals. Researchers have used this style of architecture in an image processor [3]. The image processor implemented both matrix operations and convolutions. The data path was parallel for the matrix operations and pipelined for the convolutions.

The architecture consists of five connected functional units. Each functional unit has its own local control unit, local execution unit, and local parameter file. There are four pipeline busses (Y1–Y4) that connect each local execution unit with its downstream neighbor. There is also a broadcast bus (G) that allows global data to broadcast simultaneously to all of the functional units. The broadcast bus likewise allows each of the functional units to broadcast data back to the global data storage. The global control sends meta-instructions to the functional units. These meta-instructions relate to the task that the system currently works on. The local control units return their status to the global control unit. This way the global control unit can send a meta-instruction to perform a task. The local control unit responds when it completes the task or requires additional information.

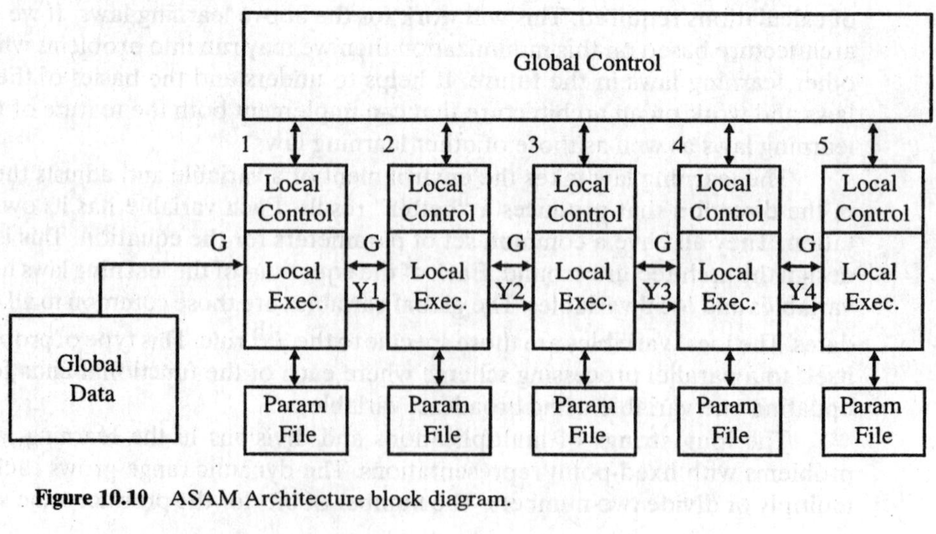

Figure 10.10 ASAM Architecture block diagram.

10.6 ASAM CHIP

An ASAM chip based on the architecture shown in Figure 10.10 is practical with today's silicon technology. The architectural concepts combine current microcode, parallel processing, and signal processing techniques. The optimization of the architecture to the task offers the advantage over general purpose implementations. The design divides roughly between the data path architecture encompassing the local execution units and parameter files and the control architecture encompassing the global control unit and the local control units.

10.6.1 Data Path Architecture

Figure 10.11 shows the basic block diagram for each of the local execution units with the parameter file integrated. This architecture is practical as long as the parameter file is less than about 4K bytes. Much more than that and the cost to have the RAM implementing the parameter file on-chip and integrated into the data path exceeds the cost of an external RAM. The basic architecture is an 8-bit derivative of the venerable Am2901 architecture [1]. This architecture has formed the basis for many CPU data paths and continues to provide many capabilities. We have enhanced the basic architecture for the ASAM application by adding the multiplier, the extension of the Q register and shifter, the parameter file, and the broadcast bus G. These additions greatly increase the performance of the data path. The local execution unit consists of a register file, the ALU and multiplier functional units, the Q shift register, the parameter file, and the essential multiplexers and buses that interconnect the various units.

The register file is the core of the data path. It provides 16 registers that have 2 read ports that provide data to the other functional units and one write port to accept data from the other units. The local control unit gives the addresses and control for the register file. The "A" and "B" ports have independent addresses from the local control unit and place data on the [A] and [B] buses. The address for the write port is often the same as the "B" port but this need not be the case. These 16 registers form the majority of the temporary variables for the system. They give intermediate calculations concatenated for calculations of up to 128 bits in length or fast access to frequently used parameters.

The ALU and multiplier provide the main computation resources for the data path. The ALU provides the addition, subtraction, and complement functions. The status bits Z, C, and S reflect the status of the most recently completed operation. The Z bit shows whether the result of the previous operation was zero. The C bit is the carry-bit and acts as the carry-in for the current operation and the carry-out for the previous operation. We can also set it under control of the local control unit. The S bit reflects the sign of the last operation. This bit is essential for signed operations and fast normalization. Using both the C and the S bits gives a signed number overflow. The multiplier takes two 8-bit unsigned integers and gives an unsigned 16-bit product.

The Q shift register gives a fast single-cycle accumulator for the multiplier products. Without this 16-bit register multiplications would require 2 cycles to store the product instead of just the one. The 16-bit Q register increases the performance of the multiplier without greatly increasing the size of the data path. The other uses

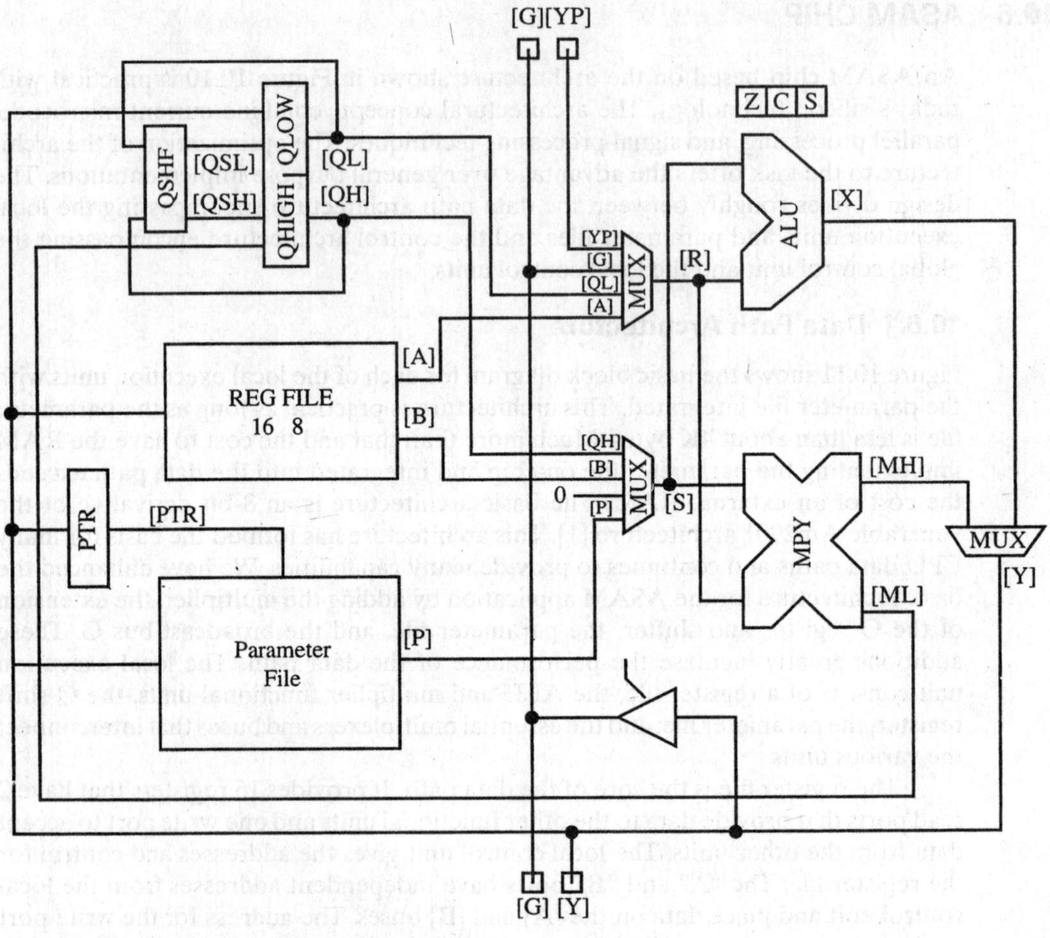

Figure 10.11 ASAM data-path block diagram.

for the Q shift register include division and exponent normalization. The number of divisions in the algorithm stays to a minimum and so there is no dedicated divider. The Q shift register speeds non-restoring division by shifting the dividend independent of the addition or subtraction of the divisor. The shift register also uses the carry status bit and treats C as the shift-in for the current operation and the shift-out for the previous operation. We can shift data in the register file without the Q shift register if we pick the desired register on both the [A] and [B] buses and add. This way algorithms can shift numbers longer than 16 bits to the left. We can also use the Q shift register operated concurrently with another addition, multiplication, or data movement operation.

The parameter file gives fast local storage for parameters. The parameter file pointer PTR provides the address for the parameter file. The pointer loads from one of the internal buses and provides the address for the parameter file. The paths between the parameter file and the global data bus "G" must go through the ALU. This is not a problem in this application as external accesses to the parameter file tend to be rare. If the size of the parameter file is larger than is practical then we may use an external RAM. Then the system will share the bus that goes into the parameter file and the bus that leaves the parameter file to minimize the number of pins required to connect the external RAM.

The multiplexers and buses route data to the blocks inside the data path. We have arranged them so that the majority of the operations and transfers takes place in a straightforward manner. There is also a buffer between the outgoing [Y] and the global [G] buses. This allows the contents of registers or parameters inside the local execution unit to pass outside to the global data storage or broadcast to all other local execution units.

10.6.2 Control Architecture

The ASAM control architecture is a variant of the two-level microcode/nanocode scheme used in the HECTOR microprocessor [8]. With HECTOR the microcode initiates a nanocode sequence that controls the operation of the execution unit. Then separate teams implement the control and data architectures. We could develop the control unit and microcode without having the nanocode and execution unit completed. We have adopted this methodology as well. Unlike the traditional cases each of the local control units need not execute the same microinstruction sequences. When the global control unit dispatches a meta-instruction then each of the local control units interprets it based on the task and phase of the current task. Sometimes we directly decode the meta-instructions into data path control signals. More often the meta-instruction starts a longer sequence of microinstructions that may themselves contain subroutines or loops to perform the task. After completion of the meta-instruction the local control unit signals the global control unit that it has finished the task. The global control unit can then provide the next meta-instruction.

Figure 10.12 shows the relationship between the meta-instructions of the global control unit and the resulting microoperations of the local control units. The meta-instruction points to a microoperation routine in each of the local control units. The 5 local control units proceed to sequence through their individual routines. The local control units finish their routine and then signal back to the global control unit to give the next meta-instruction.

The global control unit coordinates the operation of the entire device. It directly controls the global data memory and interfaces to the rest of the system. It also dispatches the meta-instruction for each of the local control units. These meta-instructions state the task as well as the phase of the task that the local control unit operates on. Depending on the granularity of the task it may be as simple as loading a register or as complex as multiplying two block floating-point numbers. It can also be far more complex and initiate multiple operations. The global control unit begins the next meta-instruction or series of instructions after each of the local control units signal they have finished their task. The global control unit serves to synchronize the parallelism in the system. This way each of the data paths in the system can perform different tasks on different data independently.

The local control unit controls the operation of the data path itself. The local control unit controls each bus, multiplexer, and operator in the data path. The data

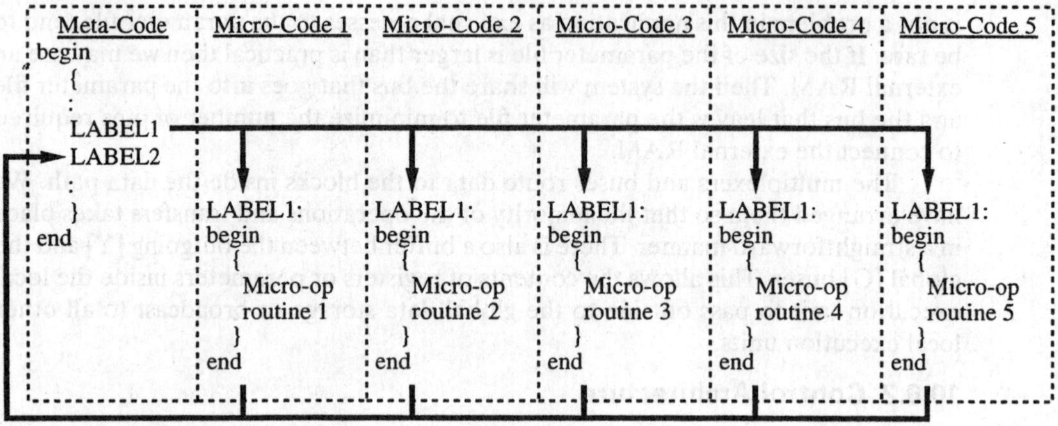

Figure 10.12 Meta-instruction dispatching.

path also gives feedback in terms of the Z, C, and S status bits. The local control units loop and branch based on the values of these status bits. These conditional operations are why the synchronizing effects of the global control unit are so important. Without global resynchronization the entire system must step through the worst-case code without regard for whatever local optimizations could take place to minimize overall execution times.

The separation of functionality between the global control unit and the local control units eases the development of code for the device. The global control unit sends meta-instructions that break down each task and phase of the system. It is not a problem to develop 5 different microinstruction sequences that operate in harmony. But it may be a big development problem to maintain some semblance of order while each of the 5 microinstruction sequences executes with some portion of conditional code. The sheer number of permutations of concurrent data path operations can be staggering. The microinstructions operate independently of one another and this helps mitigate the problem. The data transferred between data paths does so in synchrony at the beginning of the meta-instruction cycle. This is the time in a meta-instruction when we know the status of all of the functional units. When the activity that deals with data outside of a particular datapath has ended then the local control unit may control the data path autonomously. This local control is optimal to the task at hand and has no further relation to the operation of the other data paths or external resources.

10.7 ASAM OPERATION

ASAM operation has two distinct tasks: the evaluation task and the update task. The evaluation task takes the input x_t and evaluates the output $F(x_t)$. The update task modifies each of the SAM parameters based on a learning law to give an output closer to the desired output. The global control unit controls the local functional units and these units perform differently in the two phases.

10.7.1 Evaluation Task

The evaluation task consists of five pipelined phases. Each of the phases requires the output from the previous phase. Each of the local control units performs the same function once for each rule. Figure 10.13 shows the configuration of the architecture for this task and uses the Gaussian set function. The global data memory sends the value of x_t to the first functional unit at the start of each phase and then releases the bus. After all the functional units have completed their phase the global control unit has functional unit 2 send its results as a pointer to the look-up table in the global data memory. Then the global control unit has the global data memory send the translated value back to that functional unit 2 that in turn gives the value as the result of its calculations.

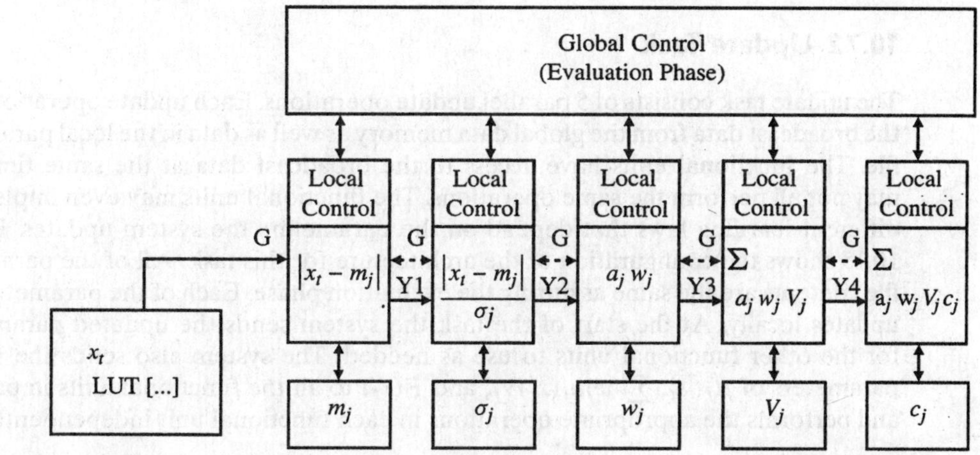

Figure 10.13 Evaluation task for a Gaussian ASAM.

The parameter file in each of the functional units deals with only one of the SAM parameters. This means there can be as many rules as there are locations in the parameter file. A 1K-byte parameter file can accommodate 1K rules. Intermediate calculations pass between the functional units by way of the pipeline bus.

The system computes $F(x_t)$ after it has evaluated all of the rules and flushed pipeline. The system copies the contents of the sum in functional unit 4 to functional unit 5. The final phase performs the final division that produces $F(x_t)$ and converts the result of the sum in functional unit 4 to the block floating-point format for later use. Functional unit 5 performs the division while functional unit 4 performs the conversion. The remaining functional units are not active and may perform other calculations. When the final phase ends the system sends $F(x_t)$ and the converted sum to the global data memory for later use.

Figure 10.14 shows the flow of operations through the pipeline. Each functional unit works on the results of the previous functional unit. The entire calculation takes place over 5 phases in the pipeline. The advantage of the pipeline over other approaches is that each functional unit is active during each phase although all of the functional units do not operate on the same rule during any phase.

1	$\lvert x_t - m_j \rvert$	$...m_{j+1}$	$...m_{j+2}$	$...m_{j+3}$	$...m_{j+4}$
2	LUT $\left[\frac{...}{\sigma_{j-1}}\right]$	LUT $\left[\frac{\lvert x_t - m_j \rvert}{\sigma_j}\right]$	LUT $\left[\frac{...}{\sigma_{j+1}}\right]$	LUT $\left[\frac{...}{\sigma_{j+2}}\right]$	LUT $\left[\frac{...}{\sigma_{j+3}}\right]$
3	$w_{j-2}...$	$w_{j-1}...$	$w_j\, a_j\,(x_t)$	$w_{j+1}...$	$w_{j+2}...$
4	$...V_{j-3}$	$...V_{j-2}$	$...V_{j-1}$	$w_j\, a_j\,(x_t) V_j$	$...V_{j+1}$
5	$...c_{j-4}$	$...c_{j-3}$	$...c_{j-2}$	$...c_{j-1}$	$\Sigma\, w_j\, a_j\,(x_t) V_j c_j$
	t_1	t_2	t_3	t_4	t_5

Figure 10.14 Gaussian ASAM evaluation phases through the pipeline.

10.7.2 Update Task

The update task consists of 5 parallel update operations. Each update operation uses the broadcast data from the global data memory as well as data in the local parameter file. The functional units have access to the broadcast data at the same time but may not all perform the same operations. The functional units may even implement different learning laws that depend on the parameters the system updates. Figure 10.15 shows the configuration of the architecture for this task. All of the parameter file contents are the same as during the evaluation phase. Each of the parameter sets updates locally. At the start of the task the system sends the updated parameters for the other functional units to use as needed. The system also sends the global parameters of μ_j, ε_t, $\sum w_j a_j(x_t) V_j$, and $F(x_t)$ to all the functional units in parallel and performs the appropriate operations in each functional unit independently.

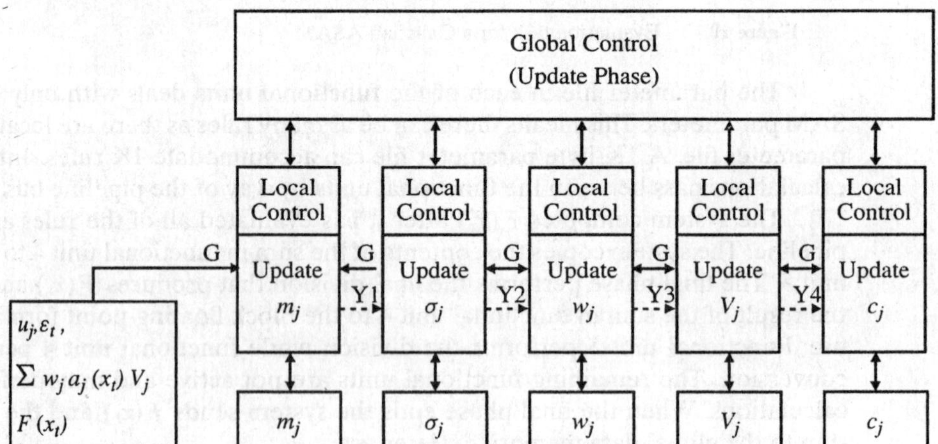

Figure 10.15 Update task for the Gaussian ASAM.

Figure 10.16 shows the concurrent operation of the task. Each functional unit updates one of the parameters during the phase. All parameters for each rule update

Functional Unit					
1	Update m_j	... m_{j+1}	... m_{j+2}	... m_{j+3}	... m_{j+4}
2	Update σ_j	... σ_{j+1}	... σ_{j+2}	... σ_{j+3}	... σ_{j+4}
3	Update w_j	... w_{j+1}	... w_{j+2}	... w_{j+3}	... w_{j+4}
4	Update V_j	... V_{j+}	... V_{j+2}	... V_{j+3}	... V_{j+4}
5	Update c_j	... c_{j+1}	... c_{j+2}	... c_{j+3}	... c_{j+4}
	t_1	t_2	t_3	t_4	t_5

Figure 10.16 Update phases through the Gaussian ASAM architecture.

at the same time. When the system has updated all parameters then the global controller moves to the next rule. The advantage of the parallel approach is that each of the parameters updates independently although they all in effect use the same data for the learning law.

10.7.3 Performance

We can implement the above ASAM architecture in a reasonably sized VLSI device. None of the requirements or architectures places special demands on the silicon. Based on current VLSI technology and libraries this architecture should run in excess of 50 MHz. This means that a new microinstruction can start every 20 ns. At this performance level the evaluation task alone can complete a SAM rule evaluation in about 300 ns or about 3 million rules per second. The complete ASAM evaluation and update tasks can evaluate and update a rule in about 4 μs or about 250,000 fully updated rules per second.

The above approach is a brute-force approach. There are many performance enhancements that do not require architectural changes. They depend on the observations of actual use rather than the strict implementation of the algorithm. The simplest enhancement has to do with the number of rules that affect the output of the SAM. We need "fire" or evaluate only those rules that affect the output and thus have a positive joint if-part fit value: $a_j(x) > 0$. In a typical system there may be only a small percentage of the rules that affect the output but this percentage can grow the system's dimensions. We determine which rules will affect the output and ignore those that do not. This achieves a major increase in speed.

For systems with more than one dimension this reduction of evaluation is even more significant. Consider a system with 6 inputs with 10 fuzzy sets per input. Fully expanding this system would give 10^6 rules to cover the graph of approximand. It would take about 4 seconds to evaluate fully the rule base at the rates we have estimated for the above architecture. Any given input x might belong to 4 or fewer if-part sets. This means that there would be roughly 4,000 rules that affect the output. Just evaluating the rules that affect the output would result in over a 200 times speed-up of the system. This is a key way to deal with rule explosion. We cannot eliminate the exponential explosion of the rules or the processing time but we can reduce its impact.

10.8 CONCLUSION

It is not hard to extend the point ASAM to the set ASAM. The biggest difference comes in how we compute $a_j(A)$ rather than $a_j(x)$. Instead of evaluating the set function as a point we evaluate it as the intersection of two sets. We can approximate the sets with triangles. In the proposed architecture the first two functional units would pick the overlap. This process is somewhat longer than the function evaluation and would slow down the evaluation task to about 1.5 million rules per second. The set intersection affects only the evaluation task. In a set ASAM implementation the architecture could evaluate and update nearly 200,000 rules per second. This gives a drop in performance of only about 20%. This is because the update task takes much more time than does the evaluation task. The same performance enhancements apply to the set SAM as well as to the scalar SAM. So fairly large set ASAM systems are practical.

The advantage of custom VLSI over general purpose VLSI is not always in the performance increase. Many of the custom chips cannot outperform workstation-class general purpose computers. The advantage of custom VLSI lies in the cost of the solution. A $4 chip might not outperform a Pentium class PC but then it does not cost $2,000. And 500 of the $4 chips will far outstrip the PC at the same cost. Future advances in silicon technology may well lead to other ASAM architectures. The silicon technology will change but the fundamental problem of exponential rule explosion will not.

REFERENCES

[1] Advanced Micro Devices, "AM2901 Four-bit Bipolar Microprocessor Slice Data Sheet," 1992, Sunnyvale, CA.

[2] American Neuralogix Inc, "NLX220 Fuzzy Logic Controller Data Sheet," Sanford, FL 1993.

[3] Aono, K., Toyokura, M., and Araki, T., "A 30 ns (600 MOPS) Image Processor with a Reconfigurable Pipeline Architecture," Custom Integrated Circuits Conference, 1989, pp 24.4.1–24.4.4.

[4] Corder, R. J., "Architectures for Custom VLSI Processor Based Embedded Fuzzy Expert Systems," *Proceedings of the 3rd IFSA Congress*, 1989.

[5] Corder, R. J., "A High-Speed Fuzzy Processor," *Proceedings of the 3rd IFSA (International Fuzzy Systems Association) Congress*, 1989.

[6] Legg, Gary, "Microcontrollers Embrace Fuzzy Logic," *Electronic Design News*, Sept. 16, 1993, pp. 100–109.

[7] Miki, T., Matsumoto, H., Ohto, K., and Yamakawa, T., "Silicon Implementation for a Novel High-Speed Fuzzy Inference Engine: Mega-Flips Analog Fuzzy Processor," *Journal of Intelligent and Fuzzy Systems*, Vol. 1, No. 1, 27–42 1993.

[8] Miller, T. K., Bhura, B. L., Barnes, R. L., Duh, J.-C., Lin, H.-B., Van den Bout, D. E., "The HECTOR Microprocessor," International Conference on Computer Design: VLSI in Computers, 1986, pp. 406–411.

[9] Wang, Jennifer, "12-Bit Fuzzy Computational Acceleration (FCA) Core," Embedded Systems Conference, 1993.

HOMEWORK PROBLEMS

10.1. SAM Learning. Estimate the time/area product for the following learning laws assuming 8-bit operators. Assume you have pre-computed $\sum_{j=1}^{m} w_j \; a_j(x) \; V_j$. Estimate the time it takes to complete each learning law with an addition taking 8 ns, a multiplication taking 20 ns, and a division taking 16 cycles. (Be sure to minimize the total number of operations and account for the necessary registers.)

(a) Gaussian

$$\sigma_j(t+1) = \sigma_j(t) + \mu_t \varepsilon_t p_j(x_t)(c_j - F(x)) \left(\frac{(x - m_j)^2}{(\sigma_j)^3} \right)$$

(b) Cauchy

$$m_j(t+1) = m_j(t) + \mu_t \varepsilon_t \frac{p_j(x_t)}{a_j(x_t)}(c_j - F(x)) \left(\frac{2d_j^2(x_t - m_j)}{[(x_t - m_j)^2 + d_j^2]^2} \right)$$

(c) Exponential

$$u_j(t+1) = u_j(t) + \mu_t \varepsilon_t p_j(x_t)(c_j - F(x_t))(v_j - x_t)$$

(d) Rule weights

$$w_j(t+1) = w_j(t) - \mu_t \varepsilon_t \frac{p_j(x_t)}{w_j(t)}[c_j - F(x_t)]$$

(e) Sign Learning

$$\sigma_j(t+1) = \sigma_j(t)(1 + k \operatorname{sgn}(\varepsilon_t) \operatorname{sgn}(c_j - F(x)))$$

10.2. Set SAM. Evaluate the $a(A)$ for these triangular sets

$$a = \{[d, 0], [e, 1], [f, 0]\} \qquad (d < e < f)$$

$$A = \{[g, 0], [h, 1], [i, 0]\} \qquad (g < h < i)$$

for these conditions:

(a) Where $g < d$ and $i = e$

(b) Where $g = e$ and $i > f$

(c) Where $g > d$ and $i < f$ and $h = e$

(d) Where $g < d$ and $i > f$ and $h > e$.

A light beam computes a Fourier transform as it passes through a thin lens. This physical computation is the emblem of the modern optical information processing. Optics excels at computing the parallel multiplies and sums in formal integration. Optics can also compute divisions and subtractions. These operations take more effort and may yield less precision.

Optics holds promise for fuzzy systems for at least two reasons. Each fuzzy output $F(x)$ in a standard additive model (SAM) depends on only multiplies and adds and one division. A light-based system can compute these terms with standard mirrors and lenses and spatial light modulators. And the vector input x fires all rules in parallel. One data light beam can fire all the stored if-part fuzzy sets. The drawback is that optical systems cannot easily store or tune the fuzzy sets.

Chapter 11 explores the device options and constraints that face optical additive fuzzy processors. The chapter shows how electro-optical designs can capture or at least approximate the operations in a SAM inference from input x to output $F(x)$. Modern optical devices restrict such optical fuzzy systems to the domain of research and laboratory test. Future advances in devices may someday allow these optical designs to compete with silicon designs. Photons can after all pass through one another without interference while electrons cannot.

The hologram is a second emblem of optical computing and one that stands for an active area of research. Van Heerden showed in 1963 that a volume holographic crystal can in theory store on the order of 10^{13} bits of information per cubic centimeter. The wavelength of light defines the side lengths of the minimum 3-D cube that can store one bit. Crystals the size of sugar cubes may someday store databases the size of small libraries. They may also store and tune and process vast banks of fuzzy rules.

11

OPTICAL ADDITIVE FUZZY SYSTEMS

Clark C. Guest

PREVIEW: SAMS IN OPTICS

Optical systems can add, multiply, distribute, and collect data. So they can implement additive fuzzy systems. This chapter presents three optical implementations of standard additive model (SAM) fuzzy systems.

The first system shows how an optical system adds, multiplies, distributes, and collects fuzzy data. The optical fuzzy systems use these operations in new ways and combinations. The optical fuzzy system can process 32 million fuzzy rules per second. The chapter ends with how future optical fuzzy systems may deal with learning and exponential rule growth.

11.1 FUZZY COMPUTATION: ELECTRONICS IN TIME VERSUS OPTICS IN SPACE

Implementations of fuzzy systems range from simple programs written in BASIC to special purpose processors using state-of-the-art technology as in Chapter 10. Optical implementations of fuzzy systems promise to extend fuzzy system processing to even higher levels of performance. Applications of optical fuzzy systems include fuzzy database operations, high-speed fuzzy control of complex systems, wireless communications and video compression, and fuzzy virtual reality and information systems.

The key advantage of optics for fuzzy systems is easy to understand. Fuzzy systems depend on fuzzy sets. The if-parts of fuzzy rules are fuzzy-set functions and so are the then-parts. These set functions are the operands of fuzzy system operations. Electronic wires, switches, and storage cells represent a voltage, a current, and a single value. Electronics represents functions as functions in *time*: Wires, switches, and storage cells represent single values of voltage and current. This slows operation. Optics represents functions as *spatial* distributions of intensity. Optical systems process two-dimensional information. One spatial dimension of an optical system can represent the fuzzy-set functions. The other spatial dimension can help process many fuzzy rules in parallel.

11.2 PARALLEL FUZZY-SET FUNCTION OPTICAL PROCESSOR

Optical fuzzy-set functions form the basis of the additive fuzzy system processor [2] shown in Figure 11.1. This additive fuzzy system [4] processes many fuzzy rules in parallel. It processes each fuzzy rule at a different vertical position in the processor. Information from the rules mixes only at the final output detector.

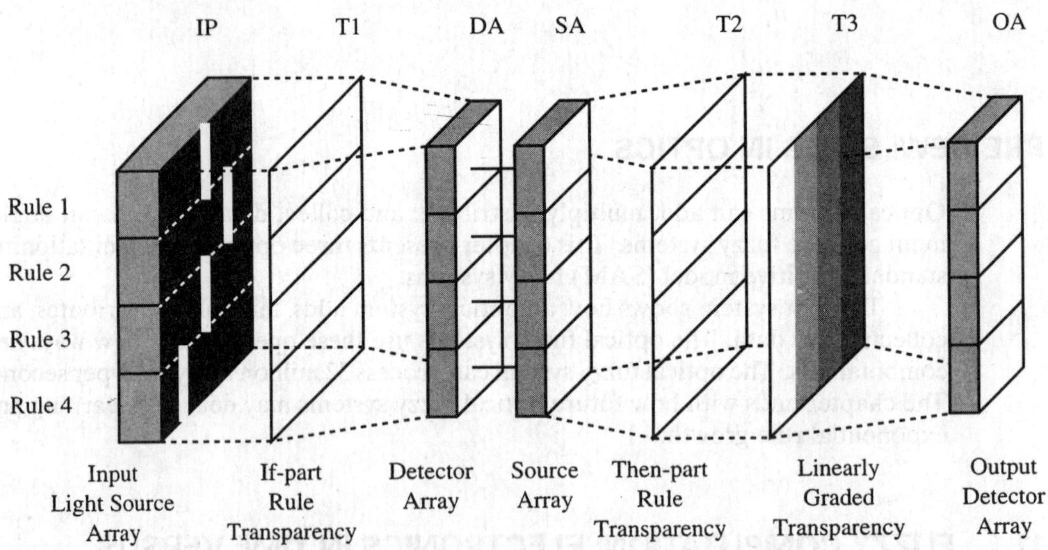

Figure 11.1 The optical parallel fuzzy set processor fires fuzzy rules in parallel.

Input for the system comes from a 2-D array of light sources labeled IP in Figure 11.1. The sources can be discrete devices such as light emitting diodes or semiconductor lasers or they can be different locations on the phosphorescent screen of a television picture tube. Each horizontal row in the array gives input for one fuzzy rule. The value of the input picks how far across the row a light source comes on. This allows for input of both nonfuzzy and fuzzy values. If an input is nonfuzzy then a single position in the row comes on at full strength. If an input value is fuzzy (as in the set SAM of Chapters 2 and 10) then the row's output intensity gives its fuzzy distribution. Figure 11.2 shows an example input array. Some rows represent nonfuzzy or point inputs and some rows represent fuzzy inputs. The optical system can approximate both a simple point SAM and a set SAM of the form $F(A) = \sum_{j=1}^{m} p(A)c_j$ as discussed in Chapter 2. The optical system approximates $F\colon R^n \to R$ by using a lens system to integrate fired then-part sets and thus approximates the total output centroid in $F(x) = \text{Centroid}(B(x))$.

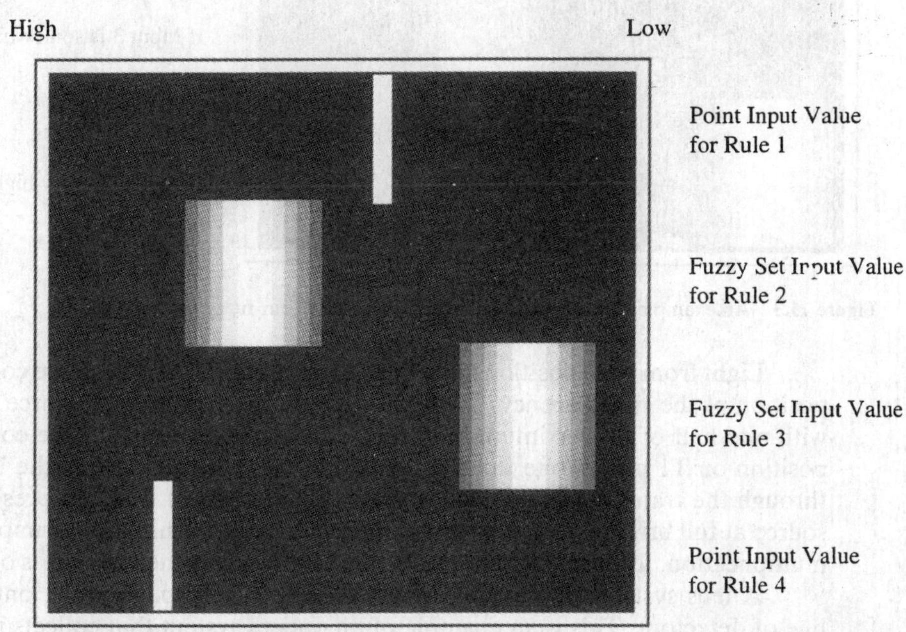

Figure 11.2 An example array of input data for the parallel fuzzy rule processor. A CRT, an array of LEDS or laser diodes, or other devices could produce this array.

Transparency T1 in Figure 11.1 shows the if-part fuzzy-set functions. Transparency T1 is an optoelectronic device that changes its pattern of transparency (its picture) in response to an electronic signal. The general name for such a transducer is a spatial light modulator (SLM). Many devices can implement them. A common SLM is a liquid crystal TV with the back of the screen removed so light can shine through. The key idea is that the device can be more transparent in some locations and less transparent in others. Horizontal patterns of transparency can define the set functions. One horizontal strip represents each fuzzy rule as with the input array IP. The horizontal strip is most transparent where the corresponding set function has a high value. The strip is opaque where the set function has a value of zero. The transparency shown in Figure 11.3 defines some sample set functions.

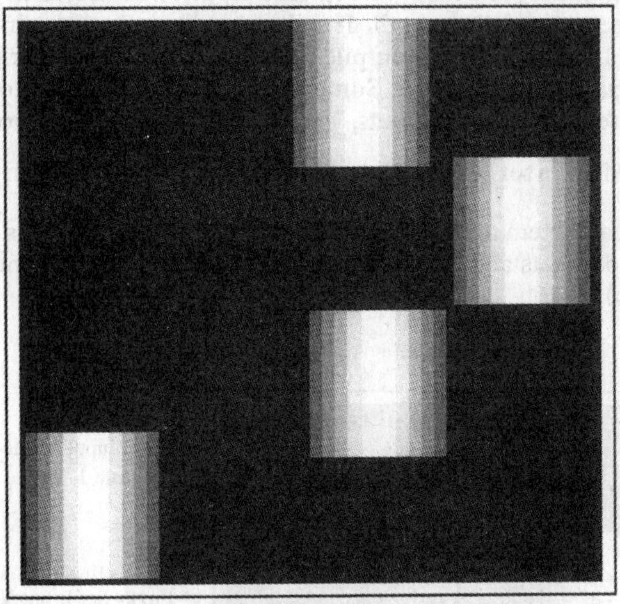

Figure 11.3 An example if-part transparency for processing four fuzzy rules in parallel.

Light from each position in the input array IP passes through the corresponding position of the transparency Tl. Suppose that a particular light source in IP shines with one half of its maximum brightness. Suppose further that the corresponding position on T1 passes one third of the light that falls on it. Then the light coming through the transparency is one sixth ($= 1/2 \times 1/3$) of the light present with the source at full brightness and with T1 fully transparent. This is an example of optical multiplication. The transparency of the SLM multiplies the brightness of the source.

A lens system focuses light passing through transparency T1 onto a vertical line of detectors. This is an example of an optical system that collects information. Collecting information in electronic circuits requires long wires that slow down the circuit. Collection of optical information with a lens has no practical speed penalty. The total light intensity falling on each detector is the sum of all light passing through the corresponding row of T1. Each detector connects to a light source. This light source can be a light emitting diode or a semiconductor laser. The light output of each source must be linearly proportional to the total light intensity that falls on the corresponding detector. The light intensity outputs of these sources are the firing level for the fuzzy rules they store.

A lens system spreads the light from each source into a horizontal line with constant intensity across its width. This is an example of optical distribution of information. Optics also performs this operation better than does electronics. These lines of light fall on transparency T2 in Figure 11.1. Transparency T2 stores the then-part sets of the fuzzy rules just as T1 stored the if-part sets. The set functions define varying levels of transparency across the rows. Figure 11.4 shows a sample T2

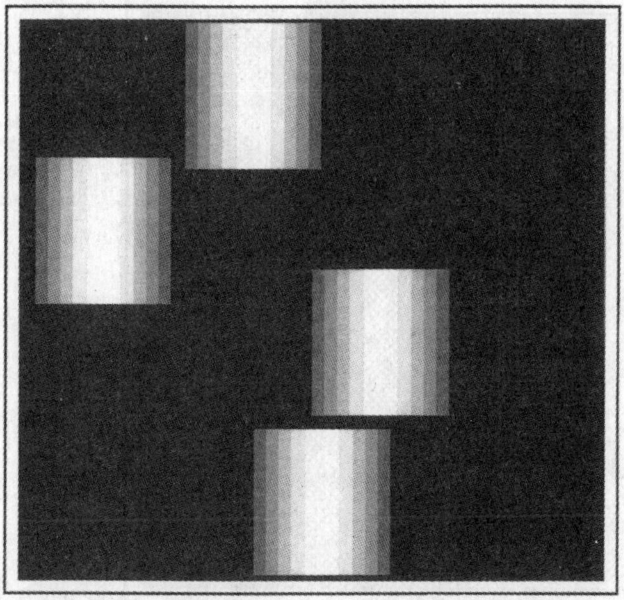

Figure 11.4 An example then-part transparency for processing four fuzzy rules in parallel.

transparency. The distribution of light intensity from each row of T2 is the then-part set function multiplied by the activation level of its rule. A spatial light modulator implements T2.

Transparency T3 and a lens system perform the centroid calculation to "defuzzify" or convert the fuzzy sets of T2 into a scalar output value for each rule. The light that T3 transmits varies across its width. T3 is opaque along one edge, transparent along the opposite edge, and varies linearly in between. Figure 11.5 shows a sample T3 transparency. The lens system following T3 sums the light intensity along each row. We can approximate the centroid in the standard additive model (SAM) system as a function $f(x)$ integrated between two limits: Centroid$(B) \approx \int_u^v yb(y)dy$. Here $f(x)$ is each of the fuzzy sets in T2, T3 gives the x that multiplies those sets, and the lens system integrates their product. The SAM approximation holds because the total incident light is constant. So the output of the position sensitive detector stays proportional to the centroid of the output set B. The normalizing integral stays constant.

The vertical line of intensities representing the centroid output for each rule falls on a set of detectors DA. Simple electronics takes the outputs of neighboring detectors, sums them, and normalizes them as groups. Groups of detectors combine to produce one output value from several rules.

The data rate of the parallel fuzzy set optical processor is extremely high. The modulation frequency of the light sources used in the input plane IP and the summing plane SP sets the speed of the system for processing a fixed set of rules. Today we can modulate semiconductor laser diodes with 1 billion values per second. A processor

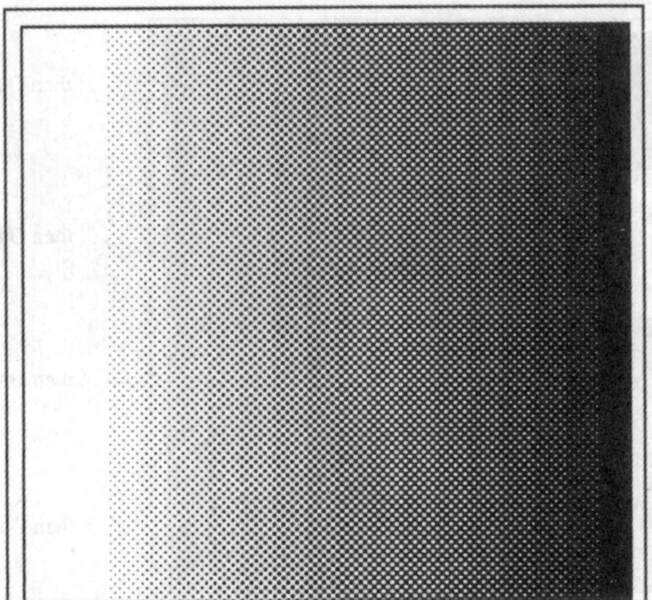

Figure 11.5 A linear graded transparency that could find the centroid of the combined output fuzzy set B in an additive fuzzy system: $F(x) = \textbf{\textit{Centroid}}(B(x))$.

made with 64 parallel channels will perform 64 billion fuzzy rule evaluations per second. The response time of the transparencies T2 and T3 fixes the rate for changing these rule bases. Current devices offer rates of 30 rule-base changes per second. Researchers are working on faster devices at Texas Instruments, Bell Laboratories, and many universities.

11.3 SCANNING LASER DIODE FUZZY SYSTEM PROCESSOR

New device technologies drive the evolution of optical processing architectures. The additive fuzzy system processor of Itoh [3] relies on new scanning laser diodes. The physics of the scanning laser diode is quite complex but the principle is simple. Two electrical currents drive the scanning laser diode. If the two currents are equal then the laser beam shoots straight out of the face of the laser. If the currents are not equal then the beam shoots out at an angle. Figure 11.6 shows this effect. Varying the ratio of the currents controls the angle of the beam.

One more technical detail of the laser diode is important. The beam projected has an intensity pattern that is high in the center and tails off symmetrically on both sides. Figure 11.6 shows this pattern. The intensity distribution is bell shaped or Gaussian (or alpha-stable as discussed in Chapters 6 and 9). This natural bell-curve intensity curve serves as the shape for the fuzzy sets.

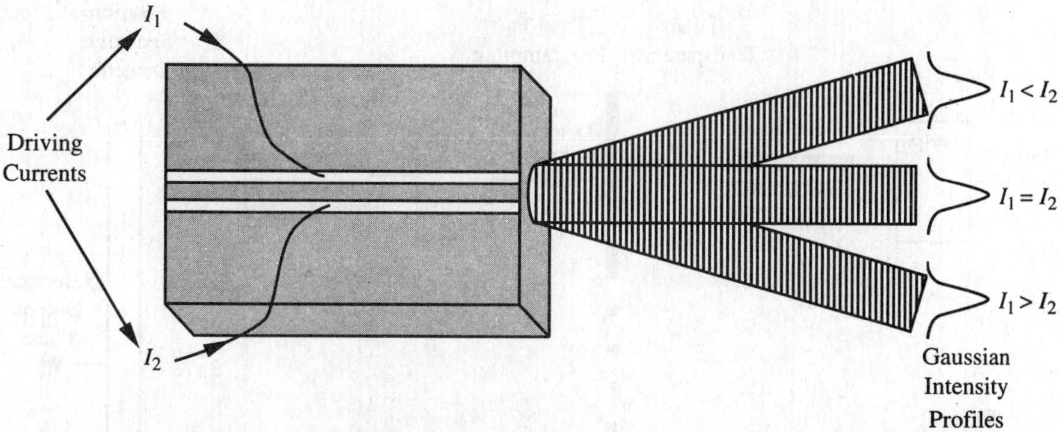

Figure 11.6 Two currents drive the beam scanning laser diode. The difference between the currents determines the angle of the output beam. The light intensity profile of the beam is approximately Gaussian.

Figure 11.7 shows the structure of the scanning laser diode fuzzy system processor. Data enter the system as electrical signal values, voltages, or currents. This system accepts only scalar inputs. Each input controls the ratio of currents that drive its scanning laser diode in SLDA1. The input signal magnitude defines the beam angle. The angle of the beam fixes the amount of light each detector receives in the array DA.

A set of electrical switches pick which detector or detectors drive the scanning laser diode in SLDA2. The if-part set of the fuzzy rule controls which switches are open and which are closed. If the rule has an if-part set "is-high" then a detector at one end of the array drives the laser. A specification of "is-low" connects the detector at the other end of the array to the laser. The method of driving differs in SLDA2 from that used in SLDA1. In SLDA2 the fuzzy rules pick the fixed angles of the beams. The electrical signals from the detectors modulate the laser intensities and not the angles.

The then-part sets of the fuzzy rules fix the angles of the beams projected from SLDA2. If the rule has the then-part set "is-high" then the corresponding beam projects to one side of the detector plane PSD (position sensitive detector). A specification of "is-low" angles the beam to the other side of PSD. Intermediate specifications call for intermediate angles.

The output detector performs the centroid defuzzification. PSD stands for position sensitive detector. This device has the property that the electrical output signal depends on both the intensity and position of light falling on the detector. A spot of light falling on one end of the detector gives a small signal. A spot with the same intensity falling on the other end gives a large signal. The total signal the detector emits is the sum of all light intensities falling on it multiplied by their respective positions. This is just the centroid of the light intensity distribution.

The speed of the position sensitive detector limits the speed of scanning laser diode fuzzy system processor. The PSD has a large area and large detectors are slower than small detectors. Existing PSD devices operate at a rate of about 1 million

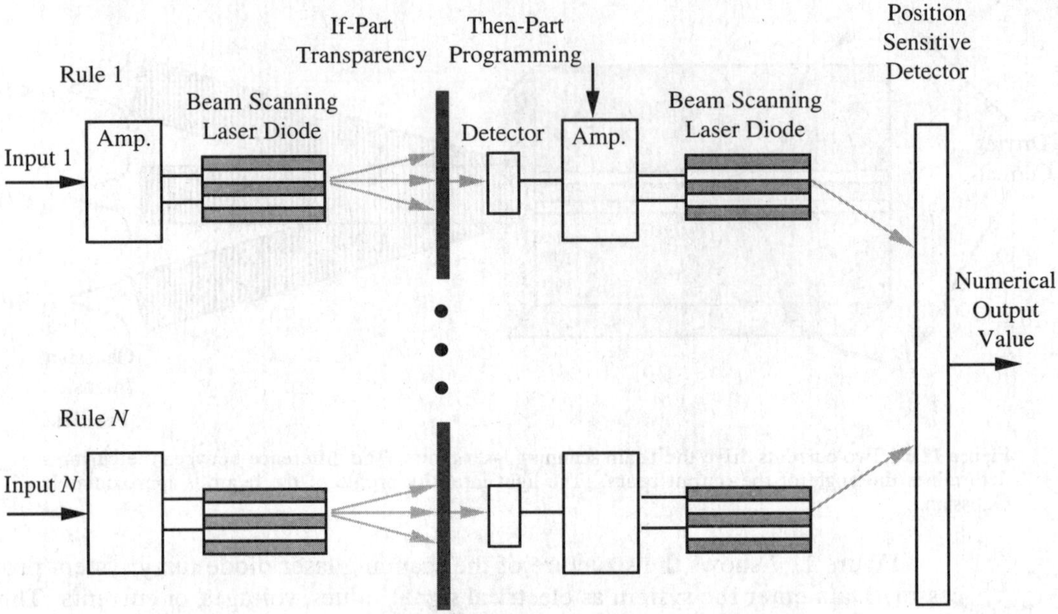

Figure 11.7 The fuzzy rule parallel processing system based on beam scanning laser diodes. Input values control the angles of the left side lasers. Photographic transparencies determine the IF-Part of each rule. The THEN-Part of each rule depends on an electrical input that fixes the angle of the right side laser. The amplitude of each right side laser output depends on the signal from its associated detector.

detections per second. Scanning laser diodes are more complex than common laser diodes. So we assume smaller arrays such as 32 devices in SLDA1 and 32 in SLDA2. Then this design can process 32 million fuzzy rules per second. The rules do not depend on SLMs but on the configuration of electrical switches and scanning laser diodes. These changes are fast. So no speed penalty results from changing the rules for each processing cycle.

11.4 DSTOP FUZZY SYSTEM PROCESSOR

The dual-scale topology optoelectronic processor (DSTOP) [5] differs from the above designs. DSTOP combines optical and electronic components to form a system for generalized matrix algebra. Generalized matrix algebra resembles conventional matrix algebra except multiplication and addition give way to other operations. The DSTOP fuzzy system processing casts the fuzzy modus ponens as an operation of generalized matrix algebra.

The fuzzy modus ponens derives from the modus ponens of traditional binary logic. Modus ponens asserts that if it is true that A implies B and if A is true then B is true. The fuzzy modus ponens is a logical extension: If it is true to some degree that A implies B and if A′ is similar to A and if A′ is true then B′ is true to some degree where B′ is similar to B [4]. Generalized matrix algebra becomes involved when A is not a single observation but a collection of them and likewise for A′, B, and B′. Vectors of fuzzy values represent these collections. The implication relations are a matrix of fuzzy values relating the vector A to the vector B. Notice that fuzzy set functions do not enter into this formulation. They determine the component values of A, B, and the implication matrix. This approach combines a vector of A′ values with the fuzzy implication matrix through generalized matrix vector multiplication to yield a B′ vector. The question is which operations to use in the generalized matrix vector multiplication. The DSTOP architecture is flexible enough that it can use any pair of the many proposed multiplication-like and addition-like functions from MIN and MAX to multiplication and addition.

The search for the optimum mix of optics and electronics to provide the fastest computation for the least energy drove development of the DSTOP architecture. With existing technology this balance uses optics for data distribution and electronics for data collection and combination. Each element of the input vector distributes optically over a row of the matrix. At each matrix location the optically entered value combines electronically with a stored matrix value. Matrix values in each column combine pairwise electronically until there is one result per column. For an *additive* fuzzy system processor the matrix elements perform multiplication and addition combines values within a column. Whether these operations occur in analog, bit-serial digital, or other circuitry depends on how we encode the data.

The core of DSTOP is that a given layout of matrix elements gives an efficient optical distribution and electrical collection. Figure 11.8 shows the layout of a DSTOP system. A1, A2, A3, and A4 are elements of the distributed vector. A lens array gives four demagnified images of the input plane. Each element of the replicated input plane falls on an optical detector in the electronic plane. Figure 11.9 shows the details of the electronic plane. The data paths connecting optical detectors form an H-tree pattern. This arrangement provides pairwise combination of column elements and short communication paths. This gives a system that has greater speed, smaller area, and lower power consumption than an all-electronic system that has the same function.

11.5 OPTICAL FUZZY SYSTEM PROCESSING: PRESENT AND FUTURE

The field of optical information processing has matured in recent years. Two developments have driven this. The first is that most researchers realize that optical processing cannot overtake progress in electronic information processing. The pace of advancement in electronic technology is too rapid. The so-called Moore's Law states that the density of circuits on a micro chip doubles every 2 years or so. Moore's Law has held for at least 3 decades. Instead optics must coexist cooperatively with electronics. Studies show that optics is best for some tasks and electronics is best for others [1]. The best system designs will be hybrids that combine optical and electronic technologies to exploit the strengths of both.

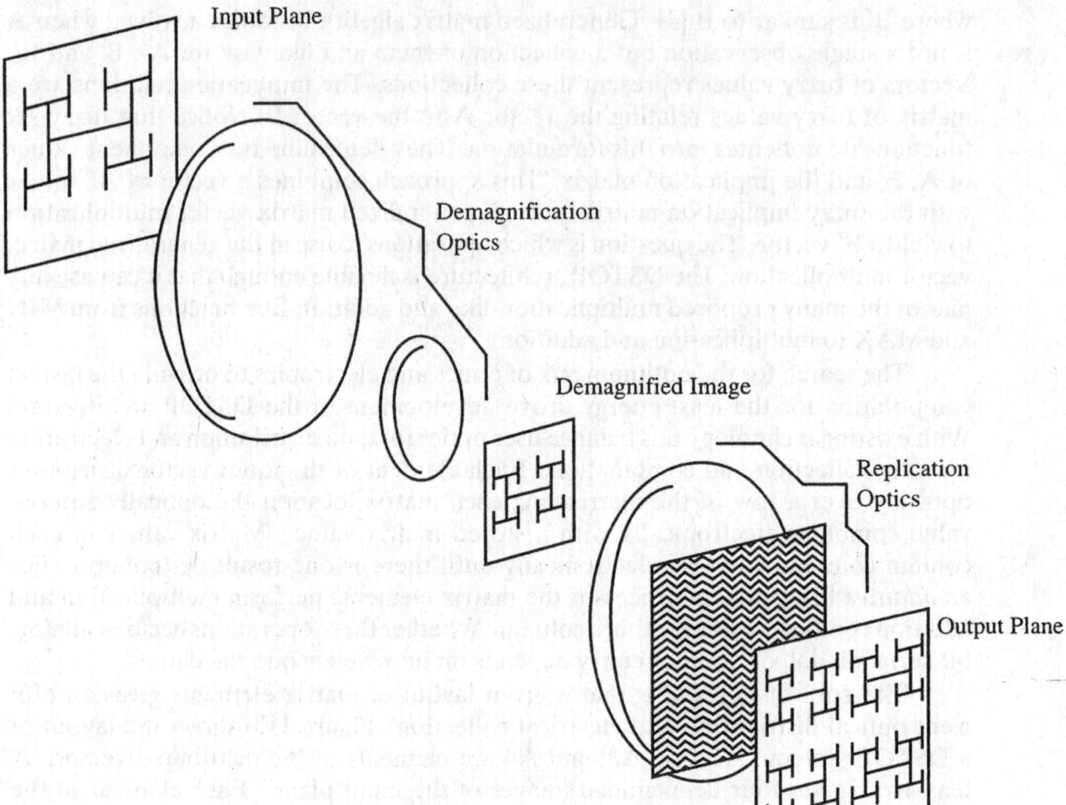

Figure 11.8 The DSTOP optoelectronic system for generalized matrix multiplication. The input plane contains an array of electronic processors and optical sources. A demagnified image of the input plane replicates onto the output plane that contains an array of electronic processors and optical detectors.

The second development is that high-performance spatial light modulator (SLM) devices are now becoming available. Architecture studies made up much of optical information processing for many years. The common claim was that "This will be a great computational architecture when good SLMs become available." Good SLMs are now close at hand. So researchers in optics can base these claims on solid physical evidence. The result is that optical information processing is getting down to business.

Successful optical fuzzy system processors will likely follow these two principles: (1) They must use both optics and electronics to their best advantage and (2) they must show significant improvement over all-electronic processors. The three optical fuzzy system processors in this chapter each possess some of these characteristics. The first system uses the parallelism of optics to store fuzzy sets. The second system uses the latest results in electro-optics to give fast reconfiguration of the fuzzy rules. The third system uses a blend of optical and electronic technology.

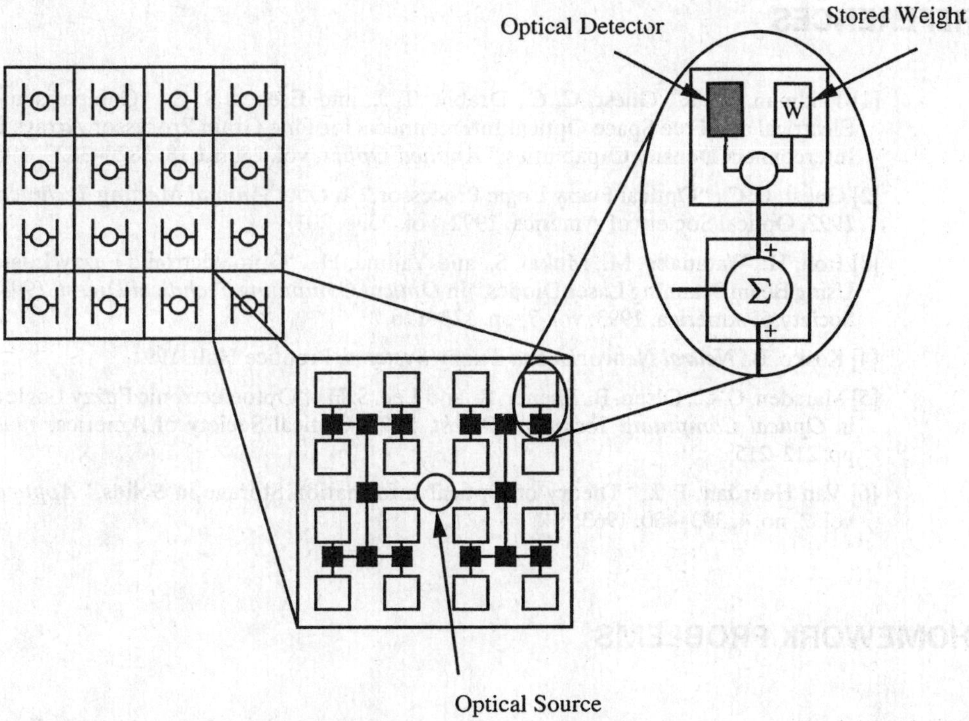

Figure 11.9 The layout of DSTOP electronic processor chips shown at three levels of detail.

These optical fuzzy system designs do not support learning. The user must code rules in advance. Learning may require a great deal of memory. Adaptive memory goes on the list of functions that electronics best provides at present.

Two key optical memory technologies are nearing commercialization: photorefractive crystals and two-photon materials. Both store data in the volume of material rather than on the surface. And both allow the system to read or write entire arrays of data simultaneously. Volume hologram crystals can in theory store on the order of 10^{13} bits per cubic centimeter [6]. In the near term adaptive optical fuzzy system processors will use electronic memory circuits. Optical memory technologies should in time offer high capacity and a high rate of data retrieval. Adaptive optical fuzzy system architectures that use the data stored in the memories remain the subject of active research.

Most problems that are interesting are also complex. They involve many inputs and many outputs and thus involve high dimension. Outputs may depend on any possible mix of inputs. As the number of factors that affect a system grows the number of possible combinations explodes exponentially. This is the curse of dimensionality. Optical systems that can process many fuzzy rules in parallel offer a partial solution to this problem of dimensionality. Adaptive systems that learn useful mixes of factors without trying all possible mixes may help complete the solution. Finding them remains the central challenge of adaptive optical fuzzy processors.

REFERENCES

[1] Feldman, M. R., Guest, C. C., Drabik, T. J., and Esener, S. C., "Comparison between Electrical and Free Space Optical Interconnects for Fine Grain Processor Arrays Based on Interconnect Density Capabilities," *Applied Optics*, vol. 28, no. 18, 3820–3829, 1989.

[2] Guest, C. C., "Optical Fuzzy Logic Processor," in *OSA Annual Meeting Technical Digest, 1992*, Optical Society of America, 1992, vol. 23, p. 201.

[3] Itoh, H., Watanabe, M., Mukai, S., and Yajima, H., "Optoelectronic Fuzzy Logic System Using Beam Scanning Laser Diodes," in *Optical Computing Technical Digest, 1993*, Optical Society of America, 1993, vol. 7, pp. 123–126.

[4] Kosko, B., *Neural Networks and Fuzzy Systems*, Prentice Hall, 1991.

[5] Marsden, G. C., Olsen, B., Esener, S., and Lee, S. H., "Optoelectronic Fuzzy Logic System," in *Optical Computing Technical Digest, 1991*, Optical Society of America, 1991, vol. 6, pp. 212–215.

[6] Van Heerdan, P. J., "Theory of Optical Information Storage in Solids," *Applied Optics*, vol. 2, no. 4, 393–400, 1963.

HOMEWORK PROBLEMS

11.1. Theory states that an optical system can carry approximately one channel of information in an area that is one wavelength on each side. How many channels of information can pass through a lens with a 2-cm diameter if one uses light with a wavelength of 500 nm?

11.2. Sketch the patterns on the T1 and T2 planes for the optical fuzzy processor for the following rules:

(a) If A is medium then B is high.

(b) If A is not low then B is very low.

(c) If A is high then B is not high.

11.3.

(a) For the fuzzy system processor that uses beam scanning laser diodes sketch the if-part transparency and specify the then-part programming voltage for each of the rules in problem 2. Assume a high voltage directs the beam toward high output values.

(b) Describe the problem you encounter in implementing rule c. Can you suggest a solution?

11.4. Suppose that A is a vector with six elements that stand for the fuzzy unit or fit value of these statements: X is high, X is medium, X is low, Y is high, Y is medium, Y is low. Suppose B is a like vector with the variables W and Z instead of X and Y.

(a) Design a matrix U containing 1s and 0s so that $B = UA$ implements these rules: "If X is high then W is medium" and "if Y is not low then Z is not high."

(b) What would be the meaning of allowing elements of U to have values between 0 and 1?

(c) What would be the meaning of allowing elements of U to be between -1 and 1?

11.5. Refer to Figure 11.9.

(a) Suppose the center-to-center spacing of leaf nodes in the H-tree is D. How far must an electrical signal travel starting at the center of a corner node and ending at the center of the optical source?

(b) What is distance traveled starting at a node that is adjacent to the source?

(c) Consider an array of 2^N by 2^N nodes. Find a general formula for distance traveled in terms of D and N.

PART VII

COMPUTING IN FUZZY CUBES

A fuzzy set can have many forms in many contexts. A fuzzy set can be a vague concept like *cool air* that has blurred borders. It can be an abstract object like *the numbers close to 5* that partially contains other objects. A fuzzy set can be a vague subset of a space like the *large mountains* in the Smoky Mountains or the set of *statistically significant random samples* in a pollster's sample space. It can be a function that maps objects in a space to the numbers between 0 and 1. The latter form is the algebraic view of a fuzzy set or the view of *sets as functions*.

A fuzzy set can also be a point in some space. This is the geometrical view of a fuzzy set or the view of *sets as points*. A continuous fuzzy set like the triangle or bell curve that stands for the set of cool air temperatures defines a point in an abstract function space of set functions.

The mind's eye cannot fully see these abstract spaces. But it can grasp the distance between two fuzzy sets as the length of the line segment that connects two points. It can grasp the neighborhood of a fuzzy set as a ball or sphere that contains the fuzzy set as the point at the ball's center. And it can grasp a changing or adapting fuzzy set as a point moving through the space.

A discrete fuzzy set has the simplest geometry. It is a point in a fuzzy cube. A fuzzy cube is a unit hypercube that has the unit interval [0,1] as each of its sides. The unit interval itself forms the simplest fuzzy cube or 1-D cube. It houses all truth values of a fuzzy or multivalued logic. The unit square houses all fuzzy subsets of two objects. The unit cube houses all fuzzy subsets of three objects and so on up to countable infinity.

Nonfuzzy sets lie at the cube corners. There and only there do they obey the either-or "laws" of bivalent logic. Long diagonals connect a bivalent set to its opposite or complement. These long diagonals must pass through the cube midpoint. The *not* operator makes the bivalent opposites hop over the midpoint to get from A to *not*-A.

Fuzzy sets fill in the cube. They get fuzzier as they approach the cube's unique midpoint. The midpoint set is the fuzziest set of all because it equals its own opposite. The midpoint set is unique to fuzzy theory. For thousands of years it has been the unseen source of many "paradoxes" of bivalent logic and set theory such as whether the Cretan lies when he says that all Cretans lie. Bivalent set theory in effect rounds off points in the cube to their closest corner. This fails for the midpoint because the midpoint is equally close to all the corners.

Chapter 12 presents the theory of fuzzy cubes in the context of information theory. All operations of discrete math have dual operations inside fuzzy cubes of perhaps huge dimension. The chapter shows that simply adding up n numbers computes the fuzzy mutual entropy of some fuzzy-set point in a fuzzy cube. These fuzzy cubes can also house information fields and dynamic flows on the fields. Chapter 12 does not deal with the standard additive model of the preceding chapters but it presents the basic theory of fuzzy cubes used in Chapter 9 and in the remaining chapters.

12

FUZZY CUBES AND FUZZY MUTUAL ENTROPY

Bart Kosko

PREVIEW

This chapter reviews and extends the theory of discrete fuzzy sets as points in unit hypercubes or "fuzzy cubes." A fuzzy cube contains all fuzzy subsets of a set X of n objects. The 2^n bivalent subsets of X lie at the 2^n corners of the n-cube $[0, 1]^n$. The continuum of fuzzy sets fill in the cube. The geometry of fuzzy cubes can give a simple way to see theorems about fuzzy sets. These theorems still hold pointwise for fuzzy subsets of continuous spaces even though we cannot picture them.

This chapter presents a measure of fuzzy mutual entropy and explores the information-theoretic structure of fuzzy cubes. The fuzzy mutual (Kullback) entropy of a fuzzy set F acts as a type of distance measure between F and its set complement F^c. It stems from the logarithm of a unique measure of the fuzziness of the set F. We first use fuzzy mutual entropy to show how operations in real space R^n can project onto fuzzy operations in fuzzy n-cubes. The first theorem shows that the basic operation of addition of real numbers is just the fuzzy mutual entropy of some fuzzy set F. The proof uses the logistic map often found in neural models. This diffeomorphism smoothly maps extended real space $[-\infty, \infty]^n$ onto the fuzzy cube $[0, 1]^n$ embedded in it. The logistic map equates the sum of a real vector's n components with the mutual entropy of some fuzzy set F and its complement F^c.

This theorem may have no direct practical value but its cube setting and proof serve as a simple introduction to the information structure of fuzzy cubes.

A deeper result shows that fuzzy mutual entropy gives back the standard Shannon entropy $H(P)$ of a probability vector P if we integrate the fuzzy mutual entropy. The set of all probability vectors of length n defines the simplex in the fuzzy n-cube. We allow the Shannon entropy to extend beyond the simplex and range over the entire fuzzy cube. Then we can compute the Shannon entropy $H(F)$ of any fuzzy set F of length n. This shows in turn that fuzzy mutual entropy has a fluid-mechanical structure and leads to the concept of an information field in a fuzzy cube. Fuzzy mutual entropy equals the negative of the divergence of Shannon entropy. Uncertainty descriptions define points in the fuzzy-cube parameter space. Versions of both extended Shannon entropy and fuzzy mutual entropy define vector fields on the fuzzy cube. The field equations show that Shannon entropy acts as the potential of the conservative mutual entropy vector field. This implies a dynamical form of the "second law of thermodynamics" for flows on the fuzzy cube: Shannon entropy can only grow in time in the fuzzy mutual entropy field. It also suggests that a simple reaction-diffusion equation may hold in fuzzy cubes. The chapter ends with a fuzzy view of the new "it from bit" thesis in physics that attempts to reduce the structure of matter and energy to the structure of information.

12.1 THE FIRST THEOREM: ADDITION AS FUZZY MUTUAL ENTROPY

Any sum of real numbers $x_1, \ldots, x_n$ equals the fuzzy mutual entropy of fuzzy set F in the unit hypercube $[0, 1]^n$:

$$\sum_{i=1}^{n} x_i = H(F/F^c) - H(F^c/F) \tag{12.1}$$

where F^c is the fuzzy set complement of F in the unit hypercube I^n. If we add two numbers then F and F^c lie in the unit square I^2. If we add three numbers then they lie in I^3 and so on up.

The proof of (12.1) maps the extended real space $\bar{R}^n = [-\infty, \infty]^n$ diffeomorphically onto the embedded unit hypercube I^n [11]. The proof views the real numbers $x_1, \ldots, x_n$ as the components of the real vector $\mathbf{x}$ in $\bar{R}^n$ and maps $\mathbf{x}$ to a unique point or fuzzy set F in the fuzzy space I^n.

The mutual entropy terms $H(F/F^c)$ and $H(F^c/F)$ stem from the logarithm of the fuzziness of F. As discussed below this fuzziness depends on how much F resembles its complement fuzzy set F^c. In this sense we can replace the two H terms in (12.1) with an entropy operator $\mathcal{H}$ applied to fuzzy set F:

$$\sum_{i=1}^{n} x_i = \mathcal{H}(F). \tag{12.2}$$

The operator $\mathcal{H}$ replaces each sum with the value of a map from fuzzy sets to real numbers.

The infinity "corners" of $-\infty$ and ∞ in $\bar{R}^n$ correspond to the 0–1 vertices in I^n. The origin in $\bar{R}^n$ corresponds to the midpoint of I^n. It is the unique fuzzy set F such that $F = F^c = F \cap F^c = F \cup F^c$. The next three sections review the needed fuzzy information theory and develop the new measure of fuzzy mutual entropy. Section 12.5 proves (12.1) as in [11]. Section 12.6 shows that the Shannon entropy measure of information theory equals the integral of fuzzy mutual entropy. Section 12.7 extends these results to derive *field* equations of fuzzy information on the fuzzy cube I^n and suggests how the results may relate to the view of modern physics that the universe is information.

12.2 FUZZY SETS AS POINTS IN HYPERCUBES: DEGREES OF SUBSETHOOD

Multivalence or "fuzziness" holds in sets and between sets. Fuzziness in a set defines *elementhood* or the degree a_i to which element x_i belongs to set A: $a_i = \text{Degree}(x_i \in A)$. A standard or bivalent or nonfuzzy set A contains elements all or none. The set value or membership degree a_i is 1 or 0, present or absent, in or out. A multivalent set contains elements to some degree. So a_i takes values in the unit interval [0,1]. Black [1] called this multivalence "vagueness" and introduced vague sets or vague lists. Zadeh [14] called these vague or multivalued sets "fuzzy" sets and worked out their algebra.

Fuzziness between sets defines *subsethood* [7–10] or the degree $S(A, B)$ to which set A belongs to or is a subset of set B: $S(A, B) = \text{Degree}(A \subset B)$. The sets A and B need not be fuzzy. If a fuzzy set A contains an element x_i to degree a_i then $S(\{x_i\}, A) = a_i$. So subsethood subsumes elementhood. In the past the subsethood operator S has defined a bivalent operator in both fuzzy and nonfuzzy set theory: $S(A, B) = 0$ or 1. The multivalued subsethood operator can also assume the values $0 < S(A, B) < 1$.

The subsethood operator arises from the unique ℓ^p extension of the Pythagorean theorem [8] in n dimensions:

$$\|A - B\|^p = \|A - B^*\|^p + \|B^* - B\|^p \tag{12.3}$$

for $p \geq 1$, n-vectors A, B, and B^*, and with the norm

$$\|A\|^p = \sum_{i=1}^{n} |a_i|^p . \tag{12.4}$$

The usual Pythagorean theorem holds if $p = 2$. For any p there are at least 2^n vectors or sets B^* that satisfy (12.3). For $p > 1$ there are only 2^n such sets B^* and $b_i^* = a_i$ or $b_i^* = b_i$. For fuzzy sets these 2^n choices reflect the 2^n choices of picking any vertex of the unit hypercube I^n as the origin or empty set. Once picked then

$b_i^* = \min(a_i, b_i)$. If A and B are bit vectors or regular nonfuzzy subsets of finite space $X = \{x_i, \ldots, x_n\}$ then this implies B^* equals $A \cap B$. The same holds for fuzzy subsets A and B of X:

$$B^* = A \cap B. \tag{12.5}$$

Suppose set or space X is finite with $X = \{x_1, \ldots, x_n\}$. Then the 2^n nonfuzzy subsets of X map to the 2^n bit vectors of length n. These map in turn to the 2^n corners of the unit hypercube I^n. This equates a set with a point in the Boolean n lattice. We can also view fuzzy subsets of X as n vectors with components in [0,1]. Then each vector component a_i of fuzzy set $A = (a_1, \ldots, a_n)$ defines a fuzzy unit or fit [7] and A defines a fit vector. The set function a: $X \to [0, 1]$ defines the n fit values $a(x_1), \ldots, a(x_n)$ for a finite space X and so gives the fit vector $A = (a_1, \ldots, a_n)$ for $a_i = a(x_i)$. Fit value a_i measures the degree to which element x_i belongs to or fits in set A. This identifies A with a point on or in the unit hypercube I^n [8]. Fuzzy sets fill in the Boolean n-cube to give the solid hypercube I^n. The midpoint of the unit cube is the fit vector $F = \left(\frac{1}{2}, \ldots, \frac{1}{2}\right)$ where each element x_i belongs to F as much as it belongs to its complement F^c. The usual set operations apply to fit vectors as Zadeh [14] proposed for fuzzy-set functions: $A \cap B = (\min(a_1, b_1), \ldots, \min(a_n, b_n))$, $A \cup B = (\max(a_1, b_1), \ldots, \max(a_n, b_n))$, $A^c = (1-a_1, \ldots, 1-a_n)$. Suppose $A = \left(\frac{1}{3}\ \frac{3}{4}\right)$ and $B = \left(\frac{1}{2}\ \frac{1}{3}\right)$. Then

$$A \cap B = \left(\frac{1}{3}\ \ \frac{1}{3}\right)$$

$$A \cup B = \left(\frac{1}{2}\ \ \frac{3}{4}\right)$$

$$A^c = \left(\frac{2}{3}\ \ \frac{1}{4}\right)$$

$$A \cap A^c = \left(\frac{1}{3}\ \ \frac{1}{4}\right)$$

$$A \cup A^c = \left(\frac{2}{3}\ \ \frac{3}{4}\right).$$

Note that $A \cap A^c \neq \emptyset$ and $A \cup A^c \neq X$ for all fuzzy sets A. Aristotle's bivalent "laws" of noncontradiction and excluded middle no longer hold. They hold only to some degree. They hold 100% only for the bit vectors at cube vertices. They hold 0% at the cube midpoint when $A = A^c$. They hold only to some degree for fit vectors between these extremes. The next section shows how the *overlap* term $A \cap A^c$ and *underlap* term $A \cup A^c$ give a unique measure of the fuzziness or entropy of A.

If A and B are not fuzzy sets then the 100% subsethood relation $A \subset B$ holds if and only if $a_i \leq b_i$ for all i. It still holds if A and B are fuzzy sets: $S(A, B) = 1$ iff $a_i \leq b_i$. Then all of B's 100% subsets define a hyperrectangle in I^n with long diagonal that runs from the origin to point B. $S(A, B) = 1$ iff A lies in or on this hyperrectangle and thus iff A lies in the fuzzy power set of B: $A \in F(2^B)$. $S(A, B) < 1$ iff A lies outside the hyperrectangle. The closer A lies to the hyperrectangle then the larger the value $S(A, B)$. The minimum distance lies between A and B^*. B^* is the 100% subset of B closest to A in any ℓ^p metric [8]. This distance gives the ℓ^p "orthogonal" projection of A onto $F(2^B)$ shown in Figure 12.1 and gives the term $\|A - B^*\|^p$ in the general ℓ^p-Pythagorean theorem (12.3).

The subsethood theorem follows from this orthogonal projection and unifies multivalued set theory. To see this first let $c(A)$ denote the *count* or cardinality of A:

$$c(A) = a_1 + a_2 + \ldots + a_n \tag{12.6}$$

$$= |a_1 - 0| + |a_2 - 0| + \ldots + |a_n - 0| \tag{12.7}$$

$$= \ell^1(\emptyset, A). \tag{12.8}$$

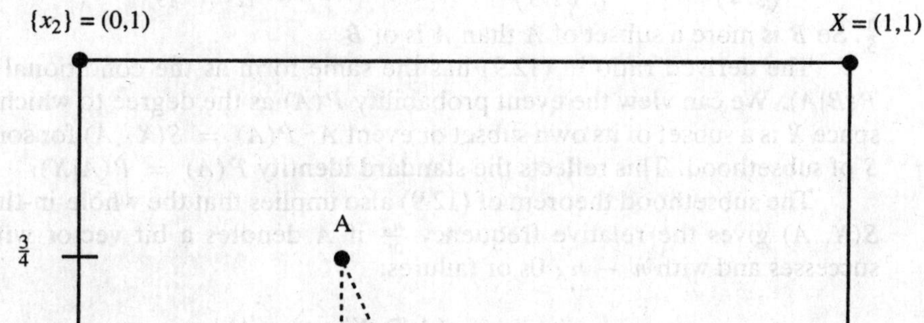

Figure 12.1 Pythagorean geometry of the subsethood theorem of discrete fuzzy sets. A like projection result holds pointwise for fuzzy subsets of continuous and more complex spaces.

If $A = (\frac{1}{3}\ \frac{3}{4})$ then $c(A) = \frac{13}{12}$. The equalities of (12.6)–(12.8) geometrize the count $c(A)$ as the ℓ^1 or fuzzy Hamming distance between A and the origin or empty set $\emptyset$. The count extends the classical counting measure of combinatorics to fuzzy sets [6]: $c(A)$ equals the counting measure of A on nunfuzzy sets A. It gives the number of elements in A if A is finite and gives ∞ if A is infinite—if A maps one-to-one to one of its proper subsets. The subsethood measure $S(A, B)$ depends on the minimal distance $d(A, B^*)$. In the fuzzy Hamming metric this means $S(A, B) = 1 - (\ell^1(A, B^*)/f(A))$. The boundary condition $S(A, \emptyset) = 0$ if $A \neq \emptyset$ [8,10] shows that $f(A) = c(A)$ since then $B^* = \emptyset$ gives $\ell^1(A, \emptyset) = c(A)$. Since $B^* = A \cap B$ then $\ell^1(A, B^*)$ equals the ℓ^1 difference $c(A) - c(A \cap B)$ shown in Figure 12.1. So $S(A, B) = 1 - (c(A) - c(A \cap B)/c(A))$. This gives the *subsethood theorem*:

$$S(A, B) = \frac{c(A \cap B)}{c(A)}. \tag{12.9}$$

If $A = (\frac{1}{3}\ \frac{3}{4})$ and $B = (\frac{1}{2}\ \frac{1}{3})$ then $S(A, B) = \frac{2}{3}/\frac{13}{12} = \frac{8}{13}$ and $S(B, A) = \frac{2}{3}/\frac{5}{6} = \frac{4}{5}$. So B is more a subset of A than A is of B.

The derived ratio in (12.9) has the same form as the conditional probability $P(B|A)$. We can view the event probability $P(A)$ as the degree to which the sample space X is a subset of its own subset or event A: $P(A) = S(X, A)$ for some measure S of subsethood. This reflects the standard identity $P(A) = P(A|X)$.

The subsethood theorem of (12.9) also implies that the whole-in-the-part term $S(X, A)$ gives the relative frequency $\frac{n_a}{n}$ if A denotes a bit vector with n_A 1s or successes and with $n - n_A$ 0s or failures:

$$S(X, A) = \frac{c(A \cap X)}{c(X)} = \frac{c(A)}{c(X)} = \frac{n_a}{n}.$$

This too confirms the relation $S(X, A) = P(A|X)$. In general $S(X, A) = \frac{c(A)}{n}$ or $c(A) = n\, S(X, A)$. So even the counting measure $c(A)$ reduces to subsethood. Subsethood depends in turn only on the metrical structure of fuzzy cubes.

The subsethood theorem (12.9) also implies $S(\{x_i\}, A) = a_i$. This holds since the singleton set $\{x_i\}$ maps to the unit bit vector $(0 \ \ldots\ 0\ 1\ 0\ \ldots\ 0)$ with a 1 in the ith slot and 0s elsewhere and since $A = (a_1, \ldots, a_n)$. Then $c(\{x_i\}) = 1$ and $c(\{x_i\} \cap A) = a_i$. So $S(\{x_i\}, A) = a_i$ and subsethood formally subsumes elementhood.

Maps between unit cubes define the simplest fuzzy systems S: $I^n \to I^p$. These fuzzy systems associate output fuzzy sets with input fuzzy sets and so generalize if-then rules. We saw in Chapter 2 that general fuzzy systems F: $R^n \to R^p$ are uniformly dense in the space of continuous funcions [9]: A fuzzy system can approximate any real continuous (or bounded Borel measurable) function on a compact set to any degree of accuracy. The fuzzy system contains m fuzzy rules of the form if $X = A$ then $Y = B$ that associate a then-part fuzzy set B with an if-part fuzzy set A. We also saw in Chapter 2 that a rule defines a fuzzy cartesian product $A \times B$ or patch in the input–output state space $X \times Y$. A fuzzy system approximates a function by covering its graph with patches and averaging patches that overlap. All the rules fire to some degree as in a neural associative memory [10]. The fuzzy approximation theorem shows that finite discretizations of A and B suffice for the covering. So the patch or fuzzy Cartesian product $A \times B$ can reduce to a fuzzy r-by-s matrix M or relation or point in I^{rs}. Then M defines the system mapping $M\colon I^r \to I^s$ and the subsethood measure in (12.9) applies to M. In the same product space each fuzzy system is a subset to some degree of all other fuzzy systems. Then (12.11) below shows that each fuzzy system has a unique numerical measure of fuzziness [5] or entropy.

12.3 FUZZINESS AND ENTROPY

The fuzziness of a fuzzy set answers a basic question: How fuzzy is a fuzzy set? A nonfuzzy set lies at a vertex of cube I^n and has 0% fuzziness. The cube midpoint M equals its own opposite ($M = M^c$) and it alone has 100% fuzziness. In between it varies. The fuzziness of set F grows as the distance falls between F and F^c and thus as F and F^c lie closer to the midpoint M.

This cube geometry motivates the ratio measure of fuzziness $E(F) = \frac{a}{b}$ [7]. Here a is the distance $\ell^1(F, F_{\text{near}})$ from F to the nearest vertex F_{near} and b is the distance $\ell^1(F, F_{\text{far}})$ from F to the farthest vertex F_{far}. A long diagonal connects F_{near} to F_{far}. The *fuzzy entropy theorem* [7] reduces this ratio of distances to a ratio of counts:

$$E(F) = \frac{c(F \cap F^c)}{c(F \cup F^c)}. \tag{12.10}$$

If $F = \left(\frac{1}{3}\ \frac{3}{4}\right)$ then $E(F) = \frac{7}{12}/\frac{17}{12} = \frac{7}{17}$. Figure 12.2 shows the fuzzy entropy theorem in the unit square.

The fuzzy entropy theorem (12.10) shows that the fuzziness of fuzzy set F depends on how much its overlap $F \cap F^c$ and underlap $F \cup F^c$ break Aristotle's laws of noncontradiction and excluded middle. The underlap $F \cup F^c$ always fully contains the overlap $F \cap F^c$: $S(F \cap F^c,\ F \cup F^c) = 1$. So we might expect $E(F)$ to involve subsethood in the converse direction $S(F \cup F^c,\ F \cap F^c)$ when the part partially contains the whole. In fact (12.9) and (12.10) reduce fuzziness to subsethood:

$$E(F) = S(F \cup F^c,\ F \cap F^c). \tag{12.11}$$

The next chapter shows that a measure of mutual subsethood or fuzzy *equivalence* has the form $\mathcal{E}(A, B) = \frac{c(A \cap B)}{c(A \cup B)}$. This theorem and (12.10) then give back the fuzziness of A as the degree to which A equals A^c: $E(A) = \mathcal{E}(A, A^c)$.

The probabilistic entropy $H(P)$ [3,5] holds for fit vectors on the simplex in I^n. Then

$$H(P) = \sum_{i=1}^{n} p_i \log \frac{1}{p_i} \tag{12.12}$$

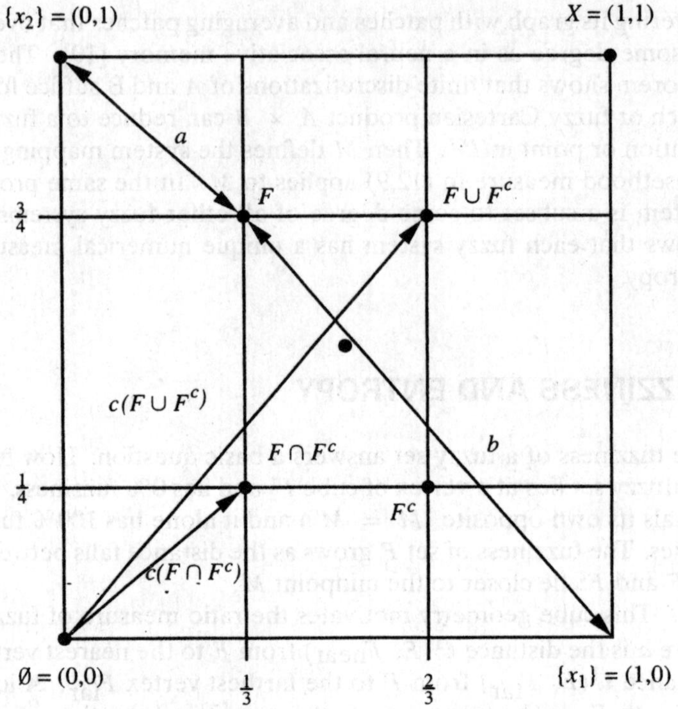

Figure 12.2 Geometry of the fuzzy entropy theorem.

and $c(P) = p_1 + \ldots + p_n = 1$. The fuzziness measure $E(P)$ differs from $H(P)$ for the same P. If no $p_j > \frac{1}{2}$ then $E(P) = \frac{1}{n-1}$ and so $E(P)$ falls to zero as the cube dimension n grows to infinity. The uniform set $\left(\frac{1}{n}, \ldots, \frac{1}{n}\right)$ belongs to this set of P vectors along with uncountably many others. If some $p_j > \frac{1}{2}$ then $E(P) < \frac{1}{n-1}$. So the uniform set maximizes $E(P)$ but does not uniquely maximize it. So E differs from H.

Now consider how E resembles H. Consider the probability element p_i and the motivation for the logarithm measure (12.12) as the average information or entropy of a message or event: "Information is inversely related to the probability of occurrence" [3]. The more improbable the event then the more informative the event if the event occurs. So information increases with $\frac{1}{p_i}$. The same intuition holds for monotone-increasing transforms of $\frac{1}{p_i}$. This includes the logarithmic transform $\log \frac{1}{p_i}$ and only the logarithmic transform in the additive case. The weighted average over the system or alphabet gives the entropy as the expected information (12.12).

In the one-fit case $E(F)$ reduces to $\frac{f}{1-f}$ if $f \leq \frac{1}{2}$ and to $\frac{1-f}{f}$ if $f \geq \frac{1}{2}$. This ratio grows to 1 as f moves to the midpoint $\frac{1}{2}$ and falls to 0 as f moves to 0 or 1. The more vague or fuzzy the event then the more informative the event if it occurs. The operator E is subadditive on fuzzy sets since in a fuzzy space all events connect to one another to some degree. Integration also shows that $\frac{f}{1-f}$ and $\frac{1-f}{f}$ define a continuous probability density on [0,1] if we normalize the integral by ln 4 − 1 (which has power series

$$\sum_{n=1}^{\infty} \frac{(-1)^{n+1}}{n(n+1)}\Bigg).$$

Note that the unit fuzziness $\frac{f}{1-f}$ is the fundamental power series of a fit value f:

$$\frac{f}{1-f} = \sum_{n=1}^{\infty} f^n \quad \text{for} \quad f < 1.$$

So far we have only reviewed fuzzy entropy. We now extend it to mutual entropy to set up the proof of the first theorem.

12.4 FUZZY MUTUAL ENTROPY

Fuzzy mutual entropy arises from a natural question: Why not take the logarithm of the unit fuzziness $\frac{f}{1-f}$? Any monotone transform will preserve its shape. So why not follow the probability example and use a logarithm? Then we can weight the log terms with the fit values and get a more proper measure of the entropy of a fuzzy set. The idea is to replace the intuition chain

$$p_i \rightarrow \frac{1}{p_i} \rightarrow \ln\frac{1}{p_i} \rightarrow \sum_i p_i \ln\frac{1}{p_i} \tag{12.13}$$

with the new fuzzy chain

$$f_i \rightarrow \frac{f_i}{1-f_i} \rightarrow \ln\frac{f_i}{1-f_i} \rightarrow \sum_i f_i \ln\frac{f_i}{1-f_i}. \tag{12.14}$$

The new fuzzy entropy term in (12.14) uses the natural logarithm to simplify the proof of the main theorem. The sum term defines a fuzzy mutual entropy.

For probability vectors P and Q in the I^n simplex define the mutual entropy $H(P/Q)$ of P given Q [12] as

$$H(P/Q) = \sum_i p_i \ln\frac{p_i}{q_i}. \tag{12.15}$$

The mutual entropy measures distance in the simplex in the rough sense that $H(P/Q) = 0$ if $P = Q$ and that $H(P/Q) > 0$ if $P \neq Q$. This follows from the Gibbs inequality [3]. Some stochastic learning automata and neural networks [4] minimize $H(P/Q)$ as the learning system's distribution P tries to estimate the distribution Q of the sampled environment. In the cube I^n the fuzzy mutual entropy term in (12.14) is the usual mutual entropy $H(F/F^c)$ defined on fit vectors.

The sum of the fuzzy information units $\ln \frac{f_i}{1-f_i}$ splits into the mutual entropies of fuzzy sets F and F^c:

Lemma:

$$\sum_{i=1}^{n} \ln \frac{f_i}{1-f_i} = H(F/F^c) - H(F^c/F). \tag{12.16}$$

Proof. Since $f_i + (1 - f_i) = 1$

$$\sum_i \ln \frac{f_i}{1-f_i} = \sum_i f_i \ln \frac{f_i}{1-f_i} + \sum_i (1-f_i) \ln \frac{f_i}{1-f_i} \tag{12.17}$$

$$= \sum_i f_i \ln \frac{f_i}{1-f_i} - \sum_i (1-f_i) \ln \frac{1-f_i}{f_i} \tag{12.18}$$

$$= H(F/F^c) - H(F^c/F). \tag{12.19}$$

Q.E.D. ■

The fuzziness measure in (12.10) shows that $E(F) = E(F^c)$. This reflects the 2^n-fold symmetry of the fuzzy cube I^n. But the mutual entropy operator is asymmetric. $H(F/F^c) = H(F^c/F)$ if $F = F^c$ and thus if F and F^c lie at the cube midpoint. The mutual entropy summands grow to infinity or zero as F and F^c move to cube vertices.

12.5 THE PROOF: DIFFEOMAPS BETWEEN REAL SPACES AND FUZZY CUBES

Fuzzy cubes map smoothly onto extended real spaces of the same dimension and vice versa. The 2^n infinite limits of extended real space $[-\infty, \infty]^n$ map to the 2^n binary corners of the fuzzy cube I^n. The real origin 0 maps to the cube midpoint. Each real point $\mathbf{x}$ maps to a unique fuzzy set F as Figure 12.3 shows.

A diffeomorphism $f: \bar{R}^n \to I^n$ is a one-to-one and onto differentiable map f with a differentiable inverse f^{-1}. Other diffeomaps can reveal other fuzzy structure of operations in real space. The theorem (12.1) follows from the choice of one of the simplest diffeomaps. We pick the logistic map used in neural models [2,10] to convert an unbounded real input x_i to a bounded signal or fit value f_i:

$$f_i = \frac{1}{1 + e^{-x_i}}. \tag{12.20}$$

In extended real space $\bar{R}^n$ the logistic map applies to each term of vector $\mathbf{x} = (x_1, \ldots, x_n)$. Note that $f_i = 0$ iff $x_i = -\infty$, $f_i = 1$ iff $x_i = \infty$, and $f_i = \frac{1}{2}$ iff $x_i = 0$. Each real $\mathbf{x}$ picks out unique dual fuzzy sets F and F^c in fuzzy space.

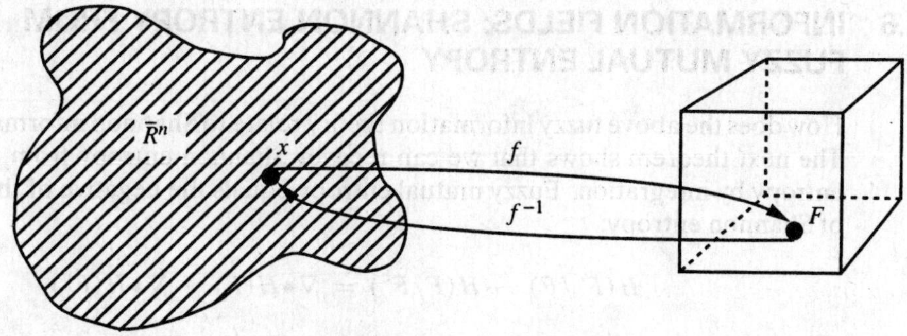

Figure 12.3 Diffeomap from extended real space to fuzzy space.

The proof of (12.1) follows from the lemma of (12.16) and from the *inverse* of the logistic map (12.20):

$$x_i = f^{-1}(f_i) = \ln \frac{f_i}{1 - f_i}. \tag{12.21}$$

So each real number is a unit of fuzzy information (12.14). This is just the logarithm of the scalar measure of fuzziness (12.10). We sum over all vector components x_i and apply the lemma (12.16) to prove (12.1) and (12.2):

$$\sum_{i=1}^{n} x_i = H(F/F^c) - H(F^c/F) \tag{12.22}$$

$$= \mathcal{H}(F) \tag{12.23}$$

in operator notation. Q.E.D.

The logistic map of (12.20) also allows a direct proof for each term x_i:

$$x_i = x_i \frac{1 + e^{-x_i}}{1 + e^{-x_i}} \tag{12.24}$$

$$= \frac{1}{1 + e^{-x_i}} \ln e^{x_i} - \frac{e^{-x_i}}{1 + e^{-x_i}} \ln e^{-x_i} \tag{12.25}$$

$$= \frac{1}{1 + e^{-x_i}} \ln \left(\frac{1}{1 + e^{-x_i}} \frac{1 + e^{-x_i}}{e^{-x_i}} \right) - \frac{e^{-x_i}}{1 + e^{-x_i}} \ln \left(\frac{e^{-x_i}}{1 + e^{-x_i}} 1 + e^{-x_i} \right) \tag{12.26}$$

$$= f_i \ln \frac{f_i}{1 - f_i} - (1 - f_i) \ln \frac{1 - f_i}{f_i} \tag{12.27}$$

since

$$f_i = \frac{1}{1 + e^{-x_i}} \quad \text{and} \quad 1 - f_i = \frac{e^{-x_i}}{1 + e^{-x_i}}.$$

12.6 INFORMATION FIELDS: SHANNON ENTROPY FROM FUZZY MUTUAL ENTROPY

How does the above fuzzy information theory relate to Shannon information theory? The next theorem shows that we can recover Shannon entropy from fuzzy mutual entropy by integration. Fuzzy mutual entropy equals the negative of the divergence of Shannon entropy:

$$H(F^c/F) - H(F/F^c) = \nabla \bullet H(F) + \nabla \bullet H(F^c) \tag{12.28}$$

for the divergence operator

$$\nabla \bullet = \frac{\partial}{\partial f_1} + \cdots + \frac{\partial}{\partial f_n}. \tag{12.29}$$

Then the theorem (12.1) implies that addition equals the negative divergence of the Shannon entropy of the dual fuzzy sets F and F^c in the fuzzy cube I^n. This result is less general than (12.28) since it depends on the logistic diffeomap to connect fuzzy space and extended real space. Other diffeomaps need not form such a bridge. The trick in (12.28) is to *extend* Shannon entropy from the simplex of probability vectors in I^n to the entire fuzzy cube I^n.

The result (12.28) holds in all cases and forms a type of bridge between fuzzy theory and standard information theory. It implies that we can in principle recover Shannon entropy H from fuzzy mutual entropy by integrating. We may have no need to do this in practice but it does give a fruitful link between the two theories.

The divergence operator from physics measures the average flow of a fluid. It is a field equation and so suggests that there is an *information field* in the fuzzy cube I^n. Each fuzzy-set point F has an amount of information or entropy as if perhaps one stood at the point F and sensed the information emitted from some source or from the cube midpoint or a vertex. The divergence suggests that we compute in an information *fluid*.

The theorem of (12.28) follows from the form of the spatial change $\frac{dH(F)}{df}$ of Shannon entropy for the two-fit case $F = (f, 1 - f)$. Then H depends on just f and so we can write it as $H(f)$:

$$H(f) = -f \ln f - (1 - f) \ln(1 - f). \tag{12.30}$$

Then $H(f)$ has the fit derivative

$$\frac{dH(F)}{df} = - \ln f - 1 + \ln(1 - f) + 1 \tag{12.31}$$

$$= \ln \frac{1 - f}{f} \tag{12.32}$$

$$= - \ln \frac{f}{1 - f}. \tag{12.33}$$

The last term is the same as the lemma summands in (12.16). So for n dimensions the lemma gives

$$\sum_{i=1}^{n} \frac{dH(f_i)}{df_i} = H(F^c/F) - H(F/F^c) \tag{12.34}$$

$$= -\mathcal{H}(F) \tag{12.35}$$

in operator notation. So we can recover each $H(f_i)$ term by integrating the fuzzy mutual entropy $\mathcal{H}(f_i)$:

$$H(f_i) = -\int \mathcal{H}(f_i) df_i. \tag{12.36}$$

This 2-D case $F = (f, 1 - f)$ has a unique property: It is the only case where a probability vector F has a complement F^c that is also a probability vector. In general the simplex in I^n is not closed under set complementation. The complement P^c lies in the cube but not in the simplex. So we can work with P only if we cannot work with P^c: $c(P) = 1$ only if $c(P^c) > 1$ for $n > 2$. The simplex of probability vectors lacks a set algebra even though for each probability measure P the underlying binary sets have such an algebra. Fuzzy mutual entropy extends the Shannon entropy measure $H(P)$ to all sets F in I^n. The fuzzy cube I^n is the maximal sigma algebra [6] of all set algebras of n-D fuzzy sets.

Now say $n > 2$. Then P and P^c do not both lie in the simplex. Call these sets F and F^c and extend them to the new *complement-coded* fuzzy set $G = [F|F^c]$ in the bigger cube I^{2n}. The fuzzy adaptive resonance system in Chapter 14 depends on just these complement-coded fit vectors. The symmetry of cubes ensures that G and F and F^c all have the same fuzzy structure. From (12.10) they have the same fuzziness:

$$E(G) = E(F) = E(F^C). \tag{12.37}$$

The extended Shannon entropy $H(G)$ is the sum of the extended Shannon entropies $H(F)$ and $H(F^c)$ of lower dimension:

$$H(G) = -\sum_{k=1}^{2n} g_k \ln g_k \tag{12.38}$$

$$= -\sum_{i=1}^{n} f_i \ln f_i - \sum_{i=1}^{n} (1 - f_i) \ln(1 - f_i) \tag{12.39}$$

$$= H(F) + H(F^c). \tag{12.40}$$

The term $H(G)$ in (12.38)–(12.40) is the fuzzy entropy measure of De Luca and Termini as Klir discusses in [5]. In turn the extended entropies $H(F)$ and $H(F^c)$ split into the sum of 2-D standard Shannon entropies $H(f_i)$:

$$H(G) = -\sum_{i=1}^{n}[f_i \ln f_i + (1 - f_i)\ln(1 - f_i)] \tag{12.41}$$

$$= \sum_{i=1}^{n} H(f_i). \tag{12.42}$$

So the partial derivative $\frac{\partial H(G)}{\partial f_i}$ is just the derivative $\frac{dH(f_i)}{df_i}$ in (12.31):

$$\frac{\partial H(G)}{\partial f_i} = \frac{dH(f_i)}{df_i} = -\ln\frac{f_i}{1 - f_i} = -\mathcal{H}(f_i). \tag{12.43}$$

Then (12.29) gives the sum of these partials as the divergence of $H(G)$ and hence of $H(F)$ and $H(F^c)$:

$$\nabla \bullet H(G) = \nabla \bullet H(F) + \nabla \bullet H(F^c) \tag{12.44}$$

$$= H(F^c/F) - H(F/F^c) \tag{12.45}$$

$$= -\mathcal{H}(F) \tag{12.46}$$

as claimed. So a multiple integral of fuzzy mutual entropy gives back extended Shannon entropy. Note that the divergence is just the directional derivative $\nabla H(G) \bullet X$ or the gradient $\nabla H(G)$ of the information flux that points to the vertex X of all 1s in the fuzzy cube I^n.

Information theorists often "derive" Shannon entropy from formal axioms [5]. This comes to no more than replacing the inverse $1/p$ with its logarithm $\ln \frac{1}{p}$ in the intuition chain in (12.13). Only a scaled logarithm turns products to sums. This is the basis of all proofs and "derivations" of the additivity of Shannon entropy:

$$H(P \times Q) = H(P) + H(Q) \tag{12.47}$$

if P and Q are independent marginal densities and if $P \times Q$ is their joint density. Of course P and P^c are not independent. Klir discusses other fuzzy measures of information in [5] that depend on logarithms.

Fuzzy mutual entropy also uses the logarithm in the intuition chain (12.14). This leads to additivity in the product of fit ratios

$$\ln \prod_{i=1}^{n} \frac{f_i}{1 - f_i} = H(F/F^c) - H(F^c/F). \tag{12.48}$$

The bridge between both types of entropy rests on the logarithm. Fuzzy entropy is more general because it holds on the whole fuzzy cube I^n and connects to a formal extension of Shannon entropy.

12.7 THE FIELD EQUATIONS OF FUZZY INFORMATION: IT FROM FIT

We can write the information or entropy field as the vector field $H(G) = (H(f_1), \ldots, H(f_n))$ for each point F in the fuzzy cube I^n (or for the dual point $\mathbf{x}$ in extended real space). Here G is the complement-coded fit vector of F and F^c in I^{2n}: $G = [F|F^c]$. The entropy field strength is maximal at the cube midpoint where $H = (\ln 2, \ldots, \ln 2)$ and minimal at the 2^n cube corners where $H = \emptyset$. Field equations for $H(G)$ should show how the information fluid compresses or rotates. The divergence and curl operators describe these fluid properties.

The vector curl depends on cross partial derivatives. The local scalars $H(f_i)$ have a zero partial derivative for distinct indices:

$$\frac{\partial H(f_i)}{\partial f_j} = 0 \quad \text{if} \quad i \neq j. \tag{12.49}$$

So the vector curl or rotational component of $H(G)$ is null. This result and (12.44)–(12.46) give the first two field equations of fuzzy mutual entropy:

$$\nabla \bullet H(G) + \mathcal{H}(F) = 0 \tag{12.50}$$

$$\nabla \times H(G) = \mathbf{0}. \tag{12.51}$$

The curl equation (12.51) may have meaning only when the dimension $n \leq 3$. So the fuzzy information fluid is compressible but irrotational.

The null curl in (12.51) implies that H is a *conservative* field:

$$H(G) = \nabla P \tag{12.52}$$

for some scalar potential $P: I^n \to R$. Integration gives P as

$$P = \sum_{i=1}^{n} (-\frac{f_i^2}{4}[2 \ln f_i - 1] + \frac{(1 - f_i)^2}{4}[2 \ln(1 - f_i) - 1] + C \tag{12.53}$$

for some constant C. Flows $F(t)$ on the entropy field H can only increase the potential: $\dot{P}(F) \geq 0$. Here the overdot stands for time differentiation. This holds for the same reason that $\dot{H}(F) \geq 0$ holds below in (12.62).

Now define the fuzzy mutual entropy field $\mathcal{H}(F)$ on the fuzzy cube I^n as

$$\mathcal{H}(F) = (-\mathcal{H}(f_1), \ldots, -\mathcal{H}(f_n)) \tag{12.54}$$

$$= \left(-f_1 \ln\frac{f_1}{1-f_1} + (1-f_1)\ln\frac{1-f_1}{f_1}, \ldots, -f_n \ln\frac{f_n}{1-f_n} + (1-f_n)\ln\frac{1-f_n}{f_n}\right). \tag{12.55}$$

So $\mathcal{H} = \emptyset$ at the midpoint and thus there the fields are orthogonal: $H \cdot \mathcal{H} = 0$.

Then relations (12.31)–(12.33) show that the extended Shannon entropy $H(G)$ in (12.41) is a potential for the fuzzy mutual entropy field $\mathcal{H}$:

$$\mathcal{H}(F) = \left(\ln\frac{1-f_1}{f_1}, \ldots, \ln\frac{1-f_n}{f_n}\right) \tag{12.56}$$

$$= \left(\frac{dH(f_1)}{df_1}, \ldots, \frac{dH(f_n)}{df_n}\right) \tag{12.57}$$

$$= \left(\frac{\partial H(G)}{\partial f_1}, \ldots, \frac{\partial H(G)}{\partial f_n}\right) \tag{12.58}$$

$$= \nabla H(G). \tag{12.59}$$

So $\mathcal{H}$ too is a conservative field and thus "irrotational" (in the 3-D case):

$$\nabla \times \mathcal{H}(F) = \mathbf{0}. \tag{12.60}$$

Differentiating the terms in (12.56) shows that $\mathcal{H}$ also has a nonzero divergence and so is compresssible:

$$\nabla \bullet \mathcal{H}(F) = -\sum_{i=1}^{n} \frac{1}{f_i(1-f_i)}. \tag{12.61}$$

In fact $\nabla \bullet \mathcal{H} \leq -4n$. Equations (12.59)–(12.61) connect the (static) field equations of fuzzy information.

The Shannon potential result [(12.59)] shows that a form of the "second law of thermodynamics" holds on the fuzzy cube:

$$\dot{H}(F(t)) \geq 0. \tag{12.62}$$

Shannon entropy can only increase in time on flows $F(t)$ on the fuzzy mutual entropy field $\mathcal{H}$.

To prove (12.62) define the *flow* $F(t) = (f_1(t), \ldots, f_n(t))$ or just $\phi(F, t)$ on $\mathcal{H}$ by (12.63)–(12.64):

$$\phi(F, 0) = F \tag{12.63}$$

$$\frac{\partial}{\partial t}\phi(F, t) = \mathcal{H}\,(\phi(F, t))\,. \tag{12.64}$$

Then $\mathcal{H} = \nabla H$ and the chain rule of differentiation give (12.62):

$$\dot{H}(\phi(t)) = \nabla H(\phi(t)) \bullet \frac{\partial \phi}{\partial t} \tag{12.65}$$

$$= \sum_{i=1}^{n} \frac{\partial H}{\partial f_i}\frac{df_i}{dt} \tag{12.66}$$

$$= \nabla H \bullet \dot{\phi} \tag{12.67}$$

$$= \nabla H \bullet \nabla H \qquad \text{from (12.59)} \tag{12.68}$$

$$= \sum_{i=1}^{n}\left(\frac{\partial H}{\partial f_i}\right)^2 \geq 0. \tag{12.69}$$

We now derive a *wave equation* for the fuzzy information fluid: The propagation of the Shannon potential H depends on the compressibility of the mutual entropy field $\mathcal{H}$. Again let $H(F, t)$ [or $H(G, t)$] stand for all the information at cube point F at time t. Let V be a homogeneous volume with closed surface $\mathcal{S}$ in the open fuzzy cube: $V \subset (0, 1)^n$. Then we can argue as in fluid mechanics that the information flux through $\mathcal{S}$ into V must balance the flux out of V through $\mathcal{S}$. Let $\mathbf{n}$ be a normal vector that points outward on the surface $\mathcal{S}$. Then the divergence theorem of vector calculus converts the normal component of the information flux ∇H (or $\mathcal{H}$) over the surface $\mathcal{S}$ to its divergence or Laplacian $\nabla^2 H$ (or $\nabla \bullet \mathcal{H}$) over the entire volume V:

$$-\int_{\mathcal{S}} c\nabla H \bullet \mathbf{n}\, d\mathcal{S} = -\int_V c\nabla \bullet \nabla H\, dV = -\int_V c\nabla^2 H\, dV \tag{12.70}$$

for some nonzero constant c that describes the flow or diffusion properties of the information fluid. The Laplacian operator ∇^2 converts a scalar map such as H back to a scalar map by summing the second partial derivatives:

$$\nabla^2 = \sum_{i=1}^{n} \frac{\partial^2}{\partial f_i^2}. \tag{12.71}$$

Now equate the volume flux change in (12.70) with the information flux per volume V on the righthand side of (12.72)

$$\frac{d}{dt}\int_V H(F)\,dV = \int_V \frac{\partial H}{\partial t}dV \tag{12.72}$$

to get the null integral

$$\int_V \left(\frac{\partial H}{\partial t} + c\nabla^2 H\right) dV = 0. \tag{12.73}$$

This holds for all volumes V in the fuzzy cube no matter how small. So the continuous integrand itself equals zero. This gives the generic form of the flux or "wave equation":

$$\frac{\partial H}{\partial t} = -c\nabla^2 H. \tag{12.74}$$

(12.74) is a simple reaction-diffusion equation. It says that the velocity of the wave depends on the concentration of information or Shannon entropy in fuzzy space. The information is minimal at the cube midpoint. It is maximal on the cube surface (which includes the 2^n vertices) where at least one fit value f_i is extremal: $f_i = 0$ or $f_i = 1$. The reaction-diffusion equation (12.74) is not a formal wave equation but a so-called heat equation of physics. The general solution of the "wave" $H(t, F)$ is a sinusoidal function in the spatial variable F that multiplies a decaying Gaussian exponential in time. The logarithmic form of Shannon's H allows us to find the F-shape of $\nabla^2 H$ in (12.77) below.

So far we have not used the logarithmic form of H. The Laplacian form of (12.74) depends on just the fluid or flux view of the fuzzy cube and the differentiability of H. We now invoke (12.56)–(12.59) and can thus rewrite (12.74) as a conservation law of entropy or information:

$$\frac{\partial H}{\partial t} + c\nabla \bullet \mathcal{H} = 0. \tag{12.75}$$

So the wave velocity depends on the compressibility of the fuzzy mutual entropy. The field is incompressible if and only if the H wave is in steady state: $\frac{\partial H}{\partial t} = 0$. We now show that this can never hold. The Shannon wave perpetually undulates.

The divergence in (12.61) and the conservation law (12.75) give the final version of the Shannon wave equation:

$$\frac{\partial H}{\partial t} = -c\nabla \bullet \mathcal{H} \tag{12.76}$$

$$= \sum_{i=1}^{n} \frac{1}{f_i(1 - f_i)}. \tag{12.77}$$

This gives a stronger statement of the second law of thermodynamics when $c > 0$: $\frac{\partial H}{\partial t} > 0$. Entropy always increases but the rate depends on the position in fuzzy space.

The inverse-logistic form of (12.77) implies that the wave velocity achieves its lower bound at the cube midpoint when all $f_i = \frac{1}{2}$:

$$\frac{\partial H}{\partial t} \geq c\,4n > 0. \tag{12.78}$$

The same form also implies that H vibrates infinitely fast on the surface of the closed fuzzy cube:

$$\frac{\partial H}{\partial t} = \infty \quad \text{if and only if} \quad f_i = 0 \quad \text{or} \quad f_i = 1 \quad \text{for at least one } f_i. \tag{12.79}$$

Suppose H cannot move faster than a finite velocity bound B (such as the speed of light). Then the system cannot achieve binary certainty:

$$\frac{\partial H}{\partial t} \leq B < \infty \quad \text{implies} \quad 0 < f_i < 1 \quad \text{for all } f_i. \tag{12.80}$$

The Shannon wave fluctuates most slowly at the cube midpoint. It fluctuates faster and faster away from the midpoint and closer to the cube surface where the certainty or information is greatest.

We can repeat the above wave arguments for the information flux $\mathcal{J}$ that comes from the gradient of the fuzzy mutual entropy potential $\mathcal{H}(F)$ in (12.35):

$$\mathcal{J} = \nabla\mathcal{H}(F) \tag{12.81}$$

Then

$$\frac{\partial \mathcal{H}}{\partial t} + d\nabla \bullet \mathcal{J} = 0 \tag{12.82}$$

holds and leads to

$$\frac{\partial \mathcal{H}}{\partial t} = d \sum_{i=1}^{n} \frac{1 - 2f_i}{f_i^2(1 - f_i)^2}. \tag{12.83}$$

So steady-state $\frac{\partial \mathcal{H}}{\partial t} = 0$ holds only at the cube midpoint. As in (12.77) the velocity is infinite if and only if some fit value is binary. But the sign of (12.83) depends on if and to what extent each fit value obeys $f_i > \frac{1}{2}$ or $f_i < \frac{1}{2}$.

These wave and "thermodynamic" results and the above information fluid equations suggest a physics interpretation. Physicist John Archibald Wheeler and others [13] have argued that physics is a branch of information theory: *It from bit.*

The entropy of thermodynamics has the same form as the Shannon entropy of information theory. Quantum wave functions depend on measurement and thus the information or brain state of observers. Maximizing Shannon entropy subject to a constraint on the mean gives the Gibbs density or partition function of quantum mechanics [5]. The Bekenstein-Hawking equations show that black holes have an entropy that acts like Shannon entropy [13].

Wheeler [13] states the thesis in terms of binary logic: "Every physical quantity, every it, derives its ultimate significance from bits, binary yes-or-no indications, a conclusion which we epitomize in the phrase, *it from bit*." Wheeler concludes that "The bit count of the cosmos, however it is figured, is ten raised to a very large power." The Bekenstein-Hawking equations give an answer if we pack all matter in the universe into a black hole. The information is proportional to the square of the mass. Then the cosmic black hole "hides" 10^{120} units of information. The universe contains only something like 10^{87} subatomic particles. This suggests that even these tiny specks of matter contain tremendous amounts of information. The bit count 10^{120} also equals the total number of chess games.

The trouble is that no one has found a statement of fact that has a binary truth value. The statements of physics are accurate to at most a few decimal places. In this sense they are fuzzy. Only statements of math and logic have the truth values 0 or 1. They are either contradictions like "$1 = 2$" or empty tautologies like "A is A." The it-from-bit view lays probability wagers on top of these unreal binary statements and grounds it all in the Shannon entropy measure.

Fuzzy mutual entropy shows how to derive the Shannon entropy measure with no assumption of binary statements or probability wagers. Fits replace bits. Fits combine both the logic and uncertainty of statements in a single unit. This in turn leads to the information field equations and suggests a more general synthesis of science and information: *It from fit*. Fuzzy mutual entropy *is* the it-from-bit gradient field ∇H.

12.8 CONCLUSIONS

Addition or counting is the most basic operation in mathematics. It equals a basic operation in fuzzy space, the entropy map $\mathcal{H}$ that assigns a real number to each fuzzy set. This equality may seem strange since we have just begun to see the unit hypercube as a fuzzy space with its own set algebra and geometry. Diffeomap projection—or in some cases the weaker homeomorphic projection—can help show the fuzzy structure of real operations and algorithms. Future research may classify diffeomaps by how they preserve basic operations such as addition or how they carve the fuzzy cube into entropy regions or balls.

Fuzzy mutual entropy also gives back the Shannon entropy of information theory. This connection is a bridge between fuzzy information theory and the standard information theory based on probability and binary logic. It suggests that an "information field" may have both physical and computational content. Information fields may also be no more than symbolic abstractions.

The fuzzy cube may extend other operations and algorithms in information theory. The fuzzy approximation theorem [9] converts continuous or measurable systems into a finite number of fuzzy patches or points in large fuzzy cubes. The cube contains both the probability simplex that describes channel transmissions and the Boolean cube that describes all binary messages of a fixed length. An algorithm can dig through the cube from binary vertex to distant binary vertex rather than hop as a gray code from vertex to local vertex in cubes of high dimension. We can also view messages as balls of entropy or fuzziness in a cube. Ball diameter falls as the ball center moves from the vague midpoint to the clear corners. Diffeomaps can map real messages or systems into signal or noise balls that overlap in the fuzzy cube. Here and elsewhere the fuzzy cube can define the framework of a fuzzy information theory.

REFERENCES

[1] Black, M., "Vagueness: An Exercise in Logical Analysis," *Philosophy of Science*, vol. 4, 427–455, 1937.

[2] Grossberg, S., *Studies of Mind and Brain*, Reidel, 1982.

[3] Hamming, R. W., *Coding and Information Theory*, 2nd ed., Prentice Hall, 1986.

[4] Hinton, G. E., and Sejnowski, T. J., "Learning and Relearning in Boltzmann Machines," *Parallel Distributed Processing*, vol. I, D. E. Rumelhart and J. L. McClelland, editors, 282–317, MIT Press, 1986.

[5] Klir, G. J., and Folger, T. A., *Fuzzy Sets, Uncertainty, and Information*, Prentice Hall, 1988.

[6] Kosko, B., "Counting With Fuzzy Sets," *IEEE Transactions on Pattern Analysis and Machine Intelligence*, vol. PAMI–8, 556–557, July 1986.

[7] Kosko, B., "Fuzzy Entropy and Conditioning," *Information Sciences*, vol. 40, 165–174, 1986.

[8] Kosko, B., "Fuzziness vs. Probability," *International Journal of General Systems*, vol. 17, no. 2, 211– 240, 1990.

[9] Kosko, B., "Fuzzy Systems as Universal Approximators," *IEEE Transactions on Computers*, vol. 43, no. 10, October 1994; an early version appears in *Proceedings of IEEE International Conference on Fuzzy Systems 1992 (FUZZ–92)*, 1153–1162, March 1992.

[10] Kosko, B., *Neural Networks and Fuzzy Systems*, Prentice Hall, 1991.

[11] Kosko, B., "Addition as Fuzzy Mutual Entropy," *Information Sciences*, vol. 73, no. 3, 273–284, October 1, 1993.

[12] Kullback, S., *Information Theory and Statistics*, Wiley, 1959.

[13] Wheeler, J. A., "Information, Physics Quantum: The Search for Links," *Complexity, Entropy, and the Physics of Information*, Santa Fe Institute Studies in the Sciences of Complexity, vol. 8, W. H. Zurek, editor, 2–28, Addison-Wesley, 1990.

[14] Zadeh, L. A., "Fuzzy Sets," *Information and Control*, vol. 8, 338–353, 1965.

HOMEWORK PROBLEMS

12.1. Use the theorem on fuzzy mutual entropy to compute the sums $1 + 2$ and $2 + (-2)$.

12.2. Show that $y = x - 1$ is the tangent line to $y = \ln x$ at the point (1,0). Graph the two functions to see the inequality

$$\ln x \leq x - 1.$$

When does strict inequality hold? Use this inequality to prove the *Gibbs inequality*

$$H(P/Q) > 0$$

if P differs from Q and $H(P/Q) = 0$ if $P = Q$. Here $P = (p_1, \ldots, p_n)$ and $Q = (q_1, \ldots, q_n)$ with unit counts $c(P) = 1$ and $c(Q) = 1$.

12.3. Show that the Shannon entropy $H(P)$ is maximized when $P = (1/n, \ldots, 1/n)$.

12.4. Show that $H(P) \leq \ln n$.

12.5. Suppose P and Q are independent probability densities. $P \times Q$ stands for their joint density. Show that their joint Shannon entropy is additive:

$$H(P \times Q) = H(P) + H(Q).$$

12.6. Show that the fuzziness $E(A)$ depends only on the distance between A and the midpoint M and has the form

$$E(A) = \frac{1 - m}{1 + m}$$

where m depends only on the distance $\ell^1(A, M)$.

12.7. Compute the fuzziness of fit vectors $A = (.3\ 1\ .2)$, $B = (.3\ .3\ .3)$, and $C = (1\ 0\ 1)$. Compute their mutual entropy. Find the subsethood values $S(A, B)$, $S(B, A)$, $S(A, C)$, $S(C, A)$, $S(B, C)$, and $S(C, B)$.

12.8. Let A be a fuzzy set or point in I^n. Let C stand for the *complement-coded* set $[A|A^c]$ in I^{2n}. Show that A and C have the same fuzziness:

$$E(A) = E(C).$$

12.9. Show that $c(A) = nS(X, A)$ for all fuzzy sets A in I^n. This fact shows that counting reduces to subsethood. It gives back the relative frequency of "event" A when A is bivalent.

12.10. Prove the ℓ^1 triangle equality of subsethood:

$$|a - b| = |a - c| + |c - b|$$

iff $a \leq c \leq b$.

12.11. Use the quadratic formula to prove the ℓ^2 triangle equality of subsethood:

$$(a - b)^2 = (a - c)^2 + (c - b)^2$$

iff $c = a$ or $c = b$.

12.12. Show that

$$u^p < (u + v)^p$$

if $p > 1$, $u > 0$, and $v > 0$.

12.13. Show that

$$u^p + v^p < (u + v)^p$$

if $p > 1$, *p is an integer*, $u > 0$, and $v > 0$. Use a derivative argument and the previous problem to prove the general case.

12.14. Use the results of problems 12 and 13 to prove the general triangle equality of subsethood:

$$\sum_{i=1}^{n} |a_i - b_i|^p = \sum_{i=1}^{n} |a_i - c_i|^p + \sum_{i=1}^{n} |c_i - b_i|^p$$

for all $p > 1$ iff $c_i = a_i$ or $c_i = b_i$.

12.15. The next problems show that the structure of the fuzzy cube still holds for continuous fuzzy subsets A of some bounded real interval $X = [a, b]$ or for more complex regions of spaces of higher dimension. Here A has set function a: $X \rightarrow [0, 1]$. Let $x_0 \in X$. Show that the fit value $a(x_0)$ comes from a delta pulse $\delta(x - x_0)$ convolved with a.

12.16. Define the count $c(A)$ on $X = [a, b]$ as the ℓ^1 distance $\ell^1(A, \emptyset)$ from the curve A to the line segment X:

$$c(A) = \int_a^b a(x)\, dx.$$

Define the fuzziness $E(A)$ as before as the ratio of the ℓ^1 distance from A to the nearest binary set over the ℓ^1 distance from A to the farthest binary set. Draw an example of A and these binary sets. Show that the fuzzy entropy theorem still holds:

$$E(A) = \frac{c(A \cap A^c)}{c(A \cup A^c)}.$$

12.17. Define the subsethood operator $S(A, B)$ on $X = [a, b]$ and state and derive the subsethood theorem. Draw the fuzzy power sets $F(2^A)$ and $F(2^B)$ and the projection geometry of $A \cap B$. Does the generalized Pythagorean theorem still hold? Why or why not? Use delta pulses to show that subsethood subsumes elementhood.

12.18. Show that $H = \nabla P$ for the potential

$$P = \sum_{i=1}^{n} \left(-\frac{f_i^2}{4}[2 \ln f_i - 1] + \frac{(1 - f_i)^2}{4}[2 \ln(1 - f_i) - 1] \right).$$

12.19. Let $\phi(F, t) = F(t)$ define a flow on the Shannon entropy field H on the fuzzy cube I^n. Show that P can only increase along H flows:

$$\frac{dP}{dt} \geq 0.$$

12.20. Show that the fuzzy mutual entropy field $\mathcal{H}$ is compressible:

$$\nabla \bullet \mathcal{H} = - \sum_{i=1}^{n} \frac{1}{f_i(1 - f_i)}.$$

12.21. Show that $\mathcal{J} = \nabla \mathcal{H}$ leads to the "wave" equation

$$\frac{\partial \mathcal{H}}{\partial t} = d \sum_{i=1}^{n} \frac{1 - 2f_i}{f_i^2(1 - f_i)^2}.$$

Most feedforward neural networks in practice are one of two types. The first type is a multi-layer feedforward perceptron. The neurons in each layer pass summed inputs to output signals with a sigmoidal or S-shaped signal function. Supervised gradient descent tunes the feedforward synapses between each layer with some form of the "backpropagation" hill-climbing algorithm.

These perceptron networks avoid the problem of rule explosion when they approximate functions. Instead they suffer from memory loss and system inscrutability. Each new input–output pattern learned changes slightly all the synapses. So the net forgets some of the patterns it has learned. There seems no formal way to know how much of which patterns it has forgotten or unlearned. The user must retest the old input–output pairs and test other samples from other parts of the input–output state space. The neurons do not stand for or map to distinct structures or features in the input–output state space. The multi-layered system is not only a black box but a tangled web of a black box. The lack of a map from neurons to state-space objects explains why there are no formal techniques to show how many layers to use for a given problem or to show how many neurons to use per layer.

The second type of neural network is a feedforward radial basis function network. The neurons have Gaussian signal functions that act in the input layer as receptive fields for subsets of the input space. Most radial basis functions coincide with Gaussian standard additive model (SAM) systems. This means that at the multi-dimensional level the Gaussian balls act as fuzzy rule patches that cover the graph of the approximand. It also means that they suffer from exponential rule explosion.

Chapter 16 extends these Gaussian SAMs with a new measure of fuzzy subsethood. The measure computes the mutual subsethood or degree of equivalence between two Gaussian fuzzy sets. This measure arises from the geometry of fuzzy power sets in fuzzy unit cubes but applies to continuous fuzzy sets like Gauss and bell curves and other integrable curves. Mutual subsethood combines with supervised learning to grow and prune and tune a radial basis network.

Most feedforward neural networks in practice are one of two types. The first type is a multi-layer feedforward perceptron. The neurons in each layer map summed inputs to output signals with a sigmoidal or S-shaped signal function. Supervised gradient descent tunes the feedforward synapses between each layer with some form of the "backpropagation" hill-climbing algorithm.

These perceptron networks avoid the problem of neuron explosion when they approximate functions. Instead they suffer from memory loss and system inscrutability. Each new input–output pattern learned changes slightly all the synapses. So the net forgets some of the patterns it has learned. There seems no formal way to know how much of which patterns it has forgotten or unlearned. The user must retest the old input–output pairs and test other samples from other parts of the input–output state space. The neurons do not stand for or map to distinct structures or features in the input–output state space. The multi-layered system is not only a black box but a tangled web of a black box. The lack of a map from neurons to state-space objects explains why there are no formal techniques to show how many layers to use for a given problem or to show how many neurons to use per layer.

The second type of neural network is a feedforward radial basis function network. The neurons have Gaussian signal functions that act in the input layer as receptive fields for subsets of the input space. Most radial basis functions coincide with Gaussian standard additive model (SAM) systems. This means that at the multi-dimensional level the Gaussian balls act as fuzzy rule patches that cover the graph of the approximand. It also means that they suffer from exponential rule explosion.

Chapter 13 extends these Gaussian SAMs with a new measure of fuzzy subsethood. The measure computes the mutual subsethood or degree of equivalence between two Gaussian fuzzy sets. This measure arises from the geometry of fuzzy power sets in fuzzy unit cubes but applies to continuous fuzzy sets like Gaussian bell curves and other integrable curves. Mutual subsethood combines with supervised learning to grow and prune and tune a radial basis network.

13

ADAPTIVE SUBSETHOOD FOR RADIAL BASIS FUZZY SYSTEMS

Chin-Teng Lin

PREVIEW

This chapter extends the fuzzy measure of subsethood to a measure of mutual subsethood or fuzzy equivalence. Gaussian fuzzy sets simplify the new measure when it applies to continuous fuzzy subsets of the real line and give closed form for the measure. The Gaussian sets give a standard additive model that has the same form as a radial basis function neural network. The bell-curve sets act as nodes in the net and the if-then rules act as links or synapses.

Supervised gradient descent tunes the Gaussian set functions. Unsupervised competitive learning initializes the gradient descent. The complete adaptive subsethood controller (ASC) system uses the network structure to store and learn and then tune the fuzzy rules. Simulations show how the ASC system can control a model car. The ASC can use other bell-curve set functions if they have a tractable measure of fuzzy equivalence.

13.1 ADAPTIVE SUBSETHOOD IN A RADIAL BASIS STANDARD ADDITIVE MODEL

Radial basis function (RBF) networks [18] are feedforward neural networks that use bell-shaped signal functions instead of the more common sigmoidal or S-shaped signal functions [17]. Chapter 2 showed that these RBF networks have the same form as Gaussian standard additive model (SAM) fuzzy systems. The Gaussian sets form the if-parts and then-parts of the fuzzy rules. Each neuron or node sums its inputs to form a scalar activation. A Gaussian or other radial basis function maps this scalar into a bounded output signal. RBF networks are well-known universal function approximators [3].

Any two Gaussian fuzzy sets give rise to a question: How much does one set contain the other? Chapter 12 gives the answer in the measure of subsethood or fuzzy set containment. This holds in both the discrete framework of fuzzy cubes where fuzzy sets define vector points and for the continuous framework where the fuzzy sets are multivalued subsets of the real line and thus points in abstract function space.

The subsethood measure extends to a measure of mutual subsethood or fuzzy *equivalence*. We derive this measure in the discrete framework of fuzzy cubes and apply it to continuous Gaussian fuzzy sets. Then we can derive simple learning laws for an adaptive subsethood scheme that picks the first Gaussian means and variances with unsupervised clustering and then uses supervised gradient descent to tune the Gaussian fuzzy sets and thus tune the rules. Adaptive subsethood helps move learned fuzzy-set points toward target fuzzy-set points.

This chapter presents a system called the Adaptive Subsethood Controller (ASC) system and shows how it can control a simulated car. The ASC extends earlier work in neural-fuzzy control [8–13] to the broader SAM framework. Section 13.2 derives the new measure of mutual subsethood or equivalence. Section 13.3 applies the new measure to continuous Gaussian set functions and derives their mutual subsethood in a tractable formula. Section 13.4 shows how mutual subsethood can help tune the rules with the supervised Gaussian SAM learning laws of Chapter 2. Then Section 13.5 tests the ASC system on Sugeno's simulated "fuzzy car."

13.2 THE MEASURE OF MUTUAL SUBSETHOOD: FUZZY EQUIVALENCE

Mutual subsethood measures the degree to which two fuzzy sets are equal or how much they are similar. It extends the subsethood measure $S(A, B)$ [15–17] that gives the degree to which fuzzy set A is a subset of fuzzy set B. Analogous to the derivation procedure of subsethood in Chapter 12 [17] we shall present an algebraic derivation and a geometric derivation of the mutual subsethood measure and discuss its properties. In the next section we will derive the equations for the mutual subsethood measure of two fuzzy sets with bell-shaped set functions and with triangular set functions. In Section 13.4 we will use these mutual subsethood equations to derive an on-line learning algorithm for adaptive SAM controllers.

Suppose $a: X \to [0, 1]$ and $b: X \to [0, 1]$ are the set functions of fuzzy sets A and B. Then fuzzy sets A and B are *equal* if $a(x) = b(x)$ for all $x \in X$. We denote this as $A = B$. If $A = B$ then $A \subseteq B$ and $B \subseteq A$. Here $A \subseteq B$ if $a(x) \le b(x)$ for every $x \in X$. This definition of equality is not fuzzy and there are only two possible cases: A is equal to B or A is not equal to B. So if $X = \{x_1, x_2, x_3\}$ and fit vectors $A = (a_1, a_2, a_3) = (.2, .7, .5)$ and $B = (.5, .7, .9)$ then $A \neq B$. If fit vector $C = (.2, .6, .6)$ then again $A \neq C$. But A is more similar to C than to B. This leads us to define a fuzzy measure of equality or *mutual subsethood measure*. The value $E(A, B)$ measures the degree to which fuzzy set A equals fuzzy set B. So

$$E(A, B) = \text{Degree}(A = B) = \text{Degree}(A \subseteq B \text{ and } B \subseteq A). \tag{13.1}$$

$E(\cdot, \cdot)$ denotes the multivalued equality measure and takes values in [0,1]. If $E(A, B)$ has a high value then A is more similar to B. In the extreme case $E(A, B) = 1$ if $A = B$.

Before we derive the equation of the mutual subsethood measure we need to define more fuzzy terms. The union of two fuzzy sets A and B is a fuzzy set $A \cup B$ such that $a \cup b(x) = \max[a(x), b(x)]$ for all $x \in X$. The intersection of two fuzzy sets A and B is a fuzzy set $A \cap B$ such that $a \cap b(x) = \min[a(x), b(x)]$ for all $x \in X$. The size or cardinality of fuzzy set A is $c(A)$. It equals the sum of the *fit* (fuzzy unit) values of A:

$$c(A) = \sum_{x \in X} a(x). \tag{13.2}$$

We replace the sum with integration for integrable set functions.

We follow the method in [17] to derive the mutual subsethood measure. We first present an algebraic derivation of the mutual subsethood measure $E(A, B)$. The algebraic derivation is a type of support-violation strategy. It studies a law by breaking it. Suppose an element $x_v \in X$ violates the equality relationship: Either $a(x_v) < b(x_v)$ or $a(x_v) > b(x_v)$. Unlike the derivation of subsethood measure in [17] we must consider both of the two cases ("greater than" and "smaller than") at once. The greater the violations in magnitude $|a(x_v) - b(x_v)|$ and in frequency then the less A is equal to B and the more A differs from B. So we can first define the degree of set difference of A and B as

$$\text{Difference}(A, B) = 1 - E(A, B). \tag{13.3}$$

We sum all the violations to count support violations in magnitude and frequency. The unnormalized sum has the form

$$\sum_{x \in X} \max(0, a(x) - b(x)) + \sum_{x \in X} \max(0, b(x) - a(x)).$$

$c(A)$ or $c(B)$ normalize the subsethood measure. The count $c(A \cup B)$ gives a simple and appropriate normalization factor for the summation. This normalization factor will become more clear in the geometric derivation. After the normalization we obtain the difference of A and B as

$$\text{Difference}(A, B) = \frac{\sum_{x \in X} \max(0, a(x) - b(x)) + \sum_{x \in X} \max(0, b(x) - a(x))}{c(A \cup B)}. \tag{13.4}$$

Then from (13.3) the mutual subsethood measure becomes

$$E(A, B) = 1 - \frac{\sum_{x \in X} \max(0, a(x) - b(x)) + \sum_{x \in X} \max(0, b(x) - a(x))}{c(A \cup B)}. \tag{13.5}$$

As in [17] the right term in (13.5) equals ratio of counts:

$$E(A, B) = \frac{c(A \cap B)}{c(A \cup B)}. \tag{13.6}$$

This is the *mutual subsethood theorem*.

The mutual subsethood measure also has geometric derivation. Figure 13.1 shows the sets-as-points geometry of the fit (fuzzy unit) vectors $A = (a_1, a_2) = (2/3, 1/4)$ and $B = (b_1, b_2) = (1/3, 3/4)$. The sets-as-points view represents a fuzzy set as a point in the cube I^n if $X = \{x_1, \ldots, x_n\}$. The cube in Figure 13.1 consists of all possible fuzzy subsets of two elements $\{x_1, x_2\}$. The four corners stand for the binary power set 2^X of $\{x_1, x_2\}$ and thus the four nonfuzzy sets. This figure also shows the points $A \cap B$ and $A \cup B$ and the cubes $F(2^A)$ and $F(2^B)$ that are the fuzzy power sets of A and B (the set of all fuzzy subsets of A or B). The sets-as-points view makes the concept of *distance* between two fuzzy sets clear and can help us develop the mutual subsethood measure. The ℓ^p distance between fuzzy sets A and B in ℓ^p is defined as

$$d(A, B) = \ell^p(A, B) = \left(\sum_{i=1}^{n} |a_i - b_i|^p \right)^{\frac{1}{p}} \tag{13.7}$$

where $1 \leq p \leq \infty$. The ℓ^2 distance is the physical Euclidean distance actually illustrated in Figure 13.1. Notice that the distances (arrows) shown in the figure stand for any ℓ^p distance. The simplest distance is the ℓ^1 or the *fuzzy Hamming distance* that sums the absolute differences. The fuzzy Hamming distance gives the size or count $c(A)$ of A as the desired ℓ^1 norm:

$$\begin{aligned} c(A) &= \sum_{i=1}^{n} a_i = \sum_{i=1}^{n} |a_i - 0| = \sum_{i=1}^{n} |a_i - \emptyset_i| \\ &= \ell^1(A, \emptyset). \end{aligned} \tag{13.8}$$

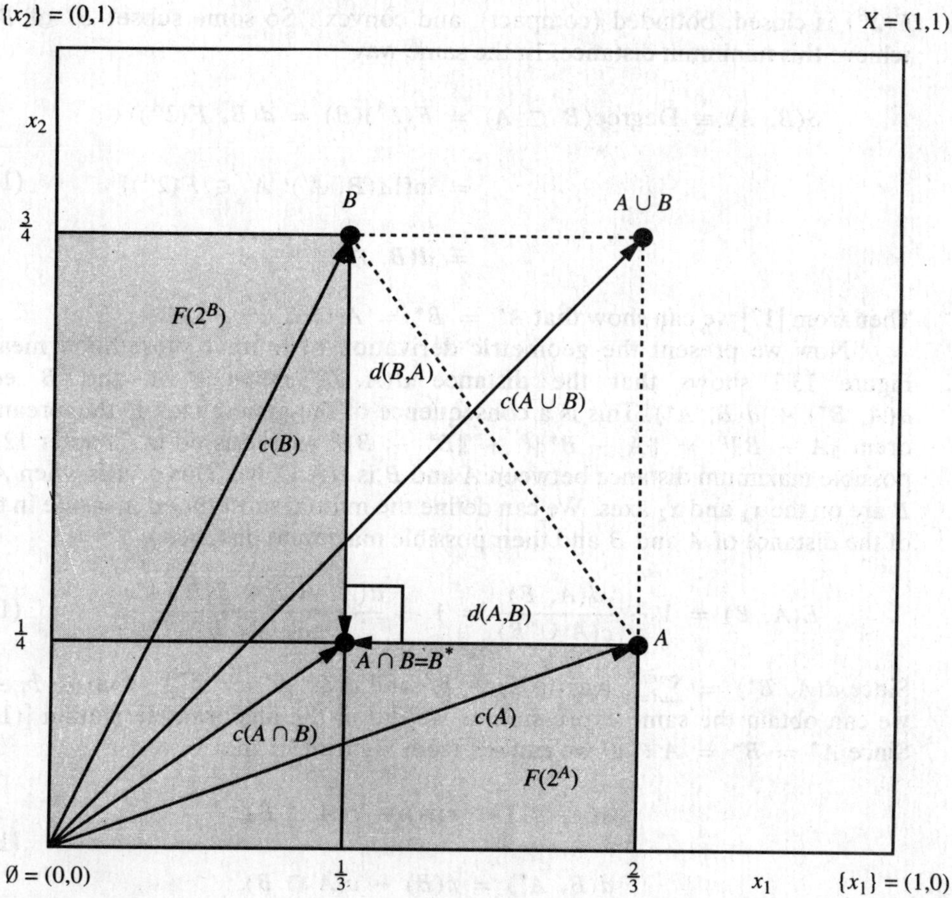

Figure 13.1 Sets-as-points geometry of the fuzzy sets A, B, $A \cup B$, and $A \cap B$. A unique Pythagorean relationship holds between the two fuzzy power sets $F(2^A)$ and $F(2^B)$. The set $A \cap B$ is the orthogonal projection of both A onto $F(2^B)$ and of B onto $F(2^A)$. This leads to both the subsethood measure $S(A, B)$ and the mutual subsethood or equivalence measure $E(A, B)$.

In Figure 13.1 $c(A \cap B)$ is the arrow from the origin (the empty set) to the fuzzy set $A \cap B$.

Now consider the mutual subsethood measure. Since $A = B$ implies $A \subseteq B$ and $B \subseteq A$ we will consider these two cases according to the subsethood theorem [17] that states that $S(A, B) = c(A \cap B)/c(A)$ if $A \neq \emptyset$. The degree to which A is a subset of B is the distance between A and the fuzzy power set $F(2^B)$ of B:

$$
\begin{aligned}
S(A, B) \equiv \text{Degree}(A \subset B) &= F(2^B)(A) = d(A, F(2^B)) \\
&= \inf\{d(A, B') : B' \in F(2^B)\} \qquad (13.9) \\
&= d(A, B^*).
\end{aligned}
$$

$F(2^B)$ is closed, bounded (compact), and convex. So some subset B^* of B will achieve this minimum distance. In the same way

$$\begin{aligned} S(B, A) \equiv \text{Degree}(B \subset A) &= F(2^A)(B) = d(B, F(2^A)) \\ &= \inf\{d(B, A'): A' \in F(2^A)\} \\ &= d(B, A^*). \end{aligned} \tag{13.10}$$

Then from [17] we can show that $A^* = B^* = A \cap B$.

Now we present the geometric derivation of mutual subsethood measure. Figure 13.1 shows that the distance $d(A, B)$ between A and B equals $d(A, B^*) + d(B, A^*)$. This is a consequence of the generalized Pythagorean theorem $\|A - B\|^p = \|A - B^*\|^p + \|B^* - B\|^p$ as discussed in Chapter 12. The possible maximum distance between A and B is $c(A \cup B)$. This occurs when A and B are on the x_1 and x_2 axes. We can define the mutual subsethood measure in terms of the distance of A and B and their possible maximum distance:

$$E(A, B) = 1 - \frac{d(A, B)}{c(A \cup B)} = 1 - \frac{d(A, B^*) + d(B, A^*)}{c(A \cup B)}. \tag{13.11}$$

Since $d(A, B^*) = \sum_{i=1}^n \max(0, a_i - b_i)$ and $d(B, A^*) = \sum_{i=1}^n \max(0, b_i - a_i)$ we can obtain the same expression as we did in the algebraic derivation [(13.5)]. Since $A^* = B^* = A \cap B$ we can see from Figure 13.1 that

$$\begin{aligned} d(A, B^*) &= c(A) - c(A \cap B), \\ d(B, A^*) &= c(B) - c(A \cap B). \end{aligned} \tag{13.12}$$

Then (13.11) becomes

$$\begin{aligned} E(A, B) &= 1 - \frac{c(A) + c(B) - 2c(A \cap B)}{c(A \cup B)} \\ &= 1 - \frac{c(A \cup B) - c(A \cap B)}{c(A \cup B)} \\ &= \frac{c(A \cap B)}{c(A \cup B)}, \end{aligned} \tag{13.13}$$

where we use the modular equality $c(A) + c(B) = c(A \cap B) + c(A \cup B)$. We obtain the same expression (13.6) of the mutual subsethood measure from the algebraic and geometric derivations. The relation between the mutual subsethood measure and the subsethood measure is the ratio

$$E(A, B) = \frac{1}{\frac{1}{S(A,B)} + \frac{1}{S(B,A)} - 1}. \tag{13.14}$$

So $E(A, B) = 1$ when $S(A, B) = S(B, A) = 1$ and thus when $A \subseteq B$ and $B \subseteq A$.

The geometry of fuzzy sets clearly describes fuzziness, defines fuzzy concepts, and proves fuzzy theorems. The geometric derivation of the mutual subsethood measure in the above is a more fundamental approach than the algebraic derivation. It also helps show why we need to use the normalization factor $c(A \cup B)$ in (13.4)–(13.6).

We now note some extreme cases of the mutual subsethood measure:

1. $E(A, B) = 1$ iff $c(A \cap B) = c(A \cup B)$ iff $\sum_{x \in X} \max(0, a(x) - b(x)) + \sum_{x \in X} \max(0, b(x) - a(x)) = 0$ iff $A = B$. This case reduces to the standard definition of equality of two fuzzy sets: $a(x) = b(x)$ for all x.
2. $E(A, B) = 0$ iff $c(A \cap B) = 0$ iff $\sum_{x \in X} \max(0, a(x) - b(x)) + \sum_{x \in X} \max(0, b(x) - a(x) = c(A \cup B)$ and thus just in case A and B do not overlap at all: $A \cap B = \phi$.
3. $E(A, B) = c(A)/c(B) < 1$ if $A \subset B$ and thus for unity subsethood: $S(A, B) = 1$. $E(A, B) = c(B)/c(A) < 1$ if $A \supset B$. The equality measure is multivalent: $0 < E(A, B) < 1$.

The subsethood measure relates to logical implication. In the same way the mutual subsethood measure relates to logical equivalence. The mutual subsethood measure reduces to the Lukasiewicz equivalence operator when viewed at the one-dimensional level of fuzzy logic (when $c(A \cup B) = 1$):

$$\begin{aligned} E(A, B) &= 1 - \max(0, a(x) - b(x)) - \max(0, b(x) - a(x)) \\ &= 1 - |a(x) - b(x)| \qquad (13.15) \\ &= A \leftrightarrow B. \end{aligned}$$

Consider an example. Define fit vectors $A = (.3, .1, .5, 0)$ and $B = (.7, .8, .1, .7)$. Then $A \cap B = (.3\ .1\ .1\ 0)$ and $A \cup B = (.7, .8, .5, .7)$. Then $c(A \cap B) = 0.5$, $c(A \cup B) = 2.7$, and $E(A, B) = 0.19$.

The mutual subsethood measure also reduces to the Tanimoto similarity measure [14] when A and B are nonfuzzy sets:

$$S_T(A, B) = \frac{n(A \cap B)}{n(A \cup B)} = \frac{n(A \cap B)}{n(A) + n(B) - n(A \cap B)} \qquad (13.16)$$

where $n(A)$ is the number of elements in binary set A and thus $n(A) = c(A)$.

The mutual subsethood measure [(13.6)] has another interesting relationship with fuzzy subsethood. It gives the measure of fuzziness (entropy) of a fuzzy set [17]. This holds when $B = A^c$ in (13.6). Here A^c is the complement of fuzzy set A with set function $a^c(x) = 1 - a(x)$:

$$\text{ENTROPY}(A) = \frac{C(A \cap A^c)}{C(A \cup A^c)} = E(A, A^c). \qquad (13.17)$$

13.3 MUTUAL SUBSETHOOD FOR BELL-SHAPED AND TRIANGULAR SET FUNCTIONS

We now use the mutual subsethood measure [(13.6)] to derive the formula for the mutual subsethood measure of two fuzzy sets with bell-shaped or Gaussian set functions. These sets are used in both radial basis networks and in fuzzy controllers. To reduce the computational complexity of the exact formula we will use an approximate formula of the mutual subsethood measure of two fuzzy sets with bell-shaped set functions in our learning algorithm. This approximate formula equals the mutual subsethood of two fuzzy sets with triangular set functions.

Consider two bell-shaped set functions for two fuzzy sets A and B:

$$a(x) = \exp\left\{-\frac{(x-m_1)^2}{\sigma_1^2}\right\} \quad \text{and} \quad b(x) = \exp\left\{-\frac{(x-m_2)^2}{\sigma_2^2}\right\}. \tag{13.18}$$

Then

$$\begin{aligned} c(A) &= \int_{-\infty}^{\infty} a(x)dx = \sigma_1\sqrt{\pi}, \\ c(B) &= \int_{-\infty}^{\infty} b(x)dx = \sigma_2\sqrt{\pi}. \end{aligned} \tag{13.19}$$

To compute $c(A \cap B)$ we set $a(x) = b(x)$ to obtain their equal-valued points. If $(1 - \sigma_1/\sigma_2)(m_1 + m_2) \geq 0$ then

$$x_1 = \frac{m_1 - \frac{\sigma_1}{\sigma_2}m_2}{1 + \frac{\sigma_1}{\sigma_2}} \quad \text{and} \quad x_2 = \frac{m_1 + \frac{\sigma_1}{\sigma_2}m_2}{1 - \frac{\sigma_1}{\sigma_2}}. \tag{13.20}$$

If $(1 - \sigma_1/\sigma_2)(m_1 + m_2) < 0$ then swap the values of x_1 and x_2. Note that $x_1 \leq x_2$. Now

$$\int_{x_1}^{x_2} \exp\left\{-\frac{(x-m)^2}{\sigma^2}\right\} dx = \sigma\sqrt{\pi}\left[G\left(\frac{\sqrt{2}(x_2 - m)}{\sigma}\right) - G\left(\frac{\sqrt{2}(x_1 - m)}{\sigma}\right)\right] \tag{13.21}$$

where $G(x)$ is the integral of the Gaussian set function that we can write in terms of the error function $erf(.)$:

$$G(x) = \frac{1}{\sqrt{2\pi}}\int_{-\infty}^{x} \exp\left\{-y^2/2\right\} dy = \frac{1}{\sqrt{2\pi}}\int_{0}^{x} \exp\left\{-y^2/2\right\} dy = \text{erf } x + \frac{1}{2}. \tag{13.22}$$

If $m_1 \geq m_2$ then

$$c(A \cap B) = \int_{-\infty}^{x_1} \exp\left\{-\frac{(x-m_2)^2}{\sigma_2^2}\right\} dx + \int_{x_2}^{\infty} \exp\left\{-\frac{(x-m_2)^2}{\sigma_2^2}\right\} dx$$

$$+ \int_{x_1}^{x_2} \exp\left\{-\frac{(x-m_1)^2}{\sigma_1^2}\right\} dx$$

$$= \sigma_2\sqrt{\pi}\left[G\left(\frac{\sqrt{2}(x_1-m_2)}{\sigma_2}\right) - 0\right]$$

$$+ \sigma_2\sqrt{\pi}\left[1 - G\left(\frac{\sqrt{2}(x_2-m_2)}{\sigma_2}\right)\right]$$

$$+ \sigma_1\sqrt{\pi}\left[G\left(\frac{\sqrt{2}(x_2-m_1)}{\sigma_1}\right) - G\left(\frac{\sqrt{2}(x_1-m_1)}{\sigma_1}\right)\right]. \tag{13.23}$$

If $m_1 < m_2$ then switch m_1 and m_2 and switch σ_1 and σ_2. So

$$E(A, B) = \frac{c(A \cap B)}{c(A) + c(B) - c(A \cap B)} \tag{13.24}$$

where x_1 and x_2 are from (13.20) and we assume that $m_1 \geq m_2$.

We use an approximate approach to avoid the computation of the function $G(x)$ and to reduce the computational complexity of the equation of $c(A \cap B)$. The area of the bell-shaped function $e^{-(x-c)^2/\sigma^2}$ is $\sigma\sqrt{\pi}$ and its height is always 1. So we can approximate it by an isosceles triangle with unity height and with length of bottom edge $2\sigma\sqrt{\pi}$. We can then compute the mutual subsethood measure of two fuzzy sets with these set functions (see Figure 13.2).

We observe all the possible relative relationships of two isosceles triangles on a horizontal axis (x-axis) to derive the equations of the mutual subsethood measure of two fuzzy sets with isosceles triangular set functions. Let $\Delta(m_i, \sigma_i)$ denote the isosceles triangle with unity height, bottom-line length $2\sigma_i\sqrt{\pi}$, and center of bottom line at m_i on the x-axis. Assume the two endpoints of the bottom line of $\Delta(m_1, \sigma_1)$ are a and b on the x-axis and the two endpoints of the bottom line of $\Delta(m_2, \sigma_2)$ are c and d on the x-axis (see Figure 13.2):

$$a = m_1 - \sigma_1\sqrt{\pi},$$

$$b = m_1 + \sigma_1\sqrt{\pi},$$

$$c = m_2 - \sigma_2\sqrt{\pi},$$

$$d = m_2 + \sigma_2\sqrt{\pi}.$$

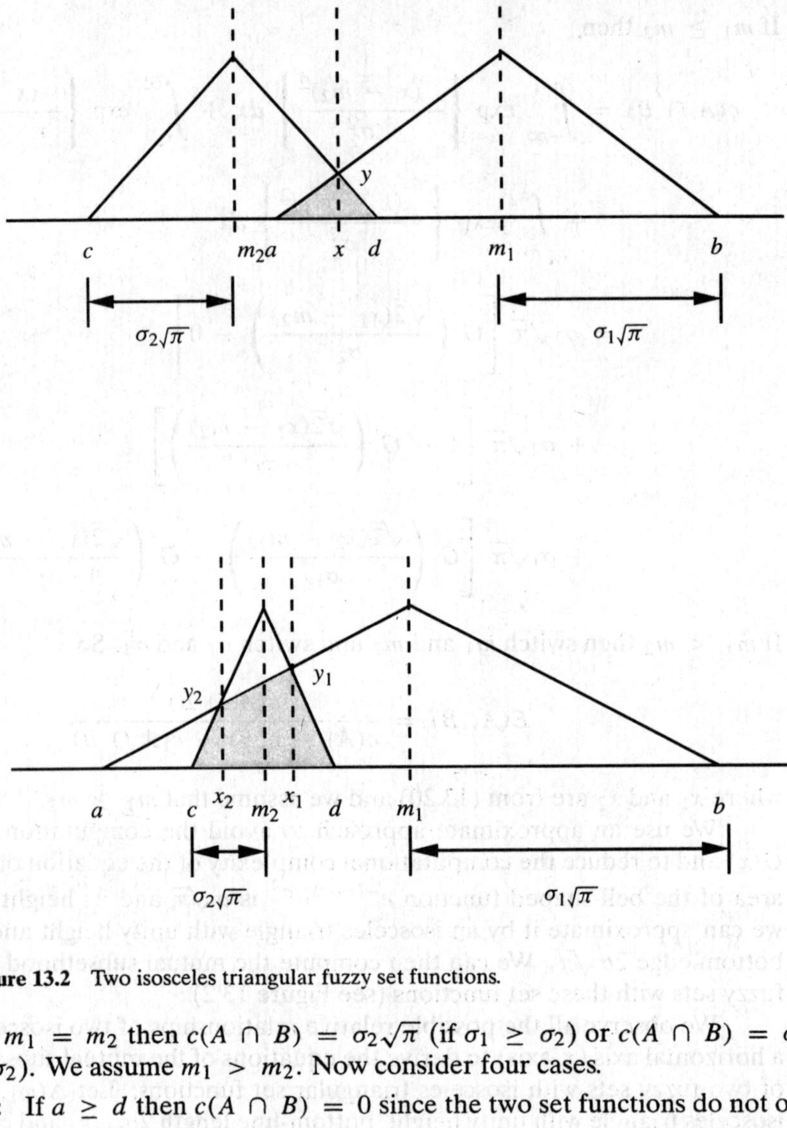

Figure 13.2 Two isosceles triangular fuzzy set functions.

Now if $m_1 = m_2$ then $c(A \cap B) = \sigma_2\sqrt{\pi}$ (if $\sigma_1 \geq \sigma_2$) or $c(A \cap B) = \sigma_1\sqrt{\pi}$ (if $\sigma_1 < \sigma_2$). We assume $m_1 > m_2$. Now consider four cases.

Case 1: If $a \geq d$ then $c(A \cap B) = 0$ since the two set functions do not overlap.

Case 2: If $b \geq d > a \geq c$ then

$$c(A \cap B) = \frac{1}{2}(d - a)y = \frac{1}{2}\frac{(m_2 - m_1 + \sigma_1\sqrt{\pi} + \sigma_2\sqrt{\pi})^2}{(\sigma_1 + \sigma_2)\sqrt{\pi}}. \qquad (13.25)$$

Case 3: If $b > d$ and $c > a$ then

$$c(A \cap B) = \frac{1}{2}(x_2 - c)y_2 + \frac{1}{2}(y_1 + y_2)(x_1 - x_2) + \frac{1}{2}(d - x_1)y_1. \qquad (13.26)$$

Then

$$\frac{1}{2}(x_2 - c)y_2 = \frac{1}{2}\sigma_2 \frac{[m_2 - m_1 + \sqrt{\pi}(\sigma_1 - \sigma_2)]^2}{\sqrt{\pi}(\sigma_1 - \sigma_2)^2},$$
$$\frac{1}{2}(d - x_1)y_1 = \frac{1}{2}\sigma_2 \frac{[m_2 - m_1 + \sqrt{\pi}(\sigma_1 + \sigma_2)]^2}{\sqrt{\pi}(\sigma_1 + \sigma_2)^2}. \tag{13.27}$$

Since

$$\frac{1}{2}(y_1 + y_2)(x_1 - x_2) = \frac{1}{2}(x_1 - a)y_1 - \frac{1}{2}(x_2 - a)y_2 \tag{13.28}$$

we get

$$\frac{1}{2}(x_1 - a)y_1 = \frac{1}{2}\sigma_1 \frac{[m_2 - m_1 + \sqrt{\pi}(\sigma_1 + \sigma_2)]^2}{\sqrt{\pi}(\sigma_1 + \sigma_2)^2},$$
$$\frac{1}{2}(x_2 - a)y_2 = \frac{1}{2}\sigma_1 \frac{[m_2 - m_1 + \sqrt{\pi}(\sigma_1 - \sigma_2)]^2}{\sqrt{\pi}(\sigma_1 - \sigma_2)^2}. \tag{13.29}$$

So the final result is

$$c(A \cap B) = \frac{1}{2}\frac{[m_2 - m_1 + \sqrt{\pi}(\sigma_1 - \sigma_2)]^2}{\sqrt{\pi}(\sigma_1 - \sigma_2)} + \frac{1}{2}\frac{[m_2 - m_1 + \sqrt{\pi}(\sigma_1 + \sigma_2)]^2}{\sqrt{\pi}(\sigma_1 + \sigma_2)}. \tag{13.30}$$

Case 4: If $b < d$ and $a > c$ then

$$c(A \cap B) = \frac{1}{2}(b - x_1)y_1 + \frac{1}{2}(x_2 - a)y_2 + \frac{1}{2}(d - x_2)y_2 - \frac{1}{2}(d - x_1)y_1. \tag{13.31}$$

Then

$$\frac{1}{2}(b - x_1)y_1 = \frac{1}{2}\sigma_1 \frac{[m_1 - m_2 + \sqrt{\pi}(\sigma_1 - \sigma_2)]^2}{\sqrt{\pi}(\sigma_1 - \sigma_2)^2},$$
$$\frac{1}{2}(x_2 - a)y_2 = \frac{1}{2}\sigma_1 \frac{[m_2 - m_1 + \sqrt{\pi}(\sigma_1 + \sigma_2)]^2}{\sqrt{\pi}(\sigma_1 + \sigma_2)^2},$$
$$\frac{1}{2}(d - x_2)y_2 = \frac{1}{2}\sigma_2 \frac{[m_2 - m_1 + \sqrt{\pi}(\sigma_1 + \sigma_2)]^2}{\sqrt{\pi}(\sigma_1 + \sigma_2)^2},$$
$$\frac{1}{2}(d - x_1)y_1 = \frac{1}{2}\sigma_2 \frac{[m_1 - m_2 + \sqrt{\pi}(\sigma_1 - \sigma_2)]^2}{\sqrt{\pi}(\sigma_1 - \sigma_2)^2}. \tag{13.32}$$

So

$$c(A \cap B) = \frac{1}{2}\frac{[m_2 - m_1 + \sqrt{\pi}(\sigma_1 + \sigma_2)]^2}{\sqrt{\pi}(\sigma_1 + \sigma_2)} + \frac{1}{2}\frac{[m_1 - m_2 + \sqrt{\pi}(\sigma_1 - \sigma_2)]^2}{\sqrt{\pi}(\sigma_1 - \sigma_2)}. \tag{13.33}$$

From the above discussion and the facts that

$$m_2 - m_1 + \sqrt{\pi}(\sigma_1 + \sigma_2) < 0$$

$$\text{implies } m_2 - m_1 + \sqrt{\pi}(\sigma_1 - \sigma_2) < 0$$

$$\text{implies } m_2 - m_1 + \sqrt{\pi}(\sigma_2 - \sigma_1) < 0$$

we can conclude a general formula for $c(A \cap B)$ as follows (assume $m_1 \geq m_2$). First we can obtain $c(A \cap B)$ as

$$\begin{aligned} c(A \cap B) &= \frac{1}{2}\frac{h^2(m_2 - m_1 + \sqrt{\pi}(\sigma_1 + \sigma_2))}{\sqrt{\pi}(\sigma_1 + \sigma_2)} \\ &+ \frac{1}{2}\frac{h^2(m_2 - m_1 + \sqrt{\pi}(\sigma_1 - \sigma_2))}{\sqrt{\pi}(\sigma_2 - \sigma_1)} \\ &+ \frac{1}{2}\frac{h^2(m_2 - m_1 - \sqrt{\pi}(\sigma_1 - \sigma_2))}{\sqrt{\pi}(\sigma_1 - \sigma_2)} \end{aligned} \tag{13.34}$$

where $h(x) = \max\{0, x\}$. Note that this holds even when $m_1 = m_2$. So the approximate mutual subsethood measure is

$$E(A, B) = \frac{c(A \cap B)}{c(A \cup B)} = \frac{c(A \cap B)}{\sigma_1\sqrt{\pi} + \sigma_2\sqrt{\pi} - c(A \cap B)}. \tag{13.35}$$

We now use this form of mutual subsethood to help tune the SAM controllers.

13.4 ADAPTIVE SUBSETHOOD FOR NEURAL FUZZY CONTROL WITH RADIAL BASIS NETWORK

This section presents an on-line *supervised* learning algorithm for building and tuning a neural fuzzy controller. We call it the *adaptive subsethood controller* (ASC). Training data lets the learning scheme learn the network *structure* and *parameters* in real time. Structure learning finds both the links and the hidden units that stand for fuzzy rules and the number of fuzzy partitions. The parameter learning adjusts the node and link values that stand for the set functions. The mutual subsethood measure drives the structure learning. Supervised learning tunes the network parameters. The supervised learning scheme builds and tunes ASC in real time. It blends *structure learning* and *parameter learning* in feedforward Gaussian SAMs or radial basis neural networks.

13.4.1 The Adaptive Subsethood Controller

This section presents the structure and functions of the Adaptive Subsethood Controller (ASC). The ASC is a feedforward multilayered radial basis net or Gaussian

SAM that acts as a fuzzy control system and that learns from input–output training data. The ASC integrates the basic elements and functions of a fuzzy system (set functions, fuzzy rules, fuzzification, defuzzification, and fuzzy implication) into a feedforward neural structure.

Figure 13.3 shows the structure of the ASC. The system has five layers. Nodes at layer 1 are input nodes (*fuzzy nodes*) that stand for input fuzzy variables. Layer 5 is the output layer. Nodes at layers 2 and 4 are *term nodes* and act as set functions that stand for the terms of each fuzzy variable. Each node at layer 3 is a rule node that stands for one fuzzy rule. So the layer 3 nodes form a fuzzy rule base. Layer 3 links define the if-parts of the rule nodes. Layer 4 links define the then-parts of the rule nodes. So for each rule node there is at most one link (perhaps none) from some term node of a fuzzy node. This is true both for if-part links and for then-part links. The links at layers 2 and 5 fully connect the fuzzy nodes and their corresponding term nodes. The arrows on the links show the normal signal flow direction. We shall later present the layer-by-layer signal propagation according to the arrow direction. Signals may flow in the reverse direction in the supervised learning process.

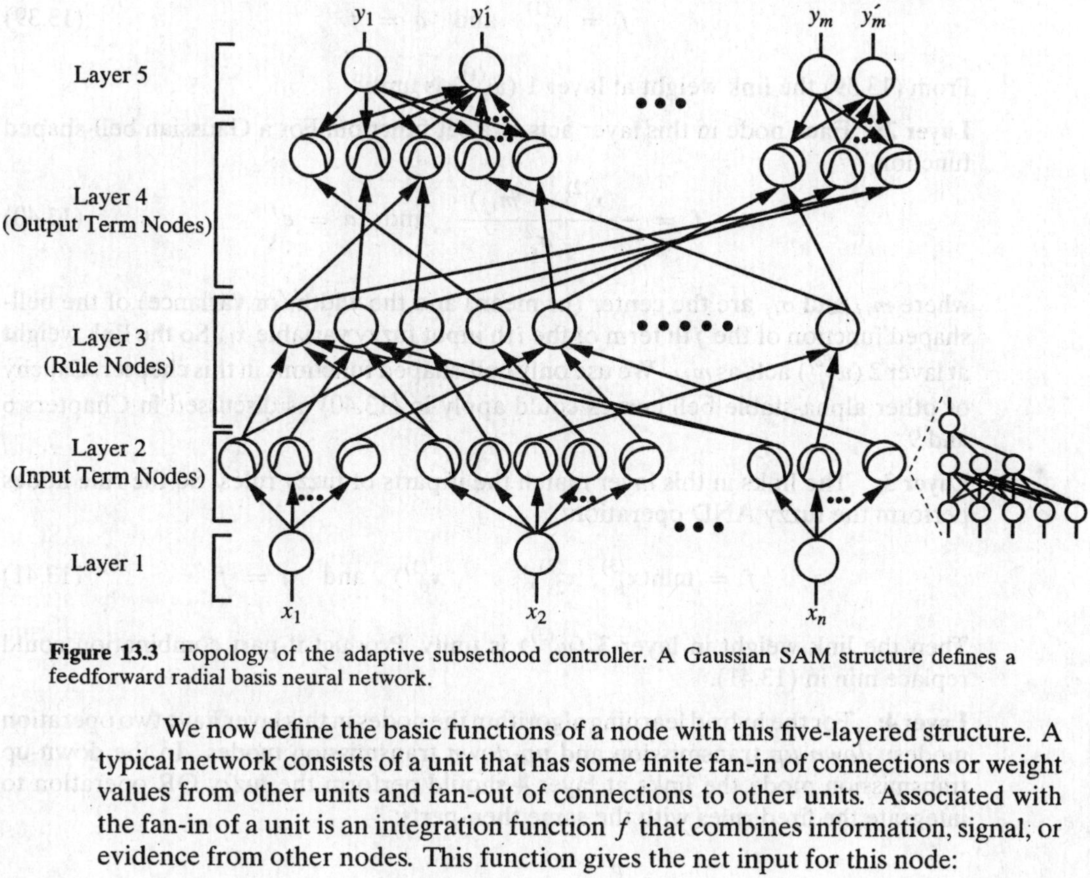

Figure 13.3 Topology of the adaptive subsethood controller. A Gaussian SAM structure defines a feedforward radial basis neural network.

We now define the basic functions of a node with this five-layered structure. A typical network consists of a unit that has some finite fan-in of connections or weight values from other units and fan-out of connections to other units. Associated with the fan-in of a unit is an integration function f that combines information, signal, or evidence from other nodes. This function gives the net input for this node:

$$\text{net-input} = f(x_1^{(k)}, x_2^{(k)}, \ldots, x_p^{(k)}; w_1^{(k)}, w_2^{(k)}, \ldots, w_p^{(k)}) \tag{13.36}$$

where $x_i^{(k)}$ stands for an ith input signal from the kth layer, w_i^k stands for the ith link weight of the kth layer, the superscript k gives the layer number, and p stands for the number of a node's input connections. We will use this notation in the following equations. A second action of each node is to emit a signal value as a function of its net input:

$$\text{output} = o_i^{(k)} = a(f). \tag{13.37}$$

where $a(\cdot)$ denotes the signal function. A logistic signal function has the form

$$f = \sum_{i=1}^{p} w_i^{(k)} x_i^{(k)} \quad \text{and} \quad a = \frac{1}{1 + e^{-f}}. \tag{13.38}$$

We next describe the functions of the nodes in each of the five layers of the ASC model.

Layer 1: The nodes in this layer transmit input values directly to the next layer:

$$f = x_i^{(1)} \quad \text{and} \quad a = f. \tag{13.39}$$

From (13.39) the link weight at layer 1 ($w_i^{(1)}$) is unity.

Layer 2: Each node in this layer acts as a set function. For a Gaussian bell-shaped function

$$f = -\frac{(x_i^{(2)} - m_{ij})^2}{\sigma_{ij}^2} \quad \text{and} \quad a = e^f \tag{13.40}$$

where m_{ij} and σ_{ij} are the center (or mean) and the width (or variance) of the bell-shaped function of the jth term of the ith input fuzzy variable x_i. So the link weight at layer 2 ($w_{ij}^{(2)}$) acts as m_{ij}. We use only bell-shaped functions in this chapter. Cauchy or other alpha-stable bell curves could apply in (13.40) as discussed in Chapters 6 and 9.

Layer 3: The links in this layer match the if-parts of fuzzy rules. So the rule nodes perform the fuzzy AND operation:

$$f = \min(x_1^{(3)}, x_2^{(3)}, \ldots, x_p^{(3)}) \quad \text{and} \quad a = f. \tag{13.41}$$

Then the link weight in layer 3 ($w_i^{(3)}$) is unity. Product if-part combination could replace min in (13.41).

Layer 4: For the hybrid learning algorithm the nodes in this layer have two operation modes: *down-up* transmission and *up-down* transmission modes. In the down-up transmission mode the links at layer 4 should perform the fuzzy OR operation to integrate the fired rules with the same then-parts:

$$f = \sum_{i=1}^{p} x_i^{(4)} \quad \text{and} \quad a = \min(1, f). \tag{13.42}$$

So the link weight $w_i^{(4)} = 1$. Other additive schemes could apply in (13.42). In the up-down transmission mode the nodes in this layer and the links in layer 5 function the same as those in layer 2. For the on-line learning algorithm the nodes in this layer operate only in the down-up transmission mode.

Layer 5: There are two kinds of nodes in this layer for the hybrid learning algorithm. The first kind of node performs the up-down transmission for the training data to feed into the network. For this kind of node

$$f = y_i \quad \text{and} \quad a = f. \tag{13.43}$$

The second kind of node performs the down-up transmission for the decision signal output. These nodes and the layer–5 links attached to them act as the defuzzifier. If $m_{ij}^{(5)}$'s and $\sigma_{ij}^{(5)}$'s are the centers and the widths of set functions then the following functions simulate the *center of gravity* or COG defuzzification method [6]:

$$f = \sum w_{ij}^{(5)} x_i^{(5)} = \sum (m_{ij}\sigma_{ij}) x_i^{(5)} \quad \text{and} \quad a = \frac{f}{\sum \sigma_{ij} x_i^{(5)}}. \tag{13.44}$$

Here the link weight at layer 5 ($w_{ij}^{(5)}$) is $m_{ij}\sigma_{ij}$. For the on-line learning algorithm we have only one kind of node performing the down-up transmission for the decision signal output.

Based on the above neural structure an off-line two-phase hybrid learning scheme can tune the ASC from the given input–output training data sets (see Appendix 13.A). This hybrid scheme is a variant of the unsupervised-supervised scheme in Chapters 3 and 4. The hybrid learning algorithm requires that we have access to sets of input and output training data off-line due to its two-phase learning scheme. The next section presents an on-line learning scheme using the mutual subsethood measure to treat learning problems whose input–output training data we can only obtain on-line. Both learning algorithms include structure and parameter learning to find the best centers (m_{ij}'s) and widths (σ_{ij}'s) of term nodes in layers 2 and 4. They learn fuzzy rules by deciding the connection types of the links at layers 3 and 4—the if-part links and then-parts links of the rule nodes. The on-line structure and parameter learning algorithm can decide the number of nodes in layer 4—the fuzzy partition of output set functions.

13.4.2 On-line Learning with Adaptive Subsethood

The mutual subsethood measure [in (13.35)] drives an on-line supervised learning algorithm for ASC based on the supervised Gaussian learning laws of Chapter 2. The system can learn on-line both the network structure and the network parameters. The learning of network structure includes deciding the proper number of output term nodes in layer 4 and the proper connections between the nodes at layers 3 and 4. This structure learning also decides the coarseness of the output fuzzy partitions and finds the fuzzy rules. The learning of network parameters also adjusts the node parameters in layers 2 and 4. This amounts to learning of if-part and then-part set functions.

The proposed structure-parameter learning algorithm uses the mutual subsethood measure to perform the structure learning. It uses the supervised algorithm to perform the parameter learning. If given the supervised training data then the proposed learning algorithm first decides whether to perform the structure learning based on the mutual subsethood measures of the then-part set functions. If it needs structure learning then the algorithm will further decide whether to add a new output term node (a new set function) and it will also change the then-parts of some fuzzy rules. After the structure learning process the parameter learning adjusts the current set functions. This structure-parameter learning repeats for each on-line incoming training input–output data pair as shown in the flow chart in Figure 13.4. The system runs the structure-parameter training loop and then combines rules to find the minimum node representation of the fuzzy rules.

We initiate the learning scheme with the desired coarseness of input fuzzy partitions (the size of the term set of each input fuzzy variable) and the initial guessed coarseness of output fuzzy partitions. Before we train this network the system builds an initial form of the network. Then during the learning process we add the new output term nodes or change some connections. Then after the learning process we delete or combine some nodes and links of the network to form the final structure of the network. In its initial form (see Figure 13.3) there are $\prod_i |T(x_i)|$ rule nodes with the inputs of each rule node coming from one possible combination of the terms of input fuzzy variables under the constraint that only one term in a term set can be a rule node's input. Here $|T(x_i)|$ denotes the number of terms of x_i (the number of fuzzy partitions of input state fuzzy variable x_i). So we divide the state space into $|T(x_1)| \times |T(x_2)| \times \ldots \times |T(x_n)|$ nodes (or fuzzy associative memory (FAM) cells) that stand for the if-parts of fuzzy rules. This is a form of *product space clustering* [17]. There is only one link between a rule node and an output fuzzy variable. This link connects to some term node of the output fuzzy variable. An expert can give the initial candidate (term node) of the then-part of a rule node or we can assign it at random. The system chooses a suitable term in each output fuzzy variable's term set for each rule node after the learning process.

We need to initialize the structure system and parameters before entering the training loop (see Figure 13.4). The structure initialization gives the connection types between layers 3 and 4. It decides the initial then-part set of each fuzzy rule. The parameter initialization decides the initial set functions of input–output fuzzy variables. One way to do this is to use the same set functions such that their domains cover the region of corresponding input–output space *evenly* according to the initial coarseness of the fuzzy partitions [17].

After the initialization process the learning algorithm enters the training loop where each loop corresponds to a set of training input data $x_i(t)$, $i = 1, \ldots, n$ and the desired output value $y_i(t)$, $i = 1, \ldots, m$ at a specific time t. Supervised gradient descent finds the errors of node outputs in each layer. The mutual subsethood measure analyzes these errors and adjusts the structure or parameters. The goal is to minimize the error or cost function

$$C = \frac{1}{2}(y(t) - \hat{y}(t))^2 \tag{13.45}$$

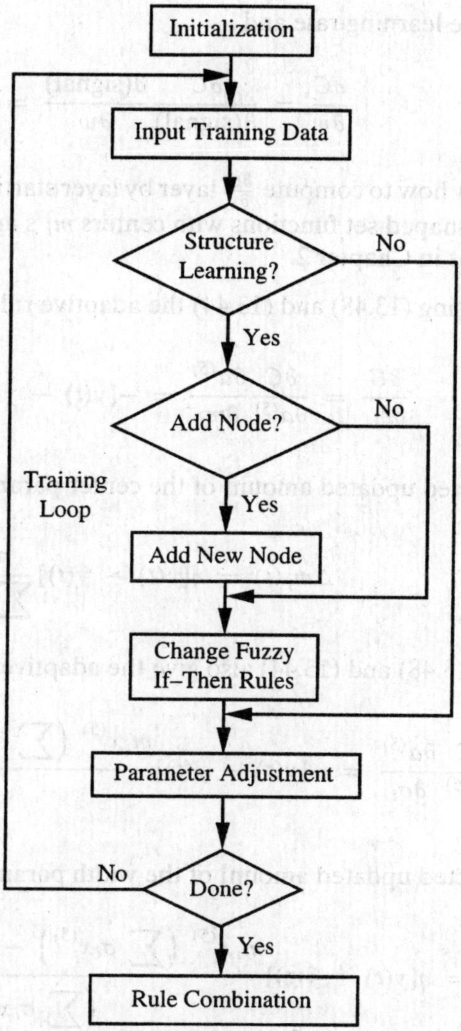

Figure 13.4 Flow chart of on-line supervised structure-parameter learning algorithm.

where $y(t)$ is the desired output and $\hat{y}(t)$ is the current output. We start each training data set at the input nodes. A forward pass computes the activity levels of all the nodes in the network. Then starting at the output nodes a backward pass computes $\frac{\partial C}{\partial y}$ for all the "hidden" nodes. If w is the adjustable parameter in a node (such as the center of a set function) then the general learning rule is the gradient-descent rule:

$$\Delta w \propto -\frac{\partial C}{\partial w}, \tag{13.46}$$

$$w(t+1) = w(t) + \eta\left(-\frac{\partial C}{\partial w}\right) \tag{13.47}$$

where η is the learning rate and

$$\frac{\partial C}{\partial w} = \frac{\partial C}{\partial(\text{signal})} \frac{\partial(\text{signal})}{\partial w} = \frac{\partial C}{\partial a} \frac{\partial a}{\partial w}. \tag{13.48}$$

We now show how to compute $\frac{\partial C}{\partial y}$ layer by layer starting at the output nodes. We will use the bell-shaped set functions with centers m_i's and widths σ_i's as the adjustable parameters as in Chapter 2.

Layer 5: Using (13.48) and (13.44) the adaptive rule of the center m_i is

$$\frac{\partial C}{\partial m_i} = \frac{\partial C}{\partial a^{(5)}} \frac{\partial a^{(5)}}{\partial m_i} = -[y(t) - \hat{y}(t)] \frac{\sigma_i x_i^{(5)}}{\sum \sigma_i x_i^{(5)}}. \tag{13.49}$$

So the expected updated amount of the center parameter is

$$\Delta m_i(t) = \eta[y(t) - \hat{y}(t)] \frac{\sigma_i x_i^{(5)}}{\sum \sigma_i x_i^{(5)}}. \tag{13.50}$$

Equations (13.48) and (13.44) also give the adaptive rule of the width σ_i as

$$\frac{\partial C}{\partial \sigma_i} = \frac{\partial C}{\partial a^{(5)}} \frac{\partial a^{(5)}}{\partial \sigma_i} = -[y(t) - \hat{y}(t)] \frac{m_i x_i^{(5)} \left(\sum \sigma_i x_i^{(5)}\right) - \left(\sum m_i \sigma_i x_i^{(5)}\right) x_i^{(5)}}{\left(\sum \sigma_i x_i^{(5)}\right)^2}. \tag{13.51}$$

So the expected updated amount of the width parameter is

$$\Delta \sigma_i(t) = \eta[y(t) - \hat{y}(t)] \frac{m_i x_i^{(5)} \left(\sum \sigma_i x_i^{(5)}\right) - \left(\sum m_i \sigma_i x_i^{(5)}\right) x_i^{(5)}}{\left(\sum \sigma_i x_i(5)\right)^2} \tag{13.52}$$

as in Chapter 2 (equations 2.17–2.28). The error "propagated" to the preceding layer is

$$\delta^{(5)} = \frac{-\partial C}{\partial a^{(5)}} = y(t) - \hat{y}(t). \tag{13.53}$$

Mutual Subsethood Measure: The system decides whether to change the current structure according to the expected updates of the center and width parameters [(13.50) and (13.52)]. To do this we compute the expected center and width as

$$\begin{aligned} m_{i-\text{new}} &= m_i(t) + \Delta m_i(t), \\ \sigma_{i-\text{new}} &= \sigma_i(t) + \Delta \sigma_i(t). \end{aligned} \tag{13.54}$$

We search current set functions of output fuzzy variables to find the one that is the most similar to the expected set function by measuring their mutual subsethoods. Let $M(m_i, \sigma_i)$ stand for the bell-shaped set function with center m_i and width σ_i. Define

$$\begin{aligned} \text{degree}(i, t) &= E[M(m_{i-\text{new}}, \sigma_{i-\text{new}}), M(m_{i-\text{closest}}, \sigma_{i-\text{closest}})] \\ &= \max_{1 \le j \le k} E[M(m_{i-\text{new}}, \sigma_{i-\text{new}}), M(m_j, \sigma_j)] \end{aligned} \tag{13.55}$$

where $k = |T(y)|$ is the size of the fuzzy partition of the output fuzzy variable $y(t)$. We find the most similar set function $M(m_{i-\text{closest}}, \sigma_{i-\text{closest}})$ to the expected set function $M(m_{i-\text{new}}, \sigma_{i-\text{new}})$ and then make the following adjustment:

IF degree$(i, t) < \alpha(t)$
 THEN
  create a new node $M(m_{i-\text{new}}, \sigma_{i-\text{new}})$ in layer 4
   and denote this new node as the i – closest node
  do the structure learning process
 ELSE IF $M(m_{i-\text{closest}}, \sigma_{i-\text{closest}}) \neq M(m_i, \sigma_j)$
  THEN
   do the structure learning process
  ELSE
   do the following parameter adjustments in layer 5:
    $m_i(t + 1) = m_{i-\text{new}}$
    $\sigma_i(t + 1) = \sigma_{i-\text{new}}$
   skip the structure learning process.

The term $\alpha(t)$ is a monotonically increasing scalar similarity criterion such that *lower similarity* is allowed in the initial stages of learning. We first compare the degree(i, t) to the similarity criterion. If the similarity is too low then we build a new term node (new set function) with the expected parameters. In this case all the current set functions differ too much from the expected one. We need a new node for the expected set function. Then we change the output connections of some just firing rule nodes to point to this new term node through the structure learning process. If the system needs no new term node then it will then check if the ith term node is the i-closest node. If this is false then it means that some just-fired fuzzy rules should have the i-closest (term) node instead of the original ith term node as their then-part. In this case the structure learning process changes the current structure. If the ith term node is the i-closest node then the system needs no structural change and it performs only the parameter learning. The following algorithm is the structure learning process.

Structure Learning: Entering this process means that the system has misassigned the ith term node in layer 4 as the then-part of some fuzzy rule that has just fired *strongly*. A better then-part for these fuzzy rules is the i – closest node. We set a *firing strength threshold* β to find the rules whose then-parts should change. We treat only the rules whose firing strengths are higher than this threshold as actual fired rules. We consider changing the then-parts of only the fired rules since only these rules fired strongly enough to contribute to the above results. Suppose that the term node $M(m_i, \sigma_i)$ in layer 4 has inputs from rule nodes 1, . . . , l in layer 3 whose corresponding firing strengths are $a_i^{(3)}$'s, i: 1, . . . , l. Then

$$\text{IF } a_i^{(3)}(t) \geq \beta \text{ THEN change the then-part of the } i\text{th rule node}$$

$$\text{from } M(m_i, \sigma_i) \text{ to } M(m_{i-\text{closest}}, \sigma_{i-\text{closest}}).$$

The following fine tuning uses the error signal more efficiently. Let

$$k_1 = \min\left(1, \sum_{a_i^{(3)} \geq \beta} a_i^{(3)}\right) \quad \text{and} \quad k_2 = \min\left(1, \sum_{a_i^{(3)} < \beta} a_i^{(3)}\right). \tag{13.56}$$

Then

$$(\Delta m_i)_{\text{extra}} = \eta' k_2 \, \Delta \, m_i = \eta' k_2 \left[\eta(y(t) - \hat{y}(t)) \frac{\sigma_i x_i^{(5)}}{\sum \sigma_i x_i^{(5)}}\right]$$

$$(\Delta \sigma_i)_{\text{extra}} = \eta' k_2 \, \Delta \, \sigma_i = \eta' k_2 \left[\eta(y(t) - \hat{y}(t)) \frac{m_i x_i^{(5)} \left(\sum \sigma_i x_i^{(5)}\right) - \left(\sum m_i \sigma_i x_i^{(5)}\right) x_i^{(5)}}{\left(\sum \sigma_i x_i^{(5)}\right)^2}\right],$$

$$\begin{aligned} (\Delta m_{i-\text{closest}})_{\text{extra}} &= \eta' k_1 (m_{i-\text{new}} - m_{i-\text{closest}}) \\ (\Delta \sigma_{i-\text{closest}})_{\text{extra}} &= \eta' k_1 (\sigma_{i-\text{new}} - \sigma_{i-\text{closest}}) \end{aligned} \tag{13.57}$$

where $0 \leq \eta' < \eta < 1$. The subscript *extra* in the above equations means the additional updated amount in addition to those calculated from other possible error signals.

Layer 4: In the down-up mode there is no parameter to adjust in this layer. We consider only the error signals ($\delta_i^{(4)}$'s). The error signal $\delta_i^{(4)}$ comes from

$$-\delta_i^{(4)} = \frac{\partial C}{\partial a_i^{(4)}} = \frac{\partial C}{\partial a^{(5)}} \frac{\partial a^{(5)}}{\partial a_i^{(4)}} \tag{13.58}$$

where from (13.44)

$$\frac{\partial a^{(5)}}{\partial a_i^{(4)}} = \frac{\partial a^{(5)}}{\partial x_i^{(5)}} = \frac{m_i\sigma_i\left(\sum \sigma_i x_i^{(5)}\right) - \left(\sum m_i\sigma_i x_i^{(5)}\right)\sigma_i}{\left(\sum \sigma_i x_i^{(5)}\right)^2} \tag{13.59}$$

and from (13.53)

$$\frac{\partial C}{\partial a^{(5)}} = -\delta^{(5)} = -[y(t) - \hat{y}(t)]. \tag{13.60}$$

Hence the error signal is

$$\delta_i^{(4)}(t) = [y(t) - \hat{y}(t)]\frac{m_i\sigma_i\left(\sum \sigma_i x_i^{(5)}\right) - \left(\sum m_i\sigma_i x_i^{(5)}\right)\sigma_i}{\left(\sum \sigma_i x_i^{(5)}\right)^2}. \tag{13.61}$$

In the multiple-output case the computations in layers 5 and 4 are the same as the above and proceed independently for each output fuzzy variable.

Layer 3: As in layer 4 we compute only the error signals. Equation (13.42) gives this error signal as

$$-\delta_i^{(3)} = \frac{\partial C}{\partial a_i^{(3)}} = \frac{\partial C}{\partial a_i^{(4)}}\frac{\partial a_i^{(4)}}{\partial a_i^{(3)}} = -\delta_i^{(4)}\frac{\partial a_i^{(4)}}{\partial x_i^{(4)}} = -\delta_i^{(4)}. \tag{13.62}$$

Hence the error signal is $\delta_i^{(3)} = \delta_i^{(4)}$. If there are multiple outputs then the error signal becomes $\delta_i^{(3)} = \sum_k \delta_k^{(4)}$ where the sum is over the then-parts of a rule node. So the error of a rule node is the sum of the errors of its then-parts.

Layer 2: Equations (13.48) and (13.40) give the adaptive rule of m_{ij} as

$$\frac{\partial C}{\partial m_{ij}} = \frac{\partial C}{\partial a_i^{(2)}}\frac{\partial a_i^{(2)}}{\partial m_{ij}} = \frac{\partial C}{\partial a_i^{(2)}} e^{f_i}\frac{2(x_i^{(2)} - m_{ij})}{\sigma_{ij}^2} \tag{13.63}$$

where from (13.62)

$$\frac{\partial C}{\partial a_i^{(2)}} = \sum_k \frac{\partial C}{\partial a_k^{(3)}}\frac{\partial a_k^{(3)}}{\partial a_i^{(2)}} \tag{13.64}$$

$$\frac{\partial C}{\partial a_k^{(3)}} = -\delta_k^{(3)} \tag{13.65}$$

and from (13.41)

$$\frac{\partial a_k^{(3)}}{\partial a_i^{(2)}} = \frac{\partial a_k^{(3)}}{\partial x_i^{(3)}} = \begin{cases} 1 & \text{if } x_i^{(3)} = \min \text{ (inputs of rule node } k). \\ 0 & \text{otherwise.} \end{cases} \tag{13.66}$$

Hence

$$\frac{\partial C}{\partial a_i^{(2)}} = \sum_k q_k \tag{13.67}$$

where the sum is over the rule nodes that $a_i^{(2)}$ feeds into and

$$q_k = \begin{cases} -\delta_k^{(3)} & \text{if } a_i^{(2)} \text{ is minimum in } k\text{th rule node's inputs.} \\ 0 & \text{otherwise.} \end{cases} \tag{13.68}$$

So the adaptive rule of m_{ij} is

$$m_{ij}(t+1) = m_{ij}(t) - \eta \frac{\partial C}{\partial a_i^{(2)}} e^{f_i} \frac{2(x_i^{(2)} - m_{ij})}{\sigma_{ij}^2}. \tag{13.69}$$

in accord with the Gaussian SAM law (2.227) in Chapter 2. And (13.48), (13.40), and (13.64)–(13.68) give the adaptive rule of σ_{ij}

$$\frac{\partial C}{\partial \sigma_{ij}} = \frac{\partial C}{\partial a_i^{(2)}} \frac{\partial a_i^{(2)}}{\partial \sigma_{ij}} = \frac{\partial C}{\partial a_i^{(2)}} e^{f_i} \frac{2(x_i^{(2)} - m_{ij})^2}{\sigma_{ij}^3}. \tag{13.70}$$

So the adaptive rule of σ_{ij} becomes

$$\sigma_{ij}(t+1) = \sigma_{ij}(t) - \eta \frac{\partial C}{\partial a_i^{(2)}} e^{f_i} \frac{2(x_i^{(2)} - m_{ij})^2}{\sigma_{ij}^3} \tag{13.71}$$

again in accord with the Gaussian SAM law (2.228) in Chapter 2.

The ASC combines rules to reduce the rule number after it finds the then-parts of the rule nodes. Three criteria decide if we need to combine a set of rule nodes into a single rule node: (1) they have the same then-parts, (2) some if-parts are common to all the rule nodes in this set, or (3) the union of other if-parts of these rule nodes equals the whole term set of some if-part fuzzy variables. If a set of nodes meets these criteria then a new rule node with only the same if-parts replaces this set of rule nodes. Figure 13.5 shows an example. The supervised learning scheme combines the structure learning and the parameter learning at the same time and on-line.

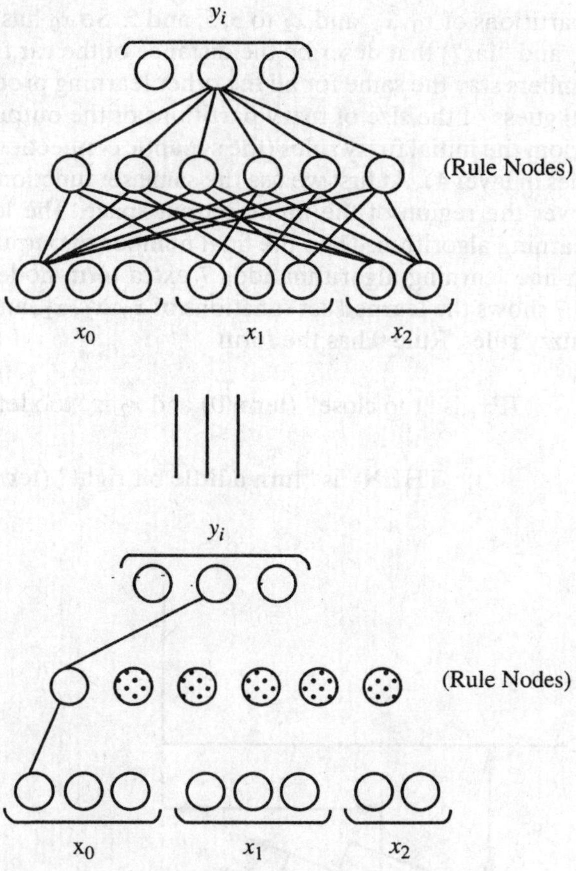

Figure 13.5 Example of combination of rule nodes.

13.5 A TEST CASE: THE FUZZY CAR

This example uses the ASC to control the "fuzzy car" of Sugeno [23]. The car can learn from examples to move along a track with rectangular turns where we know in advance the path for the car to follow. The simulation needs no path planning. The goal is to show that the ASC car can learn from past driving experiences of a skilled driver. Then the car can run on its own for similar road conditions as if a skilled driver drives it.

The input fuzzy variables are x_0, x_1, and x_2 and stand for the distance of the car from the side boundary of the track, the distance of the car from the turning point of a corner, and the current steering angle (see Figure 13.6). The output fuzzy variable y is the next steering angle. The constraints of these variables are $0 \leq x_0 \leq 250$ cm, $0 \leq x_1 \leq 700$ cm, -65 degrees $\leq x_2, y \leq 65$ degrees . The magnitude of the steering angle difference between two successive time steps cannot exceed 20 degrees . We obtain the training data in the process when an operator guides the fuzzy car along the track as shown in Figure 13.11. In the simulations we set the size of fuzzy partitions of x_0, x_1, and x_2 to 3, 5, and 5. So x_0 has three fuzzy sets ("close," "normal," and "far") that describe the distance of the car from the side of the track. These numbers stay the same for all the other learning processes as discussed below. The initial guess of the size of fuzzy partitions of the output fuzzy variable is 3. We set at random the initial fuzzy rules (the synaptic connections between nodes in layer 3 and nodes in layer 4). At first we use the same set functions such that their domains evenly cover the region of the input–output space. The learning process used the on-line learning algorithm. Then the final number of output fuzzy partitions was 10. So the on-line learning algorithm adds 7 extra term nodes to layer 4 of the ASC. Figure 13.7 shows the learned set functions of x_0, x_1, x_2, and y. Table 13.1 shows the learned fuzzy rules. Rule 0 has the form

IFx_1 is "too close" (term 0) and x_2 is "too left" (term 0)

THENy is "turn a little bit right" (term 1).

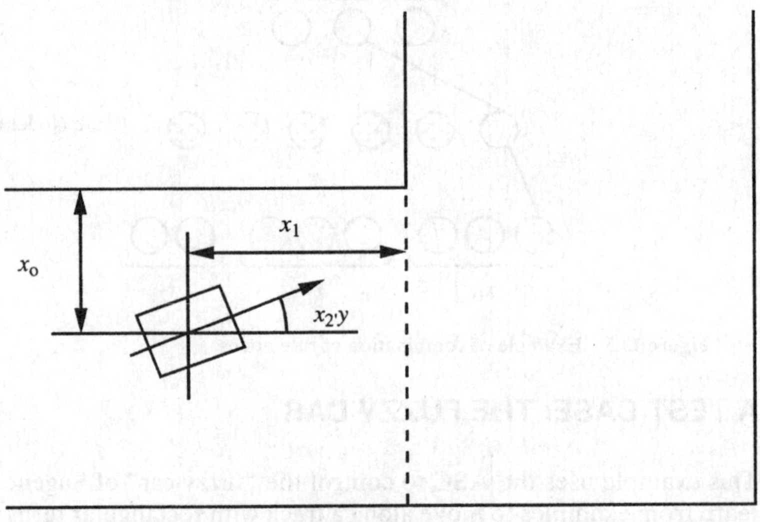

Figure 13.6 The state variables in the fuzzy car example.

These fuzzy rules also show the "hidden"-layered structure of the radial basis net as discussed in Subsection 13.4.1. Figure 13.8 shows the structure of ASC after learning. The solid curve [curve (iii)] in Figure 13.9 is the learning curve of the mean error with respect to the learning time for this simulation. We can see a large learning error in the beginning of the learning process due to our random choice of initial network structure and parameters. For comparison we show the learning curves of the two-phase hybrid learning schemes in Appendix 13.A. These are the two dashed curves [curves (i) and (ii)] in Figure 13.9. In the two-phase hybrid learning process the user must set the number of output fuzzy partitions. Here we use two values 10 and 15. The dashed curve (i) in Figure 13.9 is the learning curve when the number of

Learned Fuzzy IF-THEN Rules				
	IF-PART			THEN-PART
Rule	x_0	x_1	x_2	y
0	-	0	0	1
1	-	1	0	1
2	0	2	0	5
3	1	2	0	0
4	2	2	0	0
5	0	3	0	0
6	1	3	0	7
7	2	3	0	0
8	-	4	0	0
9	-	0	1	2
10	-	1	1	1
11	0	2	1	1
12	1	2	1	1
13	2	2	1	0
14	-	3	1	0
15	-	4	1	0
16	-	0	2	2
17	0	1	2	2
18	1	1	2	2
19	2	1	2	1
20	-	2	2	1
21	-	3	2	1
22	-	4	2	0

Learned Fuzzy IF-THEN Rules				
	IF-PART			THEN-PART
Rule	x_0	x_1	x_2	y
23	0	0	3	3
24	1	0	3	4
25	2	0	3	4
26	0	1	3	2
27	1	1	3	2
28	2	1	3	4
29	0	2	3	2
30	1	2	3	4
31	2	2	3	4
32	-	3	3	4
33	0	4	3	4
34	1	4	3	4
35	2	4	3	6
36	0	0	4	3
37	1	0	4	9
38	2	0	4	4
39	-	1	4	4
40	-	2	4	4
41	-	3	4	5
42	0	4	4	4
43	1	4	4	4
44	2	4	4	8

TABLE 13.1: Learned fuzzy rules in the fuzzy car simulation.

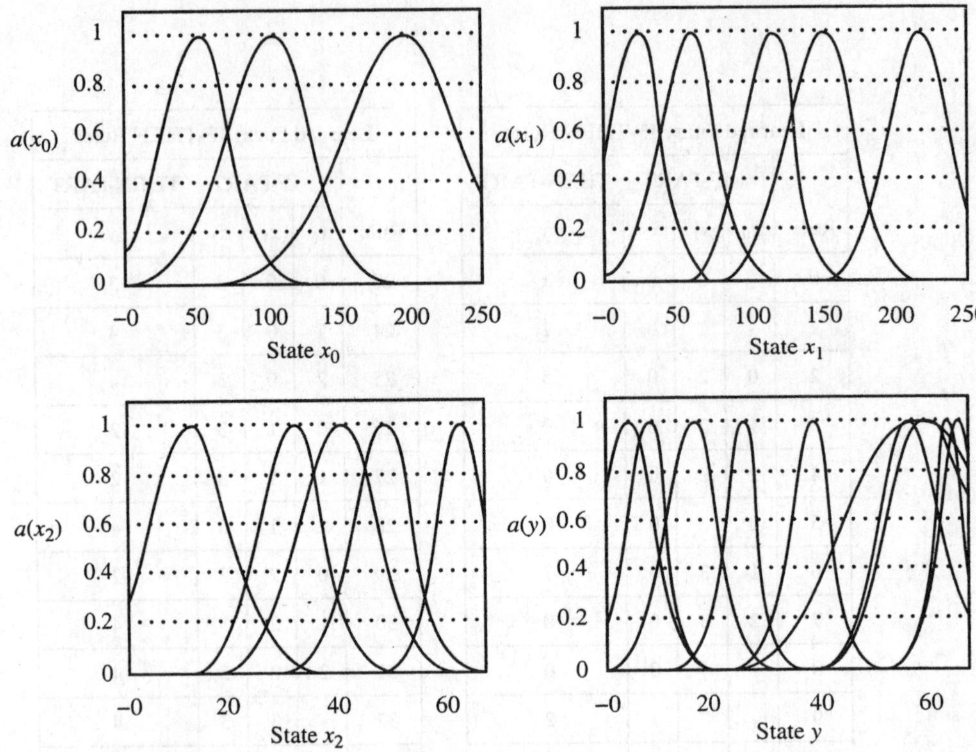

Figure 13.7 Learned set functions in the fuzzy car simulation.

output fuzzy partitions is 10. This is the final number that we obtained from on-line learning. The dashed curve (ii) in Figure 13.9 is the learning curve when the number of output fuzzy partitions is 15. The dotted curve (iv) in Figure 13.9 is the learning curve when the supervised learning algorithm replaces the second phase learning (the supervised learning) of the two-phase hybrid learning. For this modified two-phase learning algorithm if in the phase 1 learning competitive learning decides the proper initial fuzzy rules and set functions for the phase 2 learning then the on-line structure-parameter learning algorithm (used here as a phase 2 learning) gives less error as the dotted curve (iv) shows in Figure 13.9.

To show the dynamic increase of output term nodes we use three different initial guesses of the number of output fuzzy partitions. Figure 13.10 shows the growing curves of the number of output term nodes with respect to the learning time. They all reach the final number of 10. The final ASC method controls the fuzzy car. We keep the speed of the car constant and assume there are sensors on the car to measure the state variables x_0, x_1, and x_2 that feed into the controller to derive the next steering angle. Figure 13.11 shows the simulated results. The thick straight lines in this figure define the shape of the roads. We have numbered x-axis and y-axis in this figure in centimeters. The fuzzy car runs in the S-shaped road where we trained it and in the U-shaped road without further training. The curves in Figure 13.11 are the running traces of the fuzzy car under the control of the learned ASC. The

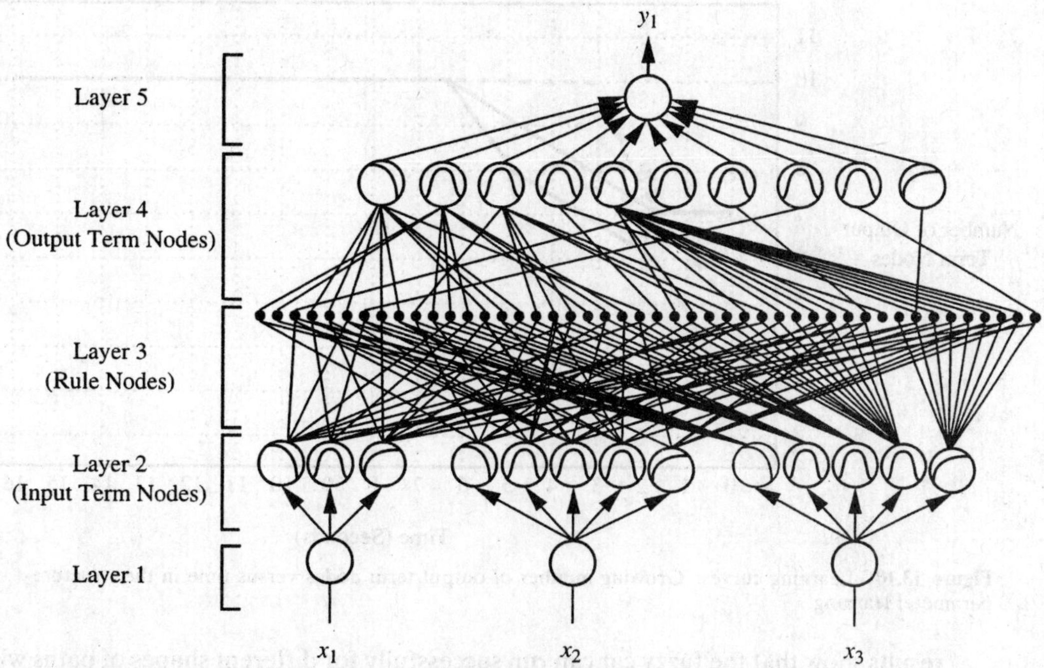

Figure 13.8 Topology of the adaptive subsethood controller after on-line learning in the fuzzy car simulation. The Gaussian SAM acts as a multilayered radial basis function neural network.

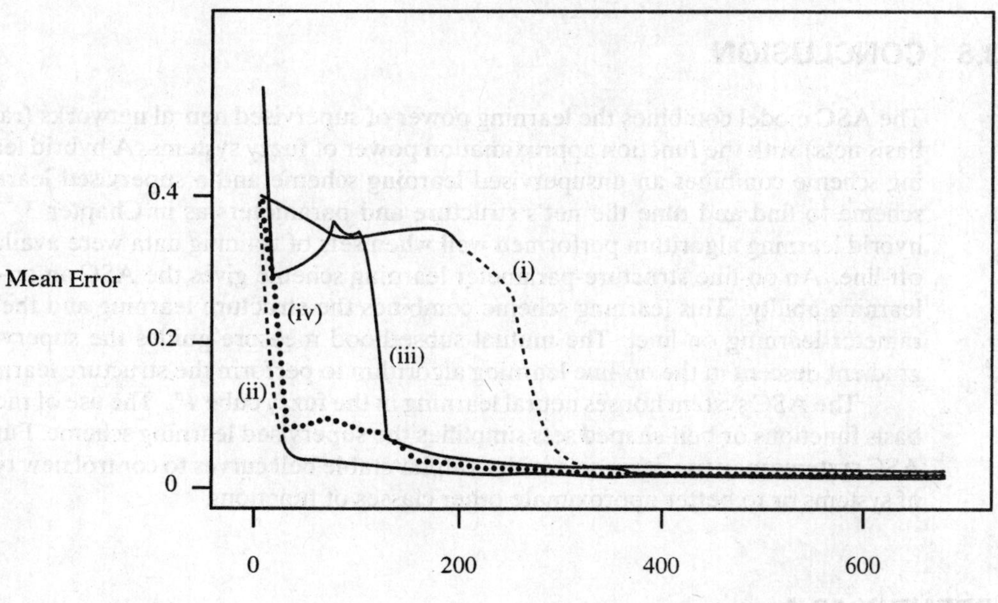

Figure 13.9 Learning curves: Mean squared error versus time in some fuzzy car simulations.

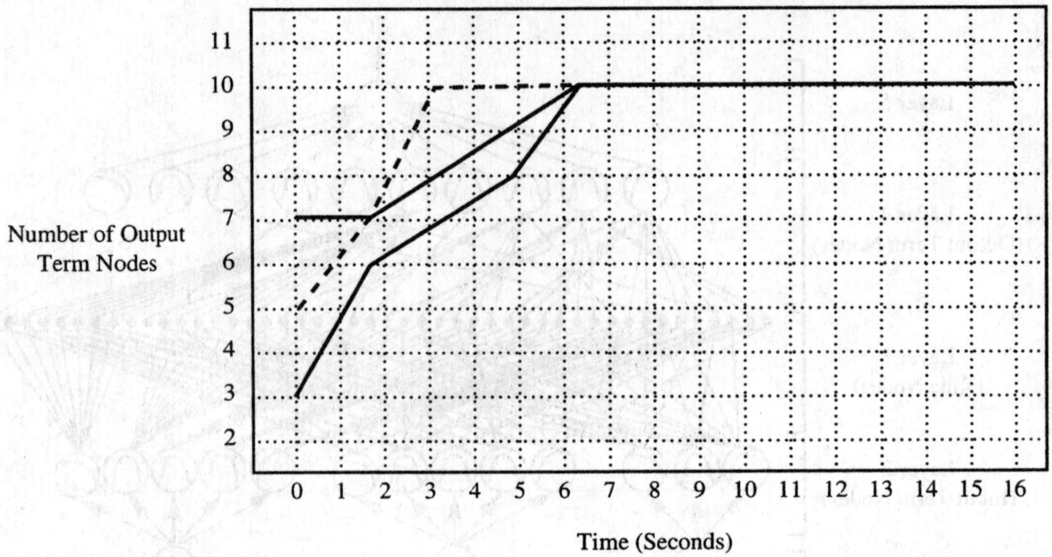

Figure 13.10 Learning curves: Growing number of output term nodes versus time in the structure-parameter learning.

results show that the fuzzy car can run successfully for different shapes of paths with different starting locations and different starting angles.

13.6 CONCLUSION

The ASC model combines the learning power of supervised neural networks (radial basis nets) with the function approximation power of fuzzy systems. A hybrid learning scheme combines an unsupervised learning scheme and a supervised learning scheme to find and tune the net's structure and parameters as in Chapter 3. The hybrid learning algorithm performed well when sets of training data were available off-line. An on-line structure-parameter learning scheme gives the ASC an on-line learning ability. This learning scheme combines the structure learning and the parameter learning on-line. The mutual subsethood measure guides the supervised gradient descent in the on-line learning algorithm to perform the structure learning.

The ASC system houses neural learning in the fuzzy cube I^n. The use of radial-basis functions or bell-shaped sets simplifies the supervised learning scheme. Future ASC systems may use other sets such as alpha-stable bell curves to control new types of systems or to better approximate other classes of functions.

APPENDIX 13.A

This Appendix presents an off-line two-phase learning scheme for realizing the ASC from the given input–output training data. This hybrid learning scheme has the same

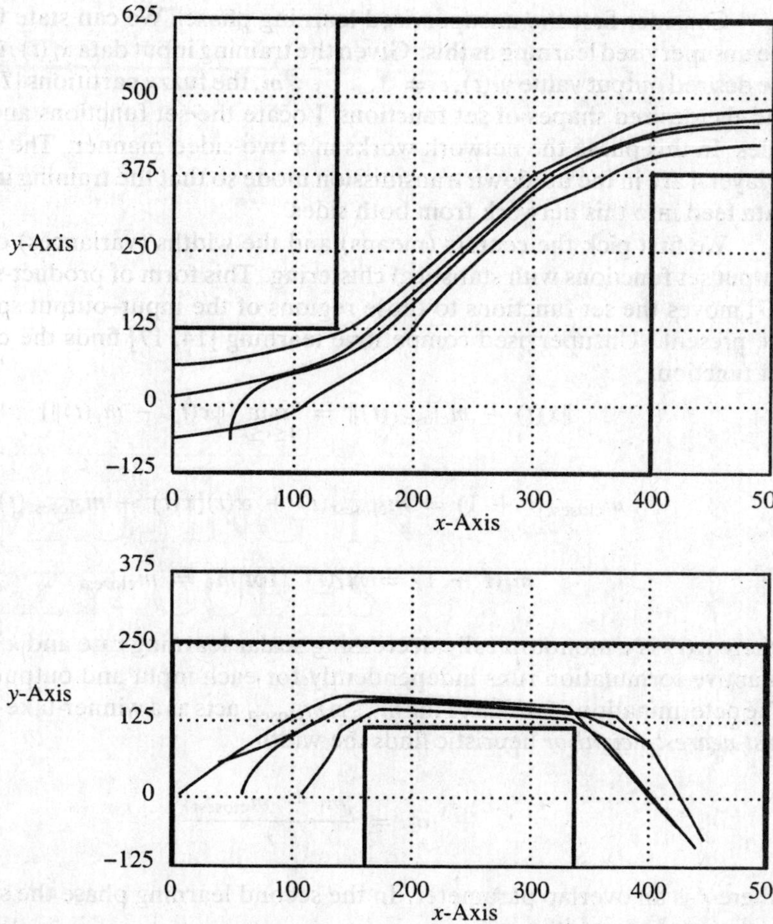

Figure 13.11 Simulation results of the fuzzy car running under the adaptive subsethood controller.

unsupervised/supervised form as the hybrid scheme in Chapters 3 and 4. In phase 1 unsupervised clustering picks the initial if-part and then-part set functions and finds the fuzzy rules. In phase 2 supervised learning tunes both if-part and then-part set functions for desired outputs. Before we train this network we construct an initial form of the network. Then during the learning process some nodes and links of this initial network combine or disappear to form the final structure of the network. In its initial form (see Figure 13.3) there are $\prod_i |T(x_i)|$ rule nodes with the inputs of each rule node coming from one possible combination of the terms of the if-part fuzzy variables under the constraint that only one term in a term set can be a rule node's input. Here $|T(x_i)|$ denotes the number of terms of x_i (the number of fuzzy partitions of input state fuzzy variable x_i). So we initially divide the state space into $|T(x_1)| \times |T(x_2)| \times \ldots \times |T(x_n)|$ nodes (or FAM cells) that stand for the if-parts of fuzzy rules. Initially the links between the rule nodes and the output term nodes are fully connected. Learning prunes the network.

Consider first the unsupervised learning phase. We can state the problem for the unsupervised learning as this: Given the training input data $x_i(t), i = 1, \ldots, n$, the desired output value $y_i(t), i = 1, \ldots, m$, the fuzzy partitions $|T(x)|$ and $|T(y)|$, and the desired shapes of set functions. Locate the set functions and find the fuzzy rules. In this phase the network works in a two-sided manner. The nodes and links at layer 4 are in the up-down transmission mode so that the training input and output data feed into this network from both sides.

We first pick the centers (means) and the widths (variances) of the input and output set functions with statistical clustering. This form of product-space clustering [17] moves the set functions to those regions of the input–output space where data are present. Unsupervised competitive learning [14, 17] finds the center m_i of the set function:

$$\|x(t) - m_{\text{closest}}(t)\| = \min_{1 \le i \le k} \{\|x(t) - m_i(t)\|\} \tag{13.72}$$

$$m_{\text{closest}}(t+1) = m_{\text{closest}}(t) + \alpha(t)[x(t) - m_{\text{closest}}(t)] \tag{13.73}$$

$$m_i(t+1) = m_i(t) \quad \text{for } m_i \neq m_{\text{closest}} \tag{13.74}$$

where $\alpha(t)$ is a monotonically decreasing scalar learning rate and $k = |T(x)|$. This adaptive formulation runs independently for each input and output fuzzy variable. The determination of which of the m_i's is m_{closest} acts as a winner-take-all method. The *first-nearest-neighbor* heuristic finds the width

$$\sigma_i = \frac{|m_i - m_{\text{closest}}|}{r} \tag{13.75}$$

where r is an overlap parameter. In the second learning phase the system tunes the centers and the widths.

After we find the parameters of the set functions the signals from both external sides can reach the output points of term nodes at layer 2 and layer 4 (see Figure 13.3). The outputs of term nodes at layer 2 transmit to rule nodes through the initial architecture of layer–3 links. This gives the firing strength of each rule node. Based on these rule firing strengths [denoted as $o_i^{(3)}(t)$'s] and the outputs of term nodes at layer 4 [denoted as $o_j^{(4)}(t)$'s] we want to decide the correct then-part links (layer 4 links) of each rule node to find the existing fuzzy rule by competitive learning [17]. As stated in Section 13.5 the links at layer 4 are at first fully connected. We denote the weight of the link between the ith rule node and the jth output term node as w_{ij}. The following competitive learning law updates these weights for each training data set:

$$\dot{w}_{ij}(t) = o_j^{(4)}(-w_{ij} + o_i^{(3)}). \tag{13.76}$$

Here $o_j^{(4)}$ acts as a win–loss index of the jth term node at layer 4 [17]. In the extreme case if $o_j^{(4)}$ is a 0–1 threshold function then the above law says *learn only if win.*

After competitive learning processes the whole training data set then the link weights at layer 4 stand for the strength of the corresponding rule then-part. Among the links which connect a rule node and the term nodes of an output fuzzy node we have at most one link with maximum weight and delete the rest. Hence only one term in an output fuzzy variable's term set can become one of the then-parts of a fuzzy rule. If all the link weights between a rule node and the term nodes of an output fuzzy node are very small then we delete all the corresponding links. So this rule node has little or no relation to this output fuzzy variable. If we delete all the links between a rule node and the layer 4 nodes then we can delete this rule node since it does not affect the outputs. After we pick the then-parts of rule nodes the rule combination scheme described in Subsection 13.4.2 reduces the number of rules.

After we find the fuzzy rules we have derived the whole network structure and the network then enters the second learning phase to adjust the parameters of the set functions. We can state the problem for the supervised learning as this: Given the training input data $x_i(t)$, $i = 1, \ldots, n$, the desired output value $y_i(t)$, $i = 1, \ldots, m$, the fuzzy partitions $|T(x)|$ and $|T(y)|$ and the fuzzy rules. Adjust the parameters of the set functions to minimize the mean-squared error. These fuzzy rules come from the first-phase learning or perhaps from experts. In the second-phase learning the network works in the feedforward manner. The nodes and the links at layer 4 are in the down-up transmission mode. Gradient descent performs the supervised learning. In the same way that we derived the on-line learning algorithm we derive the detailed learning rules of the hybrid learning algorithm as follows. Here we use the same cost function C as in (13.45) and the same general parameter learning rules as in (13.46)–(13.48).

Layer 5: Using (13.48) and (13.44), the adaptive rule of the center m_i is

$$\frac{\partial C}{\partial m_i} = \frac{\partial C}{\partial a^{(5)}} \frac{\partial a^{(5)}}{\partial m_i} = -[y(t) - \hat{y}(t)] \frac{\sigma_i x_i^{(5)}}{\sum \sigma_i x_i^{(5)}}. \tag{13.77}$$

So update the center parameter by

$$m_i(t+1) = m_i(t) + \eta[y(t) - \hat{y}(t)] \frac{\sigma_i x_i^{(5)}}{\sum \sigma_i x_i^{(5)}}. \tag{13.78}$$

And (13.48) and (13.44) and the SAM theorem give the adaptive rule of the width σ_i as

$$\frac{\partial C}{\partial \sigma_i} = \frac{\partial C}{\partial a^{(5)}} \frac{\partial a^{(5)}}{\partial \sigma_i} = -[y(t) - \hat{y}(t)] \frac{m_i x_i^{(5)} \left(\sum \sigma_i x_i^{(5)}\right) - \left(\sum m_i \sigma_i x_i^{(5)}\right) x_i^{(5)}}{\left(\sum \sigma_i x_i^{(5)}\right)^2}. \tag{13.79}$$

So update the width parameter by

$$\sigma_i(t+1) = \sigma_i(t) + \eta[y(t) - \hat{y}(t)] \frac{m_i x_i^{(5)} \left(\sum \sigma_i x_i^{(5)}\right) - \left(\sum m_i \sigma_i x_i^{(5)}\right) x_i^{(5)}}{\left(\sum \sigma_i x_i^{(5)}\right)^2}. \tag{13.80}$$

The error "propagated" to the preceding layer is

$$\delta^{(5)} = \frac{-\partial C}{\partial a^{(5)}} = y(t) - \hat{y}(t). \tag{13.81}$$

Layer 4–Layer 2: Layer 1: Since here the system performs only the parameter learning the learning rules are the same as in (13.58)–(13.71).

The flow chart in Figure 13.12 summarizes the proposed two-phase hybrid learning procedure. We found that the convergence speed of the supervised learning in the phase 2 learning is superior to the normal or blind supervised learning scheme since the unsupervised learning process in phase 1 has done much of the learning work in advance. The unsupervised learning gives a better-than-random initialization of the gradient descent.

REFERENCES

[1] H. Bersini, J.-P. Nordvik, and A. Bonarini, "A Simple Direct Adaptive Fuzzy Controller Derived from Its Neural Equivalent," *Proc. 1993 IEEE Int. Conf. on Neural Networks*, San Francisco, CA, pp. 345–350, 1993.

[2] I. Enbutsu, K. Baba, and N. Hara, "Fuzzy Rule Extraction from a Multilayered Neural Network," *Proc. 1991 IEEE Int. Joint Conf. on Neural Networks*, pp. II–461–465, 1991.

[3] E. Hartman, J. D. Keeler, and J. Kowalski, "Layered Neural Networks with Gaussian Hidden Units as Universal Approximators," *Neural Computation*, Vol. 2, 210–215, 1990.

[4] J. Hertz, A. Krogh, and R. G. Palmer, *Introduction to the Theory of Neural Computation*, Addison-Wesley, 1991, pp. 188–189.

[5] J.-S. Jang, "Self-Learning Fuzzy Controllers Based on Temporal Back Propagation," *IEEE Trans. on Neural Networks*, Vol. 3, No. 5, 741–723, 1992.

[6] C. C. Lee, "Fuzzy Logic in Control Systems: Fuzzy Logic Controller: Parts I and II," *IEEE Trans. Syst., Man, and Cybern.*, Vol. SMC–20, No. 2, 404–435, 1990.

[7] C. C. Lee and H. R. Berenji, "An Intelligent Controller Based on Approximate Reasoning and Reinforcement Learning," *Proc. of IEEE Intelligent Machines*, pp. 200–205, 1989.

[8] C. T. Lin and C. S. G. Lee, "Neural-Network-Based Fuzzy Logic Control and Decision System," *IEEE Trans. on Computers*, Vol. C–40, No. 12, 1320–1336, December 1991.

[9] C. T. Lin and C. S. G. Lee, "Real-Time Supervised Structure/Parameter Learning for Fuzzy Neural Networks," *Proc. of 1992 IEEE Int. Conf. on Fuzzy Systems*, San Diego, CA, pp. 1283–1290, March 8–12, 1992.

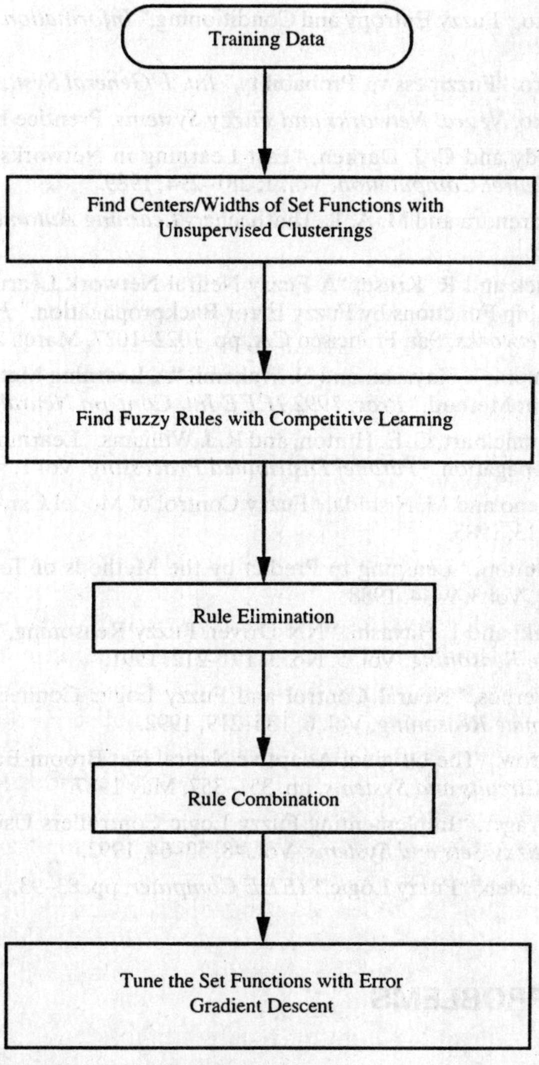

Figure 13.12 Flow chart of the hybrid learning scheme.

[10] C. T. Lin and C. S. G. Lee, "Reinforcement Structure/Parameter Learning for Neural-Network-Based Fuzzy Logic Control Systems," *IEEE Trans. on Fuzzy Systems*, Vol. 2, No. 1, 46–63, 1994.

[11] C. T. Lin, *Neural Fuzzy Control Systems with Structure and Parameter Learning*, World Scientific, Inc., 1994.

[12] C. T. Lin, "A Neural Fuzzy Control System," *Fuzzy Sets and Systems*, vol. 70 183–212, 1995.

[13] C. T. Lin, C. J. Lin, and C. S. G. Lee, "Fuzzy Adaptive Learning Control Network with On-Line Neural Learning," *Fuzzy Sets and Systems*, vol. 71, 25–45, 1995.

[14] T. Kohonen, *Self-Organization and Associative Memory*, Springer-Verlag, 1988, p. 132.

[15] B. Kosko, "Fuzzy Entropy and Conditioning," *Information Sciences*, Vol. 40, pp. 165–174, 1986.

[16] B. Kosko, "Fuzziness vs. Probability," *Int. J. General Syst.*, Vol. 17, No. 2, 211–240, 1990.

[17] B. Kosko, *Neural Networks and Fuzzy Systems*, Prentice Hall, 1991.

[18] J. Moody and C. J. Darken, "Fast Learning in Networks of Locally-Tuned Processing Units," *Neural Computation*, Vol. 1, 281–294, 1989.

[19] K. S. Narendra and M. A. L. Thathachar, *Learning Automata: An Introduction*, Prentice Hall, 1989.

[20] D. Nauck and R. Kruse, "A Fuzzy Neural Network Learning Fuzzy Control Rules and Membership Functions by Fuzzy Error Backpropagation," *Proc. 1993 IEEE Int. Conf. on Neural Networks*, San Francisco CA, pp. 1022–1027, March 28–April 1, 1993.

[21] H. Nomura, I. Hayashi, and N. Wakami, "A Learning Method of Fuzzy Inference Rules by Descent Method," *Proc. 1992 IEEE Int. Conf. on Neural Networks*, pp. 203–210, 1992.

[22] D. E. Rumelhart, G. E. Hinton, and R. J. Williams, "Learning Internal Representations by Error Propagation," *Parallel Distributed Processing*, Vol 1, pp. 318–362, MIT Press, 1986.

[23] M. Sugeno and M. Nishida, "Fuzzy Control of Model Car," *Fuzzy Sets and Systems*, Vol 16, 103–113, 1985.

[24] R. S. Sutton, "Learning to Predict by the Methods of Temporal Difference," *Machine Learning*, Vol 3, 9–44, 1988.

[25] T. Tagaki and I. Hayashi, "NN-Driven Fuzzy Reasoning," *International Journal of Approximate Reasoning*, Vol. 5, No. 3, 191–212, 1991.

[26] P. J. Werbos, "Neural Control and Fuzzy Logic: Connections and Designs," *Int. J. of Approximate Reasoning*, Vol. 6, 185–219, 1992.

[27] B. Widrow, "The Original Adaptive Neural Net Broom-Balancer," *Proc. of Int. Symposium on Circuits and Systems*, pp. 351–357, May 1987.

[28] R. R. Yager, "Implementing Fuzzy Logic Controllers Using a Neural Network Framework," *Fuzzy Sets and Systems*, Vol. 48, 53–64, 1992.

[29] L. A. Zadeh, "Fuzzy Logic," *IEEE Computer*, pp. 83–93, April 1988.

HOMEWORK PROBLEMS

13.1. Define discrete fuzzy sets A and B as $A = (.2\ .9\ .7\ .6\ .1)$ and $B = (.3\ 1\ .5\ .4\ .1)$. Compute **(a)** the subsethood value $S(A, B)$ and $S(B, A)$, **(b)** their equivalence or mutual subsethood measure $E(A, B)$, and **(c)** their difference $\text{Difference}(A, B)$.

13.2. Repeat the calculations in problem 1 for the continuous fuzzy sets A and B:

$$a(x) = \begin{cases} 2 - x & \text{if } 1 \le x < 2 \\ x & \text{if } 0 \le x < 1 \\ 0 & \text{else} \end{cases}$$

$$b(x) = \begin{cases} 3 - 2x & \text{if } 1 \le x < 1.5 \\ 1 & \text{if } -2 \le x < 1 \\ 0.5x + 2 & \text{if } -4 \le x < -2 \\ 0 & \text{else.} \end{cases}$$

13.3. Define discrete sets $A = (\frac{1}{4}\ \frac{1}{3})$ and $B = (\frac{1}{2}\ \frac{3}{5})$. Draw the set-as-points geometry of $A, B, A \cup B, A \cap B, F(2^A), F(2^B), c(A), c(B), c(A \cup B), c(A \cap B)$, and $d(A, B)$. Then use this figure to compute the values of $c(A), c(B), c(A \cup B), c(A \cap B)$, and $d(A, B)$ for the ℓ^2 norm.

13.4. For two fuzzy sets A and B prove that $A^* = B^* = A \cap B$ where A^* and B^* are as in (13.9) and (13.10).

13.5. Prove the modular equality of fuzzy counting $c(A) + c(B) = c(A \cap B) + c(A \cup B)$.

13.6. Prove for fuzzy sets A and B that equivalence reduces to subsethood:

$$E(A, B) = \frac{1}{\frac{1}{S(A,B)} + \frac{1}{S(B,A)} - 1}.$$

13.7. Show that $0 \leq E(A, B) \leq 1$.

13.8. Prove that the mutual subsethood measure reduces to the Lukasiewicz equivalence operator when viewed at the one-dimensional level of fuzzy logic [prove (13.15)].

13.9. Prove that the mutual subsethood measure $E(A, B)$ equals the Tanimoto similarity measure $S_T(A, B)$ when A and B are nonfuzzy sets.

13.10. Show that the fuzziness or entropy $c(A \cap A^c)/c(A \cup A^c)$ is the degree that A equals A^c:

$$\text{ENTROPY}(A) = E(A, A^c).$$

13.11. Prove that the formula for $c(A \cap B)$ in (13.34) holds when $m_1 = m_2$.

13.12. Explain why (13.44) can approximate the *center of gravity defuzzification* method.

13.13. Draw one possible structure of a fuzzy connected ASC with three input variables x_1, x_2, x_3 and two output variables y_1 and y_2 where $|x_1| = 2$, $|x_2| = 3$, $|x_3| = 2$, $|y_1| = 5$, $|y_2| = 3$.

13.14. Consider an ASC with isosceles triangular input and output set functions. Derive **(a)** the second phase of hybrid (off-line) learning algorithm and **(b)** the on-line learning algorithm.

13.15. Combine the rules on the following rule set and show the final set if $|x_1| = 3, |x_2| = 2, |x_3| = 2$, and $|y| = 3$.

IF x_1 is A_1 AND x_2 is B_1 AND x_3 is C_1 THEN y is D_1

IF x_1 is A_2 AND x_2 is B_1 AND x_3 is C_1 THEN y is D_3

IF x_1 is A_3 AND x_2 is B_1 AND x_3 is C_1 THEN y is D_1

IF x_1 is A_1 AND x_2 is B_2 AND x_3 is C_1 THEN y is D_1

IF x_1 is A_2 AND x_2 is B_2 AND x_3 is C_1 THEN y is D_3

IF x_1 is A_3 AND x_2 is B_2 AND x_3 is C_1 THEN y is D_2

IF x_1 is A_1 AND x_2 is B_1 AND x_3 is C_2 THEN y is D_1

IF x_1 is A_2 AND x_2 is B_1 AND x_3 is C_2 THEN y is D_2

IF x_1 is A_3 AND x_2 is B_1 AND x_3 is C_2 THEN y is D_2

IF x_1 is A_1 AND x_2 is B_2 AND x_3 is C_2 THEN y is D_1

IF x_1 is A_2 AND x_2 is B_2 AND x_3 is C_2 THEN y is D_3

IF x_1 is A_3 AND x_2 is B_2 AND x_3 is C_2 THEN y is D_2

13.16. Consider the discrete fuzzy sets

$$\text{“HIGH”} = (0\ .2\ .5\ .8\ .9\ 1),$$

$$\text{“VERY HIGH”} = (0\ .04\ .25\ .64\ .81\ 1),$$

$$\text{“MEDIUM”} = (.2\ .6\ 1\ 1\ .6\ .2),$$

$$\text{“LOW”} = (1\ .8\ .6\ .3\ .1\ 0),$$

$$\text{“VERY LOW”} = (1\ .64\ .36\ .09\ .01\ 0),$$

What is the best match for A if $A = (.4\ .7\ .9\ .5\ .2\ .2)$?

PART VIII

FEEDBACK IN FUZZY CUBES

The adaptive resonance theory (ART) model is one of the triumphs of modern neural network theory. The model combines the concepts of feedback stability between two layers of competing neurons with learning convergence of the synaptic pathways that connect the two layers. The key idea is that the synapses learn only if the two fields resonate in a dynamic equilibrium.

Neurons change faster than synapses change. So a global equilibrium requires a delicate dynamical balance. Each neural field stabilizes or equilibrates to a spatial pattern of signals. Then the synapses slowly change to encode these patterns. But this very change changes the bidirectional flow of neural signals and tends to disturb the stable patterns at each field. ART models balance neural and synaptic change with design features that ensure that the synapses learn only if the neurons resonate or equilibrate.

Real brains may use some form of this dynamic balancing. Basic psychology seems to depend on it. If the first or bottom layer codes sensory patterns then the second or upper layer codes expected patterns. The system senses a raw pattern and then in turn expects to see a pattern that matches past experience. The old buffers the new just as when we first hear the start of a new song or first see the cowboy draw his gun in a film.

We learn or relearn a new pattern when it matches a stored or expected pattern. The new pattern tends to differ slightly from the stored pattern and so the new pattern slightly changes future expectations. Thus when we look at ourselves in a mirror we may feel concern but not shock. The pattern is slightly new but largely expected.

The pure ART model is a set of design concepts. These concepts give rise to many ART models when we translate the concepts to math. The models range from pattern classifiers and real brain models to bidirectional associative memories and models of eye and robotic function. The Fuzzy ART model is one such translation.

The Fuzzy ART model approximates surfaces in a fuzzy cube of high dimension. Each pattern is a discrete fuzzy set. The ART system uses the degree of subsethood to control how well fuzzy patterns match. The system covers the decision surface with hyper-boxes much as a feedforward fuzzy system covers a function's graph with rule patches. All such graph covers lead to hyper-box or rule explosion as the dimension of the space grows. The Fuzzy ART model uses subsethood to control the fineness or grid size of the hyper-box cover in a fixed fuzzy cube. The result blends the core concepts of fuzzy cubes and feedback neural networks in a novel and powerful architecture.

14

FUZZY ADAPTIVE RESONANCE THEORY[1]

Gail A. Carpenter and Stephen Grossberg

PREVIEW

Adaptive resonance theory (ART) models are real-time neural networks for category learning, pattern recognition, and prediction. Unsupervised fuzzy ART and supervised fuzzy ARTMAP synthesize fuzzy set theory and ART networks. They exploit the formal similarity between the computations of fuzzy subsethood and the dynamics of ART category choice, search, and learning. Fuzzy ART self-organizes stable recognition categories in response to arbitrary sequences of analog or binary input patterns. It generalizes the binary ART 1 model and replaces the set-theoretic intersection (∩) with the fuzzy intersection (∧) or pairwise minimum inside a fuzzy cube. A normalization procedure called complement coding leads to a symmetric theory in which the fuzzy intersection and the fuzzy union (∨) or pairwise maximum play complementary roles. Complement coding preserves individual feature amplitudes

[1] This research was supported in part by the Advanced Research Projects Agency (ARPA) (ONR N00014–92–J–4015), the National Science Foundation (NSF IRI 90–00530), and the Office of Naval Research (ONR N00014–91–J–4100).

while normalizing the input vector and prevents a potential category proliferation problem. Adaptive weights start equal to one and can only decrease in time.

A geometric interpretation of fuzzy ART represents each category as a hyperbox in a fuzzy cube that increases in size as weights decrease. A matching criterion controls the search and decides how close an input and a learned representation must be for a category to accept the input as a new exemplar. A vigilance parameter (ρ) sets the matching criterion and determines how finely or coarsely an ART system will partition inputs. High vigilance creates fine categories or small boxes in a fuzzy cube. Learning stops when boxes cover the input space. Such a graph cover implies that the system suffers from the curse of dimensionality: The number of boxes it takes to cover a surface in a fuzzy cube grows exponentially with the dimension of the fuzzy cube. Learning stabilizes after just one presentation of each input for fast learning, fixed vigilance, and any input set. A fast-commit slow-recode option allows rapid learning of rare events yet buffers memories against recoding by noisy inputs.

Fuzzy ARTMAP unites two fuzzy ART networks to solve problems of supervised learning and prediction. A minimax learning rule controls ARTMAP category structure. It both minimizes predictive error and maximizes code compression. Low vigilance maximizes compression but may thus cause different inputs to make the same prediction. If this coarse grouping strategy causes a predictive error then an internal match tracking control process increases the vigilance just enough to correct the error. ARTMAP automatically constructs a minimal number of recognition categories (or "hidden units") to meet accuracy criteria. An ARTMAP voting strategy improves prediction by training the system several times using different orderings of the input set. Voting assigns confidence estimates to competing predictions given small, noisy, or incomplete training sets. ARPA benchmark simulations show fuzzy ARTMAP dynamics. This chapter also compares fuzzy ARTMAP to Salzberg's Nested Generalized Exemplar (NGE) and to Simpson's Fuzzy Min-Max Classifier (FMMC). It concludes with a summary of ART and ARTMAP applications.

14.1 MATCH-BASED LEARNING AND ERROR-BASED LEARNING

14.1.1 When Is a Dog Not a Dog?

A stable learning system incorporates crucial new data into an existing memory system without destroying old memories. We remember effortlessly that a dog is still a dog even as we learn that this dog is a Dalmatian named Spot. In a complex world new information often directly contradicts the old but both are important and "correct."

An ART network constructs new memories based on the success or failure of old memories as they guide the system in the world. The network encounters examples and some categories become coarse ($\bar{\text{dog}}$) (Figure 14.1a) or fine (Dalmatian) as needed. If I expect to hear "dog" but the answer is "Rover" then I am startled into paying attention to features that I had before ignored (Figure 14.1b). If I learn

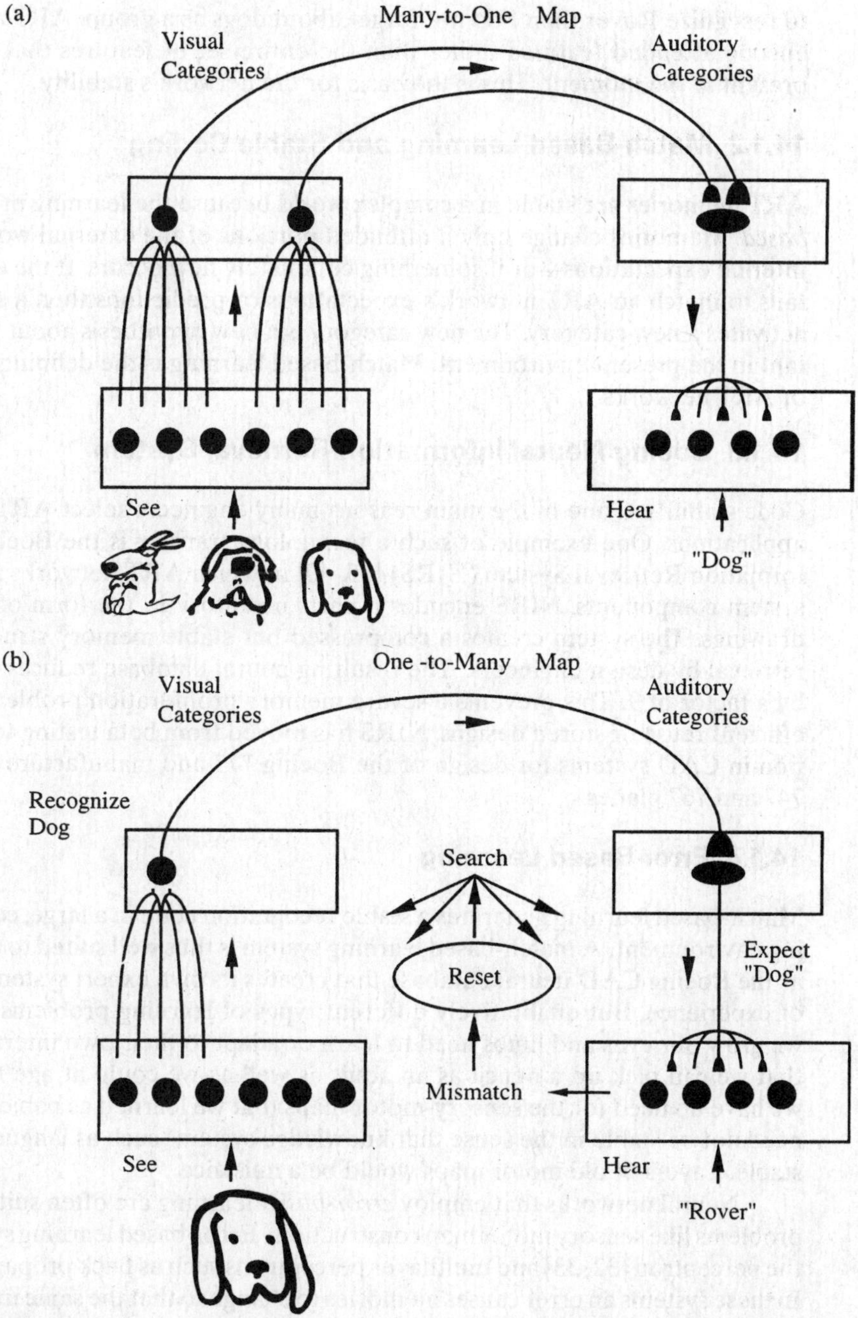

Figure 14.1 (a) An ART network creates a coarse category (dog) by a two-step code compression process. The network first groups dogs based on their shared features. Then these dissimilar categories learn the shared prediction "dog." (b) An ART network that makes a bad prediction learns a new category identification ("Rover") but at the same time preserves the coarse category representation ("dog").

to recognize Rover then I do not forget about dogs as a group. ART memories also encode attended features rather than the entire set of features that happen to be present at the moment. This is the basis for the network's stability.

14.1.2 Match-Based Learning and Stable Coding

ART memories are stable in a complex world because the learning process is *match based*. Memories change only if attended portions of the external world match our internal expectations—or if something completely new occurs. If the external world fails to match an ART network's expectations or predictions then a search process activates a new category. The new category is a new hypothesis about what is important in the present environment. Match-based learning is the defining characteristic of ART networks.

14.1.3 Boeing Neural Information Retrieval System

Code stability is one of the main reasons many engineers select ART networks for applications. One example of such a technology transfer is the Boeing Neural Information Retrieval System (NIRS) [15, 40] in which ART networks are the critical system components. NIRS encodes a parts inventory in the form of 2-D and 3-D drawings. The system creates a compressed but stable memory structure for later retrieval by design engineers. The resulting neural database reduces inventory size by a factor of 9. This prevents a severe memory proliferation problem and permits efficient reuse of stored designs. NIRS has moved from beta testing to implementation in CAD systems for design of the Boeing 777 and manufacture of the Boeing 747 and 767 planes.

14.1.4 Error-Based Learning

Match-based learning generates a stable recognition code in a large, complex, evolving environment. A match-based learning system is thus well suited to problems such as the Boeing CAD neural database that creates its own expert system as a function of experience. But qualitatively different types of learning problems also exist. As we grow our eyes and limbs need to learn or adapt to their own internal changes so that we can pick up a pencil as an adult as well as we could at age two. As adults we have no need for the sensory-motor maps that we learned as babies. These codes need not be stable in the sense that knowledge systems such as language need to be stable. Layers of old motor maps would be a nuisance.

Neural networks that employ *error-based* learning are often suited to *adaptive* problems like sensory motor map construction. Error-based learning systems include the perceptron [32, 33] and multilayer perceptrons such as back propagation [34, 42]. In these systems an error causes memories to change so that the same input seen again would give an answer that was closer to the "correct" one. If I see a dog and know it is a dog but then hear that it is a Dalmatian then a serious error has occurred—if we define error as the difference between the correct or target outcome and the system's actual or predicted outcome. An error-based network shifts the weights in such a way that the next response will be toward Dalmatian and away from dog. If this happens several times in a row then I will learn to respond "Dalmatian" but will forget that a dog is still a dog. So error-based learning is subject to catastrophic *forgetting*. I may desire this kind of forgetting if the error signal tells me that I have reached too far to touch the pencil.

14.2 ART AND FUZZY CUBES

Stephen Grossberg [21] introduced adaptive resonance as a theory of human cognitive information processing. The cognitive theory has led to an evolving series of real-time adaptive resonance theory (ART) neural network models for unsupervised and supervised category learning and pattern recognition. These models form stable recognition categories in response to arbitrary input sequences with either fast or slow learning. Unsupervised ART networks include the ART 1 [41] system that stably learns to categorize binary input patterns presented in an arbitrary order, the ART 2 [5] system that stably learns to categorize either analog or binary input patterns presented in an arbitrary order, and the ART 3 [6] system that carries out parallel search or hypothesis testing of distributed recognition codes in a multilevel network hierarchy. Many of the ART papers appear in the anthology *Pattern Recognition by Self-Organizing Neural Networks* [7].

A supervised network architecture called ARTMAP self-organizes arbitrary categorical mappings between m-dimensional input vectors and n-dimensional output vectors. ARTMAP's internal control mechanisms create stable recognition categories of optimal size by maximizing code compression while minimizing predictive error in an on-line setting. Binary ART 1 computations are the foundation of the first ARTMAP network [10] that learns binary maps. Fuzzy ART [11] extends ART 1 to learn stable recognition categories in response to analog and binary input patterns. If fuzzy ART replaces ART 1 in an ARTMAP system then the resulting fuzzy ARTMAP architecture [9] rapidly learns stable categorical mappings between analog or binary input and output vectors. Fuzzy ARTMAP learns to classify inputs by a fuzzy set of features or a pattern of fuzzy set-function values between 0 and 1 that indicate the extent to which each feature is present. Where set-theoretic operations describe ART 1 dynamics fuzzy set-theoretic operations [26, 44] describe fuzzy ART dynamics. The use of finite fuzzy sets or *fit* (fuzzy unit) vectors embeds the fuzzy ART systems in fuzzy hypercubes.

Simulations in this chapter show how fuzzy ARTMAP performs. Simulation inputs are 2-D analog patterns that we need not view as fuzzy sets but that show properties of the system. ARTMAP fast learning often leads to different adaptive weights and recognition categories for different orderings of a given training set even when the overall predictive accuracy of each simulation is similar. The different category structures cause the set of test set items where errors occur to vary from one simulation to the next. A *voting strategy* uses an ARTMAP system that trains many times on one input set with different orderings. The final prediction for a given test set item is the one that the largest number of simulations makes. The set of items making erroneous predictions varies from one simulation to the next. So voting both cancels many of the errors and assigns confidence estimates to competing predictions. This chapter concludes with some fuzzy ART and ARTMAP applications.

14.3 ART DYNAMICS

Fuzzy ART uses the basic features of all ART systems. These include pattern matching between bottom-up input and top-down learned prototype vectors. This matching process leads either to a resonant state that focuses attention and triggers stable prototype learning or to a self-regulating parallel memory search. If the search ends with the selection of an established category then the system may refine the category's prototype to include new information in the input pattern. If the search ends by choosing an untrained node then the ART network grows a new category.

Figure 14.2 shows the main parts of an ART 1 network and Figure 14.3 shows an ART search cycle. During ART search an input vector **I** registers itself as a pattern **X** of activity across level F_1 (Figure 14.3a). Multiple converging and diverging $F_1 \rightarrow F_2$ adaptive filter pathways multiply the vector **S** by a matrix of adaptive weights or long term memory (LTM) traces to generate a net input vector **T** to level F_2. The internal competitive dynamics of F_2 contrast-enhance vector **T** and create a compressed activity vector **Y** across F_2. In ART 1 strong competition selects the F_2 node that receives the maximal $F_1 \rightarrow F_2$ input. Only one component of **Y** is nonzero after this choice takes place. Activation of such a *winner-take-all* node defines the category or symbol of the input pattern **I**. Such a category stands for all the inputs **I** that maximally activate the corresponding node.

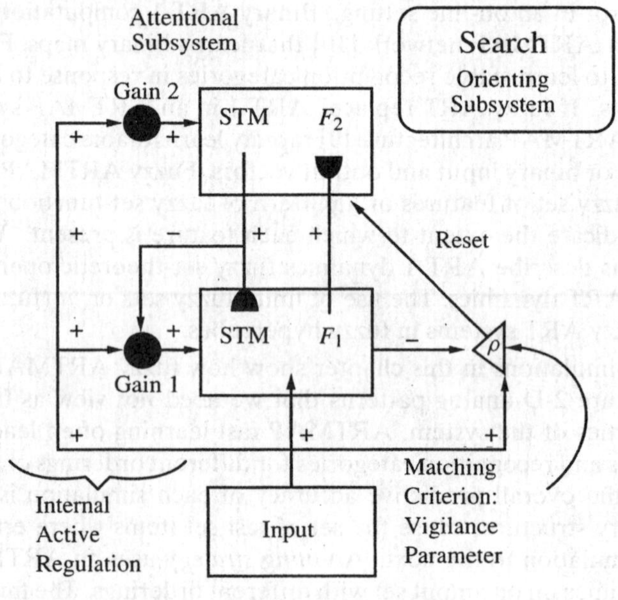

Figure 14.2 Typical ART 1 neural network [4].

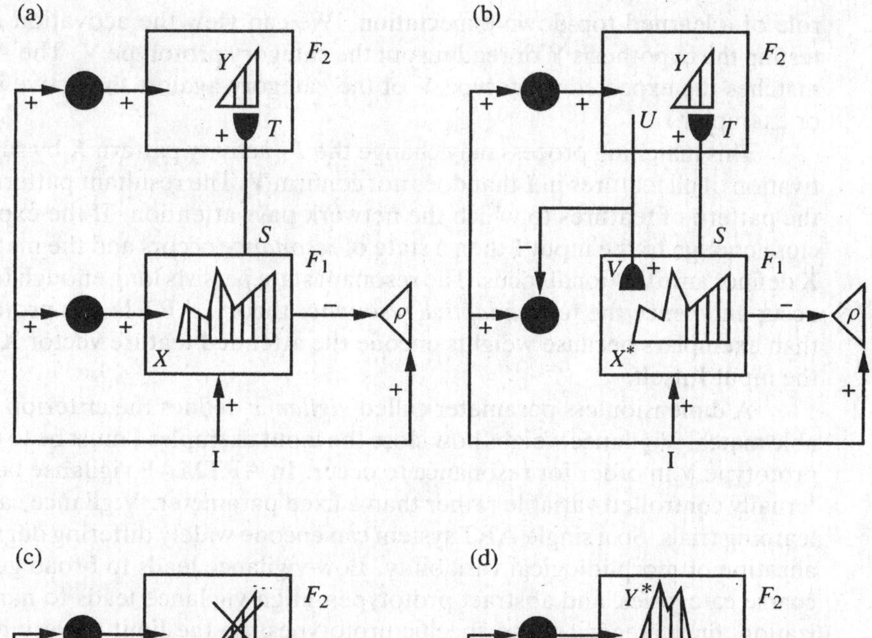

Figure 14.3 ART search for an F_2 code: (a) The input pattern **I** creates the STM activity pattern **X** at F_1 as it nonspecifically activates the orienting subsystem. Pattern **X** both inhibits A and creates the output signal pattern **S**. An adaptive filter transforms the signal pattern **S** into the pattern **T** that activates the STM pattern **Y** across F_2. (b) Pattern **Y** creates the signal pattern **U** and a top-down adaptive filter transforms **U** into the prototype pattern **V**. If **V** mismatches **I** then F_1 registers a new STM activity pattern $\mathbf{X}^*$. This reduction of total STM reduces the total inhibition from F_1 to A. (c) If the ART matching criterion fails then A releases a nonspecific signal that resets the STM pattern **Y** at F_2. (d) Reset inhibits **Y** so it also eliminates the top-down prototype signal **T**. This lets F_1 reinstate **X**. Enduring traces of the prior reset allow **X** to activate a different STM pattern $\mathbf{Y}^*$ at F_2. If the top-down prototype due to $\mathbf{Y}^*$ also mismatches **I** at F_1 then the search for an F_2 code that satisfies the matching criterion continues.

We can view the activation of an F_2 node as making a hypothesis about an input **I**. An F_2 vector generates a signal vector **U** sent top-down through the $F_2 \rightarrow F_1$ adaptive filter. After multiplication by the adaptive weight matrix of the top-down filter a vector **V** becomes the $F_2 \rightarrow F_1$ input (Figure 14.3b). Vector **V** plays the role of a learned top-down expectation. We can view the activation of **V** by **Y** as testing the hypothesis **Y** or reading out the category prototype **V**. The ART network matches the expected prototype **V** of the category against the active input pattern or exemplar **I**.

This matching process may change the F_1 activity pattern **X** by suppressing activation of all features in **I** that does not confirm **V**. The resultant pattern $\mathbf{X}^*$ encodes the pattern of features to which the network pays attention. If the expectation **V** is close enough to the input **I** then a state of *resonance* occurs and the matched pattern **X** defines an attentional focus. The resonant state persists long enough for learning to occur and hence the term *adaptive resonance* theory. ART learns prototypes rather than exemplars because weights encode the attended feature vector $\mathbf{X}^*$ rather than the input **I** itself.

A dimensionless parameter called *vigilance* defines the criterion of an acceptable match. Vigilance weighs how close the input exemplar **I** must be to the top-down prototype **V** in order for resonance to occur. In ARTMAP vigilance becomes an internally controlled variable rather than a fixed parameter. Vigilance can vary across learning trials. So a single ART system can encode widely differing degrees of generalization or morphological variability. Low vigilance leads to broad generalization, coarse categories, and abstract prototypes. High vigilance leads to narrow generalization, fine categories, and specific prototypes. In the limit of very high vigilance prototype learning reduces to exemplar learning. Varying vigilance levels allow a single ART system to recognize both abstract categories of faces and dogs and individual faces and dogs.

ART memory search or hypothesis testing begins when the top-down expectation **V** decides that the bottom-up input **I** is too novel or unexpected to satisfy the vigilance criterion. Search leads to selection of a better recognition code, symbol, category, or hypothesis to represent input **I** at level F_2. An *orienting subsystem A* controls the search process. The orienting subsystem works with the attentional subsystem (as in Figures 14.3c and 14.3d) to let the attentional subsystem learn about novel inputs without risking unselective forgetting of its past knowledge.

ART search prevents associations from forming between **Y** and $\mathbf{X}^*$ if $\mathbf{X}^*$ differs too much from **I** to satisfy the vigilance criterion. The search process resets **Y** before such an association can form. If the search ends upon a familiar category then that category's prototype may change in light of new information that carries **I**. If **I** differs too much from the previously learned prototypes then the search ends upon an uncommitted F_2 node and this begins a new category.

An ART *choice parameter* controls how deeply the search proceeds before selecting an uncommitted node. As learning self-stabilizes all inputs that a category codes access it directly and the search automatically stops. So the system selects that category whose prototype gives the globally best match to the input pattern. Stable on-line learning proceeds when familiar inputs directly activate their categories and novel inputs trigger adaptive searches until the network's memory reaches its capacity. Simulations show fuzzy ART dynamics in a parameter range called the *conservative limit*. In this limit the choice parameter α (Figure 14.4) is very small. Then an input first selects a category whose weight vector is a fuzzy subset of the input if such a category exists. No weight change occurs during learning for such a

Art 1 (Binary)	Fuzzy Art (Analog)
Category Choice	
$T_j = \dfrac{\lvert \mathbf{I} \cap \mathbf{w}_j \rvert}{\alpha + \lvert \mathbf{w}_j \rvert}$	$T_j = \dfrac{\lvert \mathbf{I} \wedge \mathbf{w}_j \rvert}{\alpha + \lvert \mathbf{w}_j \rvert}$
Match Criterion	
$\dfrac{\lvert \mathbf{I} \cap \mathbf{w}_J \rvert}{\lvert \mathbf{I} \rvert} \geq \rho$	$\dfrac{\lvert \mathbf{I} \wedge \mathbf{w}_J \rvert}{\lvert \mathbf{I} \rvert} \geq \rho$
Fast Learning	
$\mathbf{w}_J^{(new)} = \mathbf{I} \cap \mathbf{w}_J^{(old)}$	$\mathbf{w}_J^{(new)} = \mathbf{I} \wedge \mathbf{w}_J^{(old)}$
$\cap$ = Logical AND Intersection	$\wedge$ = Fuzzy AND Minimum

Figure 14.4 Analogy between ART 1 and fuzzy ART. The $\mathbf{w}_j$ in ART 1 denotes the index set of top-down LTM traces that exceed a prescribed positive threshold value.

choice and hence the name conservative limit. The system conserves learned weights wherever possible.

14.4 FUZZY ART

Fuzzy ART inherits the design features of other ART models. Figure 14.4 shows how the ART 1 operations of category choice, matching, search, and learning translate into fuzzy ART operations when the intersection operator ($\cap$) of ART 1 replaces the fuzzy intersection or pointwise minimum operator ($\wedge$). Despite this close formal homology this chapter summarizes fuzzy ART as an algorithm rather than as a locally defined neural model. Carpenter, Grossberg, and Rosen [12] describe a neural network realization of fuzzy ART. The computations of fuzzy ART are the same as those of the ART 1 neural network for the special case of binary input vectors and fast learning.

14.4.1 Fast-Learn Slow-Recode and Complement Coding Options

Many applications of ART 1 use fast learning where adaptive weights fully converge to equilibrium values in response to each input pattern. Fast learning lets a system adapt quickly to inputs that occur only rarely but that may require immediate accurate performance. Remembering many details of an exciting movie is a typical example of fast learning. Fast learning destabilizes the memories of feedforward error-based models like backpropagation. If the difference between actual output and target output defines error then present inputs drive out past learning since fast learning zeroes the error on each input trial. This feature of backpropagation restricts its domain to off-line applications with a slow learning rate. It also lacks the key feature of competition. A backpropagation system tends to average rare events with similar frequent events that have different consequences.

Some applications benefit from a fast-commit slow-recode option that combines fast initial learning with a slower rate of forgetting. Fast commitment retains the advantage of fast learning—the ability to respond to important inputs that occur only rarely. Slow recoding then prevents features in a category's prototype from being wrongly deleted in response to noisy or partial inputs. Only a statistically persistent change in a feature's relevance to an established category can delete it from the prototype of the category.

Complement coding is a preprocessing step that normalizes input patterns. Complement coding solves a potential fuzzy ART category proliferation problem [11, 30]. In neurobiological terms complement coding uses both on-cells and off-cells to represent an input pattern and preserves individual feature amplitudes while normalizing the total on-cell/off-cell activity. The on-cell portion of a prototype encodes features that are critically present in category exemplars while the off-cell portion encodes features that are critically absent. Small weights in both on-cell and off-cell portions of a prototype encode as uninformative those features that are sometimes present and sometimes absent. In set-theoretic terms complement coding leads to a symmetric ART theory in which the fuzzy intersection ($\wedge$) and the fuzzy union ($\vee$) play complementary roles. Complement coding allows a geometric interpretation of fuzzy ART recognition categories as box-shaped regions of the input fuzzy space. Fuzzy intersections and unions iteratively define the corners of each box. Simulations in this chapter show the fuzzy ART geometry for an example where inputs are two-dimensional and so the boxes are rectangles in the fuzzy 2-cube. The geometric formulation allows comparison between fuzzy ART and aspects of Salzberg's [35] Nested Generalized Exemplars (NGEs) and Simpson's [39] Fuzzy Min-Max Classifier (FMMC).

14.5 FUZZY ARTMAP

Each ARTMAP system includes a pair of adaptive resonance theory modules (ART_a and ART_b) that create stable recognition categories in response to arbitrary sequences of input patterns (Figure 14.5). During supervised learning ART_a receives a stream $\{\mathbf{a}^{(p)}\}$ of input patterns and ART_b receives a stream $\{\mathbf{b}^{(p)}\}$ of input patterns where $\mathbf{b}^{(p)}$ is the correct prediction given $\mathbf{a}^{(p)}$. An associative learning network and an internal controller link these modules to make the ARTMAP system operate in real time. The controller creates the minimal number of ART_a recognition categories or hidden units it needs to meet accuracy criteria. A minimax learning rule enables ARTMAP to learn quickly, efficiently, and accurately as it jointly minimizes predictive error and maximizes code compression. This scheme automatically links

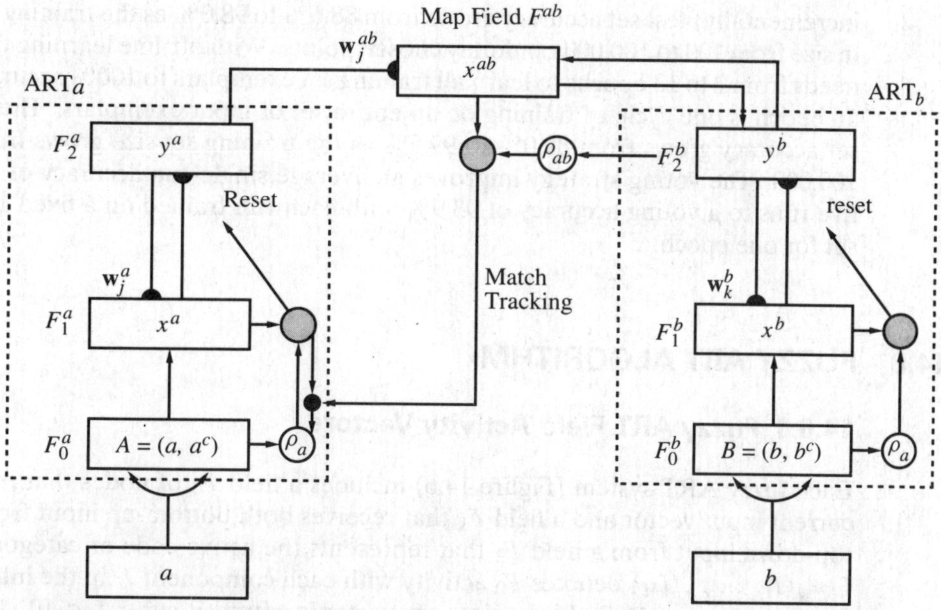

Figure 14.5 Fuzzy ARTMAP architecture. The $\boldsymbol{ART_a}$ complement coding preprocessor transforms the $M_a - D$ fuzzy unit vector $\mathbf{a}$ into the $2M_a - D$ fit vector $\mathbf{A} = (\mathbf{a}, \mathbf{a}^c)$ at the $\boldsymbol{ART_a}$ field F_0^a. $\mathbf{A}$ is the input vector to the $\boldsymbol{ART_a}$ field F_1^a. The input to F_1^b is also the $2M_b - D$ fit vector $(\mathbf{b}, \mathbf{b}^c)$. When $\boldsymbol{ART_b}$ disconfirms a prediction of $\boldsymbol{ART_a}$ then map field inhibition induces the match tracking process. Match tracking raises the $\boldsymbol{ART_a}$ vigilance (ρ_a) to just above the F_1^a-to-F_0^a match ratio $|\mathbf{x}^a|/|\mathbf{A}|$. This triggers an $\boldsymbol{ART_a}$ search that leads to activation of either an $\boldsymbol{ART_a}$ category that correctly predicts $\mathbf{b}$ or to a previously uncommitted $\boldsymbol{ART_a}$ category node.

predictive success to category size on a trial-by-trial basis using only local operations. It works by increasing the ART$_a$ vigilance parameter (ρ_a) by the minimal amount it needs to correct a predictive error at ART$_b$.

An ART$_a$ *baseline vigilance* parameter $\overline{\rho_a}$ calibrates the minimum confidence that ART$_a$ needs to accept a chosen category rather than search for a better one through automatically controlled search. Lower values of $\overline{\rho_a}$ let larger categories form and maximize code compression. Initially $\rho_a = \overline{\rho_a}$. During training a predictive failure at ART$_b$ increases ρ_a by the minimum amount needed to trigger ART$_a$ search through a feedback control mechanism called *match tracking* [10]. Match tracking sacrifices the minimum amount of compression needed to correct the predictive error. Hypothesis testing selects a new ART$_a$ category. This focuses attention on a cluster of $\mathbf{a}^{(p)}$ input features that can better predict $\mathbf{b}^{(p)}$. With fast learning match tracking allows a single ARTMAP system to learn a different prediction for a rare event rather than for a cloud of similar frequent events that occur with it.

An ARPA benchmark simulation called circle-in-the-square shows the fuzzy ARTMAP dynamics. The task is to learn to identify which points lie inside and which lie outside a circle. During training parts of the ART$_a$ input $\mathbf{a}$ are the x- and y-coordinates of a point in the unit square or fuzzy 2-cube. The ART$_b$ input equals 0 or 1 to label $\mathbf{a}$ as inside or outside the circle. As fuzzy ARTMAP learns on-line (or incrementally) test set accuracy grows from 88.6% to 98.0% as the training set grows in size from 100 to 100,000 randomly chosen points. With off-line learning the system needs from 2 to 13 epochs to learn all training set exemplars to 100% accuracy. Here an epoch is one cycle of training on an entire set of input exemplars. Then the test set accuracy grows from 89.0% to 99.5% as the training set size grows from 100 to 100,000. The voting strategy improves an average single-run accuracy of 90.5% on five runs to a voting accuracy of 93.9% with each run trained on a fixed 1,000-item set for one epoch.

14.6 FUZZY ART ALGORITHM

14.6.1 Fuzzy ART Field Activity Vectors

Each fuzzy ART system (Figure 14.6) includes a field F_0 of nodes that represent a current input vector and a field F_1 that receives both bottom-up input from F_0 and top-down input from a field F_2 that represents the active code or category. Vector $\mathbf{I} = (I_1, \ldots, I_M)$ denotes F_0 activity with each component I_i in the interval [0,1] for $i = 1, \ldots, M$. So $\mathbf{I}$ is a point or fit vector in a fuzzy n-cube: $\mathbf{I} \in [0, 1]^n$. Vector $\mathbf{x} = (x_1, \ldots, x_M)$ denotes F_1 activity and $\mathbf{y} = (y_1, \ldots, y_N)$ denotes F_2 activity. The number of nodes in each field is arbitrary.

14.6.2 Weight vector

Associated with each F_2 category node $j (j = 1, \ldots, N)$ is a vector $w_j \equiv (w_{j1}, \ldots, w_{jM})$ of adaptive weights or long-term memory (LTM) traces. At first

$$w_{j1}(0) = \ldots = w_{jM}(0) = 1. \tag{14.1}$$

Then each category is *uncommitted*. After a category codes its first input it becomes *committed*. Each component w_{ji} can decrease but never increase during learning. Thus each weight vector $\mathbf{w}_j(t)$ converges to a limit. The fuzzy ART weight or prototype vector $\mathbf{w}_j$ subsumes both the bottom-up and top-down weight vectors of ART 1 (Figure 14.2).

14.6.3 Parameters

A choice parameter $\alpha > 0$, a learning rate parameter $\beta \in [0, 1]$, and a vigilance parameter $\rho \in [0, 1]$ fix the fuzzy ART dynamics.

14.6.4 Category Choice

For each input $\mathbf{I}$ and F_2 node j the *choice function* T_j has the form

$$T_j(\mathbf{I}) = \frac{|\mathbf{I} \wedge \mathbf{w}_j|}{\alpha + |\mathbf{w}_j|} \tag{14.2}$$

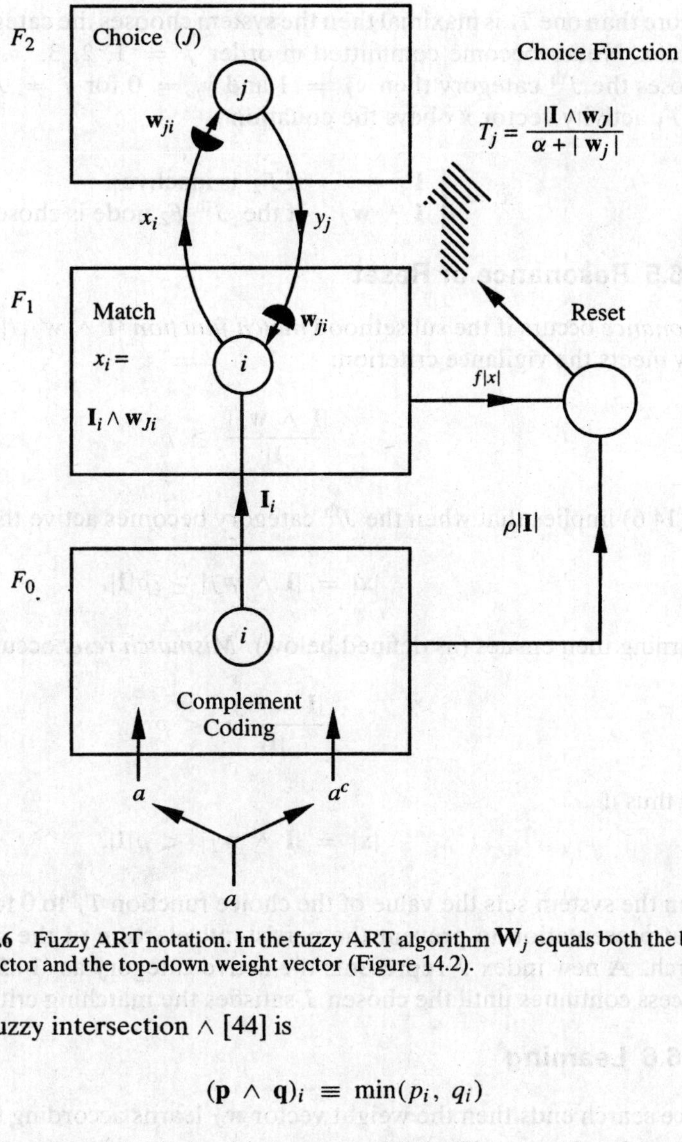

Figure 14.6 Fuzzy ART notation. In the fuzzy ART algorithm $\mathbf{W}_j$ equals both the bottom-up weight vector and the top-down weight vector (Figure 14.2).

where the fuzzy intersection $\wedge$ [44] is

$$(\mathbf{p} \wedge \mathbf{q})_i \equiv \min(p_i, q_i) \tag{14.3}$$

and where the norm $|\cdot|$ is the fuzzy count $c(\mathbf{P})$ in Chapter 12 or just

$$|\mathbf{p}| \equiv \sum_{i=1}^{M} |p_i|. \tag{14.4}$$

Note that (14.2) reduces to the fuzzy subsethood of Chapters 12 and 13 when $\alpha = 0$.

The system makes a *category choice* when at most one F_2 node can become active at a given time. The index J denotes the chosen category where

$$T_J = \max\{T_j : j = 1 \ldots N\}. \tag{14.5}$$

If more than one T_j is maximal then the system chooses the category with the smallest j index. Nodes become committed in order $j = 1, 2, 3, \ldots$. When the system chooses the Jth category then $y_J = 1$ and $y_j = 0$ for $j \neq J$. In a choice system the F_1 activity vector $\mathbf{x}$ obeys the equation

$$\mathbf{x} = \begin{cases} \mathbf{I} & \text{if } F_2 \text{ is inactive} \\ \mathbf{I} \wedge \mathbf{w}_J & \text{if the } J^{\text{th}}\ F_2 \text{ node is chosen.} \end{cases} \tag{14.6}$$

14.6.5 Resonance or Reset

Resonance occurs if the subsethood *match function* $|\mathbf{I} \wedge \mathbf{w}_J|/|\mathbf{I}|$ of the chosen category meets the vigilance criterion:

$$\frac{|\mathbf{I} \wedge \mathbf{w}_J|}{|\mathbf{I}|} \geq \rho. \tag{14.7}$$

So (14.6) implies that when the Jth category becomes active the resonance occurs if

$$|\mathbf{x}| = |\mathbf{I} \wedge \mathbf{w}_J| \geq \rho|\mathbf{I}|. \tag{14.8}$$

Learning then ensues (as defined below). *Mismatch reset* occurs if

$$\frac{|\mathbf{I} \wedge \mathbf{w}_J|}{|\mathbf{I}|} < \rho \tag{14.9}$$

and thus if

$$|\mathbf{x}| = |\mathbf{I} \wedge \mathbf{w}_J| < \rho|\mathbf{I}|. \tag{14.10}$$

Then the system sets the value of the choice function T_J to 0 for the duration of the input presentation to prevent the persistent selection of the same category during search. A new index J represents the active category as (14.5) selects. The search process continues until the chosen J satisfies the matching criterion (14.7).

14.6.6 Learning

Once search ends then the weight vector $\mathbf{w}_J$ learns according to the equation

$$\mathbf{w}_J^{(\text{new})} = \beta(\mathbf{I} \wedge \mathbf{w}_J^{(\text{old})}) + (1 - \beta)\mathbf{w}_J^{(\text{old})}. \tag{14.11}$$

Fast learning corresponds to setting $\beta = 1$. The learning law of the NGE system [35] equals (14.11) in the fast-learn limit with complement coding.

14.6.7 Fast-Commit and Slow-Recode

For efficient coding of noisy input sets it helps to set $\beta = 1$ when J is an uncommitted node and then take $\beta < 1$ for slower adaptation after the system already commits the category. The fast-commit and slow-recode option makes $\mathbf{w}_J^{(\text{new})} = \mathbf{I}$ the first time category J becomes active. Moore [30] introduced the learning law of (14.11) with fast commitment and slow recoding to explore many of generalized ART 1 models. Some of these models are similar to fuzzy ART but none uses complement coding. Moore describes a category proliferation problem that can occur in some analog ART systems when many random inputs erode the norm of weight vectors. Complement coding solves this problem.

14.6.8 Normalization by Complement Coding

Normalization of fuzzy ART inputs prevents category proliferation. The system normalizes the $F_0 \rightarrow F_1$ inputs if for some $\gamma > 0$

$$\sum_i I_i = |\mathbf{I}| \equiv \gamma \tag{14.12}$$

for all inputs **I**. One way to normalize each vector **a** is to divide by the fit count:

$$\mathbf{I} = \frac{\mathbf{a}}{|\mathbf{a}|}. \tag{14.13}$$

Complement coding normalizes the input but it also preserves amplitude information in contrast to (14.13). Complement coding represents both the on-response and the off-response to an input vector **a** (Figure 14.6). In its simplest form **a** represents the on-response and $\mathbf{a}^c$ (the complement of **a**) represents the off-response where

$$a_i^c \equiv 1 - a_i. \tag{14.14}$$

The complement coded $F_0 \rightarrow F_1$ input **I** is the $2M$-dimensional fit vector

$$\mathbf{I} = (\mathbf{a}, \mathbf{a}^c) \equiv (a_1, \ldots, a_M, a_1^c, \ldots, a_M^c). \tag{14.15}$$

So **I** is a point in the double-sized fuzzy cube $[0, 1]^{2M}$. Chapter 12 showed that **I** has the same fuzzy or entropy structure as does its terms **a** and $\mathbf{a}^c$. The system automatically normalizes a complement coded input because

$$\begin{aligned} |\mathbf{I}| &= |(\mathbf{a}, \mathbf{a}^c)| \\ &= \sum_{i=1}^{M} a_i + (M - \sum_{i=1}^{M} a_i) \\ &= M. \end{aligned} \tag{14.16}$$

With complement coding the initial condition

$$w_{j1}(0) = \ldots = w_{j,2M}(0) = 1 \tag{14.17}$$

replaces the fuzzy ART initial condition of (14.1).

The close link between fuzzy subsethood and ART choice/search/learning forms the foundation of the computational properties of fuzzy ART. In the conservative limit where the choice parameter $\alpha = 0^+$ the choice function T_j measures the degree to which $\mathbf{w}_j$ is a fuzzy subset of $\mathbf{I}$ [26]: $T_j = S(\mathbf{w}_j, \mathbf{I}) = \text{Degree}(\mathbf{w}_j \subset \mathbf{I})$. Then the system will select a category J for which $\mathbf{w}_J$ is a fuzzy subset of $\mathbf{I}$ first if such a category exists. Resonance depends on the degree to which $\mathbf{I}$ is a fuzzy subset of $\mathbf{w}_J$ by (14.7) and (14.9). When J is such a fuzzy subset choice then the match function value is

$$\frac{|\mathbf{I} \wedge \mathbf{w}_J|}{|\mathbf{I}|} = \frac{|\mathbf{w}_J|}{|\mathbf{I}|}. \tag{14.18}$$

Choosing J to maximize $|\mathbf{w}_j|$ among fuzzy subset choices by (14.2), maximizes the chance for resonance in (14.7). If reset occurs for the node that maximizes $|\mathbf{w}_J|$ then reset will also occur for all other subset choices.

A geometric interpretation of fuzzy ART represents each category as a box in a fuzzy Mcube $[0, 1]^M$ where M is the number of terms of input fit-vector $\mathbf{a}$. Consider an input set that consists of 2-D fit vectors $\mathbf{a}$. Complement coding gives

$$\mathbf{I} = (\mathbf{a}, \mathbf{a}^c) = (a_1, a_2, 1 - a_1, 1 - a_2). \tag{14.19}$$

Then each category j has the geometric shape of a rectangle R_j in the fuzzy cube. Following (14.19) a complement coded weight vector $\mathbf{w}_j$ takes the form

$$\mathbf{w}_j = (\mathbf{u}_j, \mathbf{v}_j^c) \tag{14.20}$$

where $\mathbf{u}_j$ and $\mathbf{v}_j$ are 2-D vectors. Vector $\mathbf{u}_j$ defines one corner of a rectangle R_j and $\mathbf{v}_j$ defines the opposite corner (Figure 14.7a). The size of R_j is

$$|R_j| \equiv |\mathbf{v}_j - \mathbf{u}_j| \tag{14.21}$$

which equals the height plus the width of R_j.

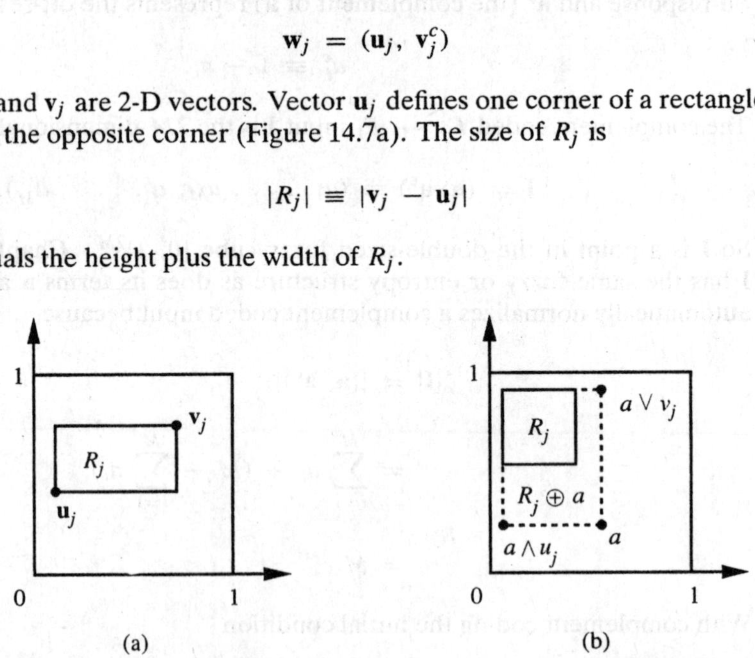

Figure 14.7 Fuzzy ART category boxes in fuzzy cubes. (a) Complement coding form with $M = 2$. Each weight vector $\mathbf{w}_j$ has the geometry of a rectangle R_j with corners $(\mathbf{u}_j, \mathbf{v}_j)$. (b) During fast learning R_J expands to $R_J \oplus \mathbf{a}$ or the smallest rectangle that includes R_J and $\mathbf{a}$ if $|R_J \oplus \mathbf{a}| \leq 2(1 - \rho)$.

Consider a fast-learn fuzzy ART system with $\beta = 1$ in (14.11). Then $\mathbf{w}_J^{(\text{new})} = \mathbf{I} = (\mathbf{a}, \mathbf{a}^c)$ when J is an uncommitted node. The corners of $R_J^{(\text{new})}$ are $\mathbf{a}$ and $(\mathbf{a}^c)^c = \mathbf{a}$. Hence $R_J^{(\text{new})}$ is just the point $\mathbf{a}$ in the fuzzy cube. Learning increases the size of R_J. This size grows as the size of $\mathbf{w}_J$ shrinks during learning. Vigilance ρ picks the maximum size of R_J with $|R_j| \leq 2(1 - \rho)$ as shown later. During each fast-learning trial R_J expands to $R_J \oplus \mathbf{a}$ or the minimum rectangle containing R_J and $\mathbf{a}$ (Figure 14.7b). The corners of $R_J \oplus \mathbf{a}$ are $\mathbf{a} \wedge \mathbf{u}_J$ and $\mathbf{a} \vee \mathbf{v}_J$ where (14.3) defines the fuzzy intersection $\wedge$ and the fuzzy union $\vee$ has the form (see [44])

$$(\mathbf{p} \vee \mathbf{q})_i \equiv \max(p_i,\ q_i) \tag{14.22}$$

Hence by (14.21) the size of $R_J \oplus \mathbf{a}$ is

$$|R_J \oplus \mathbf{a}| = |(\mathbf{a} \vee \mathbf{v}_J) - (\mathbf{a} \wedge \mathbf{u}_J)|. \tag{14.23}$$

But before R_J can expand to include $\mathbf{a}$ reset chooses another category if $|R_J \oplus \mathbf{a}|$ is too large. With fast learning R_j is the smallest rectangle that encloses all vectors $\mathbf{a}$ that have chosen category j without reset.

If $\mathbf{a}$ has dimension M then the box R_j contains the two opposing vertices $\wedge_j\mathbf{a}$ and $\vee_j\mathbf{a}$ where the ith component of each vector is

$$(\wedge_j\mathbf{a})_i = \min\{a_i\text{: category } j \text{ has coded } \mathbf{a}\} \tag{14.24}$$

and

$$(\vee_j\mathbf{a})_i = \max\{a_i\text{: category } j \text{ has coded } \mathbf{a}\} \tag{14.25}$$

(Figure 14.8). The size of R_j is

$$|R_j| = |\vee_j\ \mathbf{a} - \wedge_j\mathbf{a}| \tag{14.26}$$

and the weight vector $\mathbf{w}_j$ is

$$\mathbf{w}_j = (\wedge_j\mathbf{a},\ (\vee_j\mathbf{a})^c) \tag{14.27}$$

as in (14.20) and (14.21). Thus

$$|\mathbf{w}_j| = \sum_i (\wedge_j\mathbf{a})_i + \sum_i [1 - (\vee_j\mathbf{a})_i] = M - |\vee_j\ \mathbf{a} - \wedge_j\mathbf{a}| \tag{14.28}$$

and so the size of the box R_j is

$$|R_j| = M - |\mathbf{w}_j|. \tag{14.29}$$

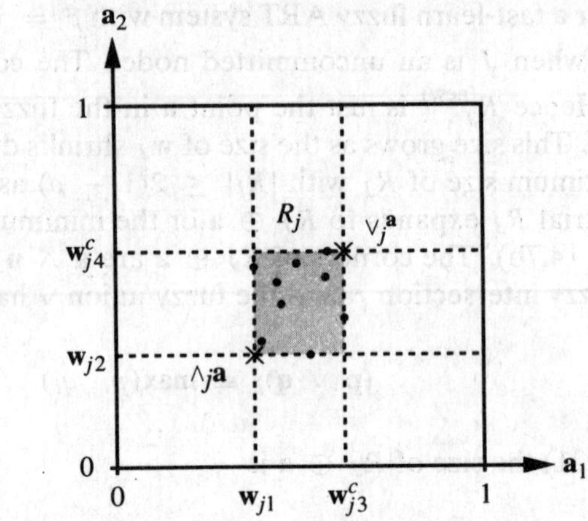

Figure 14.8 Fuzzy ART fast learning and complement coding gives the jth category rectangle R_j that includes all those fit vectors $\mathbf{a}$ in the fuzzy unit square that have activated category j without reset. The weight vector $\mathbf{w}_j$ equals $(\wedge_j \mathbf{a}, (\vee_j \mathbf{a})^c)$.

By (14.8), (14.11), and (14.16)

$$|\mathbf{w}_j| \geq \rho M. \tag{14.30}$$

By (14.29) and (14.30)

$$|R_j| \leq (1 - \rho)M. \tag{14.31}$$

Inequality (14.31) shows that high vigilance ($\rho \cong 1$) leads to small R_j while low vigilance ($\rho \cong 0$) permits large R_j. If j is an uncommitted node then $|\mathbf{w}_j| = 2M$ by (14.17). So formally $|R_j| \equiv -M$ by (14.29). The following summary of fuzzy ART dynamics combines these observations.

14.6.9 Fuzzy ART Stable Category Learning

A fuzzy ART system with complement coding, fast learning, and constant vigilance forms categories that converge to limits in response to any sequence of analog or binary input vectors. Category boxes can grow in each dimension but never shrink. The size of a box R_j equals $M - |\mathbf{w}_j|$ where $\mathbf{w}_j$ is the corresponding weight vector. The term $M(1 - \rho)$ bounds the size $|R_j|$ from above. In the conservative limit one-pass learning obtains such that no reset or additional learning occurs on subsequent presentations of any input. But if $0 \leq \rho < 1$ then the number of categories is bounded even if the number of exemplars in the training set is not bounded. Similar properties hold for the fast-learn slow-recode case except that the system may need repeated presentations of each input before stabilization occurs even in the conservative limit.

14.6.10 A Comparison of Fuzzy ARTMAP, NGE, and FMMC

The geometry of the fuzzy boxes R_j resembles parts of the Nested Generalized Exemplar (NGE) system [35] and the Fuzzy Min-Max Classifier (FMMC) system [39]. Fuzzy ARTMAP, NGE, and FMMC all use boxes in fuzzy cubes to represent category weights in a supervised learning paradigm. All three systems use some version of the learning law (14.11) to update weights when an input correctly predicts the output.

The three algorithms differ in how they respond to incorrect predictions. NGE and FMMC do not have components analogous to the ART vigilance parameter. So these algorithms do not have internal control of box size in the fuzzy cube. NGE does include a type of search process but its rules differ from those of fuzzy ARTMAP. Consider when NGE makes a predictive error. Then it searches at most two categories before creating a new one. NGE allows boxes to shrink as well as to grow. So the fuzzy ART stability properties do not hold. For the NGE system only the Greedy version (a leader algorithm that codes the first exemplar of each category) is always stable.

Stability and match tracking allow fuzzy ARTMAP to construct automatically as many categories as it needs to learn any consistent training set to 100% accuracy. Both ARTMAP and NGE rely on multilayer structures to effect their learning strategies. The FMMC is in contrast a two-layer feedforward system that allows at most one category box to represent each output class. So FMMC can learn only a limited set of category structures. Fuzzy ARTMAP can in contrast associate multiple categories with the same output. This means capital letters, script letters, and various fonts can all predict the same letter name.

14.7 FUZZY ARTMAP ALGORITHM

Fuzzy ARTMAP uses two fuzzy ART modules ART_a and ART_b that link together via an inter-ART module F^{ab} called a *map field.* The map field forms predictive associations between categories and realizes the ARTMAP *match tracking rule*. Match tracking increases the ART_a vigilance parameter ρ_a in response to a predictive error or mismatch at ART_b. Match tracking reorganizes category structure so that subsequent presentations of the input do not repeat the error. An outline of the ARTMAP algorithm follows.

14.7.1 ART_a and ART_b

First complement code the inputs to ART_a and ART_b. For ART_a: $\mathbf{I} = \mathbf{A} = (\mathbf{a}, \mathbf{a}^c)$. For ART_b: $\mathbf{I} = \mathbf{B} = (\mathbf{b}, \mathbf{b}^c)$ (Figure 14.5). We label variables in ART_a or ART_b by subscripts or superscripts a or b. For ART_a: $\mathbf{x}^a \equiv (x_1^a \ldots x_{2M_a}^a)$ denotes the F_1^a output vector. $\mathbf{y}^a \equiv (y_1^a \cdots y_{N_a}^a)$ denotes the F_2^a output vector. $\mathbf{w}_j^a \equiv (w_{j1}^a, w_{j2}^a, \ldots, w_{j,2M_a})$ denotes the jth ART_a weight vector. For ART_b: $\mathbf{x}^b \equiv (x_1^b \cdots x_{2M_b}^b)$ denotes the F_1^b output vector. $\mathbf{y}^b \equiv (y_1^b \ldots y_{N_b}^b)$ denotes the F_2^b output vector. $\mathbf{w}_k^b \equiv (w_{k1}^b, w_{k2}^b, \ldots, w_{k,2M_b}^b)$ denotes the kth ART_b weight vector. For the map field: $\mathbf{x}^{ab} \equiv (x_1^{ab}, \ldots, x_{N_b}^{ab})$ denotes the F^{ab} output vector and $\mathbf{w}_j^{ab} \equiv (w_{j1}^{ab}, \ldots, w_{jN_b}^{ab})$ denotes the weight vector from the jth F_2^a node to F^{ab}. The system resets $\mathbf{x}^a$, $\mathbf{y}^a$, $\mathbf{x}^b$, $\mathbf{y}^b$, and $\mathbf{x}^{ab}$ to $\mathbf{0}$ between input presentations.

14.7.2 Map Field Activation

The map field F^{ab} receives input from either or both of the ART$_a$ or ART$_b$ category fields. A chosen F_2^a node J sends input to the map field F^{ab} via the weights $\mathbf{w}_J^{ab}$. An active F_2^b node K sends input to F^{ab} via one-to-one pathways between F_2^b and F^{ab}. If both ART$_a$ and ART$_b$ are active then F^{ab} remains active only if ART$_a$ predicts the same category as ART$_b$. The F^{ab} output vector $\mathbf{x}^{ab}$ obeys

$$\mathbf{x}^{ab} = \begin{cases} \mathbf{y}^b \wedge \mathbf{w}_J^{ab} & \text{if the } J^{\text{th}}\ F_2^a \text{ node is active and } F_2^b \text{ is active} \\ \mathbf{w}_J^{ab} & \text{if the } J^{\text{th}}\ F_2^a \text{ node is active and } F_2^b \text{ is inactive} \\ \mathbf{y}^b & \text{if } F_2^a \text{ is inactive and } F_2^b \text{ is active} \\ \mathbf{0} & \text{if } F_2^a \text{ is inactive and } F_2^b \text{ is inactive.} \end{cases} \tag{14.32}$$

By (14.32) $\mathbf{x}^{ab} = \mathbf{0}$ if $\mathbf{y}^b$ fails to confirm the map field prediction made by $\mathbf{w}_J^{ab}$. Such a mismatch event triggers an ART$_a$ search for a better category.

14.7.3 Match Tracking

At the start of each input presentation ART$_a$ vigilance ρ_a equals a baseline vigilance parameter $\overline{\rho_a}$. If a predictive error occurs then match tracking raises ART$_a$ vigilance just enough to trigger a search for a new F_2^a coding node. ARTMAP detects a predictive error when

$$|\mathbf{x}^{ab}| < \rho_{ab}|\mathbf{y}^b| \tag{14.33}$$

where ρ_{ab} is the map field vigilance parameter. A signal from the map field to the ART$_a$ orienting subsystem causes ρ_a to track the F_1^a match. So ρ_a grows until it is slightly higher than the F_1^a match value $|\mathbf{A} \wedge \mathbf{w}_J^a||\mathbf{A}|^{-1}$. Then since

$$|\mathbf{x}^a| = |\mathbf{A} \wedge \mathbf{w}_J^a| < \rho_a|\mathbf{A}| \tag{14.34}$$

ART$_a$ fails to meet the matching criterion as in (14.10) and the search for another F_2^a node begins. The search leads to an F_2^a node J with

$$|\mathbf{x}^a| = |\mathbf{A} \wedge \mathbf{w}_J^a| \geq \rho_a|\mathbf{A}| \tag{14.35}$$

and

$$|\mathbf{x}^{ab}| = |\mathbf{y}^b \wedge \mathbf{w}_J^{ab}| \geq \rho_{ab}|\mathbf{y}^b|. \tag{14.36}$$

If no such node exists and if all F_2^a nodes are already committed then F_2^a automatically shuts down for the remainder of the input presentation.

14.7.4 Map Field Learning

Weights w_{jk}^{ab} in $F_2^a \rightarrow F^{ab}$ paths at first satisfy

$$w_{jk}^{ab}(0) = 1. \tag{14.37}$$

During resonance with the ART_a category J active $\mathbf{w}_J^{ab}$ approaches the map field vector $\mathbf{x}^{ab}$ as in (14.11). With fast learning once J learns to predict the ART_b category K that association is permanent. (Thus $w_{JK}^{ab} = 1$ for all time.)

14.8 FUZZY ARTMAP SIMULATION: CIRCLE-IN-THE-SQUARE

The circle-in-the square problem requires a system to predict which points of a square lie inside and which lie outside a circle whose area equals half that of the square (radius approximately 0.4). This task is a benchmark problem for performance evaluation in the ARPA Artificial Neural Network Technology (ANNT) Program [43]. Wilensky examined 2-n-1 backpropagation systems on this problem. He studied systems where the number (n) of hidden units ranged from 5 to 100 and the corresponding number of weights ranged from 21 to 401. Training sets ranged in size from 150 to 14,000 points identified as in or out. To avoid overfitting training ended when accuracy on the training set reached 90%. Systems with 20 to 40 hidden units reached this criterion level most quickly in about 5000 epochs. The trained systems then correctly classified approximately 90% of test set points.

Figures 14.9 and 14.10 illustrate fuzzy ARTMAP performance on the circle-in-the-square task after one training epoch. As training set size increases from 100 exemplars (Figure 14.9a) to 100,000 exemplars (Figure 14.9d) the rate of correct test set predictions grows from 88.6% to 98.0% while the number of ART_a category nodes grows from 12 to 121. Each category node j requires four learned weights $\mathbf{w}_j^a$ in ART_a plus one map field weight $\mathbf{w}_j^{ab}$ to record whether category j predicts that a point lies inside or outside the circle. Consider the case when 1-epoch trains on 100 exemplars (Figure 14.9a). This creates 12 ART_a categories and so uses 60 weights to achieve 88.6% test set accuracy. Figure 14.10 shows the ART_a category rectangles R_j^a for each simulation of Figure 14.9. As in Figure 14.7 each rectangle R_j^a corresponds to the four-dimensional weight vector $\mathbf{w}_j^a = (\mathbf{u}_j^a, (\mathbf{v}_j^a)^c)$ with $\mathbf{u}_j^a$ and $\mathbf{v}_j^a$ plotted as the lower-left and upper-right corners of R_j^a. Early in training large R_j^a estimate large areas as belonging to one or the other category (Figure 14.10a). Additional R_j^a improve accuracy especially near the boundary of the circle (Figure 14.10d). The map becomes arbitrarily accurate if the number of F_2^a nodes grows as needed.

Figure 14.11 depicts the response patterns of fuzzy ARTMAP on another series of circle-in-the-square simulations that use the same training sets as in Figure 14.9. But the system presents each input set for as many epochs as needed to achieve 100% predictive accuracy on the training set. Training in Figure 14.9 lasted for one epoch only. In each case test set predictive accuracy grows but so does the number of ART_a category nodes. Consider the case with 10,000 exemplars. Then 1-epoch training uses 50 ART_a nodes to give 96.7% test set accuracy (Figure 14.9c). The same training set after 6 epochs uses 39 more ART_a nodes to correct about half the errors and boost test set accuracy to 98.3% (Figure 14.11c).

Figure 14.9 shows how a test set error rate falls from 11.4% to 2.0% as training set size grows from 100 to 100,000 in 1-epoch simulations. Figure 14.11 shows how test set error rates fall further if learning continues for as many epochs as needed

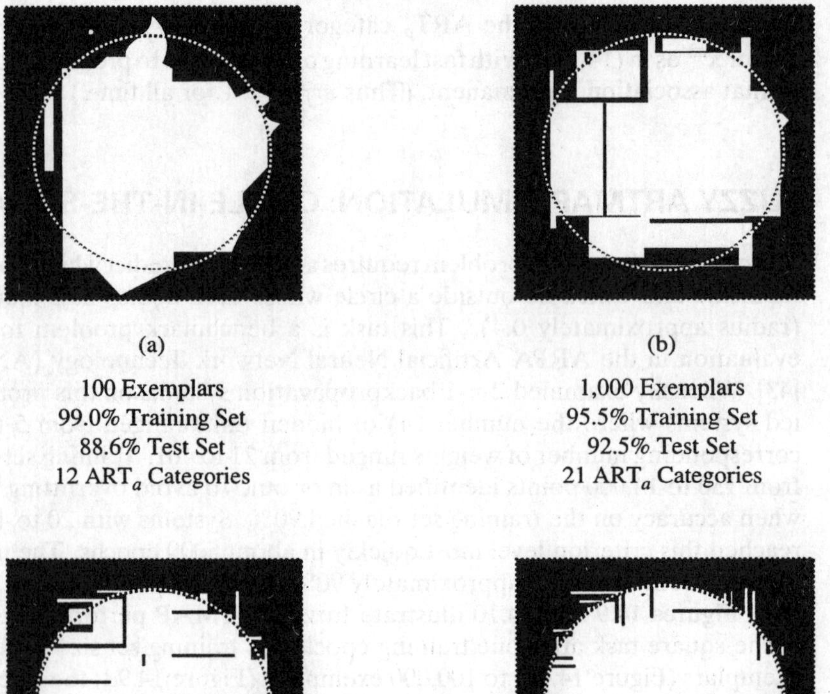

(a)
100 Exemplars
99.0% Training Set
88.6% Test Set
12 ART_a Categories

(b)
1,000 Exemplars
95.5% Training Set
92.5% Test Set
21 ART_a Categories

(c)
10,000 Exemplars
97.7% Training Set
96.7% Test Set
50 ART_a Categories

(d)
100,000 Exemplars
98.8% Training Set
98.01% Test Set
121 ART_a Categories

Figure 14.9 Circle-in-the-square test set response patterns after 1 epoch of fuzzy ARTMAP training on (a) 100, (b) 1,000, (c) 10,000, and (d) 100,000 randomly chosen training set points. The system predicts that test set points in white areas lie inside the circle and that points in black areas lie outside the circle. The test set error rate falls approximately inversely with the number of ART_a categories as the training set size grows. The box complexity of such a graph cover grows exponentially with the dimension of the fuzzy cube.

to reach 100% accuracy on each training set. The ARTMAP voting strategy gives a third way to reduce test set errors. Recall that the voting strategy assumes a fixed set of training exemplars. We randomly order inputs before each simulation. We record the prediction of each test set item after each simulation. The final prediction is the one that the largest number of simulations makes. Voting almost always reduces errors. The number of votes cast for a given outcome also attaches a measure of predictive confidence to each test set point.

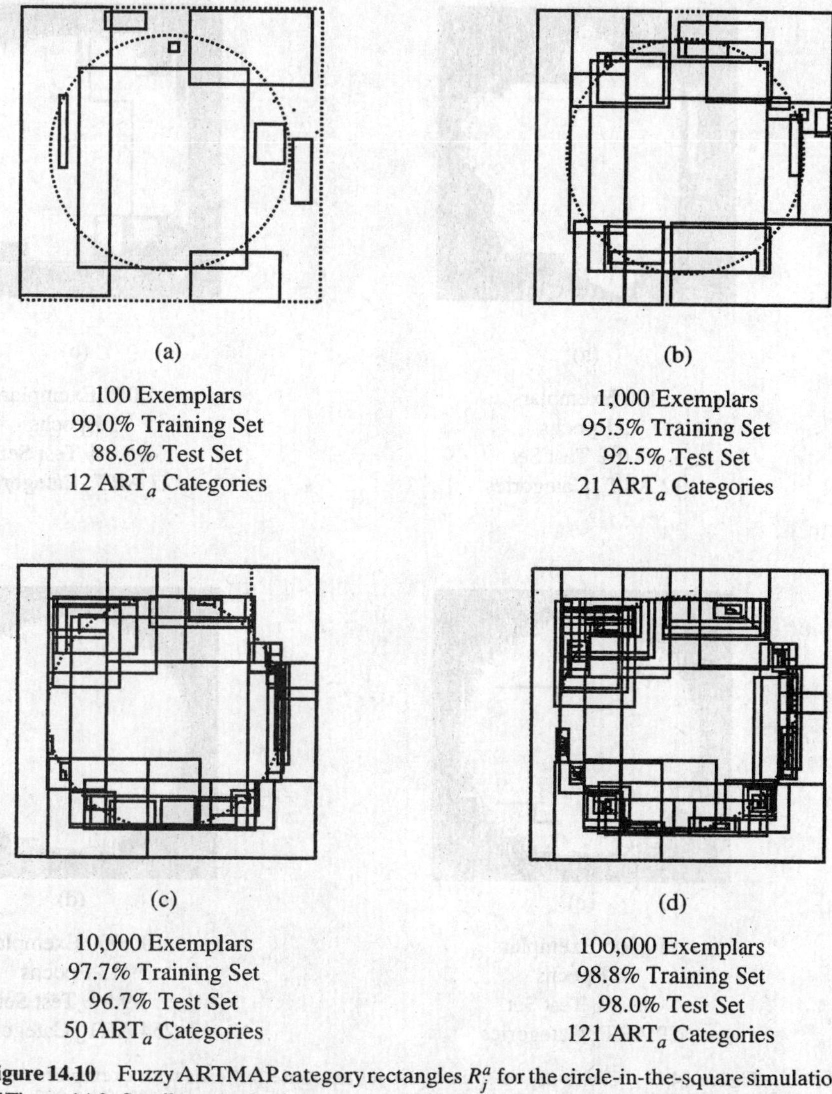

(a)

100 Exemplars
99.0% Training Set
88.6% Test Set
12 ART_a Categories

(b)

1,000 Exemplars
95.5% Training Set
92.5% Test Set
21 ART_a Categories

(c)

10,000 Exemplars
97.7% Training Set
96.7% Test Set
50 ART_a Categories

(d)

100,000 Exemplars
98.8% Training Set
98.0% Test Set
121 ART_a Categories

Figure 14.10 Fuzzy ARTMAP category rectangles R_j^a for the circle-in-the-square simulations of Figure 14.9. Small rectangles form near map discontinuities as the error rate drops toward 0.

Simulations in Figure 14.12 show how a limited training set and voting across a few input orderings improves predictive accuracy by a factor that compares well to the improvement we achieve if we increase the size of the training set by an order of magnitude. We present in Figure 14.12 a fixed set of 1,000 randomly chosen exemplars to a fuzzy ARTMAP system on five independent 1-epoch circle-in-the-square simulations. We record after each simulation inside/outside predictions on a 1,000-item test set. Accuracy on individual simulations ranges from 85.9% to 93.4% and averages 90.5%, and uses from 15 to 23 ART_a nodes. Voting by the five simulations improves test set accuracy to 93.9% (Figure 14.12c). So test set errors fall from an

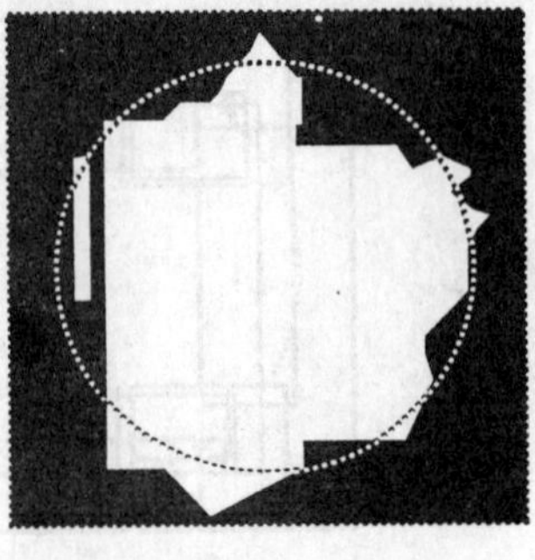

(a)

100 Exemplars
2 Epochs
89.0% Test Set
12 ART_a Categories

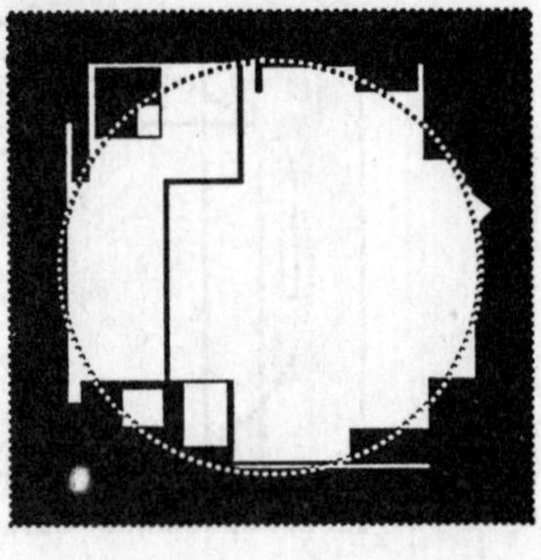

(b)

1,000 Exemplars
3 Epochs
95.0% Test Set
27 ART_a Categories

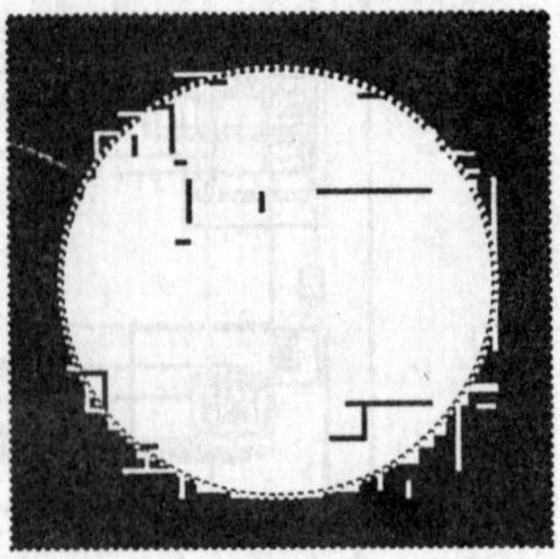

(c)

10,000 Exemplars
6 Epochs
98.3% Test Set
89 ART_a Categories

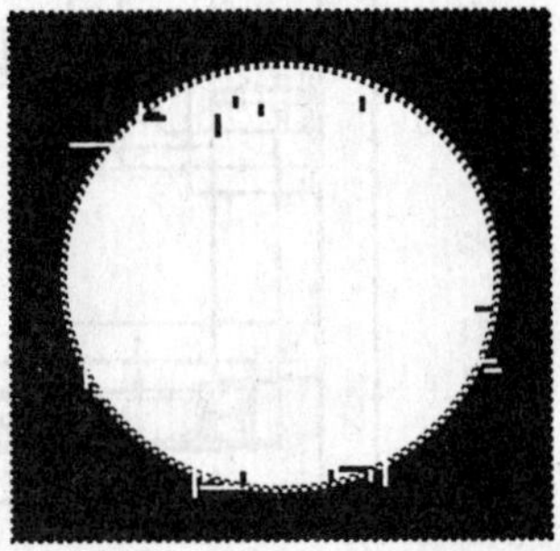

(d)

100,000 Exemplars
13 Epochs
99.5% Test Set
254 ART_a Categories

Figure 14.11 Circle-in-the-square test set response patterns with exemplars repeatedly presented until the system achieves 100% correct prediction on (a) 100, (b) 1,000, (c) 10,000, and (d) 100,000 training set points. Training sets are the same as those used for Figures 14.9 and 14.10. Training to 100% accuracy requires (a) 2 epochs, (b) 3 epochs, (c) 6 epochs, and (d) 13 epochs. Additional training epochs decrease test set error rates but create additional ART_a categories compared to the 1-epoch simulation in Figure 14.7.

average lone rate of 9.5% to a voting rate of 6.1%. Figure 14.12d shows that the number of votes cast for each test set point and reflects variations in predictive confidence across different regions with confidence lowest near the border of the circle. Voting by more than five simulations maintains an error rate between 5.8% and 6.1%. Gaps in the fixed 1,000-item training set appear to cause this limit on further improvement by voting. A 10-fold increase in the size of the training set reduces the error by an amount similar to that five-simulation voting achieved.

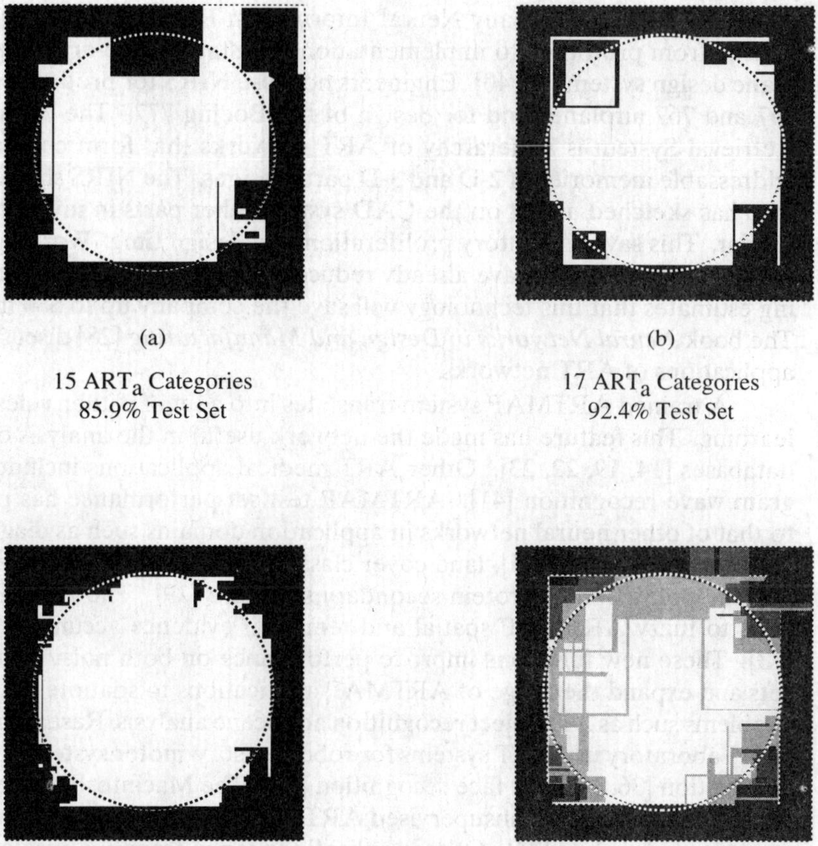

(a) 15 ART_a Categories 85.9% Test Set

(b) 17 ART_a Categories 92.4% Test Set

(c) Voting on 5 Runs 93.9% Test Set

(d) Number of Votes

Figure 14.12 Circle-in-the-square response patterns for a fixed 1,000-item training set. (a) Test set responses after training on inputs presented in random order. After 1 epoch that uses 15 ART_a nodes the test set prediction rate is 85.9% (the worst of 5 runs). (b) Test set responses after training on inputs presented in a different random order. After 1 epoch that uses 17 ART_a nodes the test set prediction rate is 92.3% (the best of 5 runs). (c) Voting strategy applied to five individual simulations. Test set prediction rate is 93.9%. (d) Cumulative test set response pattern from five 1-epoch simulations. Gray scale intensity grows with the number of votes that predict a point to lie outside the circle.

14.9 ART APPLICATIONS

The first ART network appeared in print in 1987. Scientists and engineers have since applied these systems to a variety of problems. Researchers often cite unique ART features such as code stability, speed, and incremental learning as reasons for using ART or ARTMAP instead of an error-based neural network such as a backpropagation-trained perceptron.

The Boeing Company Neural Information Retrieval System (NIRS) has advanced from prototype to implementation in a state-of-the-art computer-aided airplane design system [15, 40]. Engineers now use NIRS for production of the Boeing 747 and 767 airplanes and for design of the Boeing 777. The Neural Information Retrieval System is a hierarchy of ART networks that form compressed content-addressable memories of 2-D and 3-D parts designs. The NIRS shows to an engineer who has sketched a part on the CAD system other parts in inventory that may be similar. This saves inventory proliferation and design time. Working CAD systems that include the NIRS have already reduced parts inventories by a factor of 9. Boeing estimates that this technology will save the company up to $80 million per year. The book *Neural Networks in Design and Manufacturing* [28] discusses commercial applications of ART networks.

A trained ARTMAP system translates into a set of if-then rules at any stage of learning. This feature has made the network useful in the analysis of large medical databases [14, 19, 22, 23]. Other ART medical applications include electrocardiogram wave recognition [41]. ARTMAP test set performance has proved superior to that of other neural networks in application domains such as diagnostic monitoring of nuclear plants [25], land cover classification from remotely sensed data [20], and the prediction of protein secondary structure [29]. The ART-EMAP network adds to fuzzy ARTMAP spatial and temporal evidence accumulation capabilities [13]. These new functions improve performance on both noisy and noise-free test sets and expand the range of ARTMAP applications to spatiotemporal recognition problems such as 3-D object recognition and scene analysis. Researchers at MIT Lincoln Laboratory use ART systems for robot sensory-motor systems [1, 2], 3-D object recognition [36, 37], and face recognition [38]. The Macintosh commercial software Open Sesame! uses an unsupervised ART network to adapt the operating system to a user's work habits [24]. Other applications range from analyses of musical structure [17] to military target recognition [31]. Applications of ART networks continue to include those of the original adaptive resonance theory—to organize, clarify, and predict neural and psychological data concerning memory recognition and attention [7, 8, 16, 18].

Future fuzzy ART systems may offer new ways to approximate bounded functions and pattern decision surfaces. Other learning schemes may seek out optimal ART boxes in fuzzy cubes to deal with box explosion and to better approximate functions.

REFERENCES

[1] Bachelder, I. A., Waxman, A. M., and Seibert, M., (1993). "A Neural System for Mobile Robot Visual Place Learning and Recognition," *Proceedings of the World Congress on Neural Networks (WCNN–93)*, I–512–517, Lawrence Erlbaum Associates, 1993.

[2] Baloch, A. A., and Waxman, A. M., "Visual Learning, Adaptive Expectations, and Behavioral Conditioning of the Mobile Robot MAVIN," *Neural Networks*, vol. 4, 271–302, 1991.

[3] Carpenter, G. A., and Gjaja, M., "Fuzzy ART Choice Functions," Technical Report CAS/CNS-TR-93-060, Boston University, 1993. In the *Proceedings of the World Congress on Neural Networks (WCNN'94)*.

[4] Carpenter, G. A., and Grossberg, S., "A Massively Parallel Architecture for a Self-Organizing Neural Pattern Recognition Machine," *Computer Vision, Graphics, and Image Processing*, vol. 37, 54–115, 1987a.

[5] Carpenter, G. A., and Grossberg, S., "ART 2: Stable Self-Organization of Pattern Recognition Codes for Analog Input Patterns," *Applied Optics*, vol. 26, 4919–4930, 1987b.

[6] Carpenter, G. A., and Grossberg, S., "ART 3: Hierarchical Search Using Chemical Transmitters in Self-Organizing Pattern Recognition Architectures," *Neural Networks*, vol. 3, 129–152, 1990.

[7] Carpenter, G. A., and Grossberg, S., editors, *Pattern Recognition by Self-Organizing Neural Networks*, MIT Press, 1991.

[8] Carpenter, G. A., and Grossberg, S., "Normal and Amnesic Learning, Recognition, and Memory by a Neural Model of Cortio-Hippocampal Interactions," *Trends in Neurosciences*, vol. 16, 131–137, 1993.

[9] Carpenter, G. A., Grossberg, S., Markuzon, N., Reynolds, J. H., and Rosen, D. B., "Fuzzy ARTMAP: A Neural Network Architecture for Incremental Supervised Learning of Analog Multidimensional Maps," *IEEE Transactions on Neural Networks*, vol. 3, 698–713. Technical Report CAS/CNS-91-016, Boston University.

[10] Carpenter, G. A., Grossberg, S., and Reynolds, J. H., "ARTMAP: Supervised Real-Time Learning and Classification of Nonstationary Data by a Self-Organizing Neural Network," *Neural Networks*, vol 4, 565–588, 1991. Technical Report CAS/CNS-TR-91-001, Boston University.

[11] Carpenter, G. A., Grossberg, S., and Rosen, D.B., "Fuzzy ART: Fast Stable Learning and Categorization of Analog Patterns by an Adaptive Resonance System," *Neural Networks*, vol. 4, 759–771, 1991a. Technical Report CAS/CNS-TR-91-015, Boston University.

[12] Carpenter, G. A., Grossberg, S., and Rosen, D. B., "A Neural Network Realization of Fuzzy ART," Technical Report CAS/CNS-TR-91-021, Boston University, 1991b.

[13] Carpenter, G. A., Grossberg, S., and Ross, W. D., "ART-EMAP: A Neural Network Architecture for Learning and Prediction by Evidence Accumulation," *Proceedings of the World Congress on Neural Networks (WCNN-93)*, Lawrence Erlbaum Associates, III-649-656, 1993. Technical Reports CAS/CNS-TR-93-015 and 93-035, Boston University.

[14] Carpenter, G. A., and Tan, A.-H., "Rule Extraction, Fuzzy ARTMAP, and Medical Databases," *Proceedings of the World Congress on Neural Networks (WCNN-93)*, Lawrence Erlbaum Associates, vol. I, 501–506 1993. Technical Report CAS/CNS-TR-93-016, Boston University.

[15] Caudell, T., editor *Adaptive neural systems*. 1992 IR&D Technical Report BCS-CS-ACS-93-008, The Boeing Company, 1993.

[16] Desimone, R., "Neural Circuits for Visual Attention in the Primate Brain," *Neural Networks for Vision and Image Processing*, G. A. Carpenter and S. Grossberg, editors, MIT Press, 343–364, 1992.

[17] Gjerdingen, R. O., "Categorization of Musical Patterns by Self-Organizing Neuronlike Networks," *Music Perception*, vol. 7, 339–370, 1990.

[18] Gochin, P., "Pattern Recognition in Primate Temporal Cortex: But Is It ART?" *Proceedings of the International Joint Conference on Neural Networks (IJCNN-90)* (Washington, D.C.), I-77-80. Lawrence Erlbaum Associates, 1990.

[19] Goodman, P. H., Kaburlasos, V. G., Egbert, D. D., Carpenter, G. A., Grossberg, S., Reynolds, J. H., Rosen, D. B., and Hartz, A. J., "Fuzzy ARTMAP Neural Network Compared to Linear Discriminant Analysis Prediction of the Length of Hospital Stay in Patients with Pneumonia," *Proceedings of the IEEE International Conference on Systems, Man, and Cybernetics* (Chicago, October 1992), New York, IEEE Press, I–748–753, 1992.

[20] Gopal, S., Sklarew, D., and Lambin, E., "Fuzzy-Neural Network Classification of Land Cover Change in the Sahel," *Proceedings of the DOSES/EUROSAT Workshop on New Tools for Spatial Analysis* (Lisbon, Portugal, November 18–20, 1993), 1993.

[21] Grossberg, S., "Adaptive Pattern Classification and Universal Recoding, II: Feedback, Expectation, Olfaction, and Illusions," *Biological Cybernetics*, vol. 23, 187–202, 1976.

[22] Ham, F. M., and Han, S. W., "Quantitative Study of the QRS Complex Using Fuzzy ARTMAP and the MIT/BIH Arrhythmia Database," *Proceedings of the World Congress on Neural Networks (WCNN–93)*, Lawrence Erlbaum Associates, I–207–211, 1993.

[23] Harvey, R. M., "Nursing Diagnosis by Computers: An Application of Neural Networks," *Nursing Diagnosis*, vol. 4, 26–34, 1993.

[24] Johnson, C., "Agent Learns User's Behavior" *Electrical Engineering Times*, June 28, 1993, pp. 43, 46.

[25] Keyvan, S., Durg, A., and Rabelo, L. C., "Application of Artificial Neural Networks for Development of Diagnostic Monitoring System in Nuclear Plants," *American Nuclear Society Conference Proceedings*, April 18–21, 1993.

[26] Kosko, B., "Fuzzy Entropy and Conditioning," *Information Sciences*, vol 40, no. 2, 165–174, 1986.

[27] Kumar, S. S., and Guez, A., "ART Based Adaptive Pole Placement for Neurocontrollers," *Neural Networks*, vol. 4, 319–335, 1991.

[28] Kumara, S. R. T., Merchawi, N. S., Karmarthi, S. V., and Thazhutaveetil, M., *Neural Networks in Design and Manufacturing*, Chapman and Hall, 1993.

[29] Mehta, B. V., Vij, L., and Rabelo, L. C., "Prediction of Secondary Structures of Proteins Using Fuzzy ARTMAP," *Proceedings of the World Congress on Neural Networks (WCNN–93)*, Lawrence Erlbaum Associates, I–228–232, 1993.

[30] Moore, B., "ART 1 and Pattern Clustering," *Proceedings of the 1988 Connectionist Models Summer School*, D. Touretzky, G. Hinton, and T. Sejnowski, editors, Morgan Kaufmann Publishers, 174–185, 1989.

[31] Moya, M. M., Koch, M. W., and Hostetler, L. D., "One-Class Classifier Networks for Target Recognition Applications. *Proceedings of the World Congress on Neural Networks (WCNN–93)*, Lawrence Erlbaum Associates, III–797–801, 1993.

[32] Rosenblatt, F., "The perceptron: A Probablistic Model for Information Storage and Organization in the Brain," *Psychological Review*, vol. 65, 386–408, 1958. Reprinted in J.A. Anderson and E. Rosenfeld, editors, *Neurocomputing: Foundations of Research*, MIT Press, 18–27, 1988.

[33] Rosenblatt, F., *Principles of Neurodynamics*, Spartan Books, 1962.

[34] Rumelhart, D. E., Hinton, G., and Williams, R., "Learning Internal Representations by Error Propagation," *Parallel Distributed Processing*, D. E. Rumelhart and J. L. McClelland, editors, MIT Press, 318–362, 1986.

[35] Salzberg, S. L., *Learning with Nested Generalized Exemplars*, Kluwer Academic Publishers, 1990.

[36] Seibert, M., and Waxman, A. M., "Learning and Recognizing 3D Objects from Multiple Views in a Neural System," *Neural Networks for Perception*, vol. 1, H. Wechsler, editor, Academic Press, 426–444, 1991.

[37] Seibert, M., and Waxman, A. M., "Adaptive 3D Object Recognition from Multiple Views," *IEEE Transactions on Pattern Analysis and Machine Intelligence*, vol. 14, 107–124, 1992.

[38] Seibert, M., and Waxman, A. M., "An Approach to Face Recognition Using Saliency Maps and Caricatures," *Proceedings of the World Congress on Neural Networks (WCNN-93)*, Lawrence Erlbaum Associates, III–661–664, 1993.

[39] Simpson, P., "Fuzzy Min-Max Classification with Neural Networks," *Heuristics: The Journal of Knowledge Engineering*, vol. 4, 1–9, 1991.

[40] Smith, S. D. G., Escobedo, R., and Caudell, T. P., "An Industrial Strength Neural Network Application," *Proceedings of the World Congress on Neural Networks (WCNN-93)*, Lawrence Erlbaum Associates, I–490–494, 1993.

[41] Suzuki, Y., Abe, Y., and Ono, K., "Self-Organizing QRS Wave Recognition System in ECG Using ART 2," *Proceedings of the World Congress on Neural Networks (WCNN–93)*, Lawrence Erlbaum Associates, IV–39–42, 1993.

[42] Werbos, P., "Beyond Regression: New Tools for Prediction and Analysis in the Behavioral Sciences," Ph.D. thesis, Harvard University, 1974.

[43] Wilensky, G., "Analysis of Neural Network Issues: Scaling, Enhanced Nodal Processing, Comparison with Standard Classification," DARPA Neural Network Program Review, October 29–30, 1990.

[44] Zadeh, L., "Fuzzy Sets," *Information and Control*, vol. 8, 338–353, 1965.

HOMEWORK PROBLEMS

14.1. Fuzzy ART without complement coding Consider a fuzzy ART system without complement coding. When $M = 2$ then an input vector $\mathbf{I}$ corresponds to a point or 2-fit vector $\mathbf{a}$ in the unit square or fuzzy 2-cube.

(a) With fast learning ($\beta = 1$ in equation (14.11)) show that

$$\mathbf{w}_j = \Lambda\{\mathbf{a}\text{: } \mathbf{a} \text{ has activated category } j \text{ without reset }\}.$$

(b) Draw a typical weight vector $\mathbf{w}_j$ in the 2-cube. Shade the set of inputs $\mathbf{a}$ such that if $\mathbf{a}$ activates category j without reset then no weights will change during learning. Give a formula for this set.

(c) Pick a typical value of $\rho \in (0, 1)$. Indicate on the graph in (b) the set of points $\mathbf{a}$ in the shaded region that will not cause a reset upon choosing category j. Give a formula for this set. Call the set Δ_j.

(d) Put $\rho = 0.5$ and sketch Δ_j for (i) $\mathbf{w}_j = (0.1, 0.1)$ and (ii) $\mathbf{w}_j = (0.5, 0.5)$.

(e) Discuss the implications of the geometry of (d).

14.2. Circle-in-the-square simulations Replicate the simulations shown in Figures 14.9–14.12. Results will vary according to the random sequence of training set inputs $\mathbf{a}$ as in Figure 14.12.

14.3. ART-EMAP The ART-EMAP neural network [13] adds to fuzzy ARTMAP spatial and temporal evidence accumulation capabilities. ART-EMAP functions derive from four stages of model development. In stage 1 spatial evidence accumulation pools predictions across ART_a categories. The network trains with the fuzzy ARTMAP algorithm with choice (winner-take-all) at the ART_a category representation field F_2^a as in equation (14.5). During performance this F_2^a activity is distributed. An algebraic algorithm that approximates distributed category activations at F_2^a is

$$y_j^a = \frac{\left(T_j^a\right)^p}{\sum_{k=1}^{n}\left(T_k^a\right)^p}$$

for $p \geq 1$. Here y_j^a equals activity of the jth F_2^a node and T_j^a equals the total input from F_1^a to the jth node as in equation (14.2). During ART-EMAP performance the distributed dynamics y_j^a at F_2^a replace the fuzzy ARTMAP winner-take-all dynamics of equation (14.5). An E ("evidence") MAP field then accumulates predictive evidence across all active F_2^a categories.

(a) Sketch a graph that characterizes y_j^a as a function of T_j^a as p varies from 1 to ∞. Sketch y_j^a when $T_j^a = j$.

(b) Examine the effect of spatial evidence accumulation on circle-in-the-square simulations. Train a fuzzy ARTMAP network on 1000 randomly chosen points as in Figure 14.12. Test predictive accuracy on 1000 new test set points for (i) fuzzy ARTMAP (choice at F_2^a) and for distributed F_2^a activity using (i) $p = 1$, (ii) $p = 10$, and (iii) $p = 100$ in the above equation. Analyze the simulation results.

14.4. Attention

(a) Describe three ways an ART system (Figure 14.2) "pays attention."

(b) Illustrate each type of attention with an example from daily experience.

(c) Describe the role of each attentional process in ART dynamics.

(d) For each attentional process, show how ART dynamics would change if you removed that process from the network.

14.5. Complement coding Compare fuzzy ART dynamics with and without complement coding. For each simulation let $\alpha = 0^+$ (conservative limit) and $\beta = 1$ (fast learning).

(a) Generate a random sequence of 1,000 training set points $\mathbf{a} = (\mathbf{a}_1, \mathbf{a}_2)$ in the unit square or fuzzy 2-cube. Present each input to a fuzzy ART system for one training epoch.

(b) With complement coding plot or otherwise describe the learned category boxes R_j when vigilance $\rho = 0.2$.

(c) Repeat (b) with (i) $\rho = 0.0$, (ii) $\rho = 0.7$, and (iii) $\rho = 0.9$.

(d) Repeat (b) and (c) without complement coding.

(e) Compare fuzzy ART dynamics with and without complement coding.

14.6. Square-in-the-square calculations With fast learning ($\beta = 1$) and complement coding a 2-D fuzzy ART category box is the smallest rectangle R_j that encloses all training set points of category j (Figure 14.8). In the conservative limit ($\alpha = 0$) an input $\mathbf{a}$ that lies in a category box R_j will choose the smallest such box.

Consider the square-in-the-square problem shown in Figure 14.13. Fuzzy ARTMAP learns from a sequence of training set fit-vector inputs $\mathbf{a} = (a_1, a_2)$ each labeled as "in" or "out" with respect to the shaded square. During testing the system predicts whether a point $\mathbf{a}$ is in or out of the shaded square. Calculate fuzzy ARTMAP dynamics for this problem with fast learning, complement coding, the conservative limit, and baseline vigilance ρ_a "bar" $= 0$ as follows: For the training set of fit vectors

$$\mathbf{a}^{(1)} = (0.2,\ 0.7)$$

$$\mathbf{a}^{(2)} = (0.7,\ 0.2)$$

sketch the category box(es) R_j.

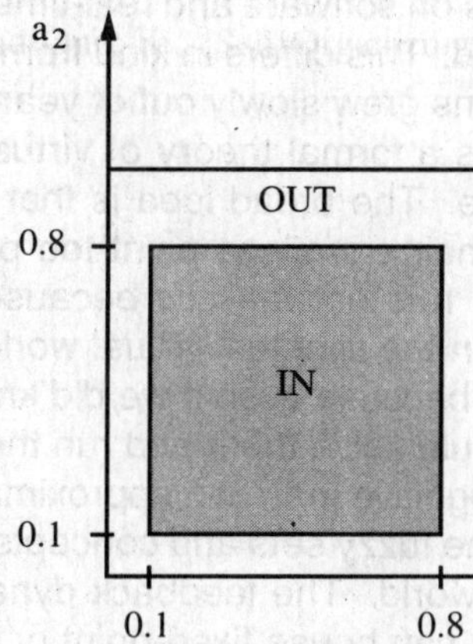

Figure 14.13

Virtual reality shares a key property with modern multimedia systems. It has far more applications than it has formal theories. It lacks a mathematical foundation and instead tends to focus on software and real-time graphics and the race to show more polygons per second. This differs in kind from the history of fuzzy and neural systems. Their applications grew slowly out of years of mathematical research.

Chapter 15 presents a formal theory of virtual reality cast in the neural-like dynamics of a fuzzy cube. The broad idea is that all math models define virtual or artificial worlds. But their equations paint too precise a picture. The result is not realistic or practical. It is not realistic because no one knows the nonlinear equations that model even the simplest virtual worlds that contain only one or two agents. It is not practical because even if we did know these equations then there is little chance that we could solve them and run them in real time.

A feedback fuzzy cognitive map can approximate a virtual world as a dynamical system. The size of the fuzzy sets and concepts defines the size or conceptual granularity of the virtual world. The feedback dynamics carve up the fuzzy cube into attractor regions that can house fixed-point or even chaotic attractors. These adaptive feedback fuzzy systems are nonlinear function approximators with even more complex dynamics than feedback neural networks. This complexity increases their approximation power but offers little chance for formal analysis.

The undersea application in Chapter 15 suggests that some future feedback fuzzy models may give up any pretense of formal analysis. These feedback systems yield a rough but novel approximation of a structured virtual world. Their novelty stems from their complex dynamics and formal inscrutability. Their ultimate value may lie as much in how well they entertain users as in how well they predict or control a physical process.

15

VIRTUAL WORLDS IN FUZZY COGNITIVE MAPS

Julie A. Dickerson and Bart Kosko

PREVIEW

Fuzzy cognitive maps (FCMs) can structure virtual worlds that change with time. An FCM links causal events, actors, values, goals, and trends in a fuzzy feedback dynamical system. An FCM lists the fuzzy rules or causal flow paths that relate events. It can guide actors in a virtual world as the actors move through a web of cause and effect and react to events and to other actors. Experts draw FCM causal pictures of the virtual world. They do not write down differential equations to change the virtual world.

Complex FCMs can endow virtual worlds with "new" or chaotic equilibrium behavior. Simple FCMs give virtual worlds with periodic behavior. They map input states to limit-cycle equilibria. An FCM limit cycle repeats a sequence of events or a chain of actions and responses. Limit cycles can control the steady-state rhythms and patterns in a virtual world. In nested FCMs each causal concept can control its own FCM or fuzzy function approximator. This gives levels of fuzzy systems that can choose goals and causal webs as well as move objects and guide actors in the webs. FCM matrices sum to give a combined FCM virtual world for any number of knowledge sources.

Adaptive FCMs change their fuzzy causal web as causal patterns change and as actors act and experts state their causal knowledge. Neural learning laws change the causal rules and the limit cycles. Actors learn new patterns and reinforce old ones. In complex FCMs the user can choose the dynamical structure of the virtual world from a spectrum that ranges from mildly to wildly nonlinear. We use an adaptive FCM to model an undersea virtual world of dolphins, fish, and sharks.

15.1 FUZZY VIRTUAL WORLDS

What is a virtual world? It is what changes in a "virtual reality" [17] or "cyberspace" [8]. A virtual world links humans and computers in a causal medium that can trick the mind or senses.

At the broadest level a virtual world is a dynamical system. It changes with time as the user or an actor moves through it. In the simplest case only the user moves in the virtual world. In general both the user and the virtual world change and they change each other.

Change in a virtual world is causal. Actors cause events to happen as they move in a virtual world. They add new patterns of cause and effect and respond to old ones. In turn the virtual world acts on the actors or on their physical or social environments. The virtual world changes their behavior and can change its own web of cause of effect. This feedback causality between actors and their virtual world makes up a complex dynamical system that can model events, actors, actions, and data as they unfold in time.

Virtual worlds are fuzzy as well as fedback. Events occur and concepts hold only to some degree. Events cause one another to some degree. In this sense virtual worlds are fuzzy causal worlds. They are fuzzy dynamical systems.

How do we model the fuzzy feedback causality? One way is to write down the differential equations that show how the virtual "flux" or "fluid" changes in time. This gives an exact model. The Navier-Stokes equations [4] used in weather models give a fluid model of how actors move in a type of virtual world. They can show how clouds or tornadoes form and dissolve in a changing atmosphere or how an airplane flies through pockets of turbulence. The inverse kinematic equations of robotics [6] show how an actor moves through or grasps in a virtual joint space. The coupled differential equations of blood glucose and insulin [1] cast the patient as a diabetic actor awash in a virtual world of sugar and hormones. Such math models are hard to find, hard to solve, and hard to run in realtime. They paint too fine a picture of the virtual world.

Fuzzy cognitive maps (FCMs) can model the virtual world in large fuzzy chunks. They model the causal web as a fuzzy directed graph [12, 13]. The nodes and edges show how causal concepts affect one another to some degree in the fuzzy dynamical system. The "size" of the nodes gives the chunk size. The causal concept node SURVIVAL THREAT can measure the degree that the fuzzy event of survival threat occurs. In a virtual world the concept nodes can stand for events, actions, values, moods, goals, or trends. The causal edges state fuzzy rules or causal flows between concepts. In a predator–prey world survival threat increases prey runaway. The degree of runaway grows or falls as the degree of threat grows or falls. The fuzzy rule states how much one node grows or falls as some other node grows or falls.

Experts draw the FCMs as causal pictures. They do not state equations. They state concept nodes and link them to other nodes. The FCM system turns each picture into a matrix of fuzzy rule weights. The system weights and adds the FCM matrices to combine any number of causal pictures. More FCMs tend to sum to a better picture of the causal web with rich tangles of feedback and fuzzy edges even if each expert gives binary (present or absent) edges. This makes it easy to add or delete actors or to change the background of a virtual world or to combine virtual worlds that are disjoint or overlap. We can also let an FCM node control its own FCM to give a nested FCM in a hierarchy of virtual worlds. The node FCM can model the complex nonlinearities between the node's input and output. It can drive the motions, sounds, actions, or goals of a virtual actor as in Figure 15.1.

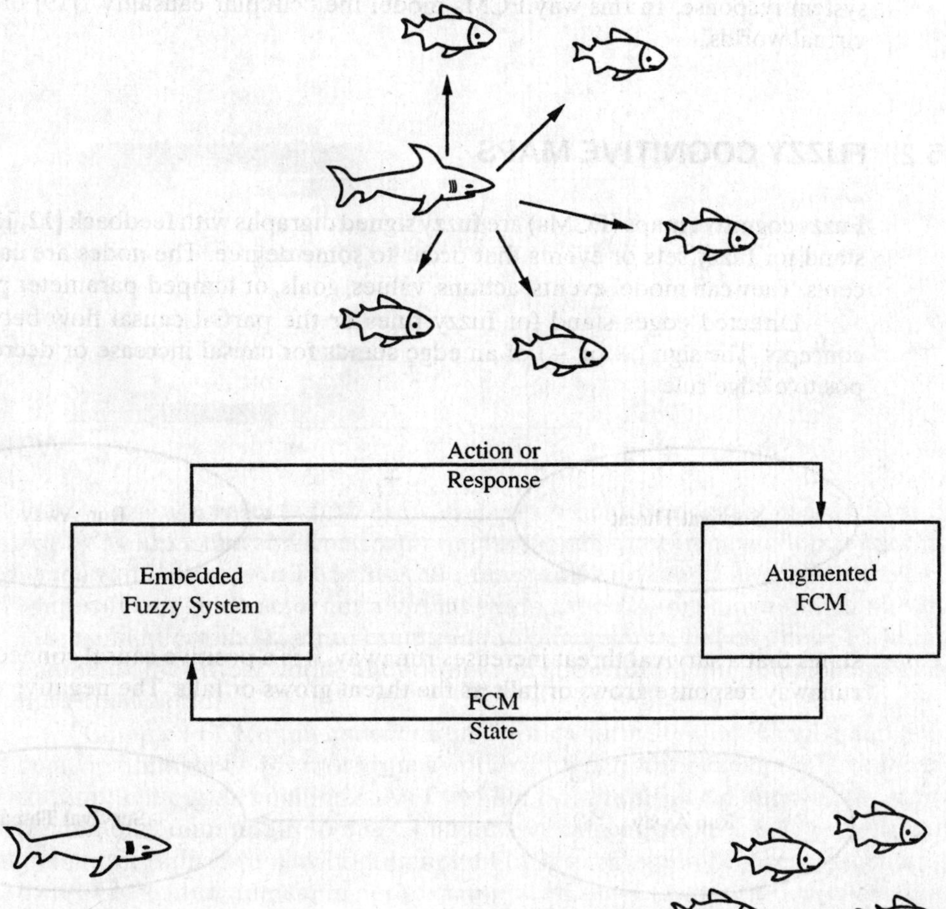

Figure 15.1 Fuzzy cognitive maps can structure virtual worlds. Embedded fuzzy systems drive lower-level fuzzy systems for animation, sounds, and other virtual world outputs. Here a shark finds a school of fish. The shark attacks and the fish flee.

The FCM itself acts as a nonlinear dynamical system. Like a neural net it maps inputs to output equilibrium states. Each input digs a path through the virtual state space. In simple FCMs the path ends in a fixed point or limit cycle. In more complex FCMs the path may end in an aperiodic or "chaotic" attractor. These fixed points and attractors represent *meta-rules* of the form "If input then attractor or fixed point." The rules are stored in the cube itself.

In contrast an artificial intelligence (AI) expert system [27] models a system as a binary rule tree with graph search. Each input fires one rule or a few rules and the search spreads down the tree branch to a leaf or leaves. The lack of feedback loops allows the tree search. But each serial inference uses only a small part of the stored knowledge. Each FCM input fires all the rules to some degree. The causal "juice" swirls through the tangles of fuzzy feedback and equilibrates in a global system response. In this way FCMs model the "circular causality" [19] of real and virtual worlds.

15.2 FUZZY COGNITIVE MAPS

Fuzzy cognitive maps (FCMs) are fuzzy signed digraphs with feedback [12, 13]. Nodes stand for fuzzy sets or events that occur to some degree. The nodes are causal concepts. They can model events, actions, values, goals, or lumped-parameter processes.

Directed edges stand for fuzzy rules or the partial causal flow between the concepts. The sign (+ or −) of an edge stands for causal increase or decrease. The positive edge rule

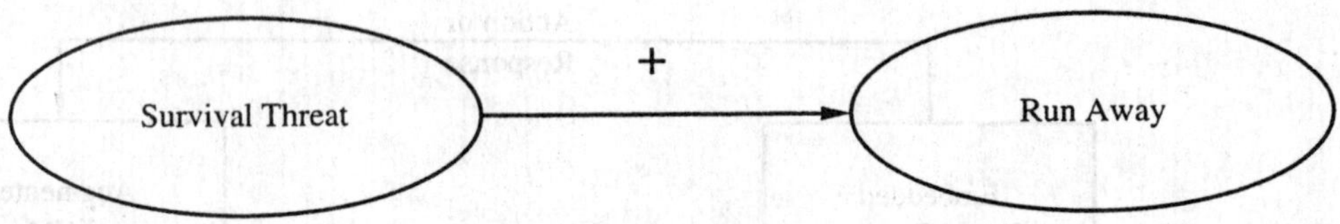

states that a survival threat increases runaway. It is a positive causal connection. The runaway response grows or falls as the threat grows or falls. The negative edge rule

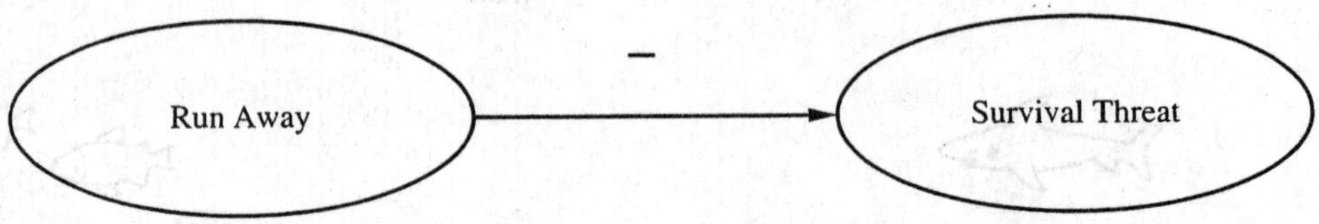

states that running away from a predator decreases the survival threat. It is a negative causal connection. The survival threat grows the less the prey runs away and falls the more the prey runs away. The two rules define a minimal feedback loop in the FCM causal web:

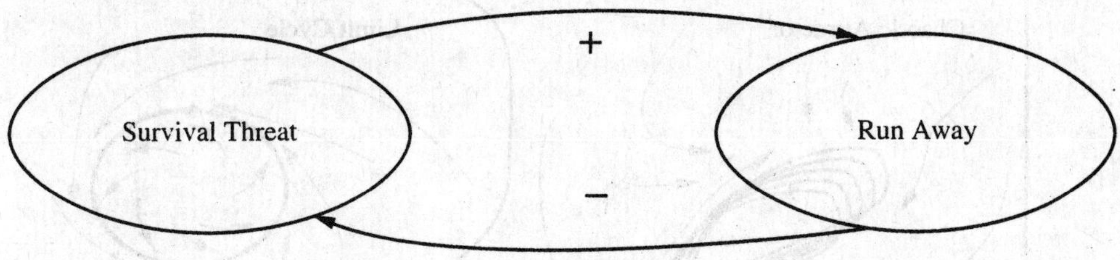

We can replace all negative rules with a positive rule if we double the number of concept nodes to add disconcepts [12] like SURVIVAL DIS-THREAT:

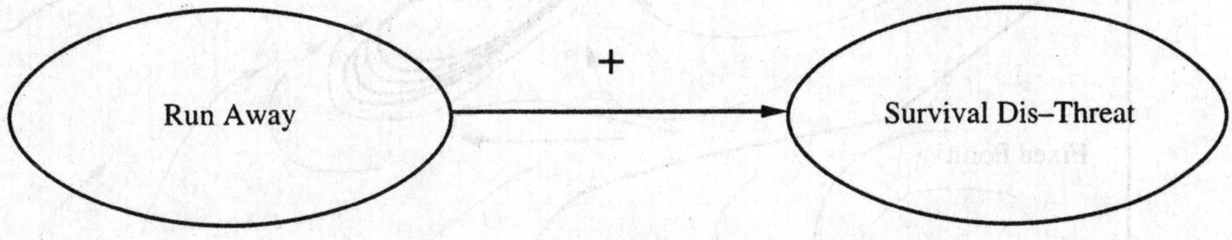

Disconcepts keep the FCM edge math in the standard fuzzy framework of concept values in the unit interval [0, 1]. We shall work with edge values in [−1, 1].

An FCM with n nodes has n^2 edges. The nodes $C_i(t)$ are fuzzy sets and so take values in [0, 1]. So an FCM state is the *fit* (fuzzy unit) vector $\mathbf{C}(t) = (C_1(t), \ldots, C_n(t))$ and thus a point in the fuzzy hypercube $I^n = [0, 1]^n$. An FCM inference is a path or point sequence in I^n. It is a fuzzy process or indexed family of fuzzy sets $\mathbf{C}(t)$. The FCM can only "forward chain" [27] to answer what-if questions. Nonlinearities do not permit reverse causality. FCMs cannot "backward chain" to answer why questions.

The FCM nonlinear dynamical system acts as a neural network. For each input state $\mathbf{C}(0)$ it digs a trajectory in I^n that ends in an equilibrium attractor **A**. The FCM quickly converges or "settles down" to a fixed point, limit cycle, limit torus, or chaotic attractor in the fuzzy cube. Figure 15.2 shows three attractors or meta-rules for a low-D dynamical FCM. The figure is only for illustration purposes.

The output equilibrium is the answer to a causal what-if question: What if $\mathbf{C}(0)$ happens? In this sense each FCM stores a set of global rules of the form "If $\mathbf{C}(0)$ then equilibrium attractor **A**."

The size of the attractor regions in the fuzzy cube governs the number of these global rules or "hidden patterns" [13]. All points in the attractor region map to the attractor. An FCM with a global fixed point has only one global rule. All input balls "roll" down its "well." FCMs can have large and small attractor regions in the fuzzy cube. The attractor types can vary in complex FCMs with highly nonlinear concepts and edges. Then one input state may lead to chaos and a more distant input state may end in a fixed point or limit cycle.

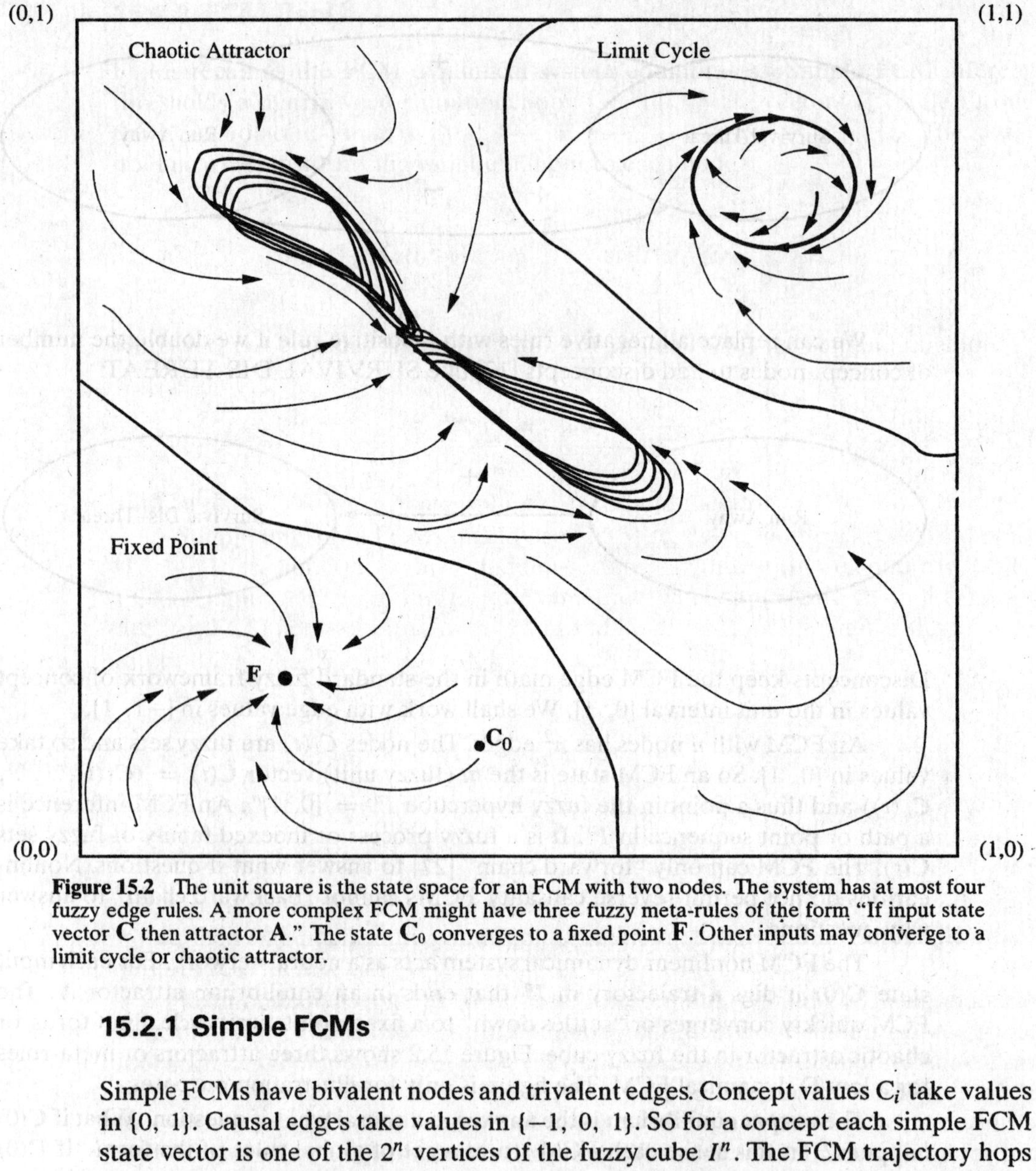

Figure 15.2 The unit square is the state space for an FCM with two nodes. The system has at most four fuzzy edge rules. A more complex FCM might have three fuzzy meta-rules of the form "If input state vector **C** then attractor **A**." The state $\mathbf{C}_0$ converges to a fixed point **F**. Other inputs may converge to a limit cycle or chaotic attractor.

15.2.1 Simple FCMs

Simple FCMs have bivalent nodes and trivalent edges. Concept values C_i take values in $\{0, 1\}$. Causal edges take values in $\{-1, 0, 1\}$. So for a concept each simple FCM state vector is one of the 2^n vertices of the fuzzy cube I^n. The FCM trajectory hops from vertex to vertex. I_t ends in a fixed point or limit cycle at the first repeated vector.

We can draw simple FCMs from articles, editorials, or surveys. Most persons can state the sign of causal flow between nodes. The hard part is to state its degree or magnitude. We can average expert responses [13, 24] as in equation (15.3) below or use neural systems to learn fuzzy edge weights from data. The expert responses can initialize the causal learning or modify it as a type of forcing function.

Figure 15.3 shows a simple FCM with five concept nodes. The connection or edge matrix **E** lists the causal links between nodes:

E =

	C_1	C_2	C_3	C_4	C_5
C_1	0	1	0	−1	0
C_2	0	0	1	0	−1
C_3	0	−1	0	1	−1
C_4	1	0	−1	0	1
C_5	−1	1	0	−1	0

The ith row lists the connection strength of the edges e_{ik} directed out from causal concept C_i. The ith column lists the edges e_{ki} directed into C_i. C_i causally increases C_k if $e_{ik} > 0$, decreases C_k if $e_{ik} < 0$, and has no effect if $e_{ik} = 0$. The causal concept C_4 causally increases concepts C_1 and C_5. It decreases C_3. Concepts C_1 and C_5 decrease C_4. Concept C_3 increases C_4.

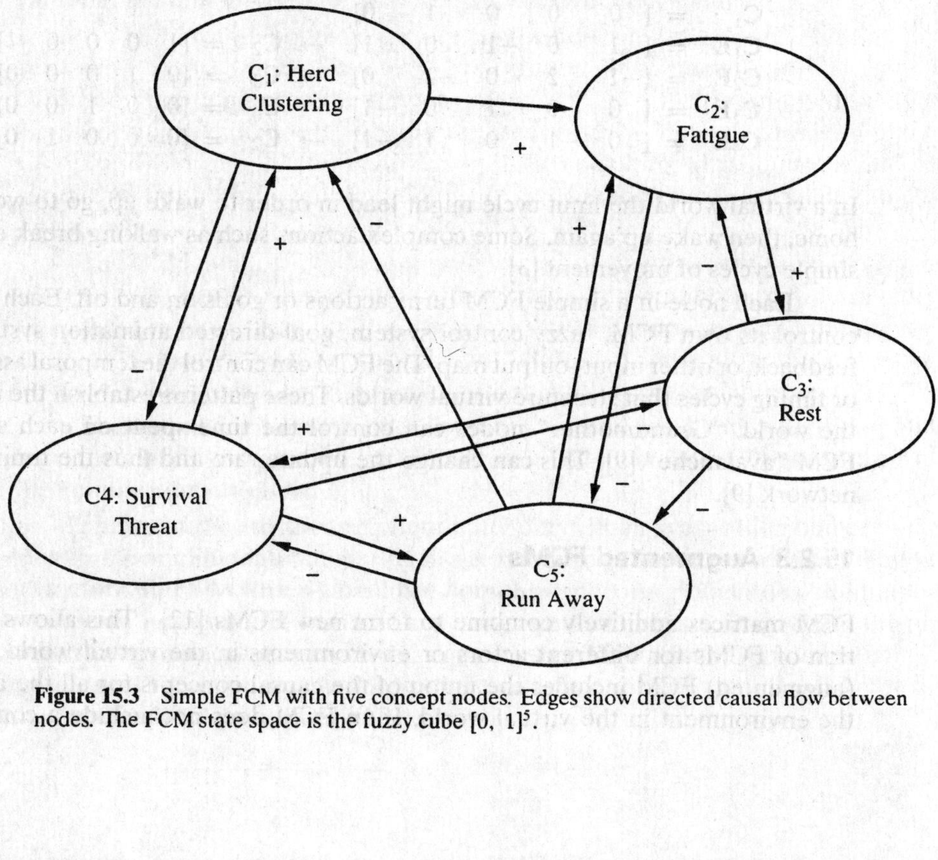

Figure 15.3 Simple FCM with five concept nodes. Edges show directed causal flow between nodes. The FCM state space is the fuzzy cube $[0, 1]^5$.

15.2.2 FCM Recall

FCMs recall as the FCM dynamical system equilibrates. Simple FCM inference thresholds a matrix-vector multiplication [13, 14]. State vectors $\mathbf{C}_n$ cycle through the FCM adjacency matrix $\mathbf{E}$: $\mathbf{C}_1 \rightarrow \mathbf{E} \rightarrow \mathbf{C}_2 \rightarrow \mathbf{E} \rightarrow \mathbf{C}_3 \ldots$. The system nonlinearly transforms the weighted input to each node C_i:

$$C_i(t_{n+1}) = S\left[\sum_{k=1}^{N} e_{ki}(t_n)\, C_k(t_n)\right]. \tag{15.1}$$

Here $S(x)$ is a bounded signal function. For simple FCMs the sigmoid function

$$S(y) = \frac{1}{1 + e^{-c(y-T)}} \tag{15.2}$$

with large $c > 0$ approximates a binary threshold function.

Simple threshold FCMs quickly converge to stable limit cycles or fixed points [13, 14]. These limit cycles show "hidden patterns" in the causal web of the FCM. The FCM in Figure 15.3 gives a three-step limit cycle when input state $\mathbf{C}_1 = [0\ 0\ 0\ 1\ 0]$ fires the FCM network. Equation (15.1) and binary thresholding gives the four-step limit cycle $\mathbf{C}_1 \rightarrow \mathbf{C}_2 \rightarrow \mathbf{C}_3 \rightarrow \mathbf{C}_4 \rightarrow \mathbf{C}_1$.

$$\begin{array}{lllll}
\mathbf{C}_1 & = [\ 0\ \ 0\ \ 0\ \ 1\ \ 0] & & & \\
\mathbf{C}_1\mathbf{E} & = [\ 1\ \ 0\ \ -1\ \ 0\ \ 1] & \rightarrow & \mathbf{C}_2 & = [1\ \ 0\ \ 0\ \ 0\ \ 1] \\
\mathbf{C}_2\mathbf{E} & = [-1\ \ 2\ \ 0\ \ -2\ \ 0] & \rightarrow & \mathbf{C}_3 & = [0\ \ 1\ \ 0\ \ 0\ \ 0] \\
\mathbf{C}_3\mathbf{E} & = [\ 0\ \ 0\ \ 1\ \ 0\ \ -1] & \rightarrow & \mathbf{C}_4 & = [0\ \ 0\ \ 1\ \ 0\ \ 0] \\
\mathbf{C}_4\mathbf{E} & = [\ 0\ \ -1\ \ 0\ \ 1\ \ -1] & \rightarrow & \mathbf{C}_2 & = [0\ \ 0\ \ 0\ \ 1\ \ 0].
\end{array}$$

In a virtual world the limit cycle might lead in order to wake up, go to work, come home, then wake up again. Some complex actions such as walking break down into simple cycles of movement [3].

Each node in a simple FCM turns actions or goals on and off. Each node can control its own FCM, fuzzy control system, goal-directed animation system, force feedback, or other input–output map. The FCM can control the temporal associations or timing cycles that structure virtual worlds. These patterns establish the rhythm of the world. "Grandmother" nodes can control the time spent on each step in an FCM "avalanche" [9]. This can change the update rate and thus the timing for the network [9].

15.2.3 Augmented FCMs

FCM matrices additively combine to form new FCMs [12]. This allows combination of FCMs for different actors or environments in the virtual world. The new (augmented) FCM includes the union of the causal concepts for all the actors and the environment in the virtual world. If an FCM does not include a concept then those rows and columns are all zero. The sum of the augmented (zero-padded) FCM matrices for each actor forms the virtual world:

$$\mathbf{F} = \sum_{i=1}^{n} w_i \mathbf{F}_i. \tag{15.3}$$

The w_i are positive weights for the ith FCM $\mathbf{F}_i$. The weights state the relative value of each FCM in the virtual world and can weight any subgraph of the FCM. Figure 15.4a shows three simple FCMs. Equation (15.3) combines these FCMs to give the new simple FCM in Figure 15.4b that has fuzzy or multivalued edges:

$$\mathbf{F} = \frac{1}{3}(\mathbf{F}_1 + \mathbf{F}_2 + \mathbf{F}_3) = \frac{1}{3} \cdot \begin{bmatrix} 0 & 2 & -1 & 0 & 0 & -1 \\ 0 & 0 & 2 & 3 & 1 & 0 \\ 0 & 0 & 0 & 2 & 1 & 0 \\ 2 & 0 & 0 & 0 & -1 & 0 \\ 0 & -2 & 0 & 0 & 0 & 0 \\ 1 & -1 & 1 & 0 & 0 & 0 \end{bmatrix} \tag{15.4}$$

The FCM sum [(15.3)] helps knowledge acquisition. Any number of experts can describe their FCM virtual world views and (15.3) will weight and combine them. In contrast an AI expert system [27] is a binary tree with graph search. Two or more trees need not combine to a tree. Combined FCMs tend to have feedback or closed loops and that precludes graph search with forward or backward "chaining." The strong law of large numbers [13] ensures that the knowledge estimate $\mathbf{F}$ in (15.3) improves with the expert sample size n if we view the experts as independent (unique) random knowledge sources with finite variance (bounded uncertainty) and identical distribution (same problem-domain focus). The sample FCM converges to the unknown population FCM as the number of experts grows.

15.2.4 Nested FCMs

FCMs can endow virtual worlds with goals and intentions as they define dynamic physical and social environments. This can give the "common representation" needed for a virtual world [2]. The FCM can combine simple actions to model "intelligent" behavior [3, 5]. Each node in turn can control its own simple FCM in a *nested* FCM. Complex actions such as walking emerge from networks of simple reflexes. Nested simple FCMs can mimic this process as a net of finite state machines with binary limit cycles.

The output of a simple FCM is a binary limit cycle that describes actions or goals. This holds even if the binary concept nodes change state asynchronously. Each output turns a function on or off as in a robotic neural net [3]. This output can control smaller FCMs or fuzzy control systems. These systems can drive visual, auditory, or tactile outputs of the virtual world. The FCM can control the temporal associations or timing cycles that structure virtual worlds. The FCM state vector drives the motion of each character as in a frame in a cartoon. Simple equations of motion can move each actor between the states.

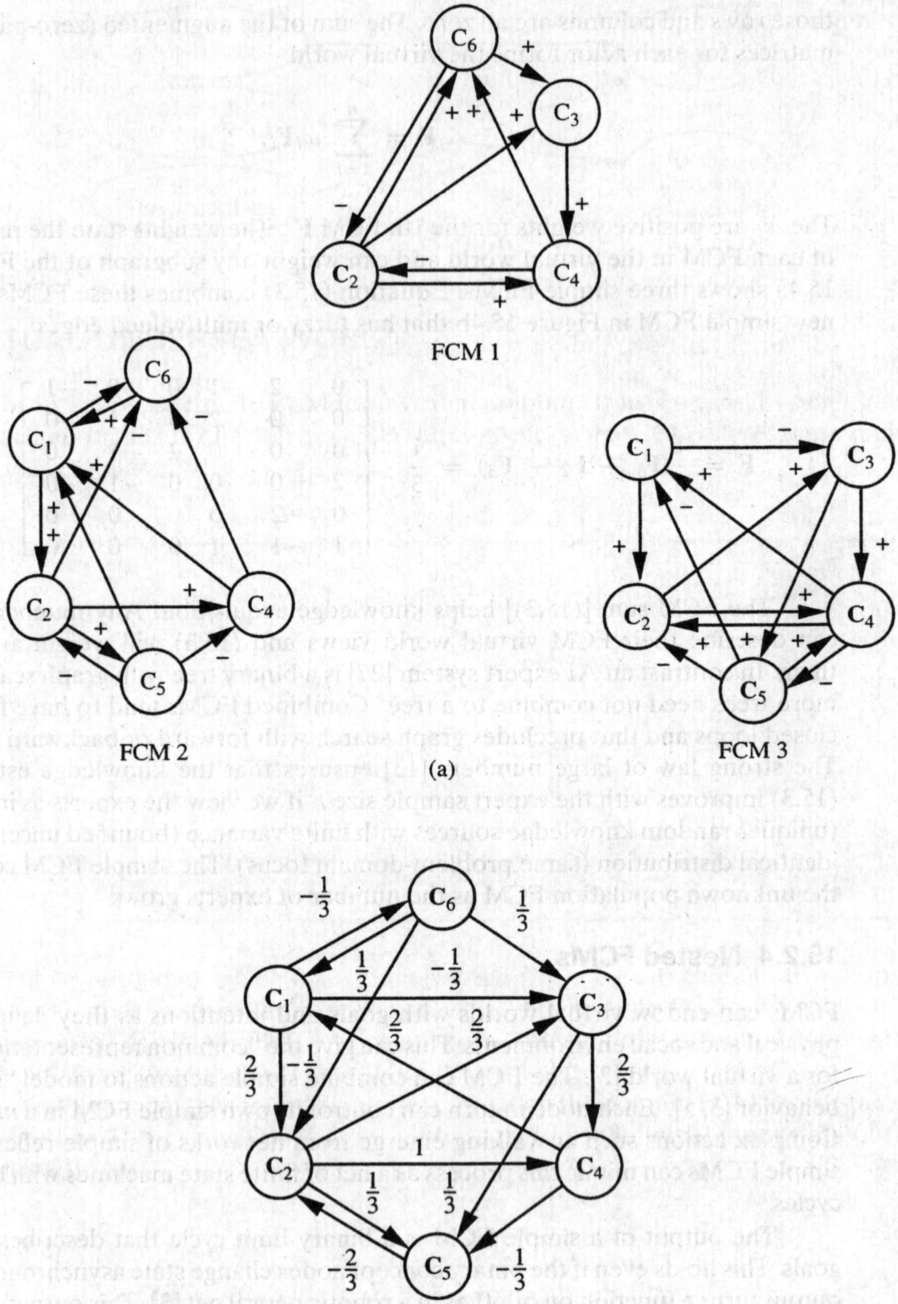

Figure 15.4 FMCs combine additively. (a) Three bivalent FMCs. (b) Augmented FCM. The augmented FCM takes the union of the causal concepts of the smaller FCMs and sums the augmented connection matrices: $\mathbf{F} = \frac{1}{3}(\mathbf{F}_1 + \mathbf{F}_2 + \mathbf{F}_3)$.

FCM nesting extends to any number of fuzzy sets for the inputs. A concept can divide into smaller fuzzy sets or subconcepts. The edges or rules link the sets. This leads to a discrete multivalued output for each node. Enough nodes allow this feed-forward system to approximate any continuous function [10] for signal functions of the form of (15.2). The subconcepts Q_{ij} partition the fuzzy concept $\mathbf{C}_j$

$$\mathbf{C}_j = \bigcup_{i=1}^{N_j} Q_{ij}. \tag{15.5}$$

Figure 15.5 shows the concept of a SURVIVAL THREAT divided into subconcepts. Each subconcept is the degree of threat.

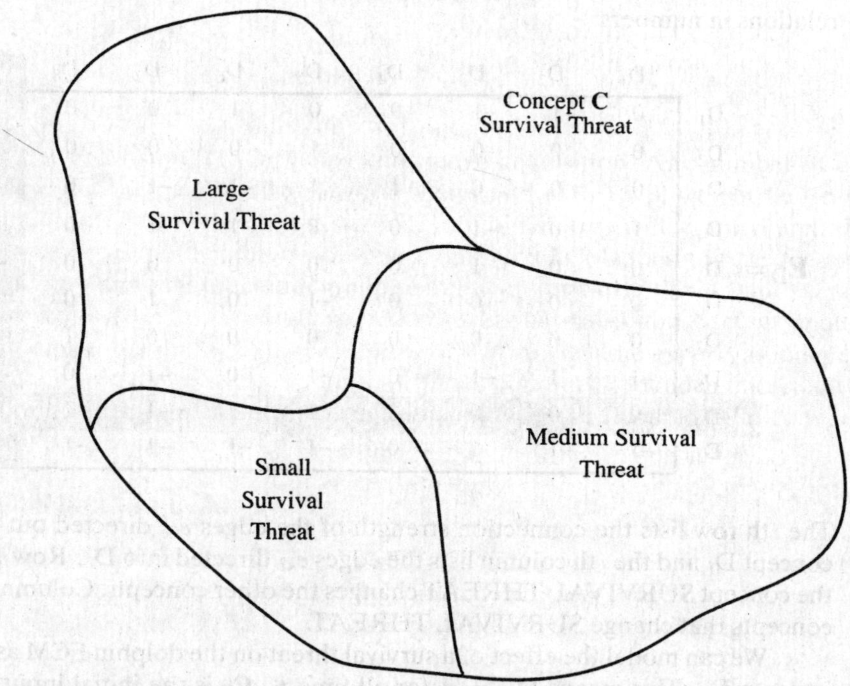

Figure 15.5 Nested FCMs can divide a concept into subconcepts.

The FCM edges or rules map one subconcept to another. These subconcept mappings form a fuzzy system or set of fuzzy if-then rules that map inputs to outputs. Each mapping is a fuzzy rule or state-space patch that links fuzzy sets. The patches cover the graph of some function in the input–output state space. The fuzzy system then averages the patches that overlap to give an approximation of a continuous function [15]. Figure 15.6 shows how subconcepts can map to different responses in the FCM. This gives a more varied response to changes in the virtual world.

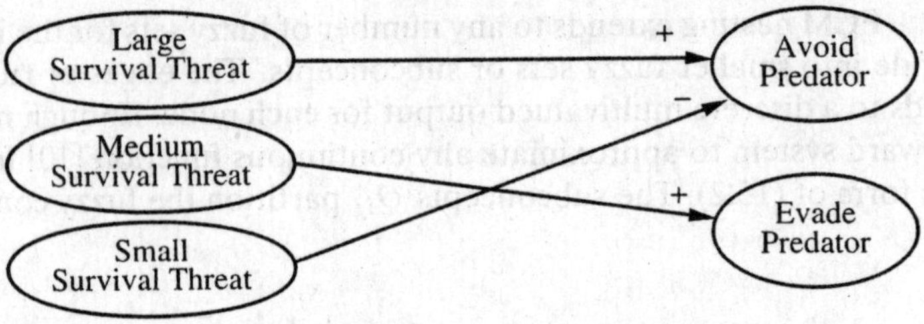

Figure 15.6 Subconcepts map to other concepts. This gives a more varied response.

15.3 VIRTUAL UNDERSEA WORLD

Figure 15.7 shows a simple FCM for a virtual dolphin. It lists a causal web of goals and actions in the life of a dolphin [22]. The connection matrix $\mathbf{E_D}$ states these causal relations in numbers:

$\mathbf{E_D} =$

	D_1	D_2	D_3	D_4	D_5	D_6	D_7	D_8	D_9	D_{10}
D_1	0	−1	−1	0	0	1	0	0	0	0
D_2	0	0	0	0	1	0	0	0	0	0
D_3	0	0	0	1	1	−1	−1	0	0	−1
D_4	1	0	−1	0	0	−1	−1	0	0	−1
D_5	0	0	1	0	0	0	0	0	−1	0
D_6	0	0	0	0	−1	0	1	0	0	0
D_7	0	0	0	0	0	0	0	1	0	0
D_8	−1	1	−1	0	1	0	−1	0	0	0
D_9	0	0	0	0	1	−1	−1	−1	0	1
D_{10}	−1	−1	1	0	−1	−1	−1	−1	−1	0

The ith row lists the connection strength of the edges e_{ik} directed out from causal concept D_i and the ith column lists the edges e_{ki} directed into D_i. Row 9 shows how the concept SURVIVAL THREAT changes the other concepts. Column 9 shows the concepts that change SURVIVAL THREAT.

We can model the effect of a survival threat on the dolphin FCM as a sustained input to D_9. This means $D_9 = 1$ for all time t_k. $\mathbf{C_0}$ is the initial input state of the dolphin FCM:

$$\mathbf{C_0} = [0 \ 0 \ 0 \ 0 \ 0 \ 0 \ 0 \ 0 \ 1 \ 0] \, .$$

Then

$$\mathbf{C_0 E_D} = [\,0 \ 0 \ 0 \ 0 \ 1 \ -1 \ -1 \ -1 \ 0 \ 1\,]$$
$$\rightarrow \mathbf{C_1} = [\,0 \ 0 \ 0 \ 0 \ 1 \ 0 \ 0 \ 0 \ 1 \ 1\,] \, .$$

The arrow stands for a threshold operation with $1/2$ as the threshold value. $\mathbf{C_1}$ keeps D_9 on since we want to study the effect of a sustained threat. $\mathbf{C_1}$ shows that when threatened the dolphins cluster in a herd and flee the threat. The negative rules in

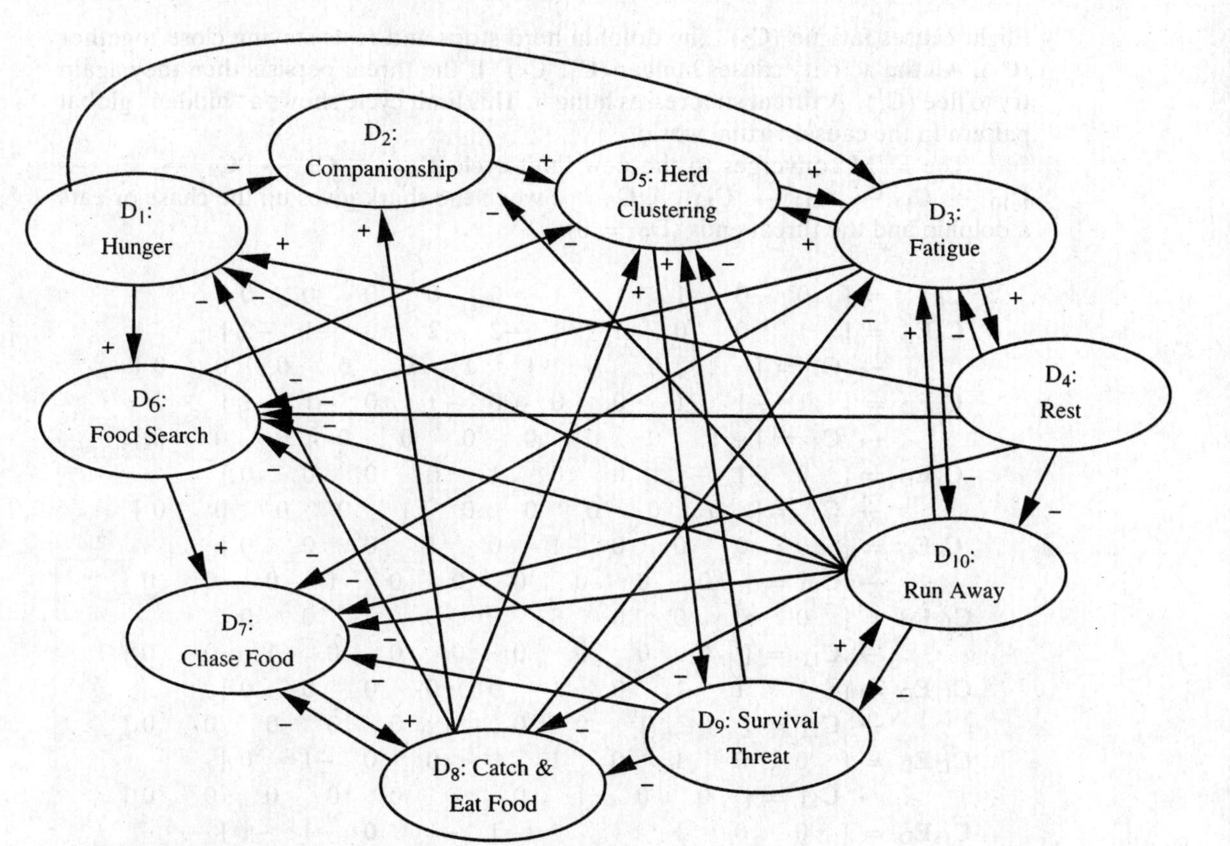

Figure 15.7 Trivalent fuzzy cognitive map for the control of a dolphin actor in a fuzzy virtual world. The rules or edges connect causal concepts in a signed connection matrix.

the ninth row of $\mathbf{E}_D$ show that a threat to survival turns off other actions. The FCM converges to the limit cycle $\mathbf{C}_1 \rightarrow \mathbf{C}_2 \rightarrow \mathbf{C}_3 \rightarrow \mathbf{C}_4 \rightarrow \mathbf{C}_5 \rightarrow \mathbf{C}_1 \ldots$ if the threat lasts:

$$
\begin{aligned}
\mathbf{C}_1\mathbf{E}_D &= [\,-1 \;\; -1 \;\; 2 \;\; 0 \;\; 0 \;\; -2 \;\; -2 \;\; -2 \;\; -2 \;\; 1\,] \\
&\rightarrow \mathbf{C}_2 = [\,0 \;\; 0 \;\; 1 \;\; 0 \;\; 0 \;\; 0 \;\; 0 \;\; 0 \;\; 1 \;\; 1\,] \\
\mathbf{C}_2\mathbf{E}_D &= [\,-1 \;\; -1 \;\; 1 \;\; 1 \;\; 1 \;\; -3 \;\; -3 \;\; -2 \;\; 1 \;\; 0\,] \\
&\rightarrow \mathbf{C}_3 = [\,0 \;\; 0 \;\; 1 \;\; 1 \;\; 1 \;\; 0 \;\; 0 \;\; 0 \;\; 1 \;\; 0\,] \\
\mathbf{C}_3\mathbf{E}_D &= [\,1 \;\; 0 \;\; 0 \;\; 1 \;\; 2 \;\; -3 \;\; -3 \;\; -1 \;\; -1 \;\; -1\,] \\
&\rightarrow \mathbf{C}_4 = [\,1 \;\; 0 \;\; 0 \;\; 1 \;\; 1 \;\; 0 \;\; 0 \;\; 0 \;\; 1 \;\; 0\,] \\
\mathbf{C}_4\mathbf{E}_D &= [\,1 \;\; -1 \;\; -1 \;\; 0 \;\; 1 \;\; -1 \;\; -2 \;\; -1 \;\; -1 \;\; 0\,] \\
&\rightarrow \mathbf{C}_5 = [\,1 \;\; 0 \;\; 0 \;\; 0 \;\; 1 \;\; 0 \;\; 0 \;\; 0 \;\; 1 \;\; 0\,] \\
\mathbf{C}_5\mathbf{E}_D &= [\,0 \;\; -1 \;\; 0 \;\; 0 \;\; 1 \;\; 0 \;\; -1 \;\; -1 \;\; -1 \;\; 1\,] \\
&\rightarrow \mathbf{C}_1 = [\,0 \;\; 0 \;\; 0 \;\; 0 \;\; 1 \;\; 0 \;\; 0 \;\; 0 \;\; 1 \;\; 1\,]\,.
\end{aligned}
$$

Flight causes fatigue ($\mathbf{C}_2$). The dolphin herd stops and rests staying close together ($\mathbf{C}_3$). All the activity causes hunger ($\mathbf{C}_4$, $\mathbf{C}_5$). If the threat persists then they again try to flee ($\mathbf{C}_1$). A threat suppresses hunger. This limit cycle shows a "hidden" global pattern in the causal virtual world.

The FCM converges to the new limit cycle $\mathbf{C}_6 \rightarrow \mathbf{C}_7 \rightarrow \mathbf{C}_8 \rightarrow \mathbf{C}_9 \rightarrow \mathbf{C}_{10} \rightarrow \mathbf{C}_{11} \rightarrow \mathbf{C}_{12} \rightarrow \mathbf{C}_{13} \rightarrow \mathbf{C}_6 \ldots$ when the shark gives up the chase or eats a dolphin and the threat ends ($D_9 = 0$):

$$
\begin{array}{rl}
\mathbf{C}_6 & = [\ 0\ \ 0\ \ 1\ \ 1\ \ 1\ \ 0\ \ 0\ \ 0\ \ 0\ \ 0\ ] \\
\mathbf{C}_6\mathbf{E}_D & = [\ 1\ \ 0\ \ 0\ \ 1\ \ 1\ \ -2\ \ -2\ \ 0\ \ -1\ \ -2\ ] \\
\rightarrow \mathbf{C}_7 & = [\ 1\ \ 0\ \ 0\ \ 1\ \ 1\ \ 0\ \ 0\ \ 0\ \ 0\ \ 0\ ] \\
\mathbf{C}_7\mathbf{E}_D & = [\ 1\ \ -1\ \ -1\ \ 0\ \ 0\ \ 0\ \ -1\ \ 0\ \ -1\ \ -1\ ] \\
\rightarrow \mathbf{C}_8 & = [\ 1\ \ 0\ \ 0\ \ 0\ \ 0\ \ 0\ \ 0\ \ 0\ \ 0\ \ 0\ ] \\
\mathbf{C}_8\mathbf{E}_D & = [\ 0\ \ -1\ \ -1\ \ 0\ \ 0\ \ 1\ \ 0\ \ 0\ \ 0\ \ 0\ ] \\
\rightarrow \mathbf{C}_9 & = [\ 0\ \ 0\ \ 0\ \ 0\ \ 0\ \ 1\ \ 0\ \ 0\ \ 0\ \ 0\ ] \\
\mathbf{C}_9\mathbf{E}_D & = [\ 0\ \ 0\ \ 0\ \ 0\ \ -1\ \ 0\ \ 1\ \ 0\ \ 0\ \ 0\ ] \\
\rightarrow \mathbf{C}_{10} & = [\ 0\ \ 0\ \ 0\ \ 0\ \ 0\ \ 0\ \ 1\ \ 0\ \ 0\ \ 0\ ] \\
\mathbf{C}_{10}\mathbf{E}_D & = [\ 0\ \ 0\ \ 0\ \ 0\ \ 0\ \ 0\ \ 0\ \ 1\ \ 0\ \ 0\ ] \\
\rightarrow \mathbf{C}_{11} & = [\ 0\ \ 0\ \ 0\ \ 0\ \ 0\ \ 0\ \ 0\ \ 1\ \ 0\ \ 0\ ] \\
\mathbf{C}_{11}\mathbf{E}_D & = [\ -1\ \ 1\ \ -1\ \ 0\ \ 1\ \ 0\ \ 0\ \ 0\ \ 0\ \ 0\ ] \\
\rightarrow \mathbf{C}_{12} & = [\ 0\ \ 1\ \ 0\ \ 0\ \ 1\ \ 0\ \ 0\ \ 0\ \ 0\ \ 0\ ] \\
\mathbf{C}_{12}\mathbf{E}_D & = [\ 0\ \ 0\ \ 1\ \ 0\ \ 1\ \ 0\ \ 0\ \ 0\ \ -1\ \ 0\ ] \\
\rightarrow \mathbf{C}_{13} & = [\ 0\ \ 0\ \ 1\ \ 0\ \ 1\ \ 0\ \ 0\ \ 0\ \ 0\ \ 0\ ] \\
\mathbf{C}_{13}\mathbf{E}_D & = [\ 0\ \ 0\ \ 1\ \ 1\ \ 1\ \ -1\ \ -1\ \ 0\ \ -1\ \ -1\ ] \\
\rightarrow \mathbf{C}_6 & = [\ 0\ \ 0\ \ 1\ \ 1\ \ 1\ \ 0\ \ 0\ \ 0\ \ 0\ \ 0\ ].
\end{array}
$$

The dolphin herd rests from the previous chase ($\mathbf{C}_6$, $\mathbf{C}_7$). Then they begin a hunt of their own ($\mathbf{C}_9$, $\mathbf{C}_{10}$). They eat ($\mathbf{C}_{11}$) and then they socialize and rest ($\mathbf{C}_{12}$, $\mathbf{C}_{13}$, $\mathbf{C}_6$). This makes them hungry and the feeding cycle repeats.

15.3.1 Augmented Virtual World

Figure 15.8 shows an augmented FCM for an undersea virtual world. It combines fish school, shark, and dolphin herd FCMs with (15.3): $\mathbf{F} = \mathbf{F}_{\text{fish}} + \mathbf{F}_{\text{shark}} + \mathbf{F}_{\text{dolphin}}$. The new links among these FCMs are those of predator and prey where the larger eats the smaller. The actors chase, flee, and eat one another. A hungry shark chases the dolphins and that leads to the limit cycle ($\mathbf{C}_1$, $\mathbf{C}_2$, $\mathbf{C}_3$, $\mathbf{C}_4$). Augmenting the FCM matrices gives a large but sparse FCM since the actors respond to each other in few ways. Figure 15.9 shows the connection matrix for the augmented FCM in Figure 15.8.

The augmented FCM moves the actors in the virtual world. The binary output states of this FCM move the actors. Each FCM state maps to equations or function approximations for movement.

We used a simple update equation for position:

$$\mathbf{p}(t_n + 1) = \mathbf{p}(t_n) + \Delta t \mathbf{v}(t_n). \tag{15.6}$$

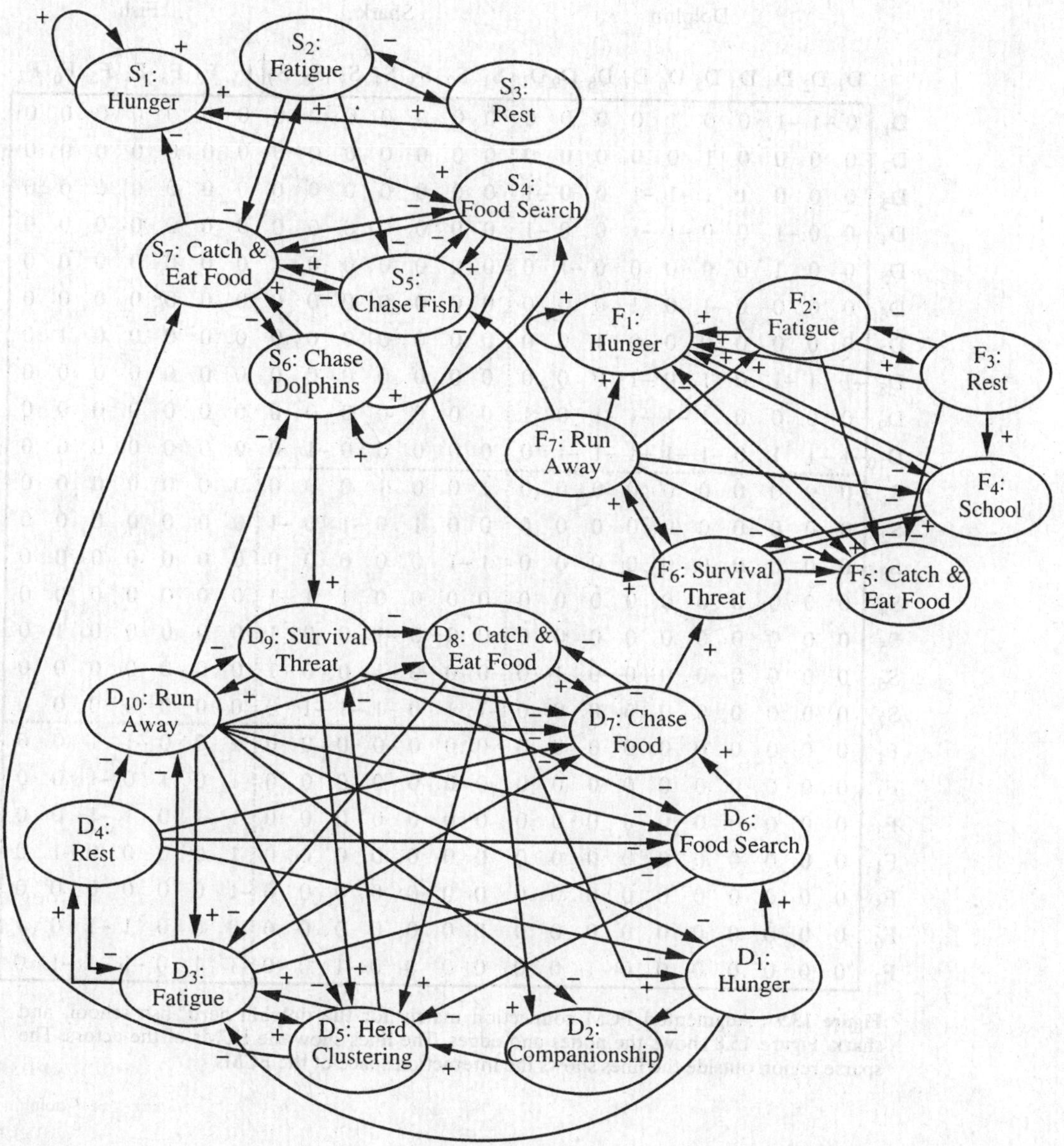

Figure 15.8 Augmented FCM for different actors in a virtual world. The actors interact through linked common causal concepts such as chasing food and avoiding a threat.

The velocity $\mathbf{v}(t)$ does not change at time step Δt. The FCM finds the direction and magnitude of movement. The magnitude of the velocity depends on the FCM state. If the FCM state is "run away" then the velocity is FAST. If the FCM state is "rest" then the velocity is SLOW. The prey choose the direction that maximizes the distance

	Dolphin										Shark							Fish						
	D_1	D_2	D_3	D_4	D_5	D_6	D_7	D_8	D_9	D_{10}	S_1	S_2	S_3	S_4	S_7	S_6	S_7	F_1	F_2	F_3	F_4	F_5	F_6	F_7
D_1	0	−1	−1	0	0	1	0	0	0	0	0	0	0	0	0	0	0	0	0	0	0	0	0	0
D_2	0	0	0	0	1	0	0	0	0	0	0	0	0	0	0	0	0	0	0	0	0	0	0	0
D_3	0	0	0	1	1	−1	−1	0	0	−1	0	0	0	0	0	0	0	0	0	0	0	0	0	0
D_4	0	0	−1	0	0	−1	−1	0	0	−1	0	0	0	0	0	0	0	0	0	0	0	0	0	0
D_5	0	0	1	0	0	0	0	0	−1	0	0	0	0	0	0	0	−1	0	0	0	0	0	0	0
D_6	0	0	0	0	−1	0	1	0	0	0	0	0	0	0	0	0	0	0	0	0	0	0	0	0
D_7	0	0	0	0	0	0	0	1	0	0	0	0	0	0	0	0	0	0	0	0	0	0	1	0
D_8	−1	1	−1	0	1	0	−1	0	0	0	0	0	0	0	0	0	0	0	0	0	0	0	0	0
D_9	0	0	0	0	1	−1	−1	−1	0	1	0	0	0	0	0	0	0	0	0	0	0	0	0	0
D_{10}	−1	−1	1	0	−1	−1	−1	−1	−1	0	0	0	0	0	0	−1	0	0	0	0	0	0	0	0
S_1	0	0	0	0	0	0	0	0	0	0	1	0	0	1	0	0	0	0	0	0	0	0	0	0
S_2	0	0	0	0	0	0	0	0	0	0	0	0	1	0	−1	0	−1	0	0	0	0	0	0	0
S_3	0	0	0	0	0	0	0	0	0	0	1	−1	0	0	0	0	0	0	0	0	0	0	0	0
S_4	0	0	0	0	0	0	0	0	0	0	0	0	0	0	1	1	−1	0	0	0	0	0	0	0
S_5	0	0	0	0	0	0	0	0	0	0	0	0	0	−1	0	0	1	0	0	0	0	0	1	0
S_6	0	0	0	0	0	0	0	0	1	0	0	0	0	−1	0	0	1	0	0	0	0	0	0	0
S_7	0	0	0	0	0	0	0	0	0	0	−1	1	0	−1	−1	−1	0	0	0	0	0	0	0	0
F_1	0	0	0	0	0	0	0	0	0	0	0	0	0	0	0	0	0	1	0	0	−1	1	0	0
F_2	0	0	0	0	0	0	0	0	0	0	0	0	0	0	0	0	0	1	0	1	0	−1	0	0
F_3	0	0	0	0	0	0	0	0	0	0	0	0	0	0	0	0	0	1	−1	0	1	−1	0	0
F_4	0	0	0	0	0	0	0	0	0	0	0	0	0	0	0	0	0	1	0	0	0	0	−1	0
F_5	0	0	0	0	0	0	0	0	0	0	0	0	0	0	0	0	0	−1	0	0	0	0	0	0
F_6	0	0	0	0	0	0	0	0	0	0	0	0	0	0	0	0	0	0	0	0	1	−1	0	1
F_7	0	0	0	0	0	0	0	−1	0	0	0	0	0	0	−1	0	0	1	1	0	−1	−1	−1	0

Figure 15.9 Augmented FCM connection matrix for the dolphin herd, fish school, and shark. Figure 15.8 shows the nodes and edges. The lines show the FCMs of the actors. The sparse region outside the lines shows the interaction space of the FCMs.

from the predator. The predator chases the prey. When a predator searches for food it swims at random [11]. Each state moves the actors through the sea.

The FCM in Figure 15.9 encodes limit cycles between the actors. For example, if we start with a hungry shark and we set the causal link between concept S4: FOOD SEARCH and S6: CHASE DOLPHINS equal to zero to look at shark interactions with the fish school then the first state $\mathbf{C}_1$ is

$$\begin{array}{rcl} \mathbf{C}_1 & = & [0\ 0\ 0\ 0\ 0\ 0\ 0\ 0\ 0\ 0\ 1\ 0 \\ & & \ 0\ 0\ 0\ 0\ 0\ 0\ 0\ 0\ 0\ 0\ 0\ 0]. \end{array}$$

This vector gives a 7-step limit cycle after four transition steps:

$$
\begin{aligned}
\mathbf{C}_1\mathbf{E}_A &= [\,0\ 0\ 0\ 0\ 0\ 0\ 0\ 0\ 0\ 0\ 1\ 0 \\
&\qquad 0\ 1\ 0\ 0\ 0\ 0\ 0\ 0\ 0\ 0\ 0\ 0\,] \\
\rightarrow \mathbf{C}_2 &= [\,0\ 0\ 0\ 0\ 0\ 0\ 0\ 0\ 0\ 0\ 1\ 0 \\
&\qquad 0\ 1\ 0\ 0\ 0\ 0\ 0\ 0\ 0\ 0\ 0\ 0\,], \\
\mathbf{C}_2\mathbf{E}_A &= [\,0\ 0\ 0\ 0\ 0\ 0\ 0\ 0\ 0\ 0\ 1\ 0 \\
&\qquad 1\ 0\ 1\ 0\ -1\ 0\ 0\ 0\ 0\ 0\ 0\ 0\,] \\
\rightarrow \mathbf{C}_3 &= [\,0\ 0\ 0\ 0\ 0\ 0\ 0\ 0\ 0\ 0\ 1\ 0 \\
&\qquad 0\ 1\ 1\ 0\ 0\ 0\ 0\ 0\ 0\ 0\ 0\ 0\,], \\
\mathbf{C}_3\mathbf{E}_A &= [\,0\ 0\ 0\ 0\ 0\ 0\ 0\ 0\ 0\ 0\ 1\ 0 \\
&\qquad 0\ 0\ 1\ 0\ 0\ 0\ 0\ 0\ 0\ 0\ 1\ 0\,] \\
\rightarrow \mathbf{C}_4 &= [\,0\ 0\ 0\ 0\ 0\ 0\ 0\ 0\ 0\ 0\ 1\ 0 \\
&\qquad 0\ 0\ 1\ 0\ 0\ 0\ 0\ 0\ 0\ 0\ 1\ 0\,], \\
\mathbf{C}_4\mathbf{E}_A &= [\,0\ 0\ 0\ 0\ 0\ 0\ 0\ 0\ 0\ 0\ 1\ 0 \\
&\qquad 0\ 0\ 0\ 0\ 1\ 0\ 0\ 0\ 1\ -1\ 1\ 1\,] \\
\rightarrow \mathbf{C}_5 &= [\,0\ 0\ 0\ 0\ 0\ 0\ 0\ 0\ 0\ 0\ 1\ 0 \\
&\qquad 0\ 0\ 0\ 0\ 1\ 0\ 0\ 0\ 1\ 0\ 1\ 1\,], \\
\mathbf{C}_5\mathbf{E}_A &= [\,0\ 0\ 0\ 0\ 0\ 0\ 0\ -1\ 0\ 0\ 0\ 1 \\
&\qquad 0\ 0\ -2\ -1\ 0\ 2\ 1\ 0\ 0\ -2\ -2\ 1\,] \\
\rightarrow \mathbf{C}_6 &= [\,0\ 0\ 0\ 0\ 0\ 0\ 0\ 0\ 0\ 0\ 0\ 1 \\
&\qquad 0\ 0\ 0\ 0\ 0\ 1\ 1\ 0\ 0\ 0\ 0\ 1\,], \\
\mathbf{C}_6\mathbf{E}_A &= [\,0\ 0\ 0\ 0\ 0\ 0\ 0\ -1\ 0\ 0\ 0\ 0 \\
&\qquad 1\ 0\ -2\ 0\ -1\ 3\ 1\ 1\ -2\ -1\ -1\ 0\,] \\
\rightarrow \mathbf{C}_7 &= [\,0\ 0\ 0\ 0\ 0\ 0\ 0\ 0\ 0\ 0\ 0\ 0 \\
&\qquad 1\ 0\ 0\ 0\ 0\ 1\ 1\ 1\ 0\ 0\ 0\ 0\,], \\
\mathbf{C}_7\mathbf{E}_A &= [\,0\ 0\ 0\ 0\ 0\ 0\ 0\ 0\ 0\ 0\ 1\ -1 \\
&\qquad 0\ 0\ 0\ 0\ 0\ 3\ -1\ 1\ 0\ -1\ 0\ 0\,] \\
\rightarrow \mathbf{C}_8 &= [\,0\ 0\ 0\ 0\ 0\ 0\ 0\ 0\ 0\ 0\ 1\ 0 \\
&\qquad 0\ 0\ 0\ 0\ 0\ 1\ 0\ 1\ 0\ 0\ 0\ 0\,], \\
\mathbf{C}_8\mathbf{E}_A &= [\,0\ 0\ 0\ 0\ 0\ 0\ 0\ 0\ 0\ 0\ 1\ 0 \\
&\qquad 0\ 1\ 0\ 0\ 0\ 2\ -1\ 0\ 0\ 0\ 0\ 0\,] \\
\rightarrow \mathbf{C}_9 &= [\,0\ 0\ 0\ 0\ 0\ 0\ 0\ 0\ 0\ 0\ 1\ 0 \\
&\qquad 0\ 1\ 0\ 0\ 0\ 1\ 0\ 0\ 0\ 0\ 0\ 0\,], \\
\mathbf{C}_9\mathbf{E}_A &= [\,0\ 0\ 0\ 0\ 0\ 0\ 0\ 0\ 0\ 0\ 1\ 0 \\
&\qquad 0\ 1\ 1\ 0\ -1\ 1\ 0\ 0\ -1\ 1\ 0\ 0\,] \\
\rightarrow \mathbf{C}_{10} &= [\,0\ 0\ 0\ 0\ 0\ 0\ 0\ 0\ 0\ 0\ 1\ 0 \\
&\qquad 0\ 1\ 1\ 0\ 0\ 1\ 0\ 0\ 0\ 1\ 0\ 0\,], \\
\mathbf{C}_1 0\mathbf{E}_A &= [\,0\ 0\ 0\ 0\ 0\ 0\ 0\ 0\ 0\ 0\ 1\ 0 \\
&\qquad 0\ 0\ 1\ 0\ 0\ 0\ 0\ 0\ -1\ 1\ 1\ 0\,] \\
\rightarrow \mathbf{C}_{11} &= [\,0\ 0\ 0\ 0\ 0\ 0\ 0\ 0\ 0\ 0\ 1\ 0 \\
&\qquad 0\ 0\ 1\ 0\ 0\ 0\ 0\ 0\ 0\ 1\ 1\ 0\,], \\
\mathbf{C}_1 1\mathbf{E}_A &= [\,0\ 0\ 0\ 0\ 0\ 0\ 0\ 0\ 0\ 0\ 1\ 0 \\
&\qquad 0\ 0\ 0\ 0\ 1\ -1\ 0\ 0\ 1\ -1\ 1\ 1\,] \\
\rightarrow \mathbf{C}_5 &= [\,0\ 0\ 0\ 0\ 0\ 0\ 0\ 0\ 0\ 0\ 1\ 0 \\
&\qquad 0\ 0\ 0\ 0\ 1\ 0\ 0\ 0\ 1\ 0\ 1\ 1\,],
\end{aligned}
$$

In this limit cycle a shark searches for food (C_1, C_2, C_3). The shark finds some fish (C_4), chases the fish (C_5), and then eats some of the fish (C_6). To avoid the shark most fish run away and then regroup as a school (C_5, C_6, C_7). Then the fish rest and eat while the shark rests (C_8, C_9). In time the shark gets hungry again and searches for fish (C_{10}, C_{11}).

The result is a complex dance among the actors as they move in a 2-D ocean. Figure 15.10 shows these movements. The forcing function is a hungry shark ($C_{11} = 1$). The shark encounters the dolphins who cluster and then flee the shark. The shark chases but cannot keep up. The shark still searches for food and finds the fish. It catches a fish and then rests with its hunger sated. Meanwhile the hungry dolphins search for food and eat more fish. Each actor responds to the actions of the other.

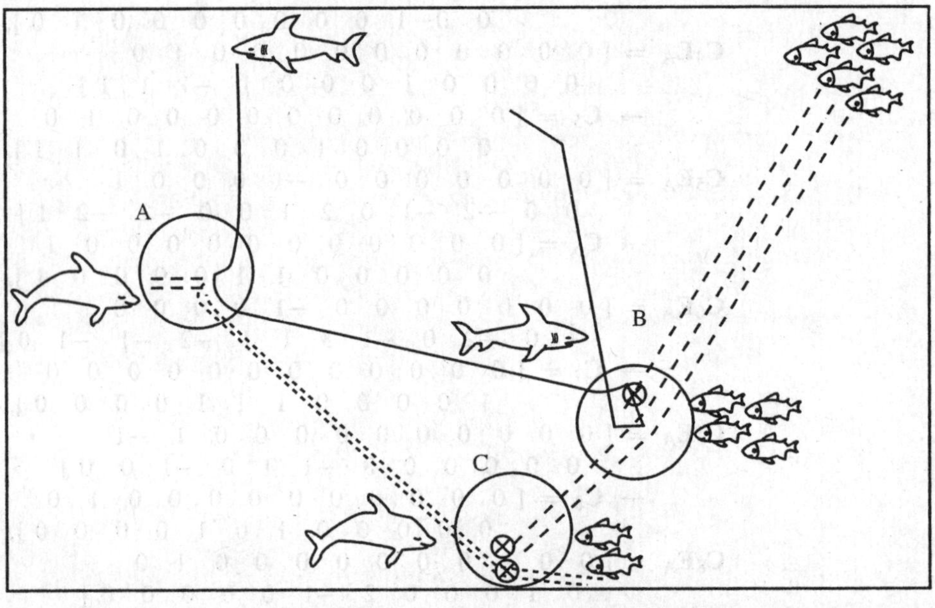

Figure 15.10 FCMs control the virtual world. The augmented FCM controls the actions of the actors. In event A the hungry shark forces the dolphin herd to run away. Each dashed line stands for a dolphin swim path. In event B the shark finds the fish and eats some. Each dashed line stands for the path of a fish in the school. The cross shows the shark eating a fish. In event C the fish run into the dolphins and suffer more losses. The solid lines are the dolphin paths. The dashes are the fish swim paths. The cross shows a dolphin eating a fish.

15.3.2 Nested FCMs for Fish Schools

In a simple FCM the threat response concepts link as a rule:

This rule does not model the effects of different threats. For that we need a nested or embedded FCM:

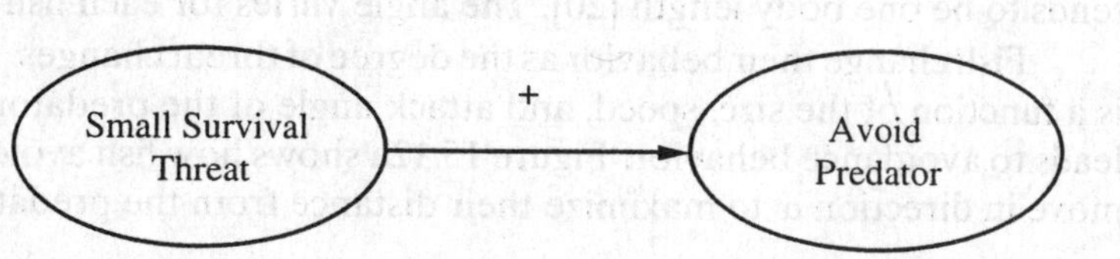

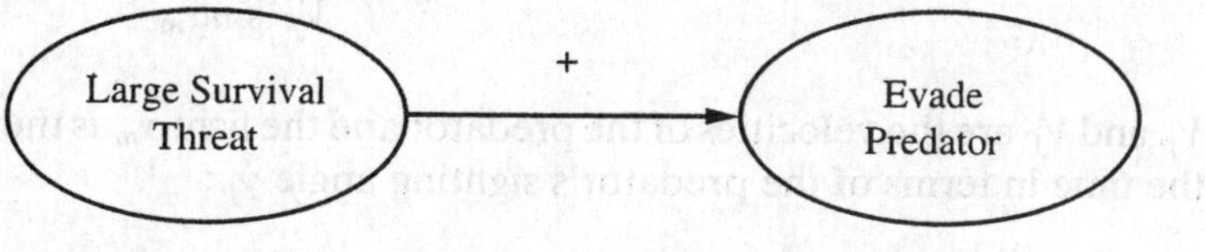

This small survival threat may be a slow-moving predator that has not seen or decided to attack the fish. The large survival threat may be a fast predator such as a barracuda or shark that swims toward the center of the school. If we insert this new sub-FCM into the Fish FCM in Figure 15.8 we get the FCM in Figure 15.11. Different limit cycles appear for different degrees of threat. For a small threat (F_6) the fish avoid the predator (F_9) as they move out of the line of sight of the predator. Large threats (F_7) cause the fish to scatter quickly to evade the predator (F_8). This leads to fatigue and rest (F_2 and F_3).

Figure 15.11 Example of a nested FCM. The concept of a survival threat divides into two subconcepts that each map to a different survival tactic.

A few simple spacing rules can model schooling behavior of fish [21]. Each fish has a preferred angle and distance from the other fish in the school. The distance tends to be one body length [20]. The angle varies for each fish species.

Fish change their behavior as the degree of threat changes. The size of the threat is a function of the size, speed, and attack angle of the predator [20]. A small threat leads to avoidance behavior. Figure 15.12a shows how fish avoid a predator. The fish move in direction α to maximize their distance from the predator [26]:

$$\cot\alpha = \cot\alpha_m + \frac{V_p}{V_f \sin\alpha_m}. \tag{15.7}$$

V_p and V_f are the velocities of the predator and the fish; α_m is the angle that minimizes the time in terms of the predator's sighting angle γ_p:

$$\tan\alpha_m = -\cot\gamma_p. \tag{15.8}$$

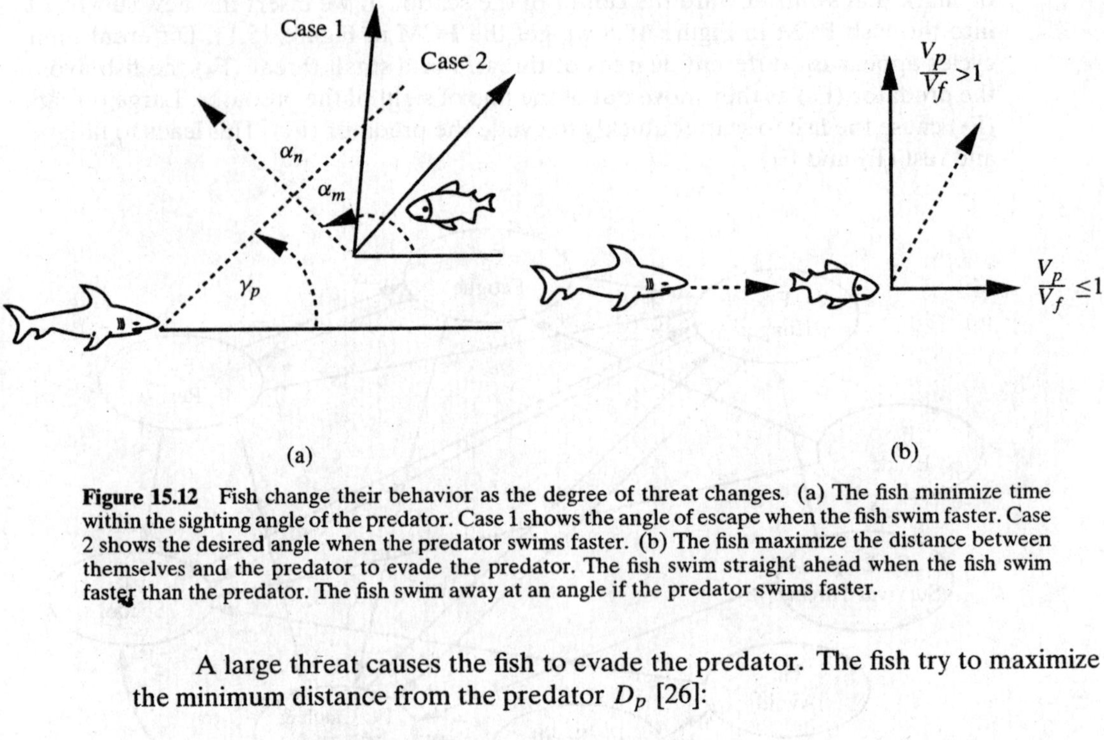

Figure 15.12 Fish change their behavior as the degree of threat changes. (a) The fish minimize time within the sighting angle of the predator. Case 1 shows the angle of escape when the fish swim faster. Case 2 shows the desired angle when the predator swims faster. (b) The fish maximize the distance between themselves and the predator to evade the predator. The fish swim straight ahead when the fish swim faster than the predator. The fish swim away at an angle if the predator swims faster.

A large threat causes the fish to evade the predator. The fish try to maximize the minimum distance from the predator D_p [26]:

$$D_p^2 = \left[(X_0 - V_p\, t) + V_f\, t \cos\alpha\right]^2 + \left(V_f\, t \sin\alpha\right)^2. \tag{15.9}$$

X_0 is the initial distance between predator and prey. α is the escape angle of the prey. V_p and V_f are the velocities of the predator and the fish. Figure 15.12b shows how fish evade a predator. The solution is a velocity ratio:

$$\cos\alpha = \begin{cases} \dfrac{V_p}{V_f} & \text{if } \dfrac{V_p}{V_f} \le 1 \\[2ex] \dfrac{V_f}{V_p} & \text{if } \dfrac{V_p}{V_f} \ge 1. \end{cases} \tag{15.10}$$

These threat responses cause the "fountain effect" and the "burst effect" in fish schools [20] as each fish tries to increase its chances of survival. The fountain effect occurs when a predator moves toward a fish school and the school splits and flows around the predator. The school re-forms behind the predator. In the burst effect the school expands in the form of a sphere to evade the predator.

15.4 ADAPTIVE FUZZY COGNITIVE MAPS

An adaptive FCM changes its causal web in time. The causal web learns from data. The causal edges or rules change in sign and magnitude. The additive scheme of (15.3) is a type of causal learning since it changes the FCM edge strengths. In general an edge e_{ij} changes with some first-order learning law:

$$\dot{e}_{ij} = f_{ij}(\mathbf{E}, \mathbf{C}) + g_{ij}(t). \tag{15.11}$$

Here g_{ij} is a forcing function. Data fire the concept nodes and in time this leaves a causal pattern in the edge. Causal learning is local in f_{ij}. It depends on just its own value and on the node signals that it connects:

$$\dot{e}_{ij} = f_{ij}(e_{ij}, C_i, C_j, \dot{C}_i, \dot{C}_j) + g_{ij}(t). \tag{15.12}$$

Correlation or Hebbian learning can encode some limit cycles in the FCMs or temporal associative memories (TAMs) [13]. It adds pairwise correlation matrices in (15.11). This method can only store a few patterns. Differential Hebbian learning encodes changes in a concept in (15.12). Both types of learning are local and light in computation.

To encode binary limit cycles in connection matrix $\mathbf{E}$ the TAM method sums the weighted correlation matrices between successive states [13]. To encode the limit cycle $\mathbf{C}_1 \rightarrow \mathbf{C}_2 \rightarrow \mathbf{C}_3 \rightarrow \mathbf{C}_1$ we first convert each binary state $\mathbf{C}_i$ into a bipolar state vector $\mathbf{X}_i$ by replacing each 0 with a -1. Then $\mathbf{E}$ is the weighted sum

$$\mathbf{E} = q_1\mathbf{X}_1^{\mathrm{T}}\mathbf{X}_2 + q_2\mathbf{X}_2^{\mathrm{T}}\mathbf{X}_3 + \cdots + q_{n-1}\mathbf{X}_{n-1}^{\mathrm{T}}\mathbf{X}_n + q_n\mathbf{X}_n^{\mathrm{T}}\mathbf{X}_1. \tag{15.13}$$

The length of the limit cycle should be less than the number of concepts or else crosstalk can occur. Proper weighting of each correlation matrix pair can improve the encoding [25] and thus increase the FCM storage capacity. Correlation learning is a form of the unsupervised signal Hebbian learning law in neural networks [16]:

$$\dot{e}_{ij} = -e_{ij} + C_i(x_i)\, C_j(x_j). \tag{15.14}$$

A virtual world can encode an event sequence with (15.13) or (15.14). A simple chase cycle might be $\mathbf{C}_1 \to \mathbf{C}_2 \to \mathbf{C}_3$:

$$\begin{aligned} \mathbf{C}_1 &= [1\ \ 0\ \ 1\ \ 0\ \ 0\ \ 0\ \ 0\ \ 0\ \ 0\ \ 1] \\ \mathbf{C}_2 &= [1\ \ 0\ \ 1\ \ 1\ \ 1\ \ 0\ \ 0\ \ 0\ \ 1\ \ 0] \\ \mathbf{C}_3 &= [1\ \ 0\ \ 0\ \ 0\ \ 1\ \ 0\ \ 0\ \ 0\ \ 1\ \ 1]. \end{aligned}$$

Then (15.13) gives the FCM connection matrix **E**:

E =

	D_1	D_2	D_3	D_4	D_5	D_6	D_7	D_8	D_9	D_{10}
D_1	3	−3	1	−1	1	−3	−3	−3	1	1
D_2	−3	3	−1	1	−1	3	3	3	−1	−1
D_3	1	−1	−1	1	3	−1	−1	−1	3	−1
D_4	−1	1	−3	−1	1	1	1	1	1	1
D_5	1	−1	−1	−3	−1	−1	−1	−1	−1	3
D_6	−3	3	−1	1	−1	3	3	3	−1	−1
D_7	−3	3	−1	1	−1	3	3	3	−1	−1
D_8	−3	3	−1	1	−1	3	3	3	−1	−1
D_9	1	−1	−1	−3	−1	−1	−1	−1	−1	3
D_{10}	1	−1	3	1	−1	−1	−1	−1	−1	−1

Then

$$\begin{aligned} \mathbf{C}_1\mathbf{E} &= [\,5\ \ -5\ \ \ 3\ \ \ 1\ \ \ 3\ \ -5\ \ -5\ \ -5\ \ \ 3\ \ -1\,] \\ &\to \mathbf{C}_2 = [\,1\ \ 0\ \ 1\ \ 1\ \ 1\ \ 0\ \ 0\ \ 0\ \ 1\ \ 0\,], \\ \mathbf{C}_2\mathbf{E} &= [\,5\ \ -5\ \ -5\ \ -7\ \ \ 3\ \ -5\ \ -5\ \ -5\ \ \ 3\ \ \ 7\,] \\ &\to \mathbf{C}_3 = [\,1\ \ 0\ \ 0\ \ 0\ \ 1\ \ 0\ \ 0\ \ 0\ \ 1\ \ 1\,], \\ \mathbf{C}_3\mathbf{E} &= [\,6\ \ -6\ \ \ 2\ \ -6\ \ -2\ \ -6\ \ -6\ \ -6\ \ -2\ \ \ 6\,] \\ &\to \mathbf{C}_1 = [\,1\ \ 0\ \ 1\ \ 0\ \ 0\ \ 0\ \ 0\ \ 0\ \ 0\ \ 1\,]. \end{aligned}$$

Correlation encoding treats negative and zero causal edges the same. It can encode "spurious" causal implications between concepts such as $e_{6,2} = 3$. This means searching for food causes a desire to socialize. Correlation encoding is a poor model of inferred causality. It says two concepts cause each other if they are on at the same time. Differential Hebbian learning encodes causal changes to avoid spurious causality. The concepts must move in the same or opposite directions to infer a causal link. They must come on and turn off at the same time or one must come on as the other turns off. Just being on does not lead to a new causal link. The patterns of turning on or off must correlate positively or negatively.

The differential Hebbian learning law [13] correlates concept changes or velocities:

$$\dot{e}_{ij} = -e_{ij} + \dot{C}_i(x_i)\ \dot{C}_j(x_j). \tag{15.15}$$

So $\dot{C}_i(x_i)\dot{C}_j(x_j) > 0$ iff concepts C_i and C_j move in the same direction. $\dot{C}_i(x_i)\ \dot{C}_j(x_j) < 0$ iff concepts C_i and C_j move in opposite directions. In this sense (15.15) learns patterns of causal change. The first-order structure of (15.15) implies that $e_{ij}(t)$ is an exponentially weighted average of paired (or lagged) changes. The most recent changes have the most weight. The *discrete* change $\Delta C_i(t) = C_i(t) - C_i(t-1)$ lies in $\{-1, 0, 1\}$. The discrete differential Hebbian learning can take the form

$$e_{ij}(t+1) = \begin{cases} e_{ij}(t) + c_t\,[\Delta C_i(x_i)\ \Delta C_j(x_j) - e_{ij}(t)] & \text{if } \Delta C_i(x_i) \neq 0 \\ e_{ij}(t) & \text{if } \Delta C_i(x_i) = 0 \end{cases} \tag{15.16}$$

Here c_t is a learning coefficient that decreases in time [14]. $\Delta C_i\ \Delta C_j > 0$ iff concepts C_i and C_j move in the same direction. $\Delta C_i\ \Delta C_j < 0$ iff concepts C_i and C_j move in opposite directions. **E** changes only if a concept changes. The changed edge slowly "forgets" the old causal changes in favor of the new ones. This causal law can learn higher-order causal relations if it correlates multiple cause changes with effect changes.

We used differential Hebbian learning to encode a feeding sequence and a chase sequence in an FCM. The concepts in the ith row learn only when $\Delta C_i(x_i)$ equals 1 or −1. We used $c_t(t_k) = 0.1\left[1 - \frac{t_k}{1.1N}\right]$. The training data came from the dolphin FCM in Figure 15.7. This gave the $\mathbf{E}_D$:

$\mathbf{E}_D =$

	D_1	D_2	D_3	D_4	D_5	D_6	D_7	D_8	D_9	D_{10}
D_1	−0.25	0.00	0.00	−0.24	−0.24	0.76	−0.51	0.00	0.00	0.00
D_2	0.00	−0.49	0.49	−0.51	0.00	0.00	0.00	0.00	0.00	0.00
D_3	−0.26	0.00	−0.25	1.00	0.75	0.00	0.00	0.00	0.00	0.00
D_4	1.00	0.00	−0.25	−0.25	−0.25	−0.50	0.00	0.00	0.00	0.00
D_5	0.51	−0.16	0.49	−0.34	−0.51	−0.33	0.00	0.00	0.00	−0.16
D_6	0.00	0.00	0.00	0.00	0.00	−0.49	1.00	−0.51	0.00	0.00
D_7	0.00	−0.51	0.00	0.00	−0.51	0.00	−0.49	1.00	0.00	0.00
D_8	0.00	1.00	−0.33	0.00	0.67	0.00	0.00	−0.67	0.00	0.00
D_9	0.00	0.00	−1.00	0.00	1.00	0.00	0.00	0.00	0.00	1.00
D_{10}	0.00	0.00	1.00	−0.51	−1.00	0.00	0.00	0.00	0.00	−0.49

This learned edge matrix $\mathbf{E}_D$ resembles the FCM matrix in Figure 15.7. The causal links it lacks between $\mathbf{D}_{10}$ and $\{\mathbf{D}_6, \mathbf{D}_7, \mathbf{D}_8\}$ were not in the training set. The diagonal links terms for self-inhibition of each concept. This occurs since each concept is on for one cycle before the matrix transitions to the next state. The hunger input

$\mathbf{CL_0} = [1\ 0\ 0\ 0\ 0\ 0\ 0\ 0\ 0\ 0]$ with a threshold of 0.51 now leads to the limit cycle:

$$
\begin{aligned}
\mathbf{CL_0E_D} &= [\,-0.25\quad 0.00\quad 0.00\quad -0.24\quad -0.24\\
&\qquad 0.76\quad -0.51\quad 0.00\quad 0.00\quad 0.00\,]\\
&\to \mathbf{CL_1} = [\,0\ \ 0\ \ 0\ \ 0\ \ 0\ \ 1\ \ 0\ \ 0\ \ 0\ \ 0\,],\\
\mathbf{CL_1E_D} &= [\,0.00\quad 0.00\quad 0.00\quad 0.00\quad 0.00\\
&\qquad -0.49\quad 1.00\quad -0.51\quad 0.00\quad 0.00\,]\\
&\to \mathbf{CL_1} = [\,0\ \ 0\ \ 0\ \ 0\ \ 0\ \ 0\ \ 1\ \ 0\ \ 0\ \ 0\,],\\
\mathbf{CL_2E_D} &= [\,0.00\quad -0.51\quad 0.00\quad 0.00\quad -0.51\\
&\qquad 0.00\quad -0.49\quad 1.00\quad 0.00\quad 0.00\,]\\
&\to \mathbf{CL_1} = [\,0\ \ 0\ \ 0\ \ 0\ \ 0\ \ 0\ \ 0\ \ 1\ \ 0\ \ 0\,],\\
\mathbf{CL_3E_D} &= [\,0.00\quad 1.00\quad -0.33\quad 0.00\quad 0.67\\
&\qquad 0.00\quad 0.00\quad -0.67\quad 0.00\quad 0.00\,]\\
&\to \mathbf{CL_1} = [\,0\ \ 1\ \ 0\ \ 0\ \ 1\ \ 0\ \ 0\ \ 0\ \ 0\ \ 0\,],\\
\mathbf{CL_4E_D} &= [\,0.51\quad -0.65\quad 0.98\quad -0.85\quad -0.51\\
&\qquad -0.33\quad 0.00\quad 0.00\quad 0.00\quad -0.16\,]\\
&\to \mathbf{CL_1} = [\,0\ \ 0\ \ 1\ \ 0\ \ 0\ \ 0\ \ 0\ \ 0\ \ 0\ \ 0\,],\\
\mathbf{CL_5E_D} &= [\,-0.26\quad 0.00\quad -0.25\quad 1.00\quad 0.75\\
&\qquad 0.00\quad 0.00\quad 0.00\quad 0.00\quad 0.00\,]\\
&\to \mathbf{CL_1} = [\,0\ \ 0\ \ 0\ \ 1\ \ 1\ \ 0\ \ 0\ \ 0\ \ 0\ \ 0\,],\\
\mathbf{CL_6E_D} &= [\,1.51\quad -0.16\quad 0.25\quad -0.59\quad -0.76\\
&\qquad -0.83\quad 0.00\quad 0.00\quad 0.00\quad -0.16\,]\\
&\to \mathbf{CL_1} = [\,1\ \ 0\ \ 0\ \ 0\ \ 0\ \ 0\ \ 0\ \ 0\ \ 0\ \ 0\,].
\end{aligned}
$$

This resembles the sequence of rest, eat, play, and rest from Section 15.3.1. Figure 15.13a shows the hand-designed limit cycle from Section 15.3.1. Figure 15.13b shows the limit cycle from FCM found with differential Hebbian learning. The DHL limit cycle is one step shorter. Both FCMs have just one limit cycle and the null fixed point in the space of 2^{10} binary state vectors. The value of $\mathbf{E_D}$ does not change over two intervals. The learning law in (15.16) learns only if there is a change in the node.

15.5 CONCLUSIONS

Fuzzy cognitive maps can model the causal web of a virtual world. The FCM can control its local and global nonlinear behavior. The local fuzzy rules or edges and the fuzzy concepts they connect model the causal links within and between events. The global FCM nonlinear dynamics give the virtual world an "arrow of time." A user can change these dynamics at will and thus change the causal processes in the virtual

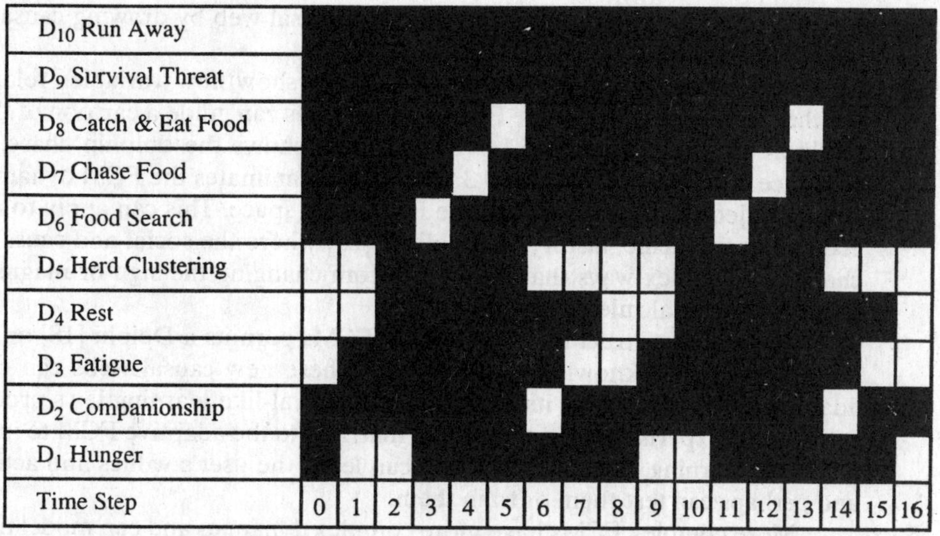

(a)

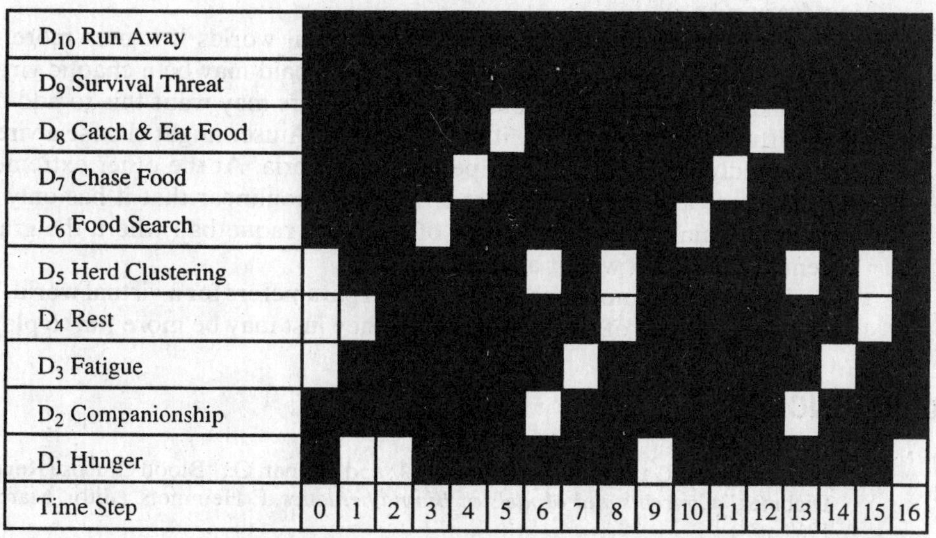

(b)

Figure 15.13 Limit cycle comparison between the hand-designed system and the FCM found with differential Hebbian learning. Each column is a binary state vector. (a) Rest, feed, play, rest limit cycle for the FCM in Figure 15.7 (b) Limit cycle for the FCM found with (15.16).

world. FCMs let experts and users choose a causal web by drawing causal pictures instead of by stating equations.

FCMs can also help visualize data. They show how variables relate to one another in the causal web. The FCM output states can guide a cartoon of the virtual world as shown in Figure 15.14. This cartoon shows the dolphin chase, rest, eat sequence described in Section 15.3. The cartoon animates the FCM dynamics as the system trajectory moves through the FCM state space. This can apply to models in economics, medicine, history, and politics [23] where the social and causal web can change in complex ways that may arise from changing the sign or magnitude of a single FCM causal rule or edge.

The additive structure of combined FCMs permits a Delphi [18] or questionnaire approach to knowledge acquisition. These new causal webs can change an adaptive FCM that learns its causal web as neural-like learning laws process time-series data. Experts can add their FCM matrices to the adaptive FCM to initialize or guide the learning. Such a causal web can learn the user's values and action habits and perhaps can test them or train them.

More complex FCMs have more complex dynamics and can model more complex virtual worlds. Each concept node can fire on its own time scale and fire in its own nonlinear way. The causal edge flows or rules can have their own time scales too and may increase or decrease the causal flow through them in nonlinear ways. This behavior does not fit in a simple FCM with threshold concepts and constant edge weights. We still seek an FCM approximator F in $\dot{x} = F(x)$ that approximates at least the qualitative attractor structure of an arbitrary dynamical system $\dot{x} = f(x)$ in a fuzzy cube.

An FCM can model these complex virtual worlds if it uses more nonlinear math to change its nodes and edges. The price paid may be a chaotic virtual world with unknown equilibrium behavior. Some users may want this to add novelty to their virtual world or to make it more exciting. A user might choose a virtual world that is mildly nonlinear and has periodic equilibria. At the other extreme the user might choose a virtual world that is so wildly nonlinear that it has only aperiodic equilibria. Think of a virtual game of tennis or raquetball where the gravitational potential changes at will or at random.

Fuzziness and nonlinearity are design parameters for a virtual world. They may give a better model of a real process. Or they just may be more fun to play with.

REFERENCES

[1] Ackerman, E., Gatewood, L., Rosevear, J., and Molnar, G., "Blood Glucose Regulation and Diabetes," *Concepts and Models of Biomathematics*, F. Heinmets, editor, Marcel Dekker, 1969.

[2] Badler, N. I., Webber, B. L., Kalita, J., and Esakov, J., "Animation from Instructions," *Making Them Move: Mechanics, Control, and Animation of Articulated Figures*, N. I. Badler, B. A. Barsky, and D. Zeltzer, editors, 51–98, Morgan Kaufmann Publishers, Inc., 1991.

[3] Brooks, R. A., "A Robot that Walks: Emergent Behaviors from a Carefully Evolved Network," *Neural Computation*, Vol. 1, No. 2, 253–262, Summer 1989.

[4] Brown, R. A., *Fluid Mechanics of the Atmosphere*, Academic Press, 1991.

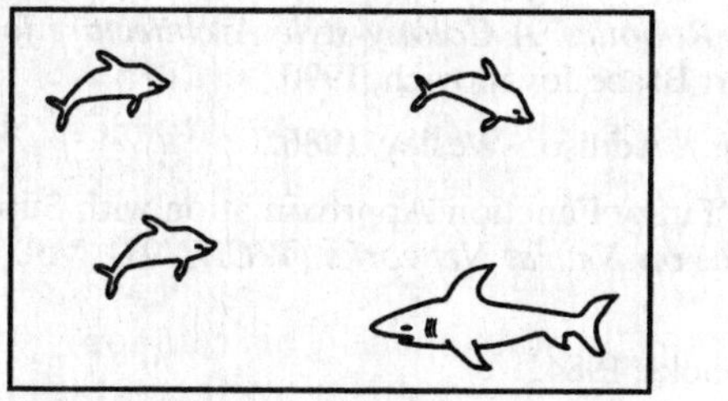

Time Step 0: Threat Appears in the Form of a Shark

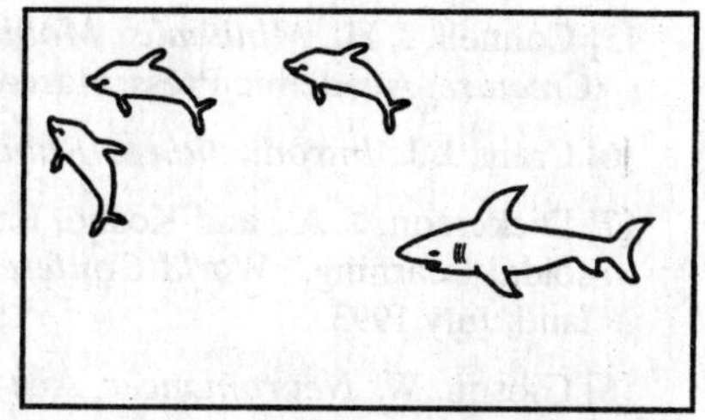

Time Steps 1&2: Dolphins Flee the Shark in a Tightly Packed Herd

Time Steps 3 &4: Dolphins Cluster Together

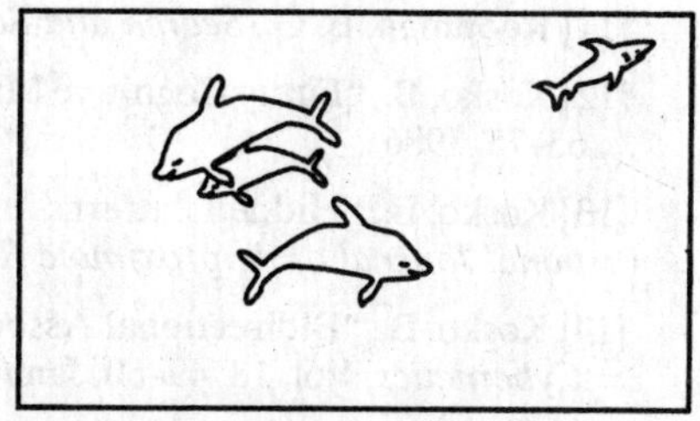

Time Steps 5 & 7: Dolphins Avoid Shark Then Rest

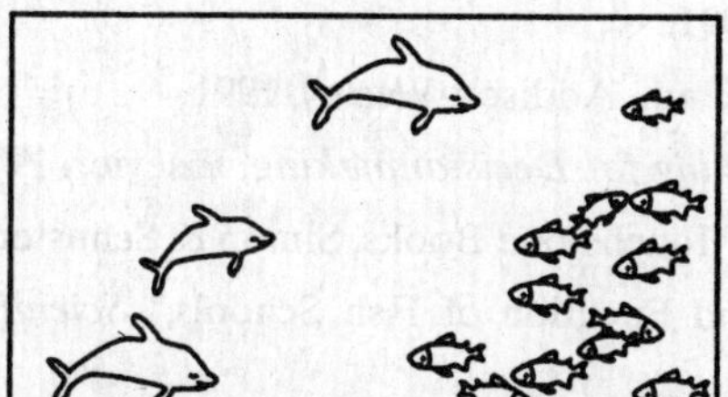

Time Steps 8 & 9: Dolphins Start a Search for Food

Time Step 10: Dolphins Find a School of Fish Then Begin to Chase Them

Time Step 11: Dolphins Catch and Eat Some Food

Time Steps 12 &13: Dolphins Then Play and Rest. Then the Cycle Begins Again

Figure 15.14 The FCM output states can guide a cartoon of the virtual world. This cartoon shows the dolphin chase, rest, eat sequence described in Section 15.3. The cartoon animates the FCM dynamics as the system trajectory moves through the FCM state space.

[5] Connell, J. H., *Minimalist Mobile Robotics: A Colony-style Architecture for an Artificial Creature*, Academic Press, Harcourt Brace Jovanovich, 1990.

[6] Craig, J. J., *Introduction to Robotics*, Addison-Wesley, 1986.

[7] Dickerson, J. A., and Kosko, B., "Fuzzy Function Approximation with Supervised Ellipsoidal Learning," *World Conference on Neural Networks (WCNN93)*, Vol. II, 9–17, Portland, July 1993.

[8] Gibson, W., *Neuromancer*, Acė Books, 1984.

[9] Grossberg, S., *Studies of Mind and Brain*, Reidel, 1982.

[10] Hornik, K., Stinchcombe, M., and White, H., "Multilayered Feedforward Networks are Universal Approximators," *Neural Networks*, Vol. 2, No. 5, 359–366, 1989.

[11] Koopman, B. O., *Search and Screening*, Pergamon Press, 1980.

[12] Kosko, B., "Fuzzy Cognitive Maps," *International Journal Man-Machine Studies*, Vol. 24, 65–75, 1986.

[13] Kosko, B., "Hidden Patterns in Combined and Adaptive Knowledge Networks," *International Journal of Approximate Reasoning*, Vol. 2, 337–393, 1988.

[14] Kosko, B., "Bidirectional Associative Memories," *IEEE Transactions Systems, Man, and Cybernetics*, Vol. 18, 49–60, Jan./Feb. 1988.

[15] Kosko, B., "Fuzzy Systems as Universal Approximators," *IEEE Transaction on Computers*, 1993; an early version appears in *Proceedings of the 1st IEEE International Conference on Fuzzy Systems (FUZZ-IEEE FUZZ92)*, 1153–1162, March 1992.

[16] Kosko, B., *Neural Networks and Fuzzy Systems: A Dynamical Systems Approach to Machine Intelligence*, Prentice Hall, 1991.

[17] Krueger, M., *Artificial Reality II*, 2nd ed., Addison-Wesley, 1991.

[18] Martino, J. P., *Technological Forecasting for Decisionmaking*, Elsevier, 1972.

[19] Minsky, M. L., *The Society of Mind*, Touchstone Books, Simon & Schuster, 1985.

[20] Partridge, B. L., "The Structure and Function of Fish Schools," *Scientific American*, Vol. 246, No. 6, 114–123, June 1982.

[21] Reynolds, C. W., "Flocks, Herds, and Schools: A Distributed Behavioral Model," *Computer Graphics*, Vol. 21, No. 4, 25–34, July 1987.

[22] Shane, S. H., "Comparison of Bottlenose Dolphin Behavior in Texas and Florida, with a Critique of Methods for Studying Dolphin Behavior," *The Bottlenose Dolphin*, S. Leatherwood and R. R. Reeves, editors, 541–558, Academic Press, 1990.

[23] Taber, W. R., "Knowledge Processing with Fuzzy Cognitive Maps," *Expert Systems with Applications*, Vol. 2, No. 1, 83–87, 1991.

[24] Taber, W. R., and Siegel, M., "Estimation of Expert Weights with Fuzzy Cognitive Maps," *Proceedings of the 1st IEEE International Conference on Neural Networks (ICNN–87)*, Vol. II, 319–325, June 1987.

[25] Wang, Y. F., Cruz, J. B., and Mulligan, J.H., "Guaranteed Recall of All Training Pairs for Bidirectional Associative Memory," *IEEE Transactions on Neural Networks*, Vol. 2, No. 6, 559–567, November 1991.

[26] Weihs, D., and Webb P. W., "Optimal Avoidance and Evasion Tactics in Predator-Prey Interactions," *Journal of Theoretical Biology*, Vol. 106, 189–206, 1984.

[27] Winston, P. H., *Artificial Intelligence*, 2nd ed., Addison-Wesley, 1984.

HOMEWORK PROBLEMS

15.1. Given the simple FCM

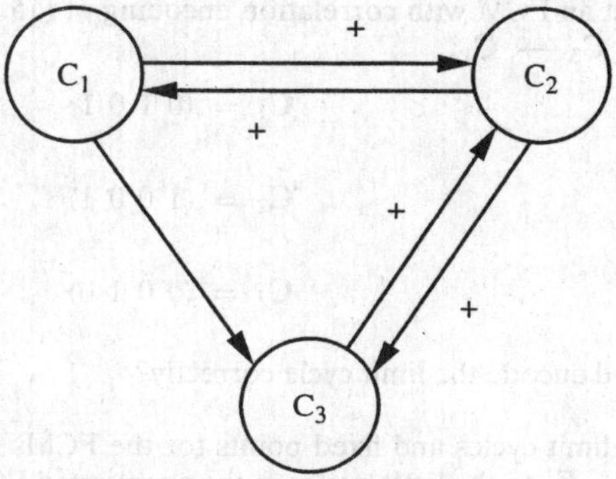

(a) Find the edge matrix.

(b) Use a threshold of 1/2. Input (110) to the dynamical system. What is the result—a fixed point or limit cycle?

(c) Input (010) to the dynamical system. What is the result?

(d) There are 8 possible input states for this FCM (000) to (111). What does each state lead to? What limit cycles and fixed points does this system have?

15.2. A simple nonlinear dynamical system is the damped pendulum

$$\dot{x} = y$$

$$\dot{y} = -ky - \sin(x).$$

Many dynamical systems transition into chaos or limit cycles depending on their starting point. Small changes in inputs can lead to vastly different outputs. Use a first-order approximation for the derivative or a routine that solves differential equations to plot x and y.

(a) Set $k = 0$. This is the *undamped* pendulum. Plot the results for $\mathbf{x} = (0.25, 0)$ and $\mathbf{x} = (0.5, 0)$. Try $\mathbf{x} = (0, 2)$. Are the trajectories periodic or aperiodic?

(b) Set $k = 0.2$. Plot the results for $\mathbf{x} = (0.25, 0)$ and $\mathbf{x} = (0, 3)$. What fixed points does the system converge to? What would be some examples of this behavior in a virtual world?

15.3. Duffing's equation for a mass on a cubic spring is

$$\dot{x} = y$$

$$\dot{y} = x - x^3 - \delta y + \gamma \cos(\omega t).$$

(a) Plot the trajectory for $\omega = 1$, $\gamma = 0.3$, and $\delta = 0.0$. Start at $\mathbf{x} = (1.5,\ 0)$.
(b) Plot the trajectory for $\omega = 1$, $\gamma = 0.3$, and $\delta = 0.50$. Start at $\mathbf{x} = (1.5,\ 0)$.
(c) Plot the trajectory for $\omega = 1$, $\gamma = 0.3$, and $\delta = -0.1$. Start at $\mathbf{x} = (1.5,\ 0)$.

Compare the trajectories.

15.4. Construct an FCM with correlation encoding of (15.13). Encode the 3-step limit cycle $C_1 \rightarrow C_2 \rightarrow C_3 \rightarrow C_1$:

$$C_1 = (0\ 1\ 0\ 1)$$

$$C_2 = (1\ 0\ 0\ 1)$$

$$C_3 = (0\ 0\ 1\ 0).$$

Did this method encode the limit cycle correctly?

15.5. Find the limit cycles and fixed points for the FCMs in Figure 15.3. Compare the limit cycles in $\mathbf{F}_1$, $\mathbf{F}_2$, $\mathbf{F}_3$ to the limit cycles in the augmented FCM. Are they similar?

15.6. Construct an FCM with differential hebbian encoding. Encode the 3-step limit cycle from problem 4. Did this method encode the limit cycle correctly if you used a threshold value of zero?

APPENDIX

HOW TO USE THE FUZZY SOFTWARE

A.1 THE FUZZY FUNCTION APPROXIMATION SOFTWARE

The FUZAPROX program on the enclosed diskette shows how a standard additive model (SAM) fuzzy system approximates one of three functions. Graphical displays show the original function or approximand, the fuzzy system output, the approximation error, and the fuzzy sets used in the rules. The second program FZCOGMAP.EXE shows a simple feedback fuzzy cognitive map with four concept nodes and allows a user to explore the dynamical system's equilibrium behavior.

The FUZAPROX program offers three types of fuzzy approximation: "Regular Rules," "Irregular Rules," and "Exact Two Rules." "Regular Rules" give a triangular SAM with evenly spaced isosceles triangles as the if-part sets. This is the simplest type of SAM approximator F. It uses the least information about the approximand function f. It also gives the poorest result in terms of the squared-error of the fuzzy function approximation. The symmetric triangular if-part sets define a SAM approximator that is piecewise linear.

"Irregular Rules" give a triangular SAM with optimal rules. The rule patches cover the extrema or bumps of the graph of the approximand f and thus they "patch the bumps." The bumps are not evenly spaced in most cases and so neither are the if-part and then-part fuzzy sets that make up the optimal rules. The rules are "irregular" only in this sense of set spacing. You can step through the system to patch one bump at a time. This triangular SAM also gives a piecewise-linear approximation of the approximand.

The SAM of "Irregular Rules" has in theory less squared error of approximation than does the like SAM of "Regular Rules." The simulation does not always show this because it simply centers the base of the if-part set at an extremum. The optimal rules in Chapter 2 require that the then-part set centroid lie at the centroid value of the approximand f when taken over the interval that the base of the if-part set covers. The simulation also treats the base of the if-part sets differently for "regular" and "irregular" sets. These artifacts appear most clearly in the case of 1-rule SAM approximation of the sinc approximand. The lone "regular" if-part defines a triangle whose base covers the whole domain and whose then-part centroid lies at the centroid of f over that domain. The lone "irregular" if-part set defines a triangle that covers the main bump of the function. Its base covers only the base of the main lobe of the sinc function and its then-part set centroid lies at the mode of f rather than at the centroid of f over that smaller interval. Patching this main bump is optimal in the sense that putting a lone if-part triangle of the same width anywhere else will have at least as large as squared error. This too assumes that the then-part set centroid lies at the centroid of f over the interval in question.

"Exact Two Rules" give an exact SAM representation of the approximand as discussed in Chapter 2. This means $F = f$ for all inputs. The SAM of "Exact Two Rules" achieves this zero-error approximation with only two rules. But this type of approximation requires total knowledge of the approximand f. It builds the structure of f into the two if-part fuzzy sets. So the if-part fuzzy sets are as complex as the function f that the SAM F represents. Here the if-part sets are scaled and translated versions of the approximand.

There are three approximand choices f that the SAM can approximate. It can approximate the sine function $f(x) = \sin x$ on [0, 2p], the sinc function $f(x) = \frac{\sin x}{x}$ on [−15, 15], or the fifth-degree polynomial f on the interval [−2, 2]:

$$f(x) = 3x^5 + 1.2x^4 - 12.27x^3 + 3.288x^2 + 7.182x.$$

This fifth-degree polynomial appears in Chapters 3 and 5. The program contains two versions packed into a single file called FUZAPROX.EXE. A different version for MS-DOS or for Microsoft Windows runs depending on how you start the program.

You can run the program directly from the diskette or you can copy the program file FUZAPROX.EXE to your hard drive first. Copy the help file FUZAPROX.HLP if you will run the software under Windows.

Be sure to read the file README.TXT for the most up-to-date information about the diskette contents.

A.1.1 Instructions for the MS-DOS Version

The MS-DOS version runs when you start the program under MS-DOS or a DOS-compatible command-line environment. This version requires color VGA graphics. It does not use a mouse. Start the program this way: Change to the drive and directory containing the program and type the command "FUZAPROX." (Do not type the quotes and remember to press Enter.)

Control the program with the menu at upper left. If the menu is not visible then press the Escape key to open it. Use the arrow keys on your keyboard to move through the menus. Select a menu command by pressing the Enter key when the command is highlighted in red. When the menu is open then you can press Escape to close it so you can see the entire display.

The Main menu has the choices for the approximation method and contains several other commands: "Information" provides useful information about the currently selected function and method. "Help" provides a brief summary of how to use the program. "About this program" gives contact information for HyperLogic Corporation (the firm that wrote this software). Exit the program by choosing the "Quit" menu item or by pressing the "q" key when the menu is open. The Function menu lists the choices for the function to approximate.

The graphical display shows results for the currently selected function and method. The window at upper right holds graphs of the target function (in green), the fuzzy system output (in yellow), and the error (in red). The window at lower right shows the input fuzzy sets. These are all one-input systems there and so have one fuzzy set per rule. The window at upper left shows the output fuzzy sets. The number of them depends on the target function and the placement of the if-part fuzzy sets.

If you choose either the "Regular Rules" or the "Irregular Rules" method then a window appears at lower left containing choices for the number of rules. (The "Exact Two Rules" method always uses two rules.) If the rule count window is visible then use the arrow keys to choose how many rules to use or press Escape to return to the menu. You can get the rule count window back by choosing the method again or by choosing a different function.

A.1.2 Instructions for the Windows Version

The Windows version runs when you start the program under Microsoft Windows 3.1 or a compatible environment as in OS/2. The program is a 16-bit Windows application with minimal memory requirements. You should run it on a system that has a color display. A mouse is helpful but you can use the program without one.

The File menu contains only the Exit command. On-line help and other information about the program are available through the Help menu. The Options menu contains all the choices for the function, approximation method, and number of rules. The menu shows quick-access command keys where they are available.

The display shows results for the currently selected function and method. The window at upper right holds graphs of the target function (in green), the fuzzy system output (in yellow), and the error (in red). The window at lower right shows the input or if-part fuzzy sets. These are all one-input systems and so there is one fuzzy set per rule. The window at upper left shows the output or then-part fuzzy sets. Their number depends on the target function and the placement of the input fuzzy sets.

A set of buttons at lower left lets you cycle the display through all the choices for function, method, and rule count. If the method is "Exact Two Rules" then the rule count is always two and the rule count button has no effect. This holds because a SAM with just two rules can represent any scalar bounded function as shown in Chapter 2 (Theorem 2.4). If you then change the method back to either "Regular Rules" or "Irregular Rules" then the rule count returns to what it was before.

A.2 THE FUZZY COGNITIVE MAP SOFTWARE

This program implements a 4-state feedback Fuzzy Cognitive Map (FCM) as described in Chapters 2 and 15. The implementation supports binary and bipolar states, optional self-connections (self-edges), and optional activation biases (constants that add to the argument of a node's signal function). Each state of the FCM is a binary or bipolar 4-vector such as (1 0 0 1) or (1 −1 −1 1) or vertex of a fuzzy cube. There are 4 concept nodes with 16 causal edges.

The program requires Microsoft Windows or a compatible environment. Start the program as you would any other with the Run menu command of Program Manager or File Manager. The program file is FZCOGMAP.EXE. It is a 16-bit Windows application with minimal memory requirements. A mouse helps but you can use the program without one.

Use the program in this sequence. First set the node states and the connection strengths (weights) between the nodes. Then use either Run or Setup to see what happens.

A.2.1 The Display

The program display shows the four state nodes as circles. The nodes number 1 through 4 starting at upper left and counting clockwise. A node with a positive state is filled with black. One with zero or negative state is white.

The causal edge strengths appear as lines or circular arcs between the nodes. Connection lines drawn in red stand for positive edge strengths. Blue lines stand for negative strengths. Short or dotted black lines stand for no causal connection or zero connection strengths. If you have a monochrome display then you can distinguish them by the width of the lines. The positive connection lines are thicker than the negative ones.

A history of the node's values appears at the right of the display. Each Run command causes the screen to display a new history (but the program retains the old history for comparison). The 4-vector states always appear in Binary (0, 1) format even if the operation mode is Bipolar. The limit cycle (if detected) appears in boldface in the history display.

A.2.2 Bipolar vs. Binary Concept Values

The FCM concept values are either binary values (0 and 1) or bipolar values (−1 and 1). Choose between these two cases with the Execute/Binary States or Execute/Bipolar States menu commands. A check mark next to the menu item shows which choice the system uses. If you change modes then the non-positive states automatically switch to the appropriate value −1 or 0.

A.2.3 Causal Edge Connection Strengths

If you want to use non-integral connection strengths then use the Connections/Connection Strengths menu command. The resulting dialog box has slots for each of the connections (not including the bias values). The program labels the grid axes "From" and "To" to help you see where to put particular values.

You can also use the mouse to set connection (edge) strengths by clicking on a connection line. The program cycles through the values +1, −1, and 0 if you repeatedly click an edge. This action overrides any fractional values you may have set with the Connections Strengths menu command.

There are special-purpose commands in the Connections menu to set certain useful configurations. You can "Remove All Connections" or "Remove All Self-connects." These commands set all the affected values to zero. You can also use "All Connections Positive" to set all connection strengths to +1 or use "All Connections Negative" to set the connection strengths to −1. The last option "Sequential Connection" forms a positive connection from each to the next node while all other connections are zero. None of these commands affects the biases.

The Connections/Biases menu command brings up a dialog that accepts a "bias" connection for each node. A bias has the effect of a constant added to the node activation rule. The program lets you enter any value between −1 and +1.

A.2.4 Concept Node Values

Use the States menu to set the current values of the FCM nodes. The States/States command brings up a dialog where you can set the output value of all four nodes. If you use Binary Mode then select either +1 or 0 as a state value. If you use Bipolar Mode then set the node states to either +1 or −1.

A "lock" checkbox lies next to each node state text box. If the lock checkbox for a node is marked with an "X" then the node is held at the same state regardless of what its FCM activation might be. (You can still change the node state manually with the mouse or the States dialog.) This gives a "stuck bit" in the binary representation of the FCM nodes. It corresponds to the continued presence of the condition that the node represents despite any influences to the contrary. Nodes show a padlock icon when in the locked state.

There are two special-purpose commands in the States menu. "All States Zero" forces all the node values to 0 (in binary mode) or −1 (in bipolar mode). "All States One" forces all the nodes to +1 regardless of the mode.

Clicking a node with the left mouse button toggles the state value. Clicking a node with the right mouse button toggles the lock for that node. If you have a single-button mouse then you can lock a node by double-clicking it.

A.2.5 Using the Fuzzy Cognitive Map

Simple FCMs converge to fixed-points or limit cycles. Fixed points are lone bit vectors like (1 0 0 1). Limit cycles chain through two or more bit vectors. You can "Run" the FCM until it reaches a limit cycle or "Single Step" it just one iteration. The commands for this are under the Execute menu. This program does not produce FCMs that have more complex attractors such as chaotic or aperiodic equilibrium states.

There are two special-purpose commands in the Execute menu. "Clear State History" clears the execution state history at the right side of the display. "Reset States" sets all the node states to the values they had at the last time you issued the Run command.

A.2.6 File and Help Menu Commands

The File menu contains only the Exit command. Use it to end the program.

The Help/Help command invokes the help file. The Help/About command brings up a dialog that contains the copyright notice and descriptive information about the program and its manufacturer. The user can contact the software developer Hyperlogic for more details:

Hyperlogic Corporation
P.O. Box 300010
Escondido, California 92030–0010
(619) 746–2765 voice
(619) 746–4089 fax.

INDEX

A

B

C

G

H

I

J

K

L

M

P

Q

R

S

T

U

V

W

X

Y

Z

LIST OF GUEST CONTRIBUTORS

Professor Gail Carpenter
Department of Cognitive and Neural Systems
Boston University
677 Beacon Street
Boston, Massachusetts 02215

Rod Corder
New Media Corporation
One Technology
Building A
Irvine, Calfornia 92718

Professor Stephen Grossberg
Department of Cognitive and Neural Systems
Boston University
677 Beacon Street
Boston, Massachusetts 02215

Professor Clark Guest
Department of Electrical And Computer Engineering
UC San Diego
La Jolla, Calfornia 92093

Professor Hiew Hong Liang
Department of Computer Science
University of Western Australia
Nedlands
West Australia 6907

Professor Chin-Teng Lin
Department of Control Engineering
National Chiao Tung University
1001 Ta Hauch Road
Hsinshu, Taiwan, R. O. C.

Professor Chi Ping Tsang
Department of Computer Science
University of Western Australia
Nedlands
West Australia 6907